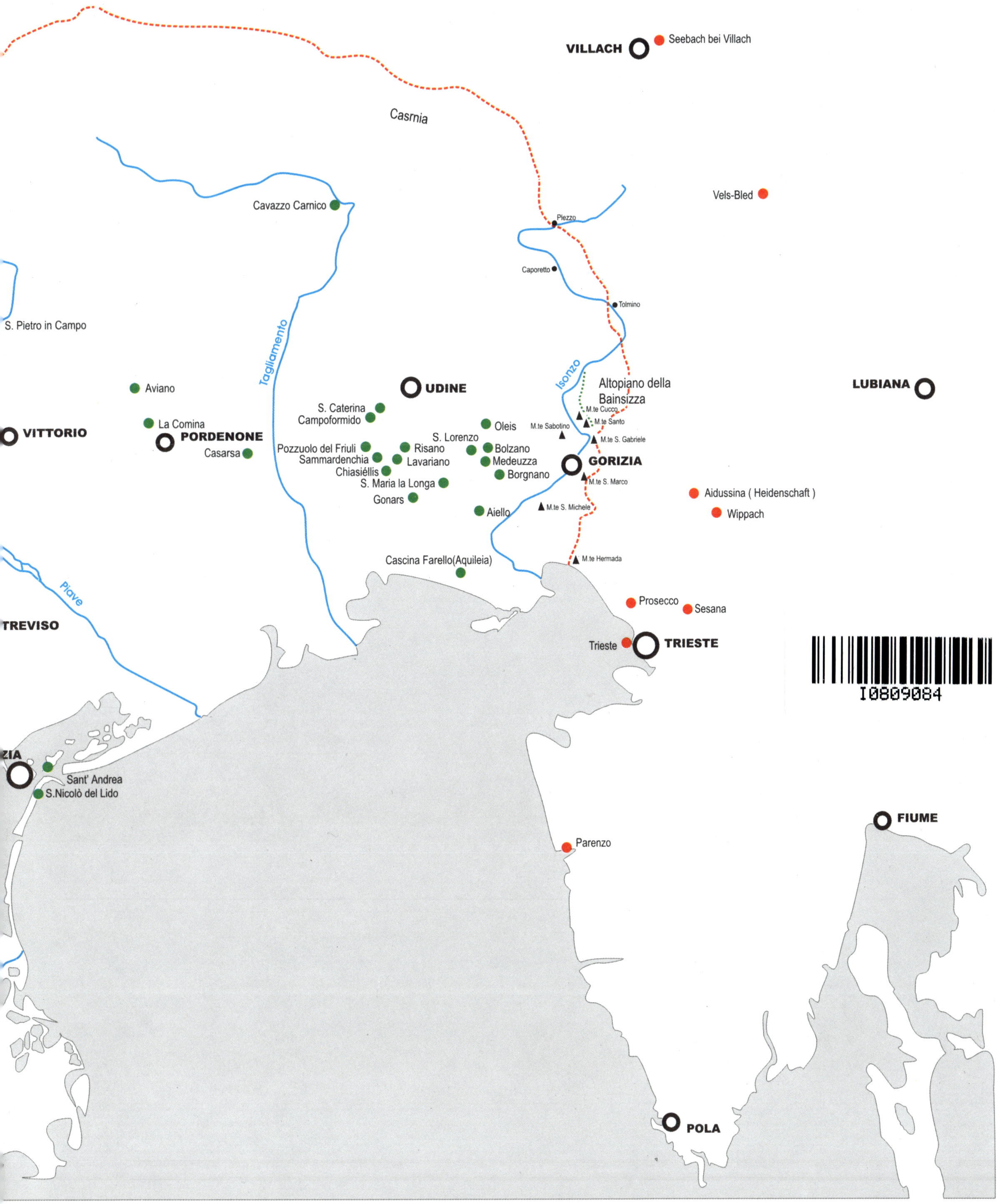
VILLACH
Seebach bei Villach
Casrnia
Vels-Bled
Cavazzo Carnico
Piezzo
Caporetto
Tolmino
S. Pietro in Campo
Tagliamento
Isonzo
Altopiano della Bainsizza
LUBIANA
Aviano
UDINE
S. Caterina
Campoformido
La Comina
VITTORIO
PORDENONE
Casarsa
Oleis
M.te Cucco
M.te Santo
M.te Sabotino
M.te S. Gabriele
S. Lorenzo
Pozzuolo del Friuli
Risano
Bolzano
Sammardenchia
Lavariano
Medeuzza
GORIZIA
Chiasiéllis
Borgnano
M.te S. Marco
S. Maria la Longa
Aidussina ( Heidenschaft )
Gonars
Aiello
M.te S. Michele
Wippach
Cascina Farello(Aquileia)
M.te Hermada
Piave
Prosecco
Sesana
TREVISO
Trieste
TRIESTE
ZIA
Sant' Andrea
S.Nicolò del Lido
FIUME
Parenzo
POLA

# *Italian Aces of World War I*
## *and their Aircraft*

Roberto Gentilli, Antonio Iozzi & Paolo Varriale

*Color aircraft profiles by Antonio Iozzi*

**Schiffer Military History**
Atglen, PA

Book design by Robert Biondi.


Library of Congress Catalog Number: 2002111964.



Printed in China.
ISBN: 0-7643-1664-8



Published by Schiffer Publishing Ltd.
4880 Lower Valley Road
Atglen, PA 19310
Phone: (610) 593-1777
FAX: (610) 593-2002
E-mail: Info@schifferbooks.com.
Visit our web site at: www.schifferbooks.com
Please write for a free catalog.
This book may be purchased from the publisher.
Please include $3.95 postage.
Try your bookstore first.

In Europe, Schiffer books are distributed by:
Bushwood Books
6 Marksbury Ave.
Kew Gardens
Surrey TW9 4JF
England
Phone: 44 (0)208 392-8585
FAX: 44 (0)208 392-9876
E-mail: Bushwd@aol.com.
Free postage in the UK. Europe: air mail at cost.
Try your bookstore first.

# Contents

# Preface and Acknowledgements

While in other nations, particularly Great Britain and the United States, a mythology was born and developed about the Aces, that produced studies and research on aviation in the Great War, and kept alive a memory, in Italy no serious study was ever undertaken, and, besides Baracca, all these heroes of the past are totally forgotten, with a few local exceptions: Cabruna in his hometown Tortona, Ancillotto in San Donà di Piave. Paradoxically, Italians know about von Richthofen, the Red Baron (thanks to *Peanuts'* Snoopy) and not about their own aces.

There are two concurring reasons for the lack of Italian research on World War I aviation: in the first quarter century after the war, while films and pulp magazines in America and England celebrated "the knights of the air", in Italy the fascist regime used aviation as a powerful propaganda tool, exploiting the success of the Schneider Trophy racers or the mass ocean flights of Balbo and celebrating the growing power of the air force, as used in Ethiopia or Spain. But in doing so, the dictatorship had to pretend that aviation was its own creation, so it belittled through silence the fact that before it, the Italian democracy had managed to create a powerful and victorious aviation in the Great War. A few good histories of the air war were written; a few pilots published their memories (Scaroni, Fucini) but that was all.

After the Second World War, in the ashes of defeat, Italy lost its entire military pride and traditions, and as the number of Great War veterans dwindled, that war faded away in history. The impressive development in aviation publishing in Italy in recent years, which, both in quality and in quantity is second only to the levels of Britain and America, dealt exclusively with World War II and the Spanish Civil War. A visit to any plastic model exhibition (or store) will show plenty of Messerschmitts, Mustangs, F-15s and F-16s but not a single SVA or Nieuport. And, as plastic modeling is the breeding ground to historical research, a vicious circle was formed: no World War I models, thus no research, thus no interest. The very fact that Italy won in World War I seems to make that war less interesting. Meanwhile, while all former World War II pilots were pestered by the interviews of enthusiasts, the fighter pilots of World War I died without anyone recording their histories.

If this book was born, it is above all thanks to the impulse that came from the Anglo-Saxon world, from people like Martin O'Connor, and Jack Herris, and journals like *Cross & Cockade, World War I Aero, Over The Front* and *Windsock.*

We want first of all to express our gratitude to the families: Allasia, Ancillotto, Avet, Baracchini, Bedendo, Bocchese, Cerutti, Chiri, Costantini, Eleuteri, Fornagiari, Fucini, Imolesi, Leonardi, Lombardi, Magistrini, Masiero, Michetti, Nicelli, Pessi, Pierozzi, Ranza, Reali, Resch, Riva, Rizzotto, Sabelli, Scaroni, Stoppani and Ticconi. They all welcomed us in their homes, generously providing to us documents, photographs, letters and mementos. Above all, they shared with us their memories, answering kindly to countless questions and giving us, after so many years, a lively image of those aces as men, and not just as pilots. It is this impression that we wish, more than anything else, to share with our readers.

This work, like all aviation research projects, depended on the collaboration of the agencies that preserve historical memories, so we want to thank the personnel of the Ufficio Storico dell'Aeronautica Militare, particularly *tenente colonnello* Fejer de Bük, *tenente* Ajello and *marescialli* Mastrangeli, Pennacchiotti and Rubbertone, the personnel of the Uffici Storici of the Marina Militare and of the Comando Generale dei Carabinieri. We are also deeply thankful of the help of Bündesarchiv, Freiburg, Germany, Kriegsarchiv, Vienna, Austria, Associazione Nazionale delle Medaglie d'Oro, Rome, Fototeca dell'Aeronautica Militare, Service Historique de l'Armée de l'Air, Paris, and Museo Baracca, Lugo.

We also thank for their support and generous help our friends Gregory Alegi, Giorgio Apostolo, August G. Blume, Luigi Buratti, Baldassarre Catalanotto, Angelo Emiliani, Gaetano Pastoressa, dottoressa Silvana Roseto, and then Andrei Alexandrov, Andrea Anesini, the Appiani family, Francesco Ballista, Maria Teresa Barbero Viazzi, Gian Marco Borea d'Olmo, Ing. Giorgio Brach Papa, Gino Burei, Ing. Ludovico Calderara, Ing. Giampiero Capparucci, Alberto Casirati, Gianni Cattaneo, Claudio Filippo Centurione Scotto, Claudio Coletta, Dott. Andrea Costantini, S.ra Adriana Crocetti, Com.te Fiorenza De Bernardi, S.ra Caterina Degli

Esposti, S.ra De Riseis, Dott. Augusto de Toro, Hermann De Wulf, Diego Ferneti, Avv. Giorgio Filippi, Franco Franci, Giulio Gauthier, S.ra Lori Gelli, Carlo Ghiglia, Cesare Gori, S.ra Patrizia Gorlero, S.ra Adriana Guglielmotti, Jon Guttman, Jack Herris, Matei Kiraly, Bernard Klaeyle, Raymond Kohn, Fiorenzo Longhi, Ing. Maurizio Longoni, Carlo Lucchini, Lucio Lucci, S.ra Anna Maria Maggi, S.ra Maia Magnetti, Dott. Natale Mancioli di Vallorsina, S.ra Alina Marsaglia Balbo, Dott. Martina, Museo di Gorizia, S.ra Paola Massa, Giovanni Massimello, Tommaso Miari Fulcis, Ten. Col. Milloch, Ermanno Molteni, Pier Luigi Moncalvo, Lucien Morareau, S.ra Maria Luisa Moscatelli Paradisi, Camillo Novellis di Coarazze, Giuseppe Osnago Gadda, Colin A. Owers, Dott. Paoletti of the Associazione Nazionale delle Medaglie d'Oro, Ing. Pier Paolo Paravicini, Prof. Gennaro Pistone, Dott. Gianclaudio Polidori, Ten. Col. Polloni, Museo della Cavalleria di Pinerolo, Wayne Ralph, Franco Ragni, Giorgio Ravazzoni, Ray Rimell, Dott. Anna Maria Rosso, Museo Leone, Generale Giuseppe Ruzzin, S.ra Salvi, Ing. Gianguido Sambonet, Prof. Alberto Santoni, Com.te Dino Sclerandi, Ing. Lucio Schiona, Umberto Scricciolo, Noel Shirley, Dott. Sobrero, Dott. Solaroli di Briona, Prof. Mariangela Spina, Augusto Spuri, Ing. Giorgio Stiavelli, S.ra Annamaria Tesei, Nevio Turella, Riccardo Udiente, S.ra Maria Rosa Valdimiro, Aaron Weaver, and we apologize for the many more whom we have overlooked.

Finally, a special thank to our friends, the historians and enthusiasts of Austria and of the nations formerly in the Hapsburg Empire, whose generous cooperation turned our job into an enterprise of brotherhood and peace: first among them, in knowledge and generosity, Karl Meindl, who freely offered us the results of his researches of thirty years in the Austrian archives, and then Zdenek Cejka, Zoltán Csizek, Tadeusz Dobrowiecki, Rudolf Höfling, Laszlo Javor, Reinhard Masajdek, Koloman Mayrhofer, Gerald Penz, Walter Schroeder, Franz Selinger, Peter Tesar.

# Introduction

The list shown opposite, issued on 1 February 1919 by the Comando Superiore d'Aeronautica, reports the Italian fighter pilots of World War I with five or more victories, the Aces. The crucial question, for a book like this, is who is an ace. The obvious answer, a pilot who has shot down at least five enemy aircraft (Five Down and Glory) is strictly limiting for a book like this: about 45 years ago, a fantastic book, "Air Aces of the 1914-1918 War" edited by Bruce Robertson, included in the Italian section a picture of Gabriele d'Annunzio, "an ace in all but victories". So there are plenty of outstanding Italian pilots, who deserve to be included. Actually, it was remarkable of the Italian air force that some bomber pilots and some reconnaissance observers (Pagliano, Gori, Palli, Locatelli etc.) gained as much recognition as fighter pilots, similarly as in World War II, when the real heroes of propaganda and the press were not fighter aces but torpedo bomber pilots. Also, there are many important pilots who may actually be aces, like Poli, Ciotti, Panero, Arrigoni, who may have scored more than the 3 or 4 victories assigned to them.

A guideline is necessary; therefore this book sticks to the official list, and considers recognized aces with five victories or more. On the other hand, a big question is raised if all confirmed victories stand up to the confirmation of facts, and to the coordinated research of historians that use data from both sides, and not just the national propaganda. This is of course a big issue that has affected the judgment of major aces in all countries, and that still raises debate.

The excellent, open and warm-hearted cooperation between Italian and Austrian historians that was launched by Rinaldo D'Ami and Karl Meindl, allows us a degree of accuracy in the control of claims that is unequalled for any other front or any other war. This clearly means that there is no confirmation for some Italian claims (and vice versa). A pattern somehow emerges that the reader can see for himself. For certain aces (Baracca, Barker and Brumowski are names that come to mind) every claim is perfectly proven, for others no such accuracy is possible. Claims were rigorously screened in some squadrons; other units seem to show a tendency to optimism. Our comment is only that this is a universal phenomenon, and that possible overclaims by Italians are very markedly less than the ones by the Royal Air Force in Italy and by the Austrian Luftfahrtruppe.

The *Provisional Instruction for the use of Fighter Squadrons*, issued by Lt. Col. Piccio, Inspector of Fighter Squadrons, on June 16, 1918 illustrates the strict criteria of Italian confirmations:

> "The pilot must.. As soon as he gets out of the airplane, if a combat took place, without talking with anybody (under the responsibility of the squadron commander) fill the form "Rapporto di Combattimento" (Combat report).
>
> Confirmation of air claims. It is indispensable that all squadrons, Groups, Aeronautic Commands, follow the same criteria to confirm or deny to their dependent pilots the claims of enemy aircraft. It must be absolutely forbidden that the pilot go personally to beg for declarations that in most cases, either for the incompetence of those who write them, and for their origin, cannot be accepted as valid. The Aeronautic Commands can rely in the armies on all the means to get the necessary information; it is their task to ask for them, if they don't arrive immediately.
>
> To have the confirmation of a claim are needed: At least 2 confirmations of artillery or balloon or first line observers, who specify in a doubtless manner that the airplane crashed in flames or that it is in the sight of the observers themselves, specifying in such case the exact place where it stands, and if our artillery is covering it.
>
> Besides these cases, exceptionally victories can be confirmed when irrefutable proofs can make certain the destruction of the enemy airplane (declarations of prisoners, uniform declarations of several other pilots who witnessed the combat – Certain vertical dive of the airplane, seen by several observers).
>
> It must be remembered that to have the confirmation of the shooting down of a draken, it must be on fire and it is not enough that it was lowered or deflated."

On the other hand, years of research in the archives of the Italian Air Force, that are remarkably well preserved, demonstrate that

# The 42 Italian "Aces"

***Supreme Command of the Royal Italian Army***
***Official list of air victories confirmed to Italian Pilots of the Army air force from 24 May 1915 to 4 November 1918***

| | | Confirmed victories | Notes |
|---|---|---|---|
| 1 | *Magg.* BARACCA, Francesco | 34 | Died 18.6.1918 |
| 2 | *Ten.* SCARON,I Silvio | 26 | |
| 3 | *T.Col.* PICCIO, Pier Ruggero | 24 | |
| 4 | *Ten.* BARACCHINI, Fulvio | 21 | |
| 5 | *Cap.* RUFFO, di Calabria Fulco | 20 | |
| 6 | *Serg.* CERUTTI, Marziale | 17 | |
| 6 | *Ten.* RANZA, Ferruccio | 17 | |
| 7 | *Ten.* OLIVARI, Luigi | 12 | Died 13.109.1917 |
| 8 | *Ten.* ANCILLLOTTO, Giovanni | 11 | |
| 8 | *Serg.* REALI, Antonio | 11 | |
| 9 | *Ten.* NOVELLI, Gastone | 8 | |
| 9 | *Ten.* AVET, Flaminio | 8 | |
| 9 | *Ten.* LOMBARDI, Carlo | 8 | |
| 9 | *S.Ten.* LEONARDI, Alvaro | 8 | |
| 9 | *S.Ten.* CABRUNA, Ernesto | 8 | |
| 9 | *Serg.* NICELLI, Giovanni | 8 | Died 5.5.1918 |
| 10 | *Cap.* RIVA, Antonio | 7 | |
| 10 | *Ten.* FUCINI, Mario | 7 | |
| 10 | *Ten.* ELEUTERI, Leopoldo | 7 | |
| 10 | *Serg.m.* FORNAGIARI, Guglielmo | 7 | |
| 10 | *Serg.* RENNELLA, Cosimo | 7 | |
| 11 | *Cap.* COSTANTINI, Bortolo | 6 | |
| 11 | *Ten.* OLIVI, Luigi | 6 | Died 17.7.1917 |
| 11 | *Ten.* PARVIS, Giuliano | 6 | |
| 11 | *Serg.* IMOLESI, Attilio | 6 | Died 3.3.1918 |
| 11 | *Serg.* STOPPANI, Mario | 6 | |
| 11 | *Serg.* NARDINI, Guido | 6 | |
| 11 | *Serg.* BOCCHESE, Aldo | 6 | |
| 11 | *Serg.* TICCONI, Romolo | 6 | |
| 11 | *Serg.* MAGISTRINI, Cesare | 6 | |
| 11 | *Serg.* RIZZOTTO, Cosimo | 6 | |
| 12 | *Cap.* LEGA, Giulio | 5 | |
| 12 | *Ten.* SABELLI, Giovanni | 5 | Died 25.10.1917 |
| 12 | *Ten.* BUZIO, Alessandro | 5 | |
| 12 | *Ten.* MASIERO, Guido | 5 | |
| 12 | *Ten.* BEDENDO, Sebastiano | 5 | |
| 12 | *Ten.* MECOZZI, Amedeo | 5 | |
| 12 | *Ten.* MICHETTI, Giorgio | 5 | |
| 12 | *S.ten.* ALLASIA, Michele | 5 | Died 30.7.1918 |
| 12 | *S.ten.* AMANTEA, Antonio | 5 | |
| 12 | *S.ten.* RESCH, Alessandro | 5 | |
| 12 | *Serg.* CHIRI, Antonio | 5 | |

200 other military pilots who shot down from 4 to 1 aircraft follow. The total number of enemy aircraft and kite balloons shot down by Italian pilots of the army aviation in the sky of our theater of war from 24 May 1915 to 4 November 1918 was 643. Our losses in the same period of time due to enemy action sum up to 128 aircraft.

Zone of War, 1 February 1919.
The General Commander of Aeronautics
*Bongiovanni*

The Nieuport 10, better known in Italy as "18 square meters" was the first real Italian fighter. These two-seaters probably belong to the 2ª Squadriglia Caccia, based at Cascina Farello in early 1916. (Photo Centro Culturale Pubblico Polivalente).

they are not 100% complete. There is the absolute certainty that some Austrian claims are accurate despite the lack of confirmation from Italian archives (for lack of documents, lost during the war, or because some documents were not filled correctly, to start with). There is no reason to doubt that the same must have happened on the enemy side.

Our conclusion is, to quote another air historian, Fernando D'Amico, that if for some claim of victory there is no confirmation, it means nothing more and nothing less that that, that there is no confirmation. We invite the reader to draw no further conclusion from this point.

The Maurice Farman, or Farman model 1914, widely produced in Italy, was used in frontline service until 1917, and in Libya it was still operational in 1919. This is aircraft 273 with De Dion engine, serving in the 10ª Squadriglia in 1915. Bought with a public subscription, it carries on its nacelle a placard with its name, "F. Crispi". *(All photos: authors' collections, except when otherwise credited).*

## Italian air victories – Official list issued by the *Comando Generale di Aeronautica* February 1919

**34 victories**
*Magg*. Baracca Francesco

**26 victories**
*Ten*. Scaroni Silvio

**24 victories**
*T.C*. Piccio Pier Ruggero

**21 victories**
*Ten*. Baracchini Flavio

**20 victories**
*Capit*. Ruffo di Calabria F.

**17 victories**
1. *serg*. Cerutti Marziale
2. *Ten*. Ranza Ferruccio

**12 victories**
*Ten*. Olivari Luigi

**11 victories**
1. *Ten*. Ancillotto Giov.
2. *serg*. Reali Antonio

**Eight victories**
1. *Ten*. Novelli Gastone
2. *Ten*. Avet Flaminio
3. *serg*. Nicelli Giovanni
4. *Ten*. Leonardi Alvaro
5. *Ten*. Lombardi Carlo
6. *S.ten*. Cabruna Ernesto

**Seven victories**
1. *S.m*. Fornagiari Gugl.
2. *Capit*. Riva Antonio
3. *serg*. Rennella Cosimo
4. *Ten*. Eleuteri Leopoldo
5. *Ten*. Fucini Mario

**Six victories**
1. *Ten*. Costantini Bortolo
2. *serg*. Ticconi Romolo
3. *serg*. Bocchese Aldo
4. *Ten*. Parvis Giuliano
5. *serg*. Imolesi Attilio
6. *serg*. Nardini Guido
7. *serg*. Stoppani Mario
8. *Ten*. Olivi Luigi
9. *serg*. Magistrini Cesare
10. *serg*. Rizzotto Cosimo

**Five victories**
1. *serg*. Chiri Antonio
2. *Ten*. Buzio Alessandro
3. *S.ten*. Amantea Antonio
4. *Ten*. Sabelli Giovanni
5. *Ten*. Masiero Guido
6. *S.ten*. Allasia Michele
7. *S.ten*. Michetti Giorgio
8. *Ten*. Mecozzi Amedeo
9. *S.ten*. Resch Alessandro
10. *Capit*. Lega Guido
11. *T*. Bedendo Sebastiano

**Four victories**
1. *Serg*. Arrigoni Bartolom.
2. *S.ten*. Conelli Carlo Alb.
3. *Serg*. Di Loreto Raimondo
4. *Sold*. Panero Clemente
5. *Ten*. Sambonet Guido
6. *Serg*. Sorrentino Ennio

**Three victories**
1. *S.ten*. Aliperta Gaetano
2. *Serg*. Bo Nicola
3. *Ten*. Bernelli Dino
4. *Ten*. Bacula Adriano
5. *Cap*. Bolognesi Domenico
6. *Serg*. Ciotti Giuseppe
7. *Serg*. Codeghini Oreste
8. *Serg*. Capparucci
9. *Ten*. Costantini Andrea
10. *Serg*. Corti Luigi
11. *Serg*. Donati Renato
12. *Asp*. Degli Esposti Alb.
13. *Ten*. De Bernardi Mario
14. *Serg*. Fanti Luigi
15. *Serg*. Ferrandi Attilio
16. *Serg*. Lucentini Edm.
17. *Ten*. Marazzani Alberto
18. *Serg*. Macchi Fermo
19. *S.ten*. Nannini Dante
20. *Serg*. Olivetti
21. *Serg*. Pereno Mario
22. *Serg*. Poli Giulio
23. *Serg*. Razzi Gastone
24. *Capit*. Serafini Filippo
25. *Ten*. Toffoletti

**Two victories**
1. *Serg*. Avon Felice
2. *S.ten*. Appiani Virginio
3. *Serg*. Astolfi
4. *Ten*. Archibugi
5. *Ten*. Brenta Giacomo
6. *Ten*. Benaglio Eugenio
7. *Serg*. Bellino
8. *Serg*. Benvenuti Paolo
9. *Ten*. Bogliolo
10. *Ten*. Censi Lodovico
11. *Ten*. Carabelli Francesco
12. *Serg*. Caselli Anselmo
13. *Capor*. Corti Carlo
14. *Serg*. D'Urso Mario
15. *Serg*. Ercoli Montaldo
16. *Sold*. Franzi Ezio
17. *Serg*. Giordano Cesare
18. *Ten*. Garrone Luigi
19. *Serg*. Gorini Goffredo
20. *Capit*. Gelmetti
21. *Ten*. Keller
22. *Capit*. Mazzucco
23. *Capor*. Morfini Eman.
24. *Sold*. Manzolini Manlio
25. *Serg*. Masaia Tommaso
26. *Ten*. Olivieri Mario
27. *Capor*. Rossi Carlo
28. *Sold*. Rustici Augusto
29. *Ten*. Sivori Camillo
30. *Serg*. Sorzini
31. *Sold*. Sartori Romeo
32. *Serg*. Tola Francesco

**One victory**
1. *S.ten*. Arone Vincenzo
2. *S.ten*. Allegri Gino
3. *Ten*. Albertozzi Bruno
4. *Serg*. Avenati Edoardo
5. *Ten*. Abbriata
6. *Serg*. Bertolini Adamo
7. *Sold*. Borgato Alessando
8. *Capor*. Baggini
9. *Serg*. Bongiovanni Mass.
10. *Ten*. Baggio
11. *Serg*. Bellaria
12. *Capor*. Belleri Lorenzo
13. *Serg*. Bocca
14. *Capor*. Breglia Salvatore
15. *Capit*. Bassani Umberto
16. *Serg*. Bellino
17. *Ten*. Benfante
18. *Ten*. Barriè
19. *Serg*. Consonni Emilio
20. *Ten*. Campanaro
21. *Serg*. Contardini Aless.
22. *Ten*. Calì
23. *Serg*. Campana
24. *Ten*. Comandone Aldo
25. *Sold*. Coccè
26. *Sold*. Cometto
27. *Serg*. Ciampitti
28. *Ten*. De Briganti Giov.
29. *Ten*. De Filippo Vinc.
30. *Serg*. Dell'Oro Arturo
31. *Serg*. Donadio Augusto
32. *Serg*. Ercoli Montaldo
33. *Ten*. Finzi Aldo
34. *Capor*. Fusco Casimiro
35. *Ten*. Frazzi
36. *S.ten*. Fraschi
37. *Ten*. Ferraris
38. *Ten*. Fochessati
39. *Capor*. Fougier Rino
40. *Ten*. Giovanardi
41. *Serg*. Giordanino Celest.
42. *Ten*. Guarisco
43. *Cap*. Gordesco Mario
44. *Sold*. Galassi Umberto
45. *Ten*. Garavoglia
46. *Ten*. Guasti
47. *Ten*. Lucci Giuseppe
48. *Ten*. Guglielmotti
49. *Sold*. Griffini Michele
50. *Serg*. Costoli Giuseppe
51. *Serg*. Gianotti Albino
52. *Cap*. Gallia Pietro
53. *Serg*. Liut Elia
54. *Serg*. Levrero Augusto
55. *S.ten*. Lamberti Carlo
56. *Capit*. Lodi
57. *Serg*. Lepore
58. *Capit*. Macchi Giacomo
59. *Ten*. Mazzini Umberto
60. *Capit*. Martinelli Mich.
61. *Ten*. Magistrelli
62. *Ten*. Moresco Alberto
63. *Serg*. Mondini
64. *Ten*. Mariani
65. *Serg*. Mori
66. *Sold*. Monetti
67. *Ten*. Negroni
68. *Capit*. Nelli Marinello
69. *Ten*. Olivi Luigi
70. *Capit*. Omizzolo Mario
71. *S.ten*. Olivero Edoardo
72. *Ten*. Pasquali
73. *S.ten*. Pascoli
74. *Serg*. Permattei
75. *Capit*. Paldaoff
76. *Ten*. Retinò Giuseppe
77. *Serg*. Rossi Arrigo
78. *Serg*. Rossi Francesco
79. *Ten*. Rondi Mario
80. *S.ten*. Restelli Eolo
81. *Ten*. Rizzi Enrico
82. *Ten*. Sommariva
83. *Capit*. Sequi Ernesto
84. *S.ten*. Sauli Gino
85. *Capor*. Sottani
86. *Serg*. Sogliani
87. *Serg*. Stoppani
88. *Capor*. Spuri
89. *Magg*. Tacchini Guido
90. *Serg*. Tesei Giuseppe
91. *Serg*. Taliani Tersilio
92. *Capor*. Tarducci
93. *Capor*. Testa
94. *Serg*. Tebaldi Andrea
95. *Ten*. Vincenzi Augusto
96. *Serg*. Vanasco
97. *Serg*. Vecco Giampiero
98. *Capit*. Vulcano Luigi
99. *Sold*. Venier
100. *S.ten*. Zamboni Rodolfo
101. *Serg*. Zingales
102. *Capit*. Zoboli Mario

This document is reproduced faithfully, with all the mistakes, spelling errors and duplications of the original (Stoppani and Olivi,for instance, appear both as a six-victories and one-victory pilots).

Fighters, probably of the 76ª Squadriglia at S. Maria la Longa: first a Nieuport 10, serial 1043, and then an unusual Nieuport 11 in silver dope finish, probably of French construction.

A Nieuport 11 of French construction in Italy, still showing its French roundels.

## Official list of the air victories confirmed to pilots of the R. Marina during the 1915-1918 War

*S.Ten.Vasc.Pilota* ARCIDIACONO, Domenico ........................ 1
*S.Ten.Vasc. Osser.* BAGGIONI, Primo ........................ 2
*S.Ten.Vasc.Pilota* BALDINI, Mario ........................ 1
*Cann.Arm. Pilota* BALLERI, Renato ........................ 4
*Cann.Sc. Osserv.* BELLINGERI, Ercole ........................ 1
*Sottonocchiere pil.* BENSA, Franco ........................ 1
*SottoCapo T.E. pil.* BONSEMPIANTE, Renato ........................ 1
*Tenente Vasc. pil.* BORTOLOZZO, Alberto ........................ 1
*2°Capo Cann. pil.* BUCARELLI, Antonio ........................ 1
*SottoCapo mecc. pil.* BURATTINI, Vincenzo ........................ 2
*S.Ten.Vasc. pil.* CALVELLO, Umberto ........................ 5
*S.Ten.Vasc. pil.* CARRERA, Alfredo ........................ 2
*Ten.Vasc. Osservatore* CENCELLI, Amedeo ........................ 1
*2°Capo Tim. pil.* CONFORTI, Mario ........................ 2
*Tenente Vasc. pil.* COPPOLA, Vincenzo ........................ 1
*S.Ten.Vasc. pil.* DALLE ORE, Giuseppe ........................ 1
*Tenente Vasc. pil.* DE PINEDO, Francesco ........................ 1
*S.Ten.Vasc. pil.* DE RISEIS, Luigi ........................ 4
*Torp.pilota* FALCONI, Dante ........................ 1
*Tenente Vasc. pil.* FIASTRI, Giorgio ........................ 1
*Tenente Vasc. oss.* GIARTOSIO, Carlo ........................ 1
*Fuochista sc. oss.* GRASSO, Salvatore ........................ 1
*Tenente R.E. oss.* GREGO, Egidio ........................ 1
*2°Capo Cann. pil.* JANNELLO, Guido ........................ 4
*2°Capo Cann. pil.* MACCHIA, Giulio ........................ 3
*Tenente Vasc. pil.* MAGALDI, Umberto ........................ 4
*S.Ten.Vasc. Oss.* MARTINENGO, Federico ........................ 5
*2°Capo Torp.E. pil.* MANZONI, Vittorio ........................ 1
*2°Nocchiere pilota* MEROLA, Antonio ........................ 1
*2°Capo R.T. pilota* MINCIOTTI, Daniele ........................ 1
*Guardiamarina pil.* MORTERRA, Paolo ........................ 1
*Volontario motonauta* NACCARI, Giovanni ........................ 4
*2°Capo Torp.Sil.pil.* ODIERNA, Edoardo ........................ 4
*Marinaio scelto pil.* PAGLIACCI, Giuseppe ........................ 4
*S°Capo Torp.Sil.pil.* PANICHI, Achille ........................ 2
*S.Tenente pilota* PARODI, Giorgio ........................ 2
*S.Ten.Vasc. pil.* PATERNITI, Giovanni ........................ 2
*2°Nocchiere pilota* PIRO, Beniamino ........................ 2
*Tenente Vasc. pil.* PIEROZZI, Orazio ........................ 7
*S.Ten.Vasc. pil.* POMILIO, Vittorio ........................ 3
*S.Tenente Macch. pil.* RAVAZZONI, Ivo ........................ 1
*S.Tenente pilota* RAVELLI, Giovanni ........................ 1
*2°Nocchiere pilota* RIVIERI, Andrea ........................ 4
*S.Ten.Vasc. pil.* SAVINO, Mario ........................ 1
*Torp.Elettr. pilota* SPADA, Renato ........................ 1
*2°Capo Torp.pilota* VALDIMIRO, Pietro ........................ 1
*S.Tenente Macch.pil.* VINER, Giulio ........................ 2
*2°Capo Tim.pilota* ZANETTI, Arturo ........................ 2

## Italian Air Claims in World War I

| | Jan | Feb | Mar | Apr | May | Jun | Jul | Aug | Sep | Oct | Nov | Dec | Total |
|---|---|---|---|---|---|---|---|---|---|---|---|---|---|
| **1915** | | | | | | | | | | | | | |
| AA fire | | | | | 1 | | | | | | 1 | 1 | 3 |
| Air combats | | | | | | | | | | | | | 0 |
| **1916** | | | | | | | | | | | | | |
| AA fire | | | 4 | 1 | 2 | 1 | 2 | | 1 | | 3 | 1 | 15 |
| Air combats | 1 | | 6 | 2 | 9 | 5 | 8 | 4 | 8 | 1 | 1 | 2 | 56 |
| **1917** | | | | | | | | | | | | | |
| AA fire | 3 | 1 | 4 | 1 | 1 | 2 | | 2 | 1 | | 3 | 5 | 23 |
| Air combats | 1 | 5 | 2 | 5 | 15 | 21 | 16 | 14 | 8 | 16 | 44 | 23 | 170 |
| **1918** | | | | | | | | | | | | | |
| AA fire | 11 | 3 | 5 | 2 | 13 | 7 | 6 | 5 | 2 | 5 | | | 59 |
| Air combats | 29 | 18 | 21 | 11 | 34 | 72 | 32 | 27 | 21 | 38 | 1 | | 304 |
| | | | | | | | | | | | | **Total** | **630** |

Regia Marina victories: 1915 ......... 8
1916 ....... 20
1917 ....... 51
1918 ....... 53
Total ..... 132

## The Official Scores of the Aces

On 11 January 1919 the 5ª Sezione (Situazione e Informazioni) of the Comando Generale di Aeronautica (General Command) wrote to the Comando Centri Formazione Squadriglie (organization command) in Turin:

> "This Command wants to establish and make public the official list of air victories scored by Italian pilots during the war.
>
> From the reports of assignments and the decorations available at this Command, for each airman result as confirmed the victories marked in their respective papers.
>
> The victories that were not confirmed for lack of any report or document are marked separately, in case there was some mistake or if it were possible to prove belatedly the victory. The interested parties who believe they have a right to a correction or who can present documents proving the shooting down of enemy aircraft, can do so, sending the required documents to this Command (Sezione 5ª) within the 20th of this month.
>
> For the Lt. Colonel Chief of Office
> Maj. Graziani".
>
> The individual letters were thus:
>
> "R. Esercito Italiano
> Comando Supremo
> Comando generale di Aeronautica
> Sezione 5ª
> General Official List of the victories scored in air combat by pilots during the war.
>
> From the documents existing at this General Command, to_______are confirmed
> N°.. air victories at the following dates:
> The following combats result as not confirmed for lack of any proof:...."

The 5ª Sezione, that is to say the intelligence section of the Comando Superiore di Aeronautica, (later Comando Generale di Aeronautica) had been charting air claims and issuing lists of confirmations at regular intervals during the war. It must be noticed that a confirmation didn't just have a moral value, it also brought a monetary reward, a prize of 1st class, 1,000 or 1,500 lire, or 2nd class, 500 lire, a rather considerable sum at that time.

It is amazing how hurriedly this investigation was conducted. In January 1919 the Italian Command had had no chance to verify Italian claims against Austrian records, nor, for that matter, to draw a reliable list using national records alone. Pilots had just a few days to get informed of the official inquiry and to collect documents and testimonials in order to specify their scores. Some of them did get the letter in time, and wrote back, for instance Mecozzi. In other cases, it was the relatives of deceased pilots who wrote pathetic letters to the Command, quoting letters written from the front, in which the fallen pilot had described his duels.

There is no way to guess why the Command was in such a hurry, and who ordered that investigation. Most likely, there had been misgivings because the Royal Air Force issued communiqués stressing the importance and the accomplishments of British flyers in Italy, and the Italian Command wanted to issue a general report telling Italians what their own airmen had done. It is obvious, from the surviving letters to specific pilots that the confirmed victories were sometimes chosen with insufficient cross-checking. Only one was confirmed to Mario Stoppani, only 21 to Baracchini, who had been credited with 31 in an official war Bulletin, and at dates apparently chosen at random.

A Nieuport 11 of the 76ª Squadriglia taking off.

Anyway, the official list was issued, and it appeared in the press in February 1919 (meaning that the deadline of January 20 was strictly adhered to, despite the fact that it was absolutely impossible that all pilots had been contacted).

Surely, some pilots did reply and clarified their records. Piccio, for instance, wrote back to the 5ª Sezione, on 20 January 1919:

> "Herewith enclosed there are the documents proving the following victories:
> 1 – 9 June 1918 – Moriago
> 2 – 29 July 1918 – Motta di Livenza
> 3 – 8 August 1918 – S.Polo
> 4 – 29 September 1918 – Montebelluna
> Regarding the victories of 29/7 – 8/8 – 29/9 I inform again that for brevity's sake and because legally sufficient only two or three confirmations of eyewitnesses of each victory are collected. For what concerns the victory of 9/6 the notes of the commander of Air Defense, 8th Army, confirm the absolute validity of the combat, validity admitted for the confirmation of claims to other pilots."

The fact that Piccio was the highest-ranking pilot, however, and commander of all fighters at the Comando Generale di Aeronautica possibly explains why he was so timely contacted and why his prompt reply was totally accepted. Dead pilots, or pilots still in hospital, didn't have the same chance.

Nobody, therefore, ever checked the accuracy of that list, nor of all the data that were hurriedly issued at the time: the list of all Italian pilots who claimed victories, the number of victories (643 including airplanes and balloons, for the Army aviation) and the ridiculously small number of losses, (128 airplanes lost due to enemy action). These data were repeated over and over, and have been accepted as the truth. The Command did issue a warning: that the list of aces doesn't take into account all the victories for which required testimonials were absent, and that were not confirmed.

The Caproni Ca.3 bomber was in many ways an exceptional airplane, as it remained in service basically unchanged from 1915 to 1927.

Overall, the list shows 42 aces, who among them shot down 391 airplanes, six pilots with 4 victories each (*serg.* Bartolomeo Arrigoni, *s.ten.* Carlo Alberto Conelli de Prosperi, *serg.* Raimondo Di Loreto, *sold.* Clemente Panero, *ten.* Guido Sambonet, *serg.* Ennio Sorrentino) 25 pilots with 3 victories, 32 pilots with two victories and 102 pilots with one victory. Statistically, this means that out of

Mixed equipment for the 71ª Squadriglia: French-built Nieuport 10s and Macchi-built Nieuport 11s.

over 1,000 active fighter pilots, 207 scored one or more victories. The 42 aces scored 60% of that total, and the ten top aces alone scored 193, meaning that 1% of all pilots scored 30% of all kills.

The list itself as it appeared, was obviously flawed: the actual total of claims was 656, not 643, (this being due to the fact, however, that the same destruction was credited to more pilots) some names appeared twice (and misspelled too), some airmen who were well known, and who had even been decorated for their victories, were absent altogether. "A higher accuracy in the compilation of these statistics would have been much desirable" was the comment of the press at the time. Another semi-official chart issued by the Command, however, reports a total of 630 claims, 530 of them to airplanes and 100 to anti-aircraft guns, with 132 more for the Regia Marina seaplanes. Finally, a typewritten list was issued, charting all the victories, giving date, place and claimant (anti-aircraft battery or air duel) for a total of just 613 claims. This "official list" is useful as a confirmation of what at the time was accepted as truth, but it is far for complete or flawless.

But for eighty years, excluding the researches of a few (Rinaldo D'Ami, James Kerr, Dr. Martin O'Connor, Karl Meindl) the question was left to rest.

So, it would be wrong to check the accuracy of this book against published sources, which uncritically relied on the 1919 official reports, disregarding the conflicting totals of 530, 613, 643 and 656 air victories. Whenever the reader finds differences between this text and well known and circulated previous works, including recent ones, please be assured that the authors are aware of such differences, and are confident in the correctness of their reconstruction, based on a constant control of original sources (squadron, group and Army air command war diaries and pilots logbooks).

However, although the utmost care was applied in trying to determine the exact course of events in each combat described, our point was not establishing for posterity who shot down whom. Our goal is only to bring out of oblivion some outstanding personalities, young men who did their duty for King and country. We were thrilled in our effort to pinpoint every air combat, to clarify as much as possible what happened during each dogfight. But what really excited us was the discovery of so many personalities: young men from all parts of our country, from big industrial cities and small hamlets, and from all conditions: wealthy princes with the noblest ancestors, and young barefoot country boys. Young men who were friends with each other ignoring their differences, in an environment of equality and enthusiasm, but not marred by cruelty or hate for the enemy.

Plenty of guns (a Vickers and a Lewis) and of roundels, including one on the fuselage deck, on this Nieuport 17, belonging to the 71ª Squadriglia, with pilot *ten.* Anesini. This plane ended up with the flying school at Bursto Arsizio.

## Italy in World War I - A Brief Overview

Italy, at the outbreak of World War I, was an ally of Germany and Austria, as member of the Treble Alliance, but as that alliance was merely defensive, the country remained neutral. All the feelings of the people and the national interest, however, were in favor of Britain and France. Austria was the traditional enemy, against which Italy had fought three wars (in 1848, 1859 and 1866) to gain its independence and national unity, and Austria still ruled two provinces with Italian population, the ones of Trento and Trieste. An invincible mixture of exalted nationalism, democratic solidarity with Belgium, France and England, and the crafty dealings of the Allied powers, brought Italy into the war against its former ally. In April 1915, a secret pact was signed in London, offering Italy territorial gains on the Dalmatian coast, besides Trento and Trieste, and a share of the colonial empires of Germany and Turkey on exchange for its entry into the war. So, despite the opposition of the Catholics, the Socialists and a sizeable part of the ruling Liberal Party, Italy entered the war on 24 May 1915.

The Italian strategy as designed by the supreme commander General Cadorna was very clear: a concentration of all efforts on the north-eastern front along the Isonzo river, aiming initially at Gorizia, then Trieste, then, climbing up the valley of the Sava River, reaching Lubjana (Laibach) and menacing Vienna from the South. This hopeful plan, worthy of a Napoleon, immediately collapsed against the reality of modern war. Austrian barbed wire and machine guns decimated Italian troops in a series of futile and bloody attacks, that were called the "battles of the Isonzo", 11 of them, the first four in 1915, costing the lives of half a million men to gain a single town, Gorizia ("damned Gorizia", as the Italian soldiers sang) and a few miles of rocky hills.

It remains incomprehensible for the modern historian how an army could lose in a single day as many soldiers as Americans died in the whole of the Vietnam War, but the slaughter on the Italian

The Farman Colombo, or MFC, was an improved version of the Farman 1914, with a rounded fuselage that served honestly in the squadrons operating over the Alps. This is aircraft serial 2661 of the 45ª Squadriglia.

front was no different than the ones on the Somme, at Yipres or at Verdun. The Austrians set up a line of resistance from the Sabotino, Podgora and San Michele mountains to the sea and, further north, around Santa Maria and Santa Lucia di Tolmino that proved impenetrable.

On 15 May 1916 the Austrian command struck back according to a plan long prepared by General Conrad, for the invasion of Italy: they struck from the northern Trentino-Tirol front, on the Asiago plateau, hoping to reach Vicenza and then pour on the plain of the Po valley. Their offensive, called "*Strafexpedition*", punishing expedition, was supported by an unprecedented barrage of artillery and initially gained some ground, but after one month of hand-to-hand fighting the Austrian thrust was stopped.

Although Cadorna, the autocratic and stern Italian supreme commander, wanted to concentrate all efforts exclusively on the main front, political considerations forced engagements on other secondary fronts. When, in 1916, Austria crushed Serbia, and the remnants of the Serbian Army fled South, Italy sent ships across the Adriatic to save as many Serbian soldiers as possible, and established a new front in Albania. It was a miserable terrain, where

A Nieuport 17 of the 71ª Squadriglia, showing a dark fuselage band and a dark tail.

The Caudron G.4, built in Italy by AER, was a twin engined version of the G.3, which served with reconnaissance units, particularly the ones operating over the Alps. This is an aircraft of the 48ª Squadriglia.

A Nieuport 17, Ni 3599 of an unidentified squadron, showing a dark engine cowling. The two-seater behind it is a Pomilio PD.

diseases took a heavier toll of combatants than bullets. In Macedonia, the Allies fighting against the Bulgarians also had a complement from Italy, both troops and aircraft. Finally, Libya was almost totally controlled by Arab rebels, who were supported by Turkish officers, while Germany could even set a submarine base in the Libyan harbor of Misurata. The desperate resistance of Libyan patriots was crushed only in 1931.

The 5th Battle of the Isonzo, 11 to 29 March 1916, was launched to engage the Austrians and stop them from sending reinforcements to the Germans at Verdun. The 6th Battle, in early August, breached through and gave to the Italians the conquest of Gorizia, but the Austrians fell back to another line of defense along another ridge of mountains: Kuk, Vodice, Monte Santo, San Gabriele, Vertoiba, Faiti. The three following battles, 14 to 17 September, 9 to 12 October, 1 to 4 November, were all to all effects fruitless. Then winter set in and the fighting slowed down.

The 10th Battle of the Isonzo, in May 1917, supported by a large preparation of artillery and aircraft, initially brought some Italian success, with the conquest of Kuk and Vodice and a penetration toward Trieste. But in June the Austrians counter-attacked and regained most lost ground, nullifying the sacrifice of 36,000 killed and 96,000 wounded Italian soldiers.

In June 1917, on the northern sector, Italy attacked on the Ortigara Mountain. Despite the incredible valor of the Alpini mountain troops, that ill-conceived secondary offensive failed, and 23,000 Alpini died on that single mountain. The hectic air actions that accompanied this battle, however, marked a turning point: the Italian air force, that had already applied its power in the 10th Battle of the Isonzo, caused serious losses to the Luftfahrtruppe and began to gain air supremacy, thanks particularly to the quality of its Nieuport and Spad fighters.

In August 1917, the 11th Battle of the Isonzo brought the conquest of the Bainsizza plateau, and the capture of a large number of prisoners and equipment. Despite the good success, the huge offensive, supported by a massive concentration of equipment and by hundreds of airplanes, was not determining either. The Italians reached the final Austrian strongpoint, the San Gabriele, and partially conquered it, they were on the point of breaking through, the Imperial army staggered under terrible punishment but the last Austrian reserves held, while Italian infantry was too much in advance of its artillery, so the battle ended with both armies bled white.

In September, fearing the influx of Austrian and German troops from the Russian front, where the revolution against the Czar had sapped the fighting spirit of the Russian Army, Cadorna ordered the Army to set on the defensive, but General Capello, chief of the 2ª Armata, disobeyed and got ready for one more offensive. Unknown to Italian intelligence, a German corps with aviation support had arrived to help the Austrians, and it struck on 24 October 1917, in a single quick penetration at Caporetto, on the higher Isonzo front. The 2ª Armata collapsed among failing leadership and confusion, as German and Austrian units quickly advanced, bypassing Italian resistance lines and spreading panic and confusion in the rear. No defense could be organized on the Tagliamento River, and the whole Italian front had to retreat, not just the crushed 2ª Armata but also the 3ª Armata, on the lower part of the front, down to the sea, and the 4ª Armata from the northern Carnia region. Over 300,000 prisoners and a huge quantity of material were captured, while thousands of refugees escaped, turning the retreat into a rout. Only at the end of November could the Italians stabilize the front, along the Piave River.

The final Austrian advance, in the fist week of December, was checked. On 5 December the Alpini put up a strenuous resistance on the Melette. The final enemy attack, on December 23, aimed at a penetration over Monte Valbella and the Frenzela valley, to pour down along the valley of the Brenta River, but again the Italians counter-attacked and on Christmas day they secured the mountain defense line.

Meanwhile, German bombers had introduced a new form of warfare, night bombardment of Italian cities, causing losses in the population. Then, as help, both of army divisions and air squad-

The reliable SAML, in its S.1 three-bay and S.2 two-bay versions, both developed from the original Aviatik design, remained in service until the early 1920s, much later than its intended successor, the Pomilio PE. This is an airplane of the 111ª Squadriglia in Macedonia.

rons, came from France and Britain, the Italian Army resisted, and slowly rebuilt its forces. On 28 January 1918 there was the first Italian attack after the retreat, and with a successful action the Col del Rosso was recaptured.

During the first half of 1918 Italy remained on the defensive. In the spring of 1918 the German moved away from the Italian front. Austria prepared a final offensive in June 1918. The hope was to break through the Piave and fan out toward the rich plains of Lombardy, but the internal situation of Austria was hopeless. Its offensive was called "the offensive of hunger" because the citizens and the troops of the Dual Monarchy were literally starving, and the rich countryside of Italy was offered to the soldiers as an attractive prize. This time, the Italian Army expected the attack, and prepared well for the defensive. Instead of concentrating their attack in a single thrust, on 15 June 1918 the Austrians struck all across the Piave River, and also in the Asiago and Mount Grappa areas, but the offensive failed. Little ground was gained across the Piave, but after fierce combats that centered on the Montello Hill, the Austrians were repulsed, while the overwhelming Italian air power harassed them as they crossed the river.

After the failure of the mid-June offensive, the war was definitely lost for Austria-Hungary, and therefore also for its ally, Germany. Still, despite the requests of the Allies, the Italian Army remained static, building up its force.

The final offensive didn't begin until 24 October 1918, the anniversary of the Caporetto defeat. The Austrian Army was not a spent force, and despite the political problems at home, where the different nationalities were openly seceding from the Empire, it resisted valiantly, with unfailing stamina. But the Italians broke through at Vittorio Veneto, and sent the enemy in a rout. After a few days of intensive fighting, the Austrian army broke down and escaped toward the Alpine passes. An armistice was signed, that went into effect on 4 November 1918, as Italian troops liberated Trento and Trieste.

The Italian victory was very important: the Austrian army was routed; an empire that had lasted 1,000 years was gone. Italy's victory and Austria's defeat could not be denied. A similar event did not happen on the Western front: Germany surrendered with its army undefeated, still solidly entrenched in enemy soil. This gave rise to a feeling of *revanche*, for an army that was stabbed in the back by traitors in the home front, a feeling on which hate mongers like Adolf Hitler could feed, stirring up hatred for a second World War.

The amount of blood that was shed for this victory was immense, Italy lost 660,000 men, and millions were wounded or mutilated. The conflict was, according to the opinions, a "useless slaughter", as the Pope said, or the 4th War of Independence, the biggest effort and the highest accomplishment of the unified Italian nation. It also was the first war in which aviation played an important part.

During the spring and summer 1917 many aircraft and pilots from second line units like the 78ª and 79ª Squadriglia were detached to the Isonzo front during offensives, sometimes directly assigned to reconnaissance units.

## Rank Equivalents

| Italian Army | Italian Navy | English (Army) | English (Navy) | Austrian |
|---|---|---|---|---|
| Generale | Ammiraglio | General | Admiral | General |
| Colonnello | Capitano di vascello | Colonel | Captain | Oberst |
| Tenente colonncllo | Capitano di fregata | Lieutenant colonel | Commander | Oberstleutnant |
| Maggiore | Capitano di corvetta | Major | Lt. Commander | Major |
| Capitano | Tenente di vascello | Captain | Lieutenant | Hauptmann |
| Tenente | Sottotenente di vas. | First lieutenant | Jr. Lieutenant | Oberleutnant |
| Sottotenente | Guardiamarina | Second lieutenant | Ensign | Leutnant |
| Aspirante | | | Midshipman cadet | Fähnrich |
| Aiutante di battaglia | | Chief w. officer | Chief w. officer | |
| Maresciallo | Capo | Warrant officer | Warrant officer | Oberfeldwebel |
| Sergente maggiore | 2° capo | Sergeant major | Chief petty officer | Stabsfeldwebel |
| Sergente | Sottocapo | Sergeant | Petty officer | Feldwebel |
| Caporal maggiore | | | | Zugsführer |
| Caporale | | Corporal | Seaman 1 Cl. | Korporal |
| Soldato | Marinaio | Private | Seaman | Soldat |

## A Brief Background to Italian Fighter Aviation

Italy had been the first country to use airplanes and airships in war, during the conflict against Turkey, in 1911-1912 that led to the conquest of Libya. Italy before the war was a country with a limited industrial basis: half its population was still employed in agriculture, and its industrial potential was a fraction of the British or German ones. Its industry produced excellent, high technology goods, like battleships or cars, but factories were active in only a limited part of the industrial north, and the country strictly dependent from coal, minerals and semi-prepared goods from abroad.

The Italian Army had an air service, the Corpo Aeronautico Militare, with two sectors, the Battaglione Aerostieri (lighter-than-air) and the Battaglione Aviatori (aircraft). The latter at the outbreak of war consisted in just a dozen squadrons, equipped with Bleriots, Farmans and Nieuport monoplanes. No special effort was made to upgrade the air service, and when Italy entered the war, on 24 May 1915, its air force consisted of the I Gruppo, with the Bleriot-equipped 1ª, 2ª, 3ª and 13ª and 14ª Squadriglia, the II Gruppo with the Nieuport-equipped 5ª, 6ª and 8ª Squadriglia, the III Gruppo with the 7ª Nieuport, 9ª, 10ª, 11ª and 12ª Farman, 15ª Caproni Parasol Squadriglia, and the autonomous 4ª Squadriglia Bleriot.

Conceived as a form of air cavalry, the Italian squadrons vanished from the front as quickly as real cavalry itself. The under-powered Bleriots and Nieuports had no performance and no war capability. By the fall of 1915, Italian aviation had almost disappeared. It received just in time new and better airplanes, the Voisin L and the Type 1914 Farman (Shorthorn for the British) and the first examples of a revolutionary weapon, the Caproni 300 hp three-engine bomber. The Austrian air force was not at all bigger nor better equipped, but it had a few efficient tractor biplanes powered by reliable inline engines, and it used them with daring, sending bombing raids to Verona, Brescia and, in February 1916, Milan and even far-away La Spezia.

The need for a fighter force had become apparent. In the summer 1915 the 2ª Squadriglia Farman was deployed at Udine as a protection for the Supreme Command headquarters. The 8ª Squadriglia was reformed flying the first Nieuport 10 biplanes, and in late August it started operations from Campoformido, near Udine. On 9 September the unit moved to Santa Caterina, near Udine, and the same day it had the first combat. *Cap.* Bolognesi with Nieuport 387 and *ten.* Baracca with Nieuport 596 intercepted an enemy plane over Udine. Baracca fired 100 rounds at the enemy, and three bullets hit his airplane. In the second half of October the unit received some French-built Nieuport 10s, 1035, 1037 and 1039. At the same time, a French squadron with Nieuport 10s had arrived in Italy, for the defense of Venice. The importance and the generosity of French help to Italy cannot be overestimated. Italy used exclusively French fighters all through the war (with the single exception of the Ansaldo Balilla): Nieuport 10, 11, 17, 24, 27, Spad 7 and 13, Hanriot Hd.1, and these fighters arrived in Italy as soon as they were available in France, always in plentiful quantities.

On 19 November 1915 all the fighters of the 8ª Squadriglia faced a large Austrian raid on Udine. Although not claimed, one of the Austrians, Albatros B.1 24.17 of Flik 14 crashed on the way back, with its observer wounded.

## Italian Air Force Order of Battle, 1 November 1915

Comando Battaglione Squadr. Aviatori Pordenone - *t.col.*Barbieri
$1^a$ Caproni Pordenone *cap.* Bailo
$2^a$ Caproni Pordenone *cap.* Carta
$3^a$ Caproni Pordenone *cap.* Piccio
$4^a$ Caproni Aviano *cap.* Visconti
Sez. Aviatik Difesa Aviano *cap.* Coppini

I Gruppo Squadriglie, S.Maria la Longa - *magg.* Gallina
$5^a$ Voisin Chiasiellis *cap.* de Rada
$7^a$ Voisin Chiasiellis *cap.* Capuzzo
$10^a$ Farman S.M.la Longa *cap.* Zappi
$11^a$ Farman Chiasiellis *cap.* Maranini

II Gruppo Squadriglie, Udine - *magg.* Gaviglio
$6^a$ Farman Chiasiellis *cap.* Olivo
$9^a$ Farman Campoformido *cap.* Beltramo

Gruppo Squadriglie Artiglieria, Medeuzza - *magg.* de Siebert
$1^a$ Caudron Oleis *cap.* Gamerra
$2^a$ Macchi Parasol Medeuzza *cap.* Santi
$3^a$ Macchi Parasol Medeuzza *cap.* Maurel
$4^a$ Caudron Gonars *cap.* Cortini
$5^a$ Caudron Oleis *cap.* Lampugnani

Autonomous units
$2^a$ Farman Campoformido *cap.* Masi
$8^a$ Nieuport Campoformido *cap.* Tacchini
$1^a$ Farman Verona *cap.* Ferraro
$12^a$ Farman Asiago *cap.* Bonazzi
$3^a$ Aviatik Brescia *cap.* Sanità
Sezione Idro Desenzano *ten.* Cesaroni

On 1 December 1915 the organization of Italian squadrons was changed, and the $8^a$ was renamed $1^a$ Squadriglia da Caccia (Ni). In February 1916 this unit received the first Nieuport 11, often called Bebé (Baby). Meanwhile, the $3^a$ Squadriglia Aviatik had been formed, active in the protection of Brescia with the German tractor biplane produced by the Italian SAML Company. In December this unit also became $3^a$ Squadriglia Caccia.

The $2^a$ Squadriglia Caccia, also equipped with Nieuports, was activated in January 1916 at La Comina, while the $4^a$ Squadriglia, equipped with a mixed component of Aviatiks and Farmans, was activated earlier, in December 1915, and operated as the defense unit of Verona. Finally the $5^a$ Squadriglia, also equipped with Aviatiks and Farmans, was activated on 16 January 1916, and de-

Homage must be paid to a gallant ally, the French. Not only was Italy timely supplied with excellent and new fighters (Nieuport 11, 17, 24, 27, Spad 7 and 13, Hanriot) but also a French fighter squadron, N.561, operated from Venice throughout the war. This is a Ni.23 marked "Pinocchio" coming in to land. *(Photo Collection SHAA N° 87-3690).*

PiNoCCHIO

ployed for the defense of Milan. In March the 2ª Squadriglia deployed to Cascina Farello, in the southern part of the front, close to the sea.

After many fruitless flights, the 1ª Squadriglia, now equipped with a dozen Nieuports, scored the first confirmed victories on 7 April 1916, facing another Austrian attack on Udine, when Francesco Baracca, flying 1451, shot down Brandenburg C.1 61.57 of Flik 19, while Brandenburg 61.59 of Flik 2, was attacked by Luigi Olivari, Bolognesi and Tacchini and also shot down.

Other air victories, not confirmed, had been claimed on 14 February 1916 by Locatelli and Ajmone Cat of the 6ª Farman, on 1 April by Ferrari and Muraro of the 5ª Voisin, on 2 April by Sala, Carta and Saletta of the 2ª Caproni at Ovcia Draga and also on 2 April by Olivari against possibly Alb.B.1 24.61 of Flik 12 and by *ten.* Barattini and gunner Moretto of the 2ª Caccia against Brandenburg C.1 26.02 of Flik 12. The Austrians themselves attribute two of their losses to Italian fighters without corresponding claims: Albatros B.1 24.04 on 25 October 1915 (maybe a victim of an airplane of the 9ª Farman), and Albatros B.1 (Ph) 24.17 on 19 November attacked by Baracca and Bolognesi.

On 16 April 1916 the organization of Italian squadrons was changed again, for the last time, creating a system that lasted until modern times. All fighter units received numbers ranging from 70, so Baracca's 1ª became 70ª, the 2ª Nieuport became the 71ª, 3ª Aviatik at Brescia the 72ª, 4ª at Verona (that also used an Austrian Lloyd 42.11 delivered by a defector) the 73ª, and the 5ª at Milan the 74ª.

In the spring of 1916 the availability of Nieuport 11 fighters, initially of French production, then license built by Nieuport Macchi at Varese, allowed the formation of more fighters squadrons that were numbered from 75 onwards. The first Italian pilots had been trained in France, at Pau, Le Bourget and Avord, but fighter conversion schools were established also in Italy at Busto Arsizio and Cascina Costa. A cursory gunnery course was offered at the San Giusto training field, near Pisa. Much later in the war, a real gunnery and combat school was created at Furbara, near Rome.

Equipped with the Nieuport, an excellent and reliable fighter, Italian pilots had more opportunities to engage Austrian intruders, particularly the redoubtable Brandenburg C.1, that was in service throughout the war, in ever-improved versions, and to score victories.

Fighter squadrons were assigned to Gruppi, groups, each supporting an Armata, or army. Each Army got to have its *Comando di Aeronautica*, aviation command, which basically established a good form of cooperation between land and air forces. From simple defense of the sensible targets, fighter units engaged in new missions, particularly escort to reconnaissance two-seaters and *crociere*, free-hunt patrol flights. The 70ª Squadriglia was based at Udine, where the Supreme Command was located, and reported directly to it, like most of the Caproni bombers strategic force. In late 1916 the first 110 hp Nieuport 17 fighters, usually equipped with a synchronized Vickers gun arrived, followed in the spring 1917 by the Macchi-built models of the same type.

## Italian Air Force Order of Battle, 1 January 1917

IV Gruppo Pordenone (Comando Supremo)
1ª Caproni, Comina - *cap.* Toccolini
2ª Caproni, Aviano
3ª Caproni, Aviano - *cap.* Falchi
4ª Caproni, Aviano - ten. Molinari
6ª Caproni, Aviano - *cap.* Sibilla
7ª Caproni, Aviano - *cap.* De Muro
8ª Caproni, Comina - *cap.* Govi
10ª Caproni, Campoformido - *cap.* De Riso
13ª Caproni, Comina - *cap.* Oppizzi

III Gruppo Verona (Aeronautica 1ª Armata)
1ª Idro FBA, Desenzano - *cap.* Amour
5ª Caproni, Verona - *cap.* Rossi
9ª Caproni, Verona - *cap.* Reggio
31ª Farman, Verona - *cap.* Straziota
32ª Farman, Villaverla - *cap.* Zoboli
71ª Nieuport, Villaverla - *cap.* Chiaperotti
72ª Aviatik, Brescia - *cap.* Matteucci
73ª Nieuport, Verona - *cap.* Sanità
75ª Nieuport, Verona - *cap.* Gordesco

VII Gruppo Nove di Bassano (Aeronautica 1ª Armata)
46ª Farman, Castenedolo and Trizzino - *cap.* Adorni
49ª Caudron, Nove di Bassano - *cap.* De Sanctis

II Gruppo Udine (Aeronautica 2ª Armata)
27ª Farman, Campoformido - *cap.* Giorgi
29ª Farman, Cavazzo Carnico - *cap.* Ferrero
30ª Farman, Chiasiellis - *cap.* Aliquò Mazzei
35ª Voisin, Campoformido - *cap.* Carignani
76ª Nieuport, S.Maria la Longa - *cap.* De Carolis

VI Gruppo Oleis (Aeronautica 2ª Armata)
41ª Caudron, Oleis - *cap.* Martucci
45ª Farman, Oleis - *cap.* Giovine

I Gruppo S. Maria la Longa (Aeronautica 3ª Armata)
25ª Voisin, Pozzuolo del Friuli - *cap.* Stanzani
26ª Voisin, S.Maria la Longa - *cap.* Minellono
28ª Farman, Chiasiellis - *cap.* Neri
36ª Farman, S. Maria la Longa - *cap.* Clementi
77ª Nieuport, Cascina Farello - *cap.* Piccio

V Gruppo Chiasottis (Aeronautica 3ª Armata)
42ª Caudron, Medeuzza - *cap.* Negro
43ª Caudron, Bolzano - *cap.* Carranza
44ª Caudron, Gonars - *cap.* Beltrami

The Savoia Pomilio SP.4 was a twin-engined derivative of the SP.3, which proved a total failure. This is aircraft serial 3197 of the 61ª Squadriglia, the only unit that used it in operations.

VIII Gruppo (Albania)
11ª Caproni, Tahyraqua - *cap.* Pascale
34ª Farman, Valona - *cap.* De Rada
Sezione Nieuport, Tahyraqua - *ten.* Sabelli

Autonomous Squadriglie
2ª Idro FBA, Grado - *cap.* Martellucci
12ª Caproni, Benghazi - *ten.* Raffaelli
33ª Farman, Campoformido - *cap.* Russi
37ª Farman, P.S. Pietro - *cap.* Oddo
47ª Farman, Saloniki - *cap.* Chiappelli
48ª Caudron, Belluno - *cap.* Porro
70ª Nieuport, S.Caterina - *cap.* Tacchini
74ª Aviatik, Trenno (Milan) - *cap.* Giorgi
78ª Nieuport, Istrana - *cap.* Bolognesi

Territorial defence units
101ª Farman, Bari - *cap.* Sisto
102ª Farman, Ancona - *ten.* Gallotti
103ª Farman, Brindisi - *ten.* Calleri
104ª Farman, Tripoli - *cap.* Suglia
105ª Voisin, Florence - *cap.* Rospigliosi
Sez. Difesa Foligno - *ten.* Debolini
Sez. Difesa Ravenna - *ten.* Cottino
Sez. Difesa Rimini - *ten.* Ponis
Sez. Difesa Comina - *ten.* De Gasperi
Sez. Difesa Benghazi - *ten.* Marchisio
Sez. Difesa Rome - *ten.* Masi
Sez. Difesa Bergamo - *ten.* Bourlot

Centro Formazione Squadriglie di Ghedi - *cap.* Ferraro
Centro Formazione Squadriglie di Arcade - *cap.* Degan

Regia Marina
Grado air station 6 L.1
Venice air station 8 L.1, 3 L.2, 9 FBA
Varano air station 5 L.1, 2 FBA
Brindisi air station 4 L.1, 6 L.2, 9 FBA
Regia Nave *Europa* 6 L.1, 1 L.2

The SIA 7b was a fast reconnaissance two-seater introduced in late 1917 and retired from service in June 1918 on account of its structural weakness. This is an aircraft of the 36ª Squadriglia at Istrana, December 1917.

Much criticized by later writers for its archaic looks, the Savoia Pomilio actually was an effective and reliable two-seater that remained in service until the end of the war. These are SP.3s of the 33ª Squadriglia in 1918. *(Photo De Vicari)*

In March 1917 France supplied its first Spad 7 that went to the 70ª Squadriglia, which in April generated the new 91ª that maintained its character of elite unit. In May 1917, during the 10th Battle of the Isonzo, the Italian air force had already established numerical superiority and was attaining supremacy over the enemy. The Italians considered the Austrian airmen gallant and dangerous foes, but their equipment had many limitations. Austrian reconnaissance biplanes were always much better than the Italian ones, that consisted in woefully obsolete Farmans, Caudrons and Voisins up to mid 1917, reliable but slow and badly defended Savoia Pomilio SP.2 and SP.3 pushers, valid but unremarkable SAML tractors since late 1916, and utterly unsafe SIA 7b and Pomilio P since late 1917. Austrian fighters, however, were generally inferior. After the early Fokker monoplanes and biplanes, the first real enemy fighter was the mediocre Brandenburg KD "star-strutter" that did not even have synchronized gun, and later the Aviatik Berg D.I with a dangerous tendency to shed its wings in flight. Only the Albatros Oeffag D.II and D.III that entered service in mid 1917 were really competitive fighters.

During the battle of the Ortigara, June 1917, and even more during the 11th Battle of the Isonzo, August 1917, Italian air supremacy became marked, as Austrian losses rose and during the peak of the battle for the Bainsizza Italian soldiers could see up to 200 friendly aircraft supporting their offensive.

## Italian Air Force Order of Battle, 1 August 1917

Comando Supremo
Raggruppamento Squadriglie da Bombardamento, Pordenone
IV Gruppo
Squadriglie 1ª, 8ª, 10ª, 13ª, 14ª Caproni
XI Gruppo
Squadriglie 2ª, 3ª, 4ª, 6ª, 7ª, 15ª Caproni
X Gruppo, Udine
Squadriglie 70ª and 82ª Nieuport, 91ª Spad S.Caterina, 114ª SP2 Campoformido

Comando Aeronautica 1ª Armata
III Gruppo, Verona
Squadriglie 5ª and 9ª Caproni, Verona, 31ª SP Castelgomberto, 46ª MF Castenedolo, 50ª MF Villaverla, 72ª SAML Brescia, 1ª Idro FBA Desenzano
IX Gruppo Squadriglie 37ª SP2 and SP3 Ponte S Pietro, 71ª Nieuport Villaverla, 73ª SAML and 75ª Nieuport Verona, 74ª MF/ SP Trenno

Comando Aeronautica 2ª Armata
II Gruppo, Udine
Squadriglie 21ª, 22ª, 33ª, 35ª and 40ª SP, 132ª P Campoformido, 27ª SP Chiasiellis
VI Gruppo, Oleis
Squadriglie 24ª SP, 29ª MF, 113ª SAML and III/83ª Nieuport Cavazzo Carnico, 41ª SP and 45ª MF Oleis, 76ª and 78ª Nieuport-Spad and 81ª Borgnano

Comando Aeronautica 3ª Armata
I Gruppo, S. Maria la Longa
Squadriglie 23ª SP and 84ª Nieuport S. Maria la Longa, 25ª Voisin Pozzuolo, 77ª Nieuport Spad and 80ª Nieuport Aiello, 112ª SAML Lavariano, 131ª P Chiasiellis, 2ª Idro FBA Grado
V Grupppo, Chiasottis
Squadriglie 28ª SP S.M.la Longa, 38ª SP Risano, 39ª SP

Sammardenchia, 42ª Caudron Medeuzza, 43ª Caudron Bolzano, 44ª Caudron Gonars

Comando Aeronautica 4ª Armata
XII Gruppo, Belluno
Squadriglie 48ª Caudron and II/83ª Nieuport Belluno, I/113ª SAML Villaverla

Comando Aeronautica 6ª Armata
VII Gruppo
Squadriglie 26ª SP Casoni, 32ª Farman S.Pietro in Gu, 49ª Caudron Nove, 79ª Nieuport Istrana

Albania
VIII Gruppo, Valona
Squadriglie 11ª Caproni, 34ª Farman, Sezione Nieuport

Macedonia
Squadriglie 47ª Farman, I/83ª Nieuport

In the summer of 1917 Italian squadrons started receiving from France the Hanriot Hd.1, which also was license built by Macchi in large quantities. The Hanriot was possibly the best fighter used in the Italian front: it was fast enough, highly maneuverable and remarkably safe, thanks to its wide wing surface. The Hanriot was described as "the fighter that turns an average pilot into a good pilot, and a good pilot into an ace".

In the late summer of 1917 the 91ª Squadriglia received also the first Spad 13 with 200-hp engine and two guns. This fighter was supplied in large numbers, and several hundreds of them were still crated when the war ended, and were used by the Italian air force for many years after the war. Meanwhile the Navy, that had to face the effective single-seat seaplane fighters of the Austrian Seeflieger, and that had to rely for escort on the Nieuports of the French squadron in Venice, started receiving from Macchi the new M single-seater, later called Macchi M.5, a formidable naval fighter.

In the autumn of 1917 the German army that was secretly sent to aid the Austrians came with its air component that included three fighters squadrons, Jasta 1, 31 and 39. When on 24 October 1917 the Central Empires struck, breaking through the Italian lines at Caporetto and putting the 2ª Armata into a rout, German airmen joined the action with the skill and aggressiveness typical of the murderous French front. German fighters scored heavily against the Italians, but German losses were just as great, both among fighters and among two-seaters.

The Caporetto retreat was terrible for Italian fighter units, because airfields had to be quickly abandoned and a large number of aircraft and equipment were lost, sometimes falling undamaged in enemy hands. Moving back from one field to the other, squadrons ran out of aircraft, and not much could come from the factories, as Nieuport 17 production was over, and Hanriot production was not yet in full swing. Help came from the Allies: the French rushed in 100 Nieuport 27 fighters, an airplane with the same engine as the Hanriot, but not with the same safety. The Nieuport 27 was quite successful in air combat, but it caused many losses in accidents.

## Italian Air Force Order of Battle, 20 November 1917

Comando Supremo:
X Gruppo
Squadriglie 91ª Padova, 70ª and 82ª Istrana
Raggruppamento squadriglie da bombardamento
XI Gruppo
Squadriglie 4ª, 5ª, 6ª Padova
IV Gruppo
Squadriglie 1ª, 8ª, 9ª and 13ª S. Pelagio
XIV Gruppo
Squadriglie 2ª, 3ª, 7ª, 10ª, 14ª and 15ª Ghedi

Aeronautica della 1ª Armata
III Gruppo
Squadriglie 37ª Brescia, 72ª, 120ª, 1ª sez./134ª Castenedolo, 75ª and 121ª Verona
IX Gruppo
Squadriglie 61ª, 71ª, 2ª sez./134ª Villaverla, 31ª Castelgomberto

Aeronautica Truppe degli Altipiani
VII Gruppo
Squadriglie 33ª, 2ª sez./139ª, 115ª Nove, 26ª and 32ª S.Pietro in Gu, 79ª Nove and Padova

Aeronautica della 4ª Armata
II Gruppo
Squadriglie 113ª, 114ª, 132ª and 133ª Casoni
XII Gruppo
Squadriglie 27ª, 48ª and 35ª Castel di Godego, 22ª and 36ª Istrana
VI Gruppo
Squadriglie 76ª, 78ª and 81 Istrana

Aeronautica della 3ª Armata
I Gruppo
Squadriglie 117ª, 131ª and 112ª Padova
V Grupppo
Squadriglie 23ª and 28ª Marcon, 38ª and 39ª Ca' Tessera
XIII Gruppo
Squadriglie 77ª, 80ª, 84ª and 2ª sez./83ª Marcon

The British sent a detachment of the Royal Flying Corps (later Royal Air Force), including three Squadrons equipped with Sopwith Camels, No. 28, No. 45 and No. 66. The British pilots showed the same aggressive spirit as the Germans, and during their stay in Italy, which lasted until the end of the war, they hit the Austrians badly, despite a certain tendency to overclaiming. The French also sent

N
3974
11324
TYPE 24

The Nieuport 27 was supplied to Italy in two batches of about 100 aircraft each. This is an airplane of the second batch, showing the removal of its French serial N 3974 and the application of the Italian one, 11324, on its rudder. It is in service in Macedonia, assigned to the 73ª Squadriglia.

The Pomilio PE was produced in large quantities as the standard two-seater for the Italian air force, but it suffered serious problems of instability and was much disliked. This is aircraft serial 18508 of the 22ª Squadriglia.

reinforcements, which include two fighter squadrons. Although this help was important, however, both on the ground and in the air the Italians managed to stop the enemy offensive by themselves. Allied troops entered service in the first lines only when the offensive was stopped, because the French Generals did not want the risk of their troops being routed with the Italian ones.

On 26 December 1917 a large German force attacked Istrana airfield that was the base of five fighter squadrons, 70ª, 82ª, 76ª, 81ª and 78ª, and of two reconnaissance units. The Italians were caught by surprise, but pilots rushed to their aircraft and took off among the falling bombs. Escorting German fighters, for unclear reasons did not intervene and the Hanriots shot down eight German bombers without losses. Later in the day another attacking German formation approached Istrana, and once more it was intercepted, losing three more bombers. This lopsided victory was a turning point in the air war, marking the end of the ascendancy briefly attained by the Central Empires.

On 28 January 1918 the Italian Army returned to the offensive attacking at Col del Rosso, and fighting in the air flared up. In the first six months of 1918 the front, now running along the Piave River, did not move, and air activity was limited. Italian squadrons were now receiving a constant stream of new fighters, particularly Hanriots, and new pilots, so they grew in size. Italian policy was not to increase the number of squadrons, but their size, that grew from an initial strength of nine pilots and aircraft in 1916 to a projected level of 18 pilots and 36 airplanes each for 1919. Nieuport 17 were no longer in service in front line squadrons, that were now equipped with Spads, Hanriots and Nieuport 27. The first two models particularly were absolutely reliable, and Italian fighter units were now fully confident and formidable.

Meanwhile, another form of conflict was being waged in the Adriatic between the two naval air services. The arrival of the Macchi M.5 fighter flying boat finally gave the Italian naval aviators air supremacy. A series of sharp encounters in which the M.5 bettered the Austrian seaplane fighters marked the turning point of the conflict, while the ever-growing Italian aircraft production ensured an absolute numerical supremacy also. The plans of the Regia Marina were far-reaching, should the war continue into 1919, and they included the formation of a large strategic bombing force --with a large number of American pilots – supported by squadrons of land fighters. Of the latter, however, only one was ready for operations before the end of the war, the 241ª Squadriglia at Venice.

In early June the Austrians, now fighting alone in Italy, as German troops had returned to the Western Front, planned the final offensive that in their hopes would break through the Piave and open all the plains of the Po River to their divisions. This time there was no surprise, so even fighter units were prepared for the onslaught. Squadrons were withdrawn from the control of the *Comandi d'Armata* and concentrated in a "*Massa da caccia*" fighter mass, under the command of *ten. col.* Piccio, consisting of Squadriglie 70ª, 82ª, 91ª, 75ª, 72ª, 76ª, 79ª and 80ª.

The Italian victory in the air was as definitive as the success on the ground. Italian fighters shot down a large number of enemy aircraft, while bombers and reconnaissance aircraft, flying without respite, had a decisive role with their unstopping attacks against gangways and boat bridges that the Austrians were trying to use to bring troops and supplies to the western bank of the Piave River. A real tempest of fire from the sky decimated Austrian troops and ensured the Italian victory. After the battle the *Massa da Caccia* was disbanded and fighter squadrons returned to their Groups.

# Italian Air Force Order of Battle, mid-June 1918

| Gruppo | Squad. | Base | Pilots | | Observers | Gunners | Aircraft | |
|---|---|---|---|---|---|---|---|---|
| | | | Officers | Troops | | | Type | Quantity |
| **COMANDO DELL'AERONAUTICA A DISPOSIZIONE – PADOVA** | | | | | | | | |
| IV | 1 | S.Pelagio | 5 | 6 | 6 | 9 | Ca.450 | 6 |
| | 5 | S.Pelagio | 3 | 5 | 6 | 9 | Ca.450 | 4 |
| | 8 | S.Pelagio | 4 | 5 | 5 | 7 | Ca.450 | 5 |
| | 13 | S.Pelagio | 4 | 4 | 7 | 8 | Ca.450 | 3 |
| | 87 | S.Pelagio | 15 | | | | SVA 17 | |
| XIV | 2 | Padova | 6 | 6 | 8 | 11 | Ca.450 | 3 |
| | 7 | Padova | 6 | 7 | 7 | 11 | Ca.450 | 5 |
| | 9 | Padova | 3 | 7 | 6 | 8 | Ca.450 | 4 |
| | 10 | Padova | 4 | 4 | 6 | 9 | Ca.450 | 4 |
| XI | 4 | Ca' degli Oppi | 6 | 4 | 7 | 10 | Ca.450 | 7 |
| | 6 | Ca' degli Oppi | 5 | 6 | 7 | 7 | Ca.450 | 5 |
| | 181 | Ghedi | 5 | 8 | 5 | 18 | Ca Triplano | 8 |
| X | 70 | Gazzo | 11 | 8 | | | HD | 18 |
| | 82 | Gazzo | 9 | 9 | | | HD | 21 |
| | 91 | Quinto di Treviso | 10 | 4 | 1 | | SPAD 13 | 17 |
| **COMANDO DELL'AERONAUTICA DELLA 1ª ARMATA** | | | | | | | | |
| III | 61 | Ganfardine | 4 | 12 | 11 | 10 | SP.4 | 8 |
| | 75 | Busiago | 9 | 5 | | | Ni-27 & HD | 15 |
| | 134 | Ganfardine | 3 | 7 | 10 | 4 | Pomilio PE | 14 |
| | 1ª | S.SVA Ganfardine | 5 | 2 | | | SVA | 6 |
| XVI | 31 | Castelgomberto | 2 | 5 | 12 | | SP.3 and SIA | 6 |
| | 71 | Castelgomberto | 5 | 8 | | | SPAD and HD | 13 |
| | 121 | Castelgomberto | 5 | 5 | 14 | | SAML | 7 |
| | 135 | Castelgomberto | 1 | 6 | 10 | 2 | Pomilio PE | 10 |
| **COMANDO DELL'AERONAUTICA DELL'8ª ARMATA** | | | | | | | | |
| XV | 78 | San Luca | 7 | 13 | | | HD | 20 |
| | 79 | San Luca | 7 | 12 | | | Ni-27 and HD | 16 |
| | 115 | San Luca | 7 | 6 | 10 | | SAML e PE | 7 |
| | 139 | San Luca | 3 | 7 | 10 | 8 | Pomilio PE | 8 |
| | 4ª S.SVA | San Luca | 4 | 2 | | | SVA | 6 |
| XIX | 23 | Istrana | 2 | 6 | 9 | | SP.3 | 5 |
| | 114 | Istrana | 3 | 9 | 19 | 2 | SAML | 6 |
| | 118 | Istrana | 3 | 8 | 17 | | SAML | 7 |
| | III/24 | Istrana | 1 | 2 | 5 | 1 | | |
| **COMANDO DELL'AERONAUTICA DELLA 3ª ARMATA** | | | | | | | | |
| I | 117 | Fossalunga | 9 | 10 | 10 | | Pomilio PE | 7 |
| | 5ª S.SVA | Fossalunga | 4 | 2 | | | SVA | 7 |
| | 28 | Marcon | 2 | 9 | 10 | | SIA 7b | 4 |
| | 38 | Malcontenta | 4 | 8 | 12 | | SIA 7b | 6 |
| | 39 | Malcontenta | 8 | 10 | 7 | | SAML | 11 |
| | 62 | Marcon | 3 | 4 | | 5 | SP.4 e PE | 4 |
| | 131 | Marcon | 2 | 7 | 11 | 3 | Pomilio PE | 10 |
| | I/24 | Marcon | | 3 | 1 | | | |
| XIII | 77 | Marcon | 7 | 3 | | | SPAD e HD | 16 |
| | 80 | Marcon | 11 | 6 | | | HD | 15 |

| Gruppo | Squad. | Base | Pilots Officers | Pilots Troops | Observers | Gunners | Aircraft Type | Aircraft Quantity |
|---|---|---|---|---|---|---|---|---|
| **COMANDO DELL'AERONAUTICA DELLA 4ª ARMATA** | | | | | | | | |
| II | 132 | Isola di Carturo | 2 | 6 | 11 | 3 | Pomilio PE | 11 |
| | 133 | Isola di Carturo | 2 | 2 | 10 | 3 | Pomilio PE | 10 |
| | 3ª S.SVA | Isola di Carturo | 5 | 1 | | | SVA | 8 |
| VI | 76 | Casoni | 8 | 6 | | | HD | 17 |
| | 81 | Casoni | 6 | 9 | | | HD e Ni-27 | 20 |
| XII | 22 | Casoni | 4 | 3 | 11 | | SIA 7b | 4 |
| | 27 | Castel di Godego | | 5 | 9 | | SIA 7b | 1 |
| | 35 | Castel di Godego | 3 | 5 | 8 | | SIA 7b | 5 |
| | 36 | Casoni | 2 | 1 | 8 | | SIA 7b | 4 |
| | 48 | Castel di Godego | 1 | 2 | 9 | 1 | Cau G.4 | 3 |
| | II/24 | Casoni | 2 | 3 | 15 | | SIA 7b | 3 |
| **COMANDO DELL'AERONAUTICA DELLA 6ª ARMATA** | | | | | | | | |
| VII | 26 | S. Pietro in Gu | 3 | 9 | 14 | 1 | SP.3 e SIA | 12 |
| | 32 | S. Pietro in Gu | 3 | 7 | 13 | | SIA 7b | 9 |
| | 33 | S. Pietro in Gu | 3 | 6 | 12 | | SP.3 | 5 |
| | 83 | Poianella | 3 | 13 | | | Ni-27 e HD | 20 |
| | 2ª S.SVA | Poianella | 6 | | | | SVA | 6 |

Below: An assortment of Italian aircraft probably at S. Pelagio airfield: on the left there are two SVAs of the 87ª Squadriglia, in the background two Caproni Ca.3s, one showing the four leaf clover insignia of the 10ª Squadriglia, and then a Nieuport 17, an Hanriot of the 82ª Squadriglia and an old Nieuport 11, serial 163?.

| Gruppo | Squad. | Base | Pilots | | Observers | Gunners | Aircraft | |
|---|---|---|---|---|---|---|---|---|
| | | | Officers | Troops | | | Type | Quantity |
| **COMANDO DELL'AERONAUTICA DELLA 7ª ARMATA** | | | | | | | | |
| IX | 72 | Busiago | 6 | 7 | | | HD | 11 |
| | 74 | Castenedolo | 5 | 14 | | | Ni-27 e HD | 17 |
| | 112 | Castenedolo | 3 | 6 | 11 | 1 | Pomilio PE | 11 |
| | 120 | Castenedolo | 2 | 4 | 13 | 2 | SAML e PE | 3 |
| | 136 | Castenedolo | 3 | 7 | 12 | | Pomilio PE | 10 |
| | 6ª S.SVA | Castenedolo | 6 | 2 | | | SVA | 7 |
| XX | 37 | Ponte S.Marco | 4 | 6 | 20 | 1 | SP.3 e SIA | 7 |
| | 40 | Pian Camuno | 5 | 6 | 10 | | SIA 7b | 2 |
| | 113 | Medole | 4 | 8 | 7 | 1 | Pomilio PE | 3 |
| **XVI CORPO D'ARMATA IN ALBANIA** | | | | | | | | |
| VIII | 11 | Tahiraqa | 3 | 4 | 3 | 5 | Ca.450 | 4 |
| | 85 | Piskupi | 12 | 1 | | | Ni-17 e HD | 16 |
| | 116 | Valona | 3 | 6 | 11 | | SAML | 9 |
| **35ª DIVISIONE SALONICCO** | | | | | | | | |
| XXI | 73 | Salonicco | 2 | 2 | | | Ni e SPAD | 11 |
| | 111 | Kremjan | 8 | 1 | 9 | 3 | SAML | 12 |
| **FRANCIA** | | | | | | | | |
| XVIII | 3 | Ochey | 5 | 5 | 5 | 5 | Ca.450 | 16 |
| | 15 | Ochey | 5 | 4 | 4 | 5 | Ca.450 | 4 |

The FBA flying boat was another French design widely built under license in Italy. These are aircraft of the 280ª Squadriglia at Milazzo, Sicily.

The large Caproni Ca.4 triplane served with only two squadrons, 181ª and 182ª Squadriglia. After an unauspicious debut in June 1918, it was then grounded for modifications and returned to action in the final week of the war.

The Ansaldo SVA was, with the Caproni, the most successful Italian airplane of the war. This is an airplane of the 87ª Squadriglia, the crack unit that flew to Vienna lead by Gabriele d'Annunzio on 9 August 1918.

After the Piave victory, the Austrian air force almost disappeared from the skies, having suffered terrible losses. It was not a spent force, however, and during the final offensive of October 1918 it was again in the air, fighting with exceptional valor and still inflicting hard losses to the Italians, but again suffering much heavier losses itself.

What are the reasons of this Italian supremacy in air combat? The reasons of Austrian weakness are many, include ethnic differences among its pilots, and also the terrible quality of its fighters: the KD was a donkey, the Berg D.1 was unreliable, shedding its fabric in flight, the Phoenix looked like a pressing iron and probably flew like one, and only the Albatros was a good fighter. The main reason of the Italian victory probably was the Hanriot, an excellent plane, who gave its pilot the highest confidence. The comparison that comes to mind is the F6F Hellcat in the Pacific War.

On 21 October 1918 General Bongiovanni, chief of the air force, gave the orders for the final battle, the offensive that started on Oct. 24 and that was called the Battle of Vittorio Veneto. The *Massa da Caccia* was reformed using Squadriglie 91ª, 71ª and 72ª at Quinto, 70ª and 82ª at Gazzo, 77ª at Marcon, with a British squadron. A *Massa da Bombardamento* was formed with the IV, XI and XIV Gruppi with nine Caproni squadrons, the new XXII Gruppo with two SVA squadrons, and the Navy two squadrons of Ca.4 triplanes. Italian fighters quickly routed the last Austrian fighters, gallant but completely outgunned, and in the final days of the war they were used for strafing Austrian troops then, as the retreat turned into a rout, they stopped shooting at the now defeated enemy. Armistice went into effect on 4 November 1918.

The Ansaldo SVA had a very long gestation, but at the end of the war it was available in very large numbers, both as a single- and a two-seater, and it remained in service for over a decade in colonial duties.This is an early SVA 10 two-seater of the 58ª Squadriglia.

An apparently flyable Nieuport 17 of the 80ª Squadriglia serves as a prop for a ceremony of decorations to airmen, 20 September 1918. The host is the Duke of Aosta, commander of the 3rd Army, with folded arms.

# Aircraft Available for War Operations, 4 November 1918

| Gruppo | Base | Squadriglia | Base | A/c | Qty. |
|---|---|---|---|---|---|
| ***Aeronautica of the Supreme Command*** | | | | | |
| ***Massa da caccia*** | | | | | |
| X | Gazzo | 70 | Gazzo | HD | 18 |
| | | 82 | Gazzo | HD | 19 |
| XVII | Quinto di Treviso | 91 | Quinto di Tv. | Spad | 12 |
| | | 71 | Quinto di Tv. | Spad | 13 |
| | | 72 | Quinto di Tv. | HD | 13 |
| | | 77 | Marcon | Spad | 13 |
| ***Massa da Bombardamento*** | | | | | |
| IV | San Pelagio | 1 | San Pelagio | Caproni 3 | 2 |
| | | 5 | San Pelagio | Caproni 3 | 2 |
| | | 8 | San Pelagio | Caproni 3 | 3 |
| | | 13 | San Pelagio | Caproni 3 | 1 |
| XIV | Arquà Petrarca | 2 | Arquá Petrarca | Caproni 3 | 1 |
| | | 7 | Arquá Petrarca | Caproni 3 | 4 |
| | | 10 | Arquá Petrarca | Caproni 3 | 3 |
| XI | Padova | 4 | Padova | Caproni 3 | 0 |
| | | 6 | Padova | Caproni 5 | 7 |
| XXII | Busiago | 89 | Busiago | SVA 6 | 10 |
| | | 90 | Busiago | SVA 5 | 8 |
| | Regia Marina | 181 | P.o Renatico | Caproni 4 | 3 |
| | | 182 | P.o Renatico | Caproni 4 | 3 |
| | | 1 S.A. | Venice | SVA, SIA, Ca | 7 |
| | | 87 | S.Pelagio | SVA | 14 |
| | | Sez.Cavalleria | S.Pelagio | SVA | 7 |
| Airships: F5, M19, M11, F6, M14 | | | | | |
| ***Comando Aeronautica della 3ª Armata - Carpenedo*** | | | | | |
| V | Carpenedo | (28) | | | |
| | | 38 | Malcontenta | PE | 6 |
| | | 39 | Malcontenta | PE | 5 |
| XIII | Marcon | 80 | Marcon | HD | 13 |
| I | Carpenedo | 131 | Marcon | PE | 7 |
| | | 5ª Sez. SVA | Marcon | SVA | 6 |
| Sp.I | Ca' Tessera | 9 | Ca'Tessera | Caproni 3 3 | |
| | | Mista | Ca'Tessera | | 3 |
| ***Comando Aeronautica della 8ª Armata - Albaredo*** | | | | | |
| XV | Paese | 115 | Fossalunga | SAML | 15 |
| | | 56 | Fossalunga | SVA | 15 |
| XIX | Falz | 23 | Istrana | PE | 6 |
| | | 114 | Istrana | PE | 10 |
| | | 118 | Istrana | PE | ? |

| Gruppo | Base | Squadriglia | Base | A/c | Qty. |
|---|---|---|---|---|---|
| XXIII | S.Luca | 78 | San Luca | HD | 16 |
| | | 79 | San Luca | HD | 18 |
| ***Comando Aeronautica della 4ª Armata - Bolzanella*** | | | | | |
| II | Castel di Godego | 27 | Cast.di Godego | PE | 8 |
| | | (48) | | | |
| VI | Casoni | 76 | Casoni | HD | 17 |
| | | 81 | Casoni | HD | 17 |
| XII | Casoni | 22 | Casoni | PE | 4 |
| | | 36 | Casoni | PE | 8 |
| | | 21 | Is. di Carturo | SAML | 7 |
| | | 57 | Is. di Carturo | SVA | 8 |
| ***Comando Aeronautica della 6ª Armata - Breganze*** | | | | | |
| VII | S.Pietro in Gu | 26 | S.Pietro in Gu | PE | 6 |
| | | 32 | S.Pietro in Gu | PE | 7 |
| | | 33 | S.Pietro in Gu | SP.3 | 6 |
| XXIV | Poianella | 83 | Poianella | Nieuport | 14 |
| | | 2ª Sez.SVA | Poianella | SVA | 2 |
| | | 136 | Villaverla | PE | 0 |
| ***Comando Aeronautica della 1ª Armata - S.Martino Buonalbergo*** | | | | | |
| III | Ganfardine | 61 | Ganfardine | PE | 4 |
| | | 75 | Ganfardine | Nieuport | 17 |
| | | 134° | Ganfardine | PE | 2 |
| | | 1° Sez.SVA | Ganfardine | SVA | 3 |
| | | 1° Sez.Idro | Desenzano | FBA | 3 |
| | | 2ª Sez.Idro | Desenzano | FBA | 3 |
| XVI | Castelgomberto | 31 | Castelgomberto | PE | 7 |
| | | 121 | Castelgomberto | SAML | 4 |
| | | 134B | Castelgomberto | PE | 2 |
| ***Comando Aeronautica della 7ª Armata - Brescia*** | | | | | |
| XX | Brescia | 74 | Ponte S.Marco | HD | 10 |
| | | 113 | Cividate Cam. | SAML | 3 |
| | | 120 | Ponte S.Marco | PE | 5 |
| IX | Castenedolo | 112 | Castenedolo | PE | 3 |
| | | 6° Sez SVA | Castenedolo | VA | 2 |
| ***Comando Aeronautica della 10ª Armata*** | | | | | |
| | | 28 | Malcontenta | PE | 6 |
| | | 34 French | | AR | 14 |
| | | 139 Sqn. RAF | | F2B | 10 |
| | | 28 Sqn. RAF | | Camel | 16 |
| | | 66 Sqn. RAF | | Camel | 18 |
| ***Comando Aeronautica della 12ª Armata*** | | | | | |
| | | 48 | Cast.di Godego | PE | 7 |
| | | 48 Sezione | | PE | 4 |
| | | 22 French A.F. | | AR | 10 |
| | | 254 French A.F. | | AR | 10 |
| | | 561 French A.F. | | Spad | 6 |

| Gruppo | Base | Squadriglia | Base | A/c | Qty. |
|---|---|---|---|---|---|
| XVIII | France | 3 | Chernizey | Caproni | 3 |
| | | 14 | Chernizey | Caproni | 3 |
| VIII | Valona | 11 | Tahyraqua | Caproni | 3 |
| | | 85 | Piskupi | Nieuport | |
| | | 116 | Valona | SAML | |
| XXI | Saloniki | 73 | Dudular | Nieuport | |
| | | 111 | Dudular | SAML | |
| | | 12 | Tripoli | Caproni | 3 |
| | | 106 | Tripoli | Farman | |

## Short Histories of the Italian Fighter Squadrons

### 70ª Squadriglia

This squadron was the 1ª Squadriglia Caccia, based at S. Caterina, Udine, and renamed after 15 April 1916. At that date, it consisted of *capitano* Guido Tacchini, commander, *cap.* Domenico Bolognesi, *ten.* Francesco Baracca, *serg.* Giulio Poli, *caporali* Anselmo Caselli and Antonio Pagliari and soldier Luigi Olivari. It was equipped with eight Nieuport 10 biplanes. It was not part of any group, being autonomous, reporting directly to the Supreme Command.

In August the unit received *ten.* Fulco Ruffo di Calabria, and it used eight Nieuport 11 and four Nieuport 10. It also had the Brandenburg C.1 shot down by Baracca on 7 April, that had been repaired and tested by Francesco Brach Papa.

The 70ª Squadriglia was much used in air patrols and escorts to Caproni bombers. In October it was supplied with Le Prieur rockets for balloon busting. It received its first Nieuport 17 110-hp fighter on 21 October 1916. The unit suffered its first casualty on 11 November, when *sergente* Renato Ghizzoni crashed to his death in a failed take-off. At the end of 1916, the 70ª Squadriglia had flown 600 missions, was engaged in 52 air combats and scored seven confirmed victories.

On 1 January 1917, its commander was still *cap.* Tacchini, and its pilots were *cap.* Baracca, *cap.* Bosio, *ten.* Ruffo, *s.ten.* Giuseppe Gentili, and *sergenti* Caselli, Gorini, Olivari, Pagliari, Poli, and Ferruccio Guicciardi. It was equipped with Nieuport 11, with serials 1609, 1639, 1641, 1659, 1662, 1685, 1698, 1764, 2127, two Ni 17, serials 2614 and 3127, and another Nieuport, serial unknown.

On 17 January 1917 *s.ten.* Gentili spun and crashed to his death. The unit was very active since February, intercepting a growing number of Austrian raids. On 15 March Olivari, now a *sottotenente*, and Ruffo ferried the first two Spad 7s, serials 4688 and 4689.

On 10 April 1917 the 70ª Squadriglia was assigned to the new X Gruppo, commanded by *mag.* Piccio. The unit flew many escort missions to Capronis, together with the fighters of the 78ª and 79ª, and also used camera-equipped Spads for photo reconnaissances.

In late April the units got new pilots, *sergente* Guido Nardini and *tenenti* Bortolo Costantini and Vittorio Melloni, while *cap.* Tacchini left command to *cap.* Buzzi Gradenigo.

On 1 May 1917, the new 91ª Squadriglia was formed, that received the Spads and all the best pilots, so the 70ª became to all effects a new unit, equipped just with 80 hp and 110 hp Nieuports, and with the single personal Spad of Group commander Piccio.

In August 1917 the 70ª included pilots Buzzi Gradenigo, commander, Caselli, Guicciardi, Pagliari, *ten.* Felice De Gasperi, *ten.* Enrico Ferreri, *s.ten.* Aldo Bognetti, *aspirante* Gioacchino Sartor, *serg.* Fernando Nelli and Angelo Pini.

After the Austro-German offensive of 24 October, this unit left S. Caterina and moved to La Comina, then on 31 October to Arcade, and finally on 9 November to Istrana. Its equipment consisted of seven Nieuport 17s that later in November were replaced by Hanriots. On 2 December Buzzi Gradenigo was replaced by *ten.* Rino Corso Fougier, future chief of staff of the Italian Air Force

The original 70ª Squadriglia at S. Caterina generated the 91ª in the spring of 1917. In this photo dated early 1917 one can see four Nieuport 11s, the last two with Colt machine guns, followed by three French built Ni.17s with the personal insignias of Poli, Ruffo and Baracca.

70ª Squadriglia, summer 1918: three Spads, two of them with horizontal stripes, then the three Hanriots of *tenenti* Guglielmotti, R esch and Eleuteri, then two Spads with vertical stripes.

during World War II. On 6 December the 70ª Squadriglia was made available to the Comando Truppe Altipiani (Command of troops on the Asiago plateau)

During the Battle of Istrana, 26 December 1917, only one of its pilots took off, *sergente* Attilio Ferrandi, while four ground troops were killed, and bombs destroyed Hanriots 6152, 6156, 6172, 6181 and 6632. At the turn of 1918, the unit was nominally commanded by *cap.* Ferdinando Martinengo, who was sick, and its pilots were *tenenti* Flaminio Avet, Felice De Gasperi, *s.ten.* Alessandro Resch, *serg.* Ferruccio Guicciardi, Ferdinando Nelli, Angelo Pini, Gioacchino Sartor. On 20 January Avet became the unit commander, and the 70ª moved to S. Pietro in Gu airfield. On 1 February there was one further command change for the troubled unit, with the arrival of *mag.* Vincenzo Lombard.

On 26 February, Austrian ace Kurt Gruber shot down and killed *sergente* Giuseppe Ghislanzoni on HD 6244.

On 14 March the unit moved back to Istrana. It now had a complement of 19 fighters. On 26 March Lombard was promoted commander of the VI Gruppo, and the new squadron commander became *cap.* Ernesto Sequi. On 21 May 1918 the unit moved to Gazzo airfield.

In June, during the Austrian offensive, it became a part of the *Massa da Caccia*. It took part in the battle with many strafing and bombing missions. In one of these, *cap.* Sequi was seriously wounded. Avet became the unit commander again. On June 19, *ten.* Ugo Capitanio was seriously wounded during a balloon-busting mission; he landed at Marcon but died of wounds the next day.

On 1 July 1918 the pilots of the unit were *ten.* Avet, temporary commander, *tenenti* Antonio Albertini, Dino Bernelli, Aldo Bognetti, Alessandro Contardini, Luigi Giovanardi, Enrico Guglielmotti, Leopoldo Eleuteri, *s.ten.* Resch, *sergenti* Aldo Bocchese, Giorgio Croci, Angelo Pini, *caporal maggiore* Aldo Torranza, soldiers Ferruccio Bucciarelli, Amedeo Guarnieri and Pietro Petitti. They flew a force of 20 Hanriots.

*Cap.* Giulio Cesare Paldaoff became commander on 24 July. In late August the 70ª Squadriglia received new aircraft, Ansaldo Balilla serials 16550, 16555, 16556 and 16558, and Spads 1360 and 1398.

On 8 October 1918 Eleuteri scored the only Balilla victory on the Italian front, flying 16558. On 23 October the 70ª was assigned again to the new *Massa da Caccia* for the final offensive.

On armistice day the unit was based at Gazzo, equipped with 18 Hanriots, 5 Spads and one Balilla, commanded by Avet, and its pilots were *tenenti* Antonio Bandino, Beduschi, Bernelli, Stefano Caselli, Eleuteri, Enrico Guglielmotti, Alessandro Resch, Carlo Tacchi Venturi, *sergenti* Bocchese, Giorgio Croci, Luigi Lucchini, *cpr.* Pietro Allegri, Giovanni Baccicchetto, Amedeo Guarnieri, Enrico Poletti, Aldo Torranza, soldiers Bagarotti, Alfredo Crotti and Ribesco.

During 1918 the unit flew 1357 combat missions, had 82 air combats and claimed 12 air victories. It was disbanded in December 1918.

## 71ª Squadriglia

The 71ª Squadriglia was the 2ª Squadriglia Caccia, renamed on 15 April 1916, when it was based at Cascina Farello. Its commander was *cap.* Giorgio Chiaperotti with pilots *ten.* Antonio Sabelli, *sergenti* Giuseppe Barattini, Roberto Ghelfi, Ferruccio Guicciardi, *caporale* Edoardo Olivero, soldier Eugenio Spano. It was equipped with Nieuports 11 658, 1249, Nieuport 10 single seater 1040 and 10 two-seaters 1263 and 1264.

On 23 May the unit was shifted to Villaverla, in the north, assigned to the XIV Corpo d'Armata, with Nieuports 1047, 1263, 1264, 1265, 1266, 1449 and 1766. On 7 June the unit had its first loss, when *serg.* Ghelfi collided with the aircraft of Chiaperotti during a mock dogfight.

On 8 July 1916 the 71ª became part of the III Gruppo. The unit was busy trying to intercept Austrian raids in the region of Vicenza, but it suffered from an inadequate alarm network, and its takeoffs and climbs are always late. In October it flew supporting an Italian offensive in the area of the Pasubio mountain. On October 30 it received its first Nieuport 17 110 hp. Its failures to intercept Aus-

The 71ª Squadriglia progress through all the types of fighters used by Italy: Ni. 10, Ni.11, Ni.17, Spad 7, Hanriot and Spad 13. These are Spad 7s in 1917.

trian aircraft were criticized, while Italian batteries shot indifferently at friend and foe. At the end of 1916, the 71ª Squadriglia had flown 523 combat missions, with 74 combats.

On 1 January 1917 the 71ª Squadriglia had *cap.* Chiaperotti as commander, with pilots Barattini, Guicciardi, Olivero, Tola, *capitano* Amerigo Notari, *s.ten.* Adriano Bacula, *sergenti* Mario Banino, Dino Menegoni and Arrigo Rossi. It was equipped with eight Nieuport 11, and one Nieuport 17 serial 3123.

Notari became commander on 17 January 1917, replacing Chiaperotti. On 21 January Nieuport 11 serial 1622 piloted by Menegoni lost its right wings in flight and crashed on the airfield. Flights of all Macchi-built Nieuports were stopped, until the bracing wires were replaced.

From January to April 1917, the unit operated sporadically, due to bad weather, in patrol and escort flights. On 10 May it was passed to the IX Gruppo, and it had as new pilots *ten.* Giuseppe Rigoni and *sergenti* Antonio Amantea and Giovanni Lenzoni.

On 15 May the 71ª received its first Spad, that was used for photo reconnaissances. During the summer, the 71ª fought against enemy aircraft in the northern sector, scoring several victories. After the Caporetto defeat, the unit sent a flight with Amantea, *s.ten.* Aldo Anesini, *sergenti* Massimiliano Bongioanni and Giampiero Vecco, as a reinforcement to the 82ª Squadriglia at Campoformido, while it received a flight of SAML two seaters of the 121ª Squadriglia commanded by *s.ten.* Leopoldo Eleuteri, that remained until November.

In mid November, as Villaverla fell within the range of Austrian artillery, the 71ª Squadriglia retreated to Sovizzo. In December, as the final effort of the Austrian offensive was spent on the Asiago plateau, the 71ª Squadriglia was engaged in ground strafing. Equipped with twelve Spads, the unit was assigned to the XVI Gruppo on 16 December 1917. On 30 December *serg.* Pietro Camperi on Spad 5413 was shot down and killed by German fighters. During 1917 the unit flew 1288 combat flights and engaged in 89 air combats.

In January 1918 the 71ª Squadriglia, commanded by Notari, had pilots Amantea, Anesini, Barattini, Bongioanni, Ercoli, Tola, Vecco, Vulcano, *cap.* Cesare Buzzi Gradenigo, *s.ten.* Agostino Consigli, *s.ten.* Goffredo Gorini and *s.ten.* Giorgio Orsini. Its Spads had serials 1401, 1530, 1531, 1669, 1683, 1686, 1688, 1715, 2474 and 5422. The unit received a SVA flight with *cap.* Natale Palli and *serg.* Giovanni Bartolomeo Arrigoni, and an Hanriot flight from the 72ª led by *ten.* Sebastiano Bedendo.

In February, while *cap.* Palli was commander of the 71ª, the Hanriot flight returned to the 72ª, but Bedendo remained. On March 1 the new commander was *ten.* Anesini. In May, Sovizzo airfield was flooded, so the squadron moved to Castelgomberto, and at the same time lost the SVA Flight.

On 28 May 1918 *cap.* Ettore Croce, an expert leader and test pilot, became the new commander, but he died three days later, crashing with a pupil during a mock dogfight. Anesini became commander again.

On June 30, the 71ª Squadriglia had a stregth of 12 Spads and three Hanriots, and its pilots were *tenenti* Anesini and Bedendo, *s.ten.* Consigli and Vecco, *sergente maggiore* Giovanni Rosso, *sergenti* Guido Consonni, Augusto Levrero, Andrea Mancini, Pereno, Tola and Vulcano. *Cap.* Salvatore Breglia became commander on July 16.

On August 28 the unit was reinforced by an Hanriot flight from the 75ª, with six pilots. On September 23 the unit was assigned to the new XVII Gruppo. On October 16 it moved to Gazzo airfield, and became part of the *Massa da Caccia* assembled for the final offensive. It sent a flight of six Spads with pilots Breglia, Bedendo, Mancini, Natalini, Pereno, Vecco and *cpr.* Giuseppe Dal Santo to form a mixed squadron with six Niueport 27s of the 75ª Squadriglia, that was active in strafings and bombings in the sector of Conegliano. The other flight was deployed at Castelgomberto with pilots Anesini, Danieli, Levrero, Spuri, Vulcano, *capitani* Arnoldo May and Arnaldo Zocca, *ten.* Luigi Dufour, *s.ten.* Agostino Consigli and. Umberto Corti and *serg. m.* Giovanni Rosso. During the final days, ground fire took a heavy toll, *cap.* Breglia was wounded, Vecco and *s.ten.*

Modesto Strazzabosco crash-landed. During 1918, the 71ª Squadriglia flew 1183 war missions and had 46 combats.

## 72ª Squadriglia

The 3ª Squadriglia Caccia, based at Brescia with *cap.* Gino Matteucci and pilots *ten.* Umberto Re, *sergenti* Angelo Alberto and Remigio Robino, *caporali* Adriano Cargiani, Umberto Ferro, Gustavo Rizzoli and soldier Mario Moccafighe, became the 72ª Squadriglia on 15 April 1916. It had four Aviatiks and two Farmans. It was rarely employed and didn't see much action. In December *cap.* Leone Tappi became its new commander. It was part of the III Gruppo.

In January 1917 it had new pilots *sergenti* Luigi Lucchini, Antonio Nava, Alfredo Tagliapietra and many gunners, including Luigi Burei, future test pilot at Aermacchi. In February it received SAMLs that replaced its Aviatiks. In March it moved from Brescia to nearby Ghedi. On 17 April *serg.* Cargiani died in a flight accident with SAML 1581.

On 11 May all the squadron aircrfact took off to contrast an Austrian raid, but SAML 1580 crashed, with the death of the unit commander, *cap.* Tappi and his gunner Dondelli. *Cap.* Teresio Ferrè became the new commander. On 20 September 1917 the numbers of squadrons were rationalized, and the 72ª, to all effects a two-seaters unit, became 120ª Squadriglia.

A new 72ª Squadriglia was formed on 22 October 1917 at Castenedolo with personnel and aircraft from the 75ª. Its commander was *ten.* Giuseppe Rigoni, it was equipped with Nieuport 17s and was part of the III Gruppo. It flew only 19 missions in October and November, then it passed to the IX Gruppo and re-equipped with Hanriots.

By the end of the year, it was formed by commander Rigoni, then replaced by *cap.* Mario Omizzolo, with *ten.* Sebastiano Bedendo, Giovanni Fasoli, Natale Palli, *sergenti* Giovanni Bartolomeo Arrigoni, Giovanni Gerardini, Teresio Marcellino, Mario Pereno, Francesco Pomponi, Arrigo Rossi, Attilio Simonetta, soldiers Gino Baldi and Ezio Franzi.

In January the unit had eleven Hanriots, two 80 hp Nieuports and two Farman Colombo, used as night fighetrs above Brescia. This Farman flight passed to the 37ª Squadriglia in March. On 19 January 1918 three pilots and fighters were deployed at Sovizzo, as reinforcement to the 71ª Squadriglia, for one month..

The unit had its first combat on 18 March, when *serg.* Franchi and *serg.* Celestino Giordanino shot down Brandenburg 269.06 of Flik 10/F. On May 13, Franzi shot down another enemy, returned to base and then took off again to take pictures of his victim. He didn't come back, and was captured by the Austrians with his intact Hanriot.

In May the unit was equipped with 18 Hanriots, one Nieuport and one Spad, 5384. Its pilots were *cap.* Pietro Marchesini, *tenenti.* Ottorino Frazzi, Arrigo Archibugi, Camillo Sommariva, Giuseppe Rolando, *sergenti* Gerardini, Giordanino, Teresio Marcellino, Serafino Monti, Pomponi, Giovanbattista Vercelli, *cpr.* Giovanni Dematteis and soldiers Vincenzo Gallo and Zanchi.

In June the unit operated from Busiago airfield as part of the *Massa da Caccia.* On 17 June *ten.* Sommariva had a fierce duel and crash landed, dying of his wounds two days later.

The 72ª Squadriglia used in 1918 the Lion of St. Mark as its unit insignia. This is Hanriot 11403 of *soldato* Ezio Franzi, forced down by engine failure behind enemy lines on 13 May 1918. *(Photo Tötschinger).*

On 1 July the 72ª Squadriglia returned to Castenedolo. Its new pilots were *ten.* Aldo Passalacqua and *sergente* Riccardo De Paolini. On July 16 *ten.* Amerigo Crocetti and *serg.* Pomponi, escorting a two-seater, had a fierce duel with Austrian pilots Navratil, Peter and Stec, of Flik 3/J. Crocetti was wounded.

On 17 July the 72ª Squadriglia was assigned to the new XXIV Gruppo. On August 4 Archibugi crashed to his death on Hanriot 646 as he was practicing aerobatics above Ghedi. On 7 September Omizzolo became the new commander of the 79ª Squadriglia, and was replaced by *cap.* Renato Tasselli. As the XXIV Gruppo was disbandend, the 72ª passed under the XX Gruppo on 25 September, then on 16 October it went to the XVII and moved to Quinto di Treviso airport.

As part of the *Massa da Caccia*, this unit took part in the final actions of the war. *Ten.* Gino Suali, on 4 November, crashed during an observation flight. This was the last casualty of the Italian Air Force in the conflict.

The 72ª Squadriglia carried as unit emblem the winged Lion of St. Mark painted in red on the fuselages. This unit flew 1358 combat missions and had 61 air combats. It was disbanded on 10 December 1918.

## 73ª Squadriglia

The 4ª Squadriglia Caccia at Verona, part of the III Gruppo, became on 15 Aprile 1916 73ª Squadriglia. Its commander was *cap.* Ferdinando Sanità and it was equipped with Aviatiks 1423, 1424, 1595, 1598 and Lloyd 42.11. On 10 May the unit moved to S.Anna d'Alfaedo, where it remained until September, then returning to Verona.

At the beginning of 1917 the 73ª Squadriglia da Difesa had commander Sanità with pilots *cap.* Arnaldo Bevagna, *s.ten.* Gino Tonietti and *serg.* Giuseppe Pascoli. It was re-equipped with SAML, and on 10 April 1917 was assigned to the new IX Gruppo. Its pilots occasionally had duels with Austrian airplanes, like *cpr.* Eugenio Beria d'Argentine on 16 April 1917 and *s.ten.* Leopoldo Eleuteri on 24 April.

On 29 May SAML 1586 of *cpr.* Carlo Brambilla crashed on landing after an interception flight, killing its pilot, gunner Arturo

Marzari was wounded. On 8 June *serg.* Alfredo Tagliapietra had an inconclusive combat against two enemy airplanes over Nogaredo. On 19 June the unit sent six of its aircraft to escort a Caproni raid, with pilots *cap.* Sanità, *ten.* Flaminio Avet, *ten.* Clemente Giorelli, *serg.*Giuseppe Pascoli, *serg.* Tagliapietra and *cpr.* Beria, who had combats with enemy airplanes.

On 7 July 1917 *cap.* Mario Van Axel Castelli became the new commander. In the summer, the pilots of the 73ª Squadriglia were *cap.* van Axel and Giorelli, *ten.* Avet and Fernando Armati, *s.ten.* Eleuteri, *aspiranti* Beria d'Argentine and Pascoli, *serg.* Giuseppe Negri, corporals Pasquale Ceccarelli and Rinaldo Bulgheroni. On 20 September 1917 the 73ª, that was to all effects a reconnaissance unit, was re-named 121ª Squadriglia.

A new 73ª Squadriglia was formed in Macedonia on 10 November 1917 from the Sezione (Flight) of the 83ª that operated at Thessaloniki in support of the 35th Division on the Italian Army in Macedonia. Its commander was *ten.* Ernesto Bonavoglia. Its other pilots were *ten.* Giovanni Righi, *serg.* Miracca and Contò. It was equipped with Nieuport 17 and 11, and also one Spad. It was based at "Hill 619" 20 km south of Monastir.

On January 1, 1918 Bonavoglia had a serious accident with a Nieuport 17, and *ten.* Righi became the commander. He was badly wounded in combat on 29 January, so *ten.* Vittorino De Biasi replaced him. In February the 73ª Squadriglia received some Nieuport 27s. On 25 May 1918 the unit became a part of the new XXI Gruppo.

In the summer the 73ª was equipped with Hanriots, receiving no less than 26 of them, for its eight pilots. In September, with new pilots, it formed a new 2nd Flight while the 1st was deployed at Uskub. In September, with the collapse of the Bulgarian Army, the war was over on that front.

The 73ª Squadriglia was not heavily engaged, and suffered for the low quality of its airplanes, particularly the Ni.27, and lack of spares. Overall, it flew 384 combat missions and had 27 air duels, without claiming any victory. It returned to Italy in August 1919 and was disbanded.

The second 73ª Squadriglia operated in Macedonia with a mix of Spads and Nieuport 27s like these ones. The first one carries the name Jane, not an Italian name and probably connected to some form of inter-Allied fraternization.

## 74ª Squadriglia

74ª became the new name of the 4ª Squadriglia Caccia for the defense of Milan, commanded by *cap.* Lombard and equipped with Aviatiks. In October *cap.* Ferdinando Giorgi became its new commander, and in November it moved to Trenno, Milan. It was equipped with Farman 1914, Aviatik and Farman Colombo, and during 1916 it flew about 300 cruises, with no particular event.

Its pilots were *ten.* Antonio Dominici, *m.llo.* Pagano, *sergenti* Leccese, Miari Fulcis, Orsi, Stillio, Strina, *sold.* Paolucci and ten gunners. In March it received new SAMLs and Savoia Pomilio SP.2. 2765, 2768, 2770. On 10 April 1917 it was assigned to the IX Gruppo.

On 6 May 1917 *cap.* Ercole Ercole became its new commander: he was a Caproni pilot decorated with the Gold Medal for Bravery for a mission in Albania. In June, now equipped only with SAML and S.P.2, the squadron had as pilots *cap.* Ercole, *tenenti* Nino Bartoli, Emilio Botturi, Francesco Petrosemolo, Antonino Serra, Eriberto Torello, *s.ten.* Corradino Raineri, Carlo Tacchi Venturi, Ruggero Tarantelli, *serg.m.* Carlo Massagli, *v.brig.* Facciocchi and *serg.* Bianchi.

On 11 August 1917 the 74ª Squadriglia deployed two SAML flights to Campoformido, to participate in the 11th Battle of the Isonzo, with pilots *ten.* Petrosemolo, Serra and Tarantelli, *s.ten.* Tacchi Venturi and *serg.* Lucchini and Tagliapietra, later joined by Ercole. At the front, the formerly under-used defense squadron fought valiantly, flying many bombing and strafing missions. On 4 Sept. 1917 the pilot *ten.*Corradino Ranieri was wounded in action, dying later of his wounds. On 12 October the SAMLs returned to base. Meanwhile the remaining flights were commanded by *cap.* Giorelli. A Caproni Flight was also formed on 24 August, with Capronis 1132 and 1164. On 20 September 1917, this unit also was re-named as 122ª Squadriglia.

A new 74ª Squadriglia was formed on 20 May 1918 at Castenedolo airfield, near Brescia, with the personnel and aircraft of the 2ª Sezione of the 83ª Squadriglia, part of the IX Gruppo. Its commander was *cap.* Marinello Nelli. A flight of the new squadron was immediately sent to Cividate Camuno airfield, in the Tonale area, and remained there until the end of the war.

The unit was equipped with seven "Super Nieuports", that is, Ni.27, and its pilots were *cap.* Nelli and Nilo Tibaldi, *ten.* Pierfausto Barelli, *serg. m.* Guido Branca, *sergenti* Carlo Campana, Eligio Cruciani, Cesare Giordano, Giovanni Riva, Francesco Sodani, *cpr.* Raffaele Tarducci, *sold.* Giosuè Lombardi, Stefano Borla and Guido Roncuzzi. On 1 June 1918 soldier Roncuzzi got lost and landed in Switzerland. He was interned until the end of the war.

*Cap.* Stefano Battaini became the new commander of the 74ª Squadriglia on 8 June, and on 24 June he forced down an Austrian airplane. Another one was shot down by Frazzi, Tarducci and Campana on July 24. On 7 July the squadron was assigned to the XXIV Gruppo. In July, the unit had three Nieuport 27s, 13 Hanriots, one S.P.2 and three Farmans, used for night flights. Battaini shot down another enemy on 29 July, while Lombardi was wounded on 24 August.

On 25 September the XXIV Gruppo was disbanded so the 74ª Squadriglia passed first to the IX, then to the XX Gruppo, moving to Ponte S. Marco. The flight at Cividate Camuno had pilots *cap.* Nelli, *sergenti* Giordano, Raffaele Capo (an Argentinian), Sodani, Angelo Lucchini and soldier Adolfo Novarese.

During the war, this squadron flew about 800 war missions and had 14 air combats. It was disbanded on 6 February 1919.

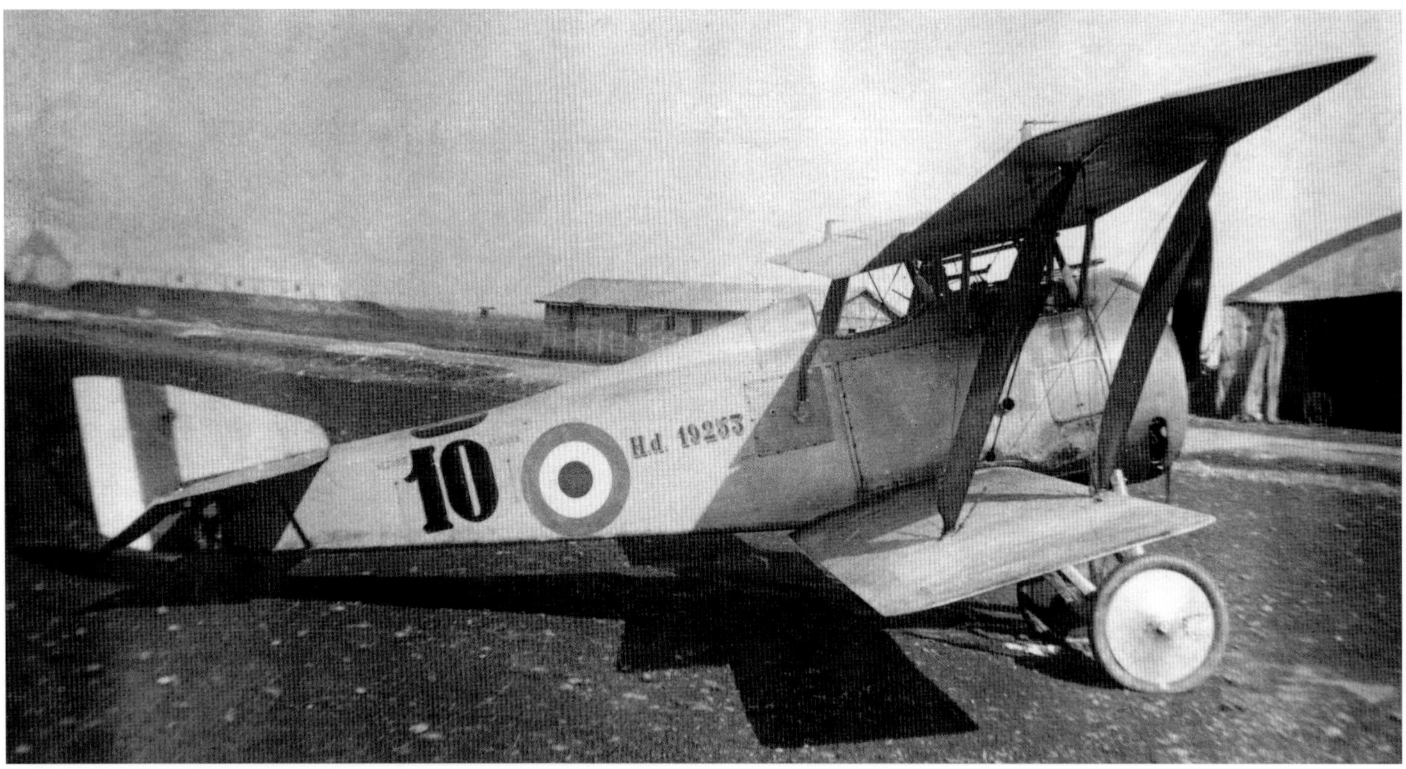


A Hanriot, serial 19253, of the 74ª Squadriglia showing its individual number 10 repeated over the fuselage deck. *(Photo Lucchini).*

The 75ª Squadriglia was not heavily engaged during the war.This Nieuport 17 of French construction shows double armament and a striped cowling. In the background there is the Nieuport 11 with large tricolor band of Buzio.

## 75ª Squadriglia

The 75ª Squadriglia was formed on Tombette airfield on 1 May 1916. It was lead by *cap.* Maffeo Scarpis and assigned to the III Gruppo for the defense of Verona. Its initial pilots were *s.ten.* Alessandro Buzio, *s.ten.* Mario De Bernardi and *caporale* Elia Liut, later *cap.* Ettore Croce, *s.ten.* Paolo Michetti, *caporale* Guido Consonni and volunteer soldier Guido Nardini. It flew Nieuports 1455, 1456, 1457, 1458 and 1459.

It sent a flight to Villaverla, with Michetti, Croce and Liut. On May 16 Buzio, Consonni and Nardini intercepted an enemy above Verona, and drove it away. The first victory was scored on 27 June, when De Bernardi, Consonni and Nardini, forced down Br. C.1 26.11 in Italian territory. On 18 August, as Scarpis was moved to the Technical Direction, *cap.* Mario Gordesco became the new unit commander. Later he was replaced by *cap.* Calori. At the end of the year the 75ª had pilots *ten.* Giovanni De Briganti, *s.ten.* Buzio, de Bernardi, Pietro Bergonzi, *serg.* Consonni, Liut, soldiers Carlo Iannace and Dante Nannini. It was equipped with Nieuport 11 serials 1646, 1647, 1649, 1654, 1688, 1690 and 1695. During 1916 this unit flew 361 combat missions with 24 combats and scored one victory.

On 8 April 1917 *cap.* Antonio Bosio became the new commander, and the squadron was assigned to the IX Gruppo. In August 1917 its 3rd flight, equipped with Spads, left Ghedi and moved to Castenedolo. In October two flights of the 75ª left the unit and formed the new 72ª Squadriglia.

The 75ª Squadriglia, that had returned within the III Gruppo, was engaged during the Caporetto retreat. On 30 October the squadron was joined by the 1ª Sezione SVA, with *ten.* Natale Palli and *sergente* Giovanni Bartolomeo Arrigoni with SVAs 6781 and 6789.

On 18 November 1917 Arrigoni was shot down in flames by aces Arigi, Maier and Kiss of Flik 55/J, but survived. At the end of 1917, the 75ª Squadriglia had pilots cdr. Bosio, *ten.* Pietro Campanaro, Pietro Danieli, Argiro Mele, Natale Palli, Giuseppe Vanasco, *s.ten.* Carlo Rovedati *serg.* Arrigoni, Consonni, Andrea Mancini and Antonio Palpacelli. It was equipped with five Ni-17, two SVA and two Farmans of a Nigth Flight with *sergenti* Antonio Coda and Luigi Recchioni. Both of them were writtten off landing in the fog in the night of December 31.

In January 1918 the unit got a re-equipment of Nieuport 27s. First combat of 1918 took place on 27 February, by Consonni and Palpacelli. On 18 March 1918 the unit moved to S. Luca airfield. In the spring it received Hanriots, serials 6199, 6220 and 6223. The Farman Flight, that also used S.P.4 twin engined biplanes, operated training the observers of the 61ª Squadriglia.

On May 14 the 75ª Squadriglia moved to Ganfardine airfield. On June 1, the squadron had pilots *cap.* Raffaele Lioce, *tenenti* Campanaro, Danieli, Argiro Mele, Carlo Rovedati, *s.ten.* Benedetto Claris Appiani and Remo Segala, *asp.* Arnaldo Brandolini, *serg.* Nazzareno Natalini, *caporali* Rolando Spuri and Carlo Novelli, soldiers Michele Gresino and Daniele Manzolini. They flew twelve Hanriots, eight Ni.27s and one Ni.17. On June 15 the unit moved to Buriago, as part of the *Massa da Caccia*, returning to Gandfardine on June 29, with the claim for one victory.

In July, *cap.* Arturo Oddo became the new commander. In late August, the unit sent a Hanriot flight to reinforce the 71ª Squadriglia, with pilots Danieli, Gresino, Natalini, Novelli, *cpr.* Rolando Spuri, *s.ten.* Pietro Cascina and *serg.* Francesco Bellino.

Mixed equipment for the 76ª Squadriglia in early 1917: a French built Nieuport 17 showing a fuselage tricolor band and a Lewis gun, and then a Nieuport 11.

On 29 October, the squadron suffered its only casualty of the war, as *serg.* Bellino was killed during a strafing mission. During the final year of the war, the squadron flew 946 combat missions, meeting the enemy in the air 30 times. It was disbanded in March 1919.

## 76ª Squadriglia

The 76ª Squadriglia was formed at La Comina on 25 May 1916, equipped with Nieuports 1042, 1043, 1805 and 1806. Its commander was *cap.* Ettore De Carolis with pilots *s.ten.* Giuseppe Rigoni, *sergente* Mario Stoppani and volunteer soldiers Giovanni Menegoni and Otello Venchiartti. On 29 May the new unit flew to S. Maria la Longa airfield, assigned to the I Gruppo. The unit was soon heavily engaged in interceptions and escorts to two-seaters and Caproni bombers.

On 24 August 1916 it lost *cpr.* Guido Iori, who ran out of gas and landed in enemy territory, his Ni.11 fighter being captured intact. On 29 September the 76ª Squadriglia was assigned to the II Gruppo. By the end of the year 1916, it had flown 624 combat missions, with 39 duels, and claiming five victories.

In February 1917, when it was equipped with a mix of Nieuport 11 and 17, its personnel included commander De Carolis and pilots *tenenti* Ernesto Bonavoglia, Luigi Olivi, Giuseppe Rigoni, *s.ten.* Menegoni, and *sergenti* Virgilio Appiani, Giovanni Bartolomeo Arrigoni, Stoppani and Venchiarutti.

On 25 February 1917 the squadron moved to Borgnano airfield, also known as Langoris, in the hottest area of the Isonzo front, so close to the frontline that it was occasionally shelled by Austrian artillery. *Ten.* Olivi became commander on 22 April.

The unit had its first casualty on 14 May, when *ten.* Francesco Broili was shot down near Gorizia, probably victim of *Zugsf.* Stefan Fejes and *Leut.* Alexander Tahy. On 17 June 1917 Olivi shot down two airplanes in one day, but, returning over the location of the combat to take pictures, he crashed to his death on Spad 5403.

The new commander was *cap.* Salvatore Calori, coming from the 81ª Squadriglia. On July 9 the unit was assigned to the VI Gruppo, in preparation for the 11th Battle of the Isonzo. On 17 July its most outstanding pilot, *ten.* Flavio Torello Baracchini scored his 10th victory, and on August 30 he was decorated with the Medaglia d'Oro al Valor Militare, Italy's highest decoration.

On 11 August the 76ª, 78ª and 81ª formed the "Sottogruppo Aeroplani di Borgnano", lead by Calori, so *ten.* Gastone Novelli became commander of the 76ª. In August this unit was equipped with the first Hanriots to reach the front. On 13 August its force included six Hanriots, six Nieuport 80 hp and three SPADs. On 28 September *ten.* Francesco di Rudinì became the new commander, but he was shot down and killed five days later by *Leut.* Schroeder, of Jasta 1.

During the Caporetto retreat, the unit left Borgnano, where many Hanriots had to be abandoned, and moved to Campoformido, then La Comina and Arcade. On 25 October *ten.* Alessandro Veronesi crashed near Jdensko, reportedly for an accident, but more likely a victim of *Feldwebel* Hippert, of Jasta 39.

On 1 November, when based at Arcade, the 76ª Squadriglia had pilots *cap.* Alberto De Bernardi, commander, *tenenti* Alessandro Buzio, Mario Fucini, Amedeo Mecozzi, *s.ten.* Giorgio Michetti and Silvio Scaroni, *asp.* Renato Donati, *sergenti* Fernando Armati, Edoardo Avenati, Raimondo Di Loreto, Luigi Fanti and Gastone Razzi. On 10 November it moved to Istrana, with a force of just four Hanriots.

On 26 December the squadron was protagonist of the "battle of Istrana", a sudden attack by forty German aircraft that was met by the fighters of all the squadrons based at Istrana, 70ª, 76ª, 78ª and 82ª, that shot down eleven Germans with no losses.

During 1917 the 76ª Squadriglia flew about 2,000 combat missions and had 300 combats.

During the winter, Istrana was heavily bombed, because the aircraft and hangars stood out in the snow, so the 76ª was forced to move, first to Isola di Carturo then, on 17 February 1918, to Casoni.

Yet another night bombing, on 27 February, hit an hangar killing two men and destroying ten of its fighters.

On 24 March Baracchini returned to his squadron, after a long period in hospital. The squadron was thus full of remarkable pilots: on 13 May, with commander De Bernardi, its pilots were *cap.* Giulio Lega, *ten.* Baracchini, Buzio, Ludovico Censi, Di Loreto, Donati, Fanti, Michetti, Razzi, Giuseppe Retinò, Scaroni, *s.ten.* Dante Nannini, *brig.* Pacifico Malfranci, *serg.* Piero Mondini and *cpr.* Luigi Boccasaiva and Umberto Baggini. It had on charge Hanriots 519, 524, 541, 546, 558, 6182, 6206, 6216, 6222, 6239, 6254, 6269, 6270, 6274, 6280, 6558, 7458 and 7460. Baracchini, however, passed to the 81ª Squadriglia on 31 May.

The 76ª Squadriglia fought very effectively during the Austrian offensive of mid June.On 12 July, its top ace, Silvio Scaroni, was seriously wounded in combat. *Cap.* Amerigo Notari became the new commander on 9 September. At the armistice the unit, still based at Casoni, with 21 fighters, under the command of Notari had pilots *cap.* Lega, *tenenti* Roberto Carrara, Di Loreto, Carlo Fornasari, Camillo Sivori, *brig.* Roberto Balandi and Malfranci, *sergenti* Baggini, Giuseppe Magnetti, Pietro Mondini, Antonio Pasta, Romolo Ticconi, *cpr.* Armando Ramberti, soldiers Vasco Giovanardi, Giovanni Randali and Renato Silvi.

During 1918 the 76° Squadriglia flew 2464 war missions and had 101 combats.

## 77ª Squadriglia

The 77ª Squadriglia Nieuport was formed on 31 May 1916 at the La Comina airfield. It was assigned to the Supreme Command and based at Istrana, where it became operational on 18 June, equipped with Nieuport 10s. Its commander was *cap.* Piccio, and its initial pilots were *ten.* Carlo Savio, *s.tenenti* Ferruccio Ranza and Giovanni Fasoli, *sergenti* Giuseppe Tesei, Domenico Piaggio and Cosimo Rizzotto, *cpr.* Giovanni Bozzetto. It soon received Nieuport 11s and had its first combat on 26 July, and its first victory claim on July 27, for Ranza and Savio.

In August 1916 it moved to Cascina Farello, near Aquileia, close to the sea, where it flew escorting Navy seaplanes. On 5 September it was assigned to the I Gruppo. It participated in the large air attack against Trieste on 13 September, and fought against Austrian seaplanes the next day.

The 77ª Squadriglia flew 109 combat flights in October, with the first Italian victory over a kite balloon, on the 18th. During the whole of 1916, this unit flew 526 combat missions, with 30 combats and three claims.

On 1 January 1917 the 77ª was partially equipped with Nieuport 17s, and in January it was shortly put under the command of *cap.* Ettore Croce, then of Ranza. In March it moved to Aiello, and received its first Spads. In May its equipment consisted of Nieuport 11 1627, 1680, 2113 and 2132, Nieuport 17 2613, 3135, 3136, 3137, and Spad 3121, 4695 and 4696. The unit emblem was a red heart, possibly inherited from its first commander, Piccio.

During the August 1917 offensive, the 11th Battle of the Isonzo, the 77ª Squadriglia flew no less than 398 combat flights, having 20

Fighters of the 77ª Squadriglia at Aiello in 1917, Spad 7s and a Nieuport 17. The red heart insigina of the unit was painted without a background on light finish aircraft such as these.

*serg.* Giuseppe Ciotti, *cpr. m.* Edmondo Lucentini, *soldati* Remo Bozzetti and Riccardo Ferrero.

In January and February it supported the first Italian counter-offensive in the Asiago sector. In March 1918 it moved to San Luca airfield, and received some Hanriots besides its "Super-Nieuports".

On 10 April its ace Attilio Imolesi crashed to his death. On May 5, also *serg.* Giovanni Nicelli, its other ace, crashed to his death in a Nieuport.

In mid June the 79ª Squadriglia was assigned to the *Massa da Caccia*. At the time of the offensive of the Piave, the unit was equipped with 12 Hanriots and five Nieuport 27s. When the *Massa da Caccia* was disbanded, on 29 June, the squadron returned to its XV Gruppo.

On 30 June 1918 its temporary commander was *cap.* Mazzini with pilots *tenenti* Federico Comirato and Giovambattista Toffoletti, *sottotenenti* Angelo Federici and Francesco Fusconi Pagnani, *sergenti* Cerutti, Giuseppe Ciotti, Carlo Gaffuri, Edmondo Lucentini, Reali, Fortunato Sicco, Umberto Soliani, *caporali* Attilio Aime and Augusto Rustici, *soldati* Bozzetti, Riccardo Ferrero, Carlo Rossi.

On 7 July 1918 the 79ª Squadriglia passed to the new XXIII Gruppo. *Ten.* Eugenio Mossi became the commander on 3 September, and five days later he was replaced by *cap.*Mario Omizzolo. The latter was shot down and killed by Austrian artillery on 8 October. *Cap.* Arturo Freddi Cavalletti then lead the unit since October 17.

On the day of the armistice the 79ª Squadriglia was based at San Luca and its pilots were *cap.* Freddi Cavalletti, with pilots Cerutti, Ciotti, Comirato, Federici, Ferrero, Fusconi Pagnani, Gaffuri, Lucentini, Melloni, Mossi, Rossi, Rustici, Soliani, *ten.* Carlo Ravedati, *s.ten.* Arnaldo Brandolini, *sergenti* Gino Andreutti and Arturo Bosio, *carab.* Franco Bianchi, *soldati* Domenico Bertone, Dino Pini and Giacomo Savioli. It was armed with18 Nieuport 27s and three Hanriots. The unit emblem was a black she-wolf. During 1918 this unit flew 2910 combat flights and had 158 air combats.

## 80ª Squadriglia

The 80ª Squadriglia became operational on 28 February 1917 at S. Maria la Longa airfield, assigned to the I Gruppo supporting the III Armata. Its commander was *cap.* Mario Gordesco, with pilots *tenenti* Raoul Da Barberino, Guido Keller, *s.ten.* Guido Sambonet, *serg. m.* Amleto Degli Esposti, *sergenti* Michele Allasia, Felice Avon, Alvaro Leonardi and Abbondio Viganò. It was equipped with Nieuports 2060, 2125, 2139, 2140, 2142, 2147, 2152, 2157, 2172 and 2174.

It detached a flight to Cascina Farello airfield, that also operated from Aiello, in escort flights. *Ten.* Keller scored the unit's first victory on 24 April 1917, when he shot down Brandenburg 129.04 of Flik 12. The whole squadron moved to Cascina Farello on 30 April. From there, close to the coastline, the 80ª Squadriglia engaged enemy seaplanes. On 16 July, the whole unit gave away its Nieuport 11s and re-equipped with twelve Nieuport 17s. By that time, with commander *cap.* Gordesco, its pilots were Allasia, Avon, Da Barberino, Degli Esposti, Leonardi, Sambonet, now a *tenente*, *ten.* Carlo Molinari and *s.ten.* Giovanni Ancillotto. The 80ª Squadriglia operated valiantly during the 11th Battle of the Isonzo, in August, strafing enemy trenches.

On 26 October, Leonardi shot down an enemy seaplane, the same day the unit received the order to retreat to Comina, and had to set on fire four unflyable fighters. On 1 November it moved to Arcade, while still fighting against the enemy air force. On 8 November the 80ª was assigned, together with its sister unit, the 77ª, to the new XIII Gruppo and set its base at Marcon. Gordesco became the Group commander, while *cap.* Da Barberino lead the 80ª Squadriglia, but he was wounded in an accident 9 days later, and Sambonet became the temporary commander.

At the end of 1917, the pilots of the 80ª Squadriglia were Sambonet and Leonardi, *tenenti* Giovanni Mazzia, Renato Mazzucchelli, *sottotenenti* Francesco Carabelli, Eolo Restelli, *brig.* Ernesto Cabruna, *sergenti* Mario Banino, Elia Liut, Nello Parlanti, Rodolfo Piermattei, Cesare Rappalli, Giovanni Riva and Francesco Rossi.

On 30 November the unit received its first Hanriots, serials 6138 and 6162. During 1917, the 80ª Squadriglia performed 2049 combat missions engaging in 128 air fights.

*Ten.* Giorgio Zoli, that reached the unit together with *ten.* Carlo Alberto Conelli De Prosperi, became commander on 6 February

Aircraft of the 79ª squadriglia in January 1918: a Nieuport 27 followed by two Ni.17s and then an Hanriot.

Nieuport 17s of the 80ª Squadriglia at Aiello, summer 1917.

1918, while *cap.* Umberto Gelmetti replaced him as commander on 30 April 1918. Totally equipped with Hanriots, the unit had a white star in a red field as its emblem.

On 7 June *cap.* Achille Pierro became the new commander, and under his leadership the squadron, assigned to the *Massa da Caccia*, took part in the operations of the Battle of the Piave, in mid June.

On 15 June 1918 a new pilot, *ten.* Guido Baricalla, got lost, and force-landed in enemy territory: his Hanriot, 6273, was re-serialled 00.72 by the Austrians.

On 1 July 1918 the pilots of the 80ª Squadriglia were *cap.* Pierro, *ten.* Zoli, Piermattei, Carlo De Rossi, Carlo Pasquinelli, Giorgio Romano, *sottotenenti* Lombardi, Ettore Dario, Giuseppe Pascoli, *sergenti* Felice Avon, Giuseppe Pizzoccheri, *caporali* Luigi Corti, Telemaco Pillotti, Ottavio Sottani, Giovenale Testa, *soldati* Vincenzo Bonanni, Giovanbattista Gaiando and Silvio Grassi.

In the last summer of the war, the squadron dropped leaflets over enemy territory. A Balilla fighter was assigned to the unit for evaluation, but it crashed on 26 September, wounding *brig.* Cabruna.

In the days of the final offensive toward Vittorio Veneto, the 80ª Squadriglia flew 277 combat missions. Overall, during the war, it flew 4637 combat missions. It was disbanded in January 1919.

## 81ª Squadriglia

The 81ª Squadriglia was formed at the Centro Formazione Squadriglie of Arcade on 20 March 1917. It moved to Borgnano (Langoris) airfield, west of Gorizia, on 20 April 1917, assigned to the II Gruppo that flew for the 2ª Armata. Equipped with twelve Nieuport 11s, its commander was *cap.* Salvatore Calori, and its pilots were *ten.* Gastone Novelli, *s.ten.* Vincenzo Arone and Flavio Torello Baracchini, *sergenti* Riccardo Ciotti, Paolo Nicosia, Ennio Sorrentino, *cpr.* Paolo Avenati and *soldato* Paolo Simoncini.

The start of this squadron was unauspicious, because *serg.* Ciotti crashed in the ferry flight to Borgnano, and died of his wounds on April 23. He was soon replaced by *ten.* Alessandro Buzio. Sorrentino and Avenati sustained the first air combat on April 24 and on May 1 Sorrentino scored the first victory, shooting down Brandenburg C.1 27.64 of Flik 2. Its leading pilot was Baracchini, who claimed four victories in late May and June 1917.

*M.llo* Arturo Guglielmi crashed to his death on take-off on 14 June 1917. On June 24, Calori was shifted to the 76ª Squadriglia, and *cap.* Mario Zoboli became the new commander of the 81ª. On July 14, also Baracchini was shifted to the 76ª.

On 9 July 1917 the 81ª Squadriglia was assigned to the VI Gruppo, supporting the 4ª Armata. At the same time, it received the

The Austrians found a large booty at Campoformido airport, including this Nieuport 17 serial 3625 of the 81ª Squadriglia, which eventually received Austrian serial 00.59. Behind it is a Pomilio PD serial 3818. *(Photo Cejka).*

new 110 hp Nieuport 17. On August 10, together with the 76ª and the 78ª, it formed the so-called "Sottogruppo Aeroplani di Borgnano", assembled for the forthcoming offensive. On 28 August *serg.* Antonio Pellanda was killed in a crash due to engine failure.

After the enemy breakthrough, the 81ª left Borgnano with its efficient fighters, and moved back to Aviano and then to Arcade. On 9 November it moved to Istrana, where it was equipped with the Nieuport 27.

On 26 December the unit took part in the battle of Istrana together with the other squadrons of its VI Gruppo, the 76ª and 78ª, with Sorrentino claiming a victory. During 1917 the 81ª Squadriglia flew 1,650 missions, had 150 air combats and claimed 15 victories.

On 1 January 1918 the force of this unit consisted of the commander *cap.* Zoboli, *tenenti* Sorrentino, Giuseppe Lucci, Lamberto Toncher, *s.ten.* Antonio Bogliolo, *sergenti maggiori* Mario Baldaccini, Alfredo Sozzini, *sergenti* Gino Allegri, Alessandro Borgato, Alberto Rettori, *caporali* Raffaele Tarducci, Attilio Venuti, *carabiniere* Lino Montello Chiaventone and *soldato* Giovanbattista Gays.

On 15 January *serg.* Rettori failed to return from a mission. His Nieuport 27 5800 had a failure and landed at S. Fior di Sopra airfield, base of the fighters of Jasta 39. A similar accident happened on 1 May, when *serg.* Silvio Mellone was lost with his Ni.27 5836, captured by the Austrians at S. Pietro in Campo. On 2 February the 81ª moved to Isola di Carturo, and two weeks later to Casoni, where it started to re-equip on the Hanriot. On 25 March Zoboli left command of the squadron to *cap.* Renato Mazzucco. On 31 May *ten.* Baracchini returned to serve in this squadron.

The 81ª was very active during the Piave battle of mid-June, losing on 20 June *serg.* Antonio Nava, shot down and killed by Austrian fighters. On 25 June Baracchini, after a combat in which he claimed a fighter, was wounded in the belly by a bullet fired from the ground. Overall, in June the 81ª put up 385 war flights, had 34 combats and claimed 14 aircraft and kite balloons shot down.

On 30 June the force of the Squadriglia included commander *cap.* Ponis, and the pilots *cap.* Arturo Freddi Cavalletti, *tenenti* Antonio Bogliolo, Manillo Zerbinati, Michele Borea d'Olmo, *serg.m.* Mario Baldaccini and Alfredo Sozzini, *sergenti* Carlo Ciampitti, Carlo Corti, Francesco Genta, Alessandro Oliveti, Ennio Sorrentino, Aldo Ungaro and *caporale* Aldo Astolfi.

On 31 July *serg.* Ciampitti and *cpr.* Astolfi had a long and difficult duel with an Aviatik D.I that was shot down near Vas. Their victim was the Austrian 27-victories ace *Oberleutnant* Frank Linke-Crawford. On 26 October the squadron was assigned to the *Massa da Caccia*, and on that day Corti scored the last victory of his unit. At the end of the war the 81ª at Casoni had 21 Hanriots and one Nieuport. During the last year of war it had flown 2,468 combat missions with 80 air combats and had scored 19 confirmed victories.

## 82ª Squadriglia

The 82ª Squadriglia was activated on 11 April 1917 and assigned to the new X Gruppo. It was equipped with nine Nieuport 11 serials 1704, 2189, 2198, 2208, 2211, 2219, 2221, 2235 and 2236. On 25 May its 1st Sezione (flight) moved to S. Caterina. Its commander was *tenente* Giuliano Pessi, an Austria-born Italian who used the fictitious name Parvis. On 9 June *ten.* Antonio Fochessati became commander.

The unit, that shared the airfield as a junior member with the famous 91ª and the 70ª, flew 96 war flights in June, with four combats. On 2 July command was assigned to *cap.* Pietro Cavadini. On 17 August a flight was deployed at Bolzano. On 7 September Fochessati returned as temporary commander. On 1 October *serg.* Gaetano Aliperta was seriously wounded in a combat with enemy fighters. During the Caporetto retreat, on 26 October *serg.* Pietro Zanatta had an accident and was then captured by the enemy while in Udine hospital. On 28 October the unit, with its equipment of eight Nieuport 11 and one Ni.17, moved back to Pordenone, and then on 5 November to Arcade.

The 82ª Squadriglia used individual letters on its Hanriots.

Late in the war the 83ª Squadriglia introduced a red diamond as its unit emblem. The aircraft are Nieuport 27s, with a Hanriot on the right.

On 7 November *serg.* Agostino Buttazzoni was shot down, probably by Buttman, of Jasta 1. He got out of his Hanriot 7211 unharmed, set it on fire and managed to return to his unit.

On 10 November the 82ª Squadriglia moved to Istrana, where it got new Hanriots. At that time its pilots were *ten.* Fochessati, Flaminio Avet, Alberto Comandone, Giovanni Pirelli, Renato Rossetti, *sottotenenti* Alessandro Resch, Michele Riello, *sergenti* Paolo Benvenuti, Alessandro Contardini, Mario D'Urso, Giulio Poli, Andrea Teobaldi and *soldato* Clemente Panero.

On 26 December the unit participated in the air battle against German airplanes raiding Istrana, and lost four of its Hanriots on the ground. During 1917 the 82ª performed 889 war missions and had 49 air combats.

On 9 January 1918 the unit moved from Istrana to S. Pietro in Gu. On 29 January *maggiore* Ferruccio Coppini became its commander. On 22 February *sold.* Panero crashed while taking off in a Hanriot, he died of wounds one month later. On 14 March the 82ª Squadriglia returned to Istrana, then on 21 May it moved to Gazzo.

On 1 June 1918 the pilots of the 82ª were *magg.* Coppini commander, *capitani* Luigi Aglietti, Salvatore Breglia, Francesco Fassi, Giulio Cesare Paldaoff, Pompeo Vaccarossi, *tenenti* Alberto Comandone, Arturo Ferrarin, Umberto Garelli, Giovanni Pirelli, Renato Rossetti, Andrea Teobaldi, *sergenti* Mario Jacazio, Gaspare Monetti, Paolo Benvenuti, Agostino Roggeri and *sold.* Guido Colombatto.

During the battle of mid June the unit, attached to the *Massa da Caccia*, was engaged in air combats and particularly in strafing attacks. In one such mision, Austrian fighters jumped a patrol of five Italian Hanriots and shot down and killed *cap.* Fassi on Hd 7452.

During the summer 1918 its pilots were often tasked with the escort of Caproni bombers. In August the unit got two Ansaldo Balilla for evaluation, and at the end of September also four Spad 7.

On 23 October the *Massa da Caccia* was assembled again for the final offensive, and the 82ª was part of it. Equiment consisted in 15 Hanriot, 6 SPAD and one Balilla. The 82ª Squadriglia was disbanded in December 1918.

## 83ª Squadriglia

The 83ª Squadriglia Nieuport was a particular unit because it was formed on 5 May 1917 with more Sezioni (flights) that operated separately. The 1a Sezione was destined to Macedonia and it sailed from Taranto on 16 May. On 1 June its first fighter, Nieuport 2273 was set up at Dudular. Personnel of this unit included *s.ten.* Giovanni Righi, commander, and *sergenti* Giovanni Contò and Giovanni Miracca. On 8 June it moved to Kremsan, where also the 47ª Squadriglia Farman was based. On 22 June there was the first mission and the first air combat for *s.ten.* Righi and *serg.* Miracca. On 1 August 1917 *ten.* Ernesto Bonavoglia became commander.

In September Miracca shot down a German Albatros, and on 5 October Bonavoglia, flying a Spad, damaged another German airplane. On 10 December 1917 the Sezione was upgraded to Squadriglia, taking the number 73ª. During its existence the 1ª Sezione of the 83ª Squadriglia made 189 war flights, had 15 air combats and claimed two victories.

The 2ª and 3ª Sezione of the 83ª were sent respectively to S. Pietro in Campo, Belluno, and to Cavazzo Carnico. The 2ª Sezione had *ten.* Vittorio Bonomi, commander, with *sergenti* Arturo Dell'Oro and Mario Pereno, flying Nieuport 11s. On 10 May 1917 there was the first combat mission. On 1 September Arturo Dell'Oro rammed intentionally an enemy aircraft and crashed together with it, losing his life. For this deed, the Medaglia d'Oro, Italy's highest decorarion, was awarded to his memory.

The 3ª Sezione moved to Cavazzo Carnico on 16 May with *ten.* Pietro Danieli and *serg.* Antonio Pellanda, and flew its first mission on 19 May, during which *serg.* Pellanda drove off an enemy two-seater. In June this flight had *ten.* Arone as commander, and the pilots *serg.* Tommaso Masala and Giuseppe Magnetti.

After the retreat, the 2ª Sezione moved back to Feltre, then to Casoni and finally to Marcon, where also the 3ª moved from Cavazzo Carnico. The two united flights were assigned to the XIII Gruppo having commander *ten.* Vittorio Bonomi and the pilots *ten.* Arone, *sergenti* Signorini, Giannotti, Magnetti and Masala. On 8 December 1917 *serg.* Giannotti shot down a kite balloon using Le Prieur rockets on his Nieuport 11.

On 15 December 1917 the 83ª Squadriglia had *ten.* Vittorio Bonomi, temporary commander, *ten.* Arone, *s.ten.* Carabelli, *sergenti* Magnetti, Masala, Buri, Giovanni Riva, Francesco Rossi. On 28 December it received the first Nieuport 27.

In January 1918 *cap.* Giulio Moroni became the commander. On 18 March the squadron left Marcon, and the XIII Gruppo, sending its 1ª sezione (formerly 3ª) to S. Pietro in Gu with *tenenti* Fougier (future chief of the Italian Air Force in World War II) and Gadda, *cpr.* Donadio, *sergenti* Caminada, Masala and Magnetti and the 2ª Sezione to Castenedolo with *ten.* Marinello Nelli, commander, then with *cap.* Serafino Battaini who lead pilots Nelli, now *capitano* and Bruno Lodolo, *ten.* Pierfausto Barelli, *serg.m.* Eligio Cruciani, *sergenti* Angelo Comolli, Giovanni Riva, Raffele Capo, Francesco Sodani. Both flights were equipped with Nieuport 27s. On 22 March Fougier became commander of the squadron.

On 23 April 1918 ten. Enrico Gadda, of the 1ª Sezione (brother of the famous writer Paolo Emilio Gadda) was killed returning from a combat flight. In mid May the 1ª Sezione included *cap.* Fougier, *ten.* Enrico Rizzi, *s.ten.* Giovanni Menegoni, *m.llo* Giulio Martinotti, *sergenti* Aprà, Magnetti, Masala, Rossi and Adamo Bortolini, *cpr.m.* Donadio and Lemmi, *cpr.* Silvio Vischioni, *soldati* Caminada, Luigi Bello and Rodolfo Maggia.

The 2ª Sezione included *capitani* Lodolo, Nelli, Battaini and Nilo Tibaldi, *serg.m.* Cruciani and Poggiali, *sergenti* Riva, Raffaele Capo, Angelo Comolli, Francesco Sodani, Cesare Giordano, Carlo Campana, Guido Branca, *cpr.m.* Roncuzzi, *cpr.* Raffaele Tarducci and *soldati* Giosuè Lombardi and Stefano Borla, (an unusual composition, four captains and eleven troopers). The 2ª Sezione on 20 May 1918 became the new 74ª Squadriglia, that on 1 June moved to Pojanella.

On 16 June 1918 during the Battle of the Piave *sergenti* Martinotti and Aprà had a combat and Aprà shot down an airplane at Riva Grassa. On 18 June *cpr.* Lemmi was hit during a strafing and he force landed Hanriot 6259, being badly wounded in the face. This was the only combat casualty of the squadron. On 22 June *serg.* Magnetti, with Rossi and Donadio, shot down an enemy at Villa Premuda. During the battle the 83ª Squadriglia flew 262 offensive missions and 122 escorts.

During the summer the unit, with a mixed complement of Nieuports and Hanriots, received more pilots: *cpr.m.* Camillo Trabucco, *sold.* Umberto Galassi and *sold.* Romeo Sartori (future Macchi test pilot) who on 13 August was attacked by a British Camel. The 83ª scored more victories in August and September.

On 4 October the 83ª joined the new XXIV Gruppo and moved from S. Pietro in Gu to Poianella. During 1918 it flew over 2,500 combat flights. It was disbanded on 25 March 1919.

## 84ª Squadriglia

The 84ª Squadriglia Nieuport was formed in June 1917, at Arcade and on July 2 it moved to S. Maria La Longa, where it became operational, assigned to the I Gruppo. Its commander was *cap.* Giovanni Fancello and its pilots *tenenti* Giovanni Mazzia and Renato Mazzucchelli, *serg.m.* Leonida Schiona, *sergenti* Albino Giannotti and Francesco Rossi. It used Nieuport 11s, among them the ones serialled 3209, 2294, 2296, 2129, 2191, 2308 and 3244. Its main missions were escorts to the Voisins of the 25° Squadriglia and to the Caudrons of the V Gruppo.

This unit then received *ten.* Felice Martinengo and *brig.* Ernesto Cabruna. Its commander *cap.* Fancello, died on 22 July 1917 when his Nieuport went into a spin. Cabruna also crashed on 6 August, but he was unharmed. It it likely that *ten.* Martinengo took over as temporary commander.

This is the only known photo of the short-lived 84ª Squadriglia, showing the fatal accident of its commander *cap.* Giovanni Fancello on 22 July 1917.

On 26 September 1917 the 84ª Squadriglia moved to Aiello. On 26 October it was temporarily attached to the X Gruppo, and it retreated to La Comina. It was still equipped with old Nieuport 11s, but its pilots were used to save the better fighters of the 77ª and the 80ª from the enemy advance, so nine Ni 11s and three new Nieuport 17s and one Spad, still crated, had to be abandoned. With few pilots and no aircraft, the 84ª Squadriglia was disbanded on 11 November 1917. A new 84ª Squadriglia Hanriot was being formed at Ponte San Pietro since 15 October 1918 but it was suppressed at the end of the month.

## 85ª Squadriglia

A Sezione Nieuport was formed on 23 September 1916 at Durazzo (Durres) for the XVI Army Corps operating in Albania. Its commander was *ten.* Giovanni Sabelli, with *sergenti* Antonio Lovadina (Fiat test pilot after the war) and Antonio Palpacelli. It moved to Tahiraqua airfield and it became part of the VIII Gruppo when this was formed, 9 December 1916. Initially it was not much active, on account of bad weather and the horrible conditions of the airfield. In May 1917 the Sezione Nieuport in Albania had pilots *ten.* Sabelli, *ten.* Adriano Bacula (later a famous test pilot with SIAI Marchetti), *serg.* Lovadina, *serg.m.* Palpacelli, *serg.* Lorenzo Cortesi and *cpr.* Guido Gerard.

This flight became the 85ª Squadriglia on 25 September 1917, with commander *cap.* Ernesto Pellegrino. In October it completely re-equipped with Nieuport 17. During 1917 it flew 97 combat missions, during which it had three combats (on 13 October *cpr.* Gerard, 15 October *ten.* Bacula, 26 October *asp.* Cena) that were inconclusive.

On 1 January 1918 the 85ª Squadriglia had on strength *cap.* Pellegrino, *tenenti* Vinicio Muraro and Prospero Freri, *s.tenenti* Balbi, Nicola Cena, Raffele Castellano, Silvio Ciappi, Giuseppe Ghislanzoni and Natale Ravasso and *serg.* Cortesi. In 16 January *cap.* Mario Sarrocchi became temporary commander. In February the unit moved from Tahiraqua to Piskupi (Peshkep).

On 17 April two enemy aircraft shot down and killed *serg.* Cortesi. In May the squadron re-equipped with Hanriots, of which no less than two dozens arrived from Italy.

On 15 June Pellegrino left for Italy, and the new commander, *cap.* Mario Ponis, arrived on 17 July. The fighters escorted the

Capronis of the 11ª Squadriglia and the FBA flying-boats of the 257ª. In September a flight was deployed at Durazzo, as soon as this town was reconquered from the Austrians, with *ten.* Arrigo Tessari (commander of the post-armistice Fascist Air Force in World War II).

During the war the 85ª Squadriglia put up 925 combat flights, had one pilot killed but claimed no victory. In December the 85° was still at Piskupi with commander *ten.* Cartoni. It flew liaison and mail flights among the towns of Albania. It was disbanded on 10 October 1919, leaving a flight attached to the Squadriglia Mista di Valona, that was active during the Albanian insurgency in June-July 1920.

The 85ª Squadriglia operated in the remote Albanian front, and its fighters had their share of colorful personal markings, like this Hanriot embalzoned with four aces.

## 91ª Squadriglia

The 91ª Squadriglia was formed on 1 May 1917 at S. Caterina di Udine, with the best pilots of the 70ª. Under *cap.* Guido Tacchini were pilots *capitano* Francesco Baracca, *ten.* Fulco Ruffo di Calabria, *ten.* Ferruccio Ranza, *s.ten.* Luigi Olivari and *sergente* Goffredo Gorini. The squadron, assigned to the X Gruppo, initially had Spad 4689, 4690, 4691, 4700, and Nieuport 17 2614, 3127 and 3138. The unit emblem was a black prancing griffon.

On May 7 the squadron detached a flight to Aiello, attached to the 77ª Squadriglia, formed by Olivari, Gorini and Ranza, for the 10th Battle of the Isonzo. In May the unit received ten more Spads, some of them equipped with cameras for photo reconnaissance and a new pilot, *ten.* Giovanni Sabelli.

The aces of this unit took part in thc heavy air fighting during the 10th Battle of the Isonzo, May 1917. On 3 June the flight detached at Aiello returned to the squadron. *Cap.* Baracca became the new commander on 6 June, and that same day the 91ª Squadriglia moved to Istrana, to support the Alpine troops in the Battle of the Ortigara. The unit returned to Santa Caterina on 2 July, and at this date it had claimed 13 victories in 100 air fights.

Ths élite unit received new pilots: *ten.* Franco Di Rudinì, a Sicilian prince, son of a former Prime Minister, *ten.* Giuliano Pessi (born in the Austrian held part of Italy, who had to fight under the false identity Parvis) and *s.ten.* Enrico Ferreri.

On 8 July 1917 the 91ª Squadriglia was attached to the Comando Aeronautico della III Armata.

During the intense fighting of the 11th battle, in August, the 91ª Squadriglia effectively escorted raids of Caproni bombers. Its

A Spad 13 of the 91ª Squadriglia late in the war. The squadron griffon appears on both sides of the fuselage, individual numbers are written in Roman numerals and as a personal marking the front cowling and rocker arms fairings are painted white.

efforts are reflected by its victory claims: seven in July and five in August. In late August, the unit received two Ansaldo SVA, 6755 and 6757, that were evaluated as fighters, but were found unsuitable for this role. The unit now had, besides the two SVAs, seventeen Spads, one Nieuport 11 and a captured Brandenburg C.1.

On 13 October Olivari, who had returned to the squadron after spending some time in Genoa cooperating in the development of the SVA, was killed in an accident. On 21 October Baracca scored a double victory against German two-seaters flying a Spad 13 with two guns. In the first day of good weather after the Austro-German attack, 25 October, the 91ª flew 17 combat missions, losing two pilots killed, Giovanni Sabelli in Spad 4699 and Ferreri on Spad 1369, and claiming six victories. On 28 October the unit left S. Caterina, burning the unserviceable aircraft and retreated to La Comina. The "squadron of aces" kept up a high level of activity throughout this period, scoring heavily against German aeroplanes. On 2 November a new pilot, *sergente* Fermo Macchi, was killed during a test flight.

On 10 November the 91ª was at Padua, where it found a quantity of new Spads. The pilots with commander Baracca were *tenenti* Costantini, Gastone Novelli, Guido Keller, Parvis, Ranza, Ruffo, *s.ten.* Giovanni Bozzetto, *aspirante* Amleto Degli Esposti, *sergenti* Carlo Caselli, Cesare Magistrini, Montù, Edoardo Olivero and Pagliari.

On 23 November there was a series of air combats, with Piccio forced to land at Istrana by fighters, Parvis and Keller shooting down a two-seater, and then getting damaged by fighters, Baracca and Novelli shooting down an Albatros D.III, probably Karl Überschär of Jasta 39. On 7 and 8 December the squadron moved to Nove di Bassano and participated in the actions against the final Austrian attack at the Melette. Then the tired Baracca finally got a furlough and Costantini became acting commander.

In the night of 4 January enemy bombers destroyed three Spads and damaged some more. Again in the night of 20 February bombers attacked Padua airfield and destroyed 13 Spads, some of them brand new. On 11 March 1918 the 91ª Squadriglia moved to Quinto di Treviso. *Sergente* Mario D'Urso, considered as an expert in night flying, remained at Padua, with the local Sezione Difesa, (Defence flight) to prevent further attacks.

In May the unit got new personnel, so it was composed by Baracca, Costantini, D'Urso, Keller, Magistrini, Novelli, Olivero, Piccio, Ranza, Ruffo, *ten.* Adriano Bacula, *ten.* Mario De Bernardi, *sergenti* Gaetano Aliperta and Guido Nardini. The new pilots were all veterans with previous victories. Baracca, after five months, scored again on 3 May and on 23 May. When the expected Austrian offensive was launched, on 15 June, Baracca scored two victories, his last ones. On 19 June he failed to return, his wingman *ten.* Osnago just lost him. His body was found a few days later on the Montello, with a single bullet hole in his head.

On 21 June *cap.* Ruffo became commander of the 91ª Squadriglia that, by royal decree, in July was named "Baracca". The unit in May and June had flown 500 combat missions for 960 flight hours shooting down 14 enemy aircraft. When the *Massa da Caccia* was disbanded the 91ª was attached to the IV Gruppo, a bomber unit, and remained at Quinto di Treviso.

On 28 June De Bernardi brought the first Ansaldo Balilla, followed by two more in August, but the pilots, having tested it, much preferred the Spad. On 8 September a typhoon swept the plains of the Po river, smashing impartially countless Italian and Austrian aeroplanes. At Quinto eight hangars were wrecked destroying 14 Spads and the three Balillas. Replacements were plentiful, and the unit was soon back in business. On 23 September the 91ª was assigned to the XVII Gruppo, of which Ruffo became commander, so Ranza was appointed as commander of the 91ª.

On 17 October *ten.* Augusto Stobia was wounded during a strafing mission, he landed at the Treviso racing track but died of wounds the next day. With the final offensive the 91ª, back in the *Massa da Caccia*, operated in strafing missions, losing an inordinate amount of fighters. Keller on Spad 5415 failed to return on 29 October, Ranza on Spad 2482 the next day was hit and force landed at San Fior airfield, just deserted by the Austrians, where he found gasoline and returned. On 31 October *ten. col.* Piccio himself, the commander of all fighter forces, flying Spad 2962 was hit and landed behind enemy lines. He was captured by the Austrians, but on 2 November he evaded, and three days later he returned, wearing a dirty enemy overcoat. Keller returned on 3 November, with his wounded leg carefully medicated by Austrian doctors.

At the end of the war the 91ª had 13 Spads and the pilots Ranza, Aliperta, Bacula, Conelli de Prosperi, De Bernardi, D'Urso, Novelli, Olivero, *ten.* Giorgio Castelli, *ten.* Arturo Bonucci, *s.ten.* Mario Ottolini and *sergente* Guido Consonni.

This outstanding unit, that was never disbanded, during the war effected 3,412 combat flights with 117 confirmed victories. Eleven of its pilots were declared aces: Baracca, Piccio, Ruffo, Ranza, Olivari, Novelli, Costantini, Magistrini, Nardini, Parvis, and Sabelli.

## 241ª Squadriglia

The 241ª Squadriglia of Hanriot land fighters of the Regia Marina was formed at Venice in the late spring of 1918, with commander *Tenente di Vascello* Luigi Bologna and pilots TV Umberto Magaldi, *s.tenenti* Giorgio Parodi and Giovanni Ravelli, *marinaio* Renato Balleri, STV Luigi de Riseis. On 24 June Bologna flew the first combat mission, then in July the unit escorted L.3 and M.8 seaplanes.

On 6 August 1918 Giorgio Parodi on Hanriot 13216 "P" shot down a kite balloon and then also Brandenburg 429.18 of Flik 22. On 24 August *capo* Edoardo Odierna was killed as his Hanriot crashed. Two days later in another accident Parodi was wounded. On 5 October now *sottocapo* Balleri shot down an Albatros D.III at Susegana, then on 8 October he attacked three fighters at Revedole and shot one down

Since 22 October the 241ª was equipped with Balillas, but only Hanriots were used on combat missions. Totally this unit flew over 80 war flights and damaged several kite balloons.

The 241ª Squadriglia of the Regia Marina operated land based Hanriot and Balilla fighters from Lido airfield, Venice. Here *STV* Giovanni Ravelli, third from left, and commander *TV* Luigi Bologna, sixth, appear with US Navy pilots attached to that unit for conversion.

## 260ª Squadriglia

The 260ª Squadriglia was formed at Venice on 1 November 1917, equipped with the new Macchi M.5 seaplane fighters and at the command of *Tenente di Vascello* Luigi Bologna, with the pilots *Sottotenente di Vascello* Fedrico Martinengo, *guardiamarina* Paolo Morterra, *2° capo* Guido Jannello (the winner of the 1919 Schneider race), Achille Panichi, Beniamino Piro, Pietro Valdimiro, Arturo Zanetti and *marinaio* Tiziano Finati. Finati was killed in a crash on 3 November.

The operational debut of this squadron was unauspicious, as on 5 November the Austrians shot down a seaplane of the 259ª Squadriglia escorted by Martinengo and Morterra.

On 17 November Morterra and Zanetti intercepted three Austrian seaplanes and together with the French *sous-lieutnant* Jean Bignon they shot down K 211. Then Morterra alighted and picked up the Austrian pilot *Fregattenleutnant* Julius Fedrigoni Edl. von Etschtal on his fighter, while an L.3 flying boat picked up the other crew member. On 19 November *2° capo* Daniele Minciotti ran out of gas and was captured with his Macchi serial 7068.

On 27 November GM Morterra was killed while practicing spins on an L.3 with Pomilio, of the 252ª Squadriglia, who was wounded. On 8 December Odierna of the 261ª had to alight as ground fire had hit his propeller, Piro alighted besides him, picked him up in his Macchi and brought him to safety. In December Bologna became leader of the seaplane fighters Group, so Martinengo became commander of the 260ª Squadriglia.

On 1 January 1918 the unit had Martinengo with pilots Jannello, Panichi, Piro, Valdimiro, Zanetti and *Guardiamarina* Umberto Calvello, flying Macchi M.5 serials 7059, 7058, 7071, 7062 and 7065. Three ISVA, the floatplane version of the SVA were also supplied, but these aircraft were under-performing and unpopular.

On 24 February *2° capo macchinista* Giorgio Parodi on Macchi 7102 had a duel with a patrol of Flik 61/J lead by the ace Franz Gräser on Albatros 153.106. The Italian pilot was wounded twice, he alighted and later his seaplane was towed back to base. After the war Parodi created the Guzzi motorcycle company.

In March TV Orazio Pierozzi became commander of the Gruppo Squadriglie Caccia, and he introduced strict combat formations and rules. Many training flights were flown using a Hynta gun cine camera to improve pilots aim. On 14 March *torpediniere* Carlo Baldini was killed on M.5 7233 that went into a spin. On 20 March, after an official visit, Giovanni Naccari on a Macchi collided with an L.3 that crashed, killing two airmen.

On 4 May the unit had a combat against several Austrian seaplane fighters from Trieste, two of which were shot down, with the capture of their pilots, and a third one was forced down. On 22 May Piro and Pierozzi had a combat with two Phönix land fighters and shot down one of them.

On 1 June Rivieri was killed, as his Macchi 7278 had to alight for an engine failure and capsized. On 8 June TV Alberto Bortolozzo replaced Martinengo as commander. During the mid-June battle the unit was very active, flying 45 combat missions. On 17 July it escorted the first daylight bombing of Pola, and had inconclusive duels with the defending fighters.

In the summer, while the ISVA were retired, the Gruppo Idrocaccia received the first Macchi M 7, and tested also a Hanriot floatplane.

On 15 September three Albatros of Flik 61/J bounced and shot down the Macchi of *sottocapo* Emilio Dri escorting a L.3. The plane burned but the pilot alighted and survived. On 22 October Pagliacci, escorting bombers over Pola, claimed the destruction of a Phönix together with Garrone and Macchi, of the "Squadra San Marco", flying SIA 9b 10818.

The 260ª Squadriglia from Venice used green and red stripes as an identification marking. This is Macchi M.5 of *marò* Giuseppe Pagliacci.

Acting against enemy troops during the final offensive, on 31 October 1918 Calvello was hit by ground fire and alighted in a marsh. Ravazzoni alighted near him, picked him up and let him the controls and the two took off amidst the rifle fire of the Austrians.

On 2 November the whole squadron flew over Trieste and performed aerobatics, but Pagliacci couldn't resist and he alighted in the port, being the first Italian to enter the city. At the end of the war the 260ª Squadriglia had Bortolozzo commander and the pilots Calvello, Dri, Pagliacci, Ravazzoni, *s.capo* Bensa, *marinaio* Agostino Barberis and *torpediniere* Renato Spada.

## 261ª Squadriglia

The 261ª Squadriglia was formed at Venice in early December 1917 at the command of *Sottotenente di Vascello* Giovambattista Bevilacqua with the pilots *Guardiamarina* Luigi de Riseis, *2° capo torpediniere* Raffaele Petta, *sottocapo cannoniere* Giulio Macchia and *marinaio* Luigi Bruzzone. Its aircraft were Macchi M 5. The unit, despite its limited forces, was very active. In December Bevilacqua was wounded in a flying accident, and STV de Riseis became commander.

At the beginning of 1918 the 261ª had eight M.5s for just four pilots, who were very active escorting bomber flying boats and attacking Austrian kite balloons at the mouth of the Piave River.

On 14 March 1918 *Tenente di Vascello* Umberto Magaldi became commander of the squadron.

On 1 May 1918 a patrol of four Macchis with Pierozzi, *fuochista* Vincenzo Burattini, de Riseis and Macchia escorting a photo recce seaplane to Pola clashed with an Austrian naval fighter, probably A 67 of Josef Niedermayer, that was seriously damaged and forced to alight. On 14 May the Seeflieger seaplanes attacked Italian ships off Pola. The pilots of the 261ª Squadriglia who contrasted them were Pierozzi, Magaldi, de Riseis, *volontario motonauta* Giovanni Naccari, *2° capo torpediniere* Edoardo Odierna and *marinaio* Renato Balleri. The Italians damaged K 217 and shot down A 70, with its pilot *Fliegermaat* Josef Gindl unharmed and A 85, that capsized killing *Seekadett* Franz Pichl. On 19 May, however, Austrian land fighters shot down Macchi 7280 of Magaldi, who was picked up by a seaplane of 251ª Squadriglia.

On 1 June the 261ª got a new commander, STV Domenico Arcidiacono, under whom were the pilots Balleri, Macchia, Odierna, STV Mario Savino, *2° capo* Attilio Buono, Giorgio Parodi and Dante Falconi, *2° nocchiere* Antonio Merola. On that day, however, *2° capo* Buono failed to return from a mission on Macchi 7274. The next day also Rivieri of the 260ª disappeared during a mission searching for the missing colleague.

During the Austrian offensive on the Piave the unit escorted bombers inland and hunted for kite balloons but on the first day of the battle, 15 June, during a bomber escort to San Donà, Austrian guns hit Macchi 7257 instantly killing *2° capo* Falconi.

On 2 July Pierozzi, Arcidiacono, Macchia and STV Vittorio Pomilio intercepted a flight of Austrian bombers and shot down K 394, its three crew members were captured. On 17 July the squadron took part in the massive day bombing of Pola. On 21 August the Macchis escorted the new SIA 9b of d'Annunzio's Squadra San Marco again attacking Pola.

On 17 September the 261ª tangled again with the famous Austrian ace Banfield. A patrol of seven M.5s lead by Savino and including Pomilio, Macchia, Merola, *Guardiamarina* Giuseppe Dalle Ore, *2° capo* Antonio Bucarelli and *2° capo* Beniamino Piro met three enemy fighters over the Tagliamento, shooting one down and

The 261ª Squadriglia from the Miraglia Naval Air Station, Sant'Andrea, Venice, used a white serrated design as identification marking.

driving away another. Banfield, on A 11, managed to escape the fire of Piro and Merola, who then went down and strafed the Austrian air station at Grado.

On 12 October while practicing mock combat Macchia, flying on M.5 13057, collided with a French Nieuport. The Italian survived but the French pilot, *maréchal* Autissier of Escadrille 561 was killed. During the final offensive the 261ª Squadriglia again operated over land with escort and strafing missions. On 3 November Pierozzi lead a flight to Trieste, and occupied the local naval air station. The pilots in service at armistice day were commander Arcidiacono, Dalle Ore, Macchia, Merola, Piro, Savino, *serg.* Umberto Guarnieri, *brig.* Arturo Simioli and *marinaio* Antonio Papola.

## 262ª Squadriglia

The 262ª Squadriglia was formed on Macchi M.5 in early 1918 in Brindisi, facing the coast of Albania, with the best pilots of the local Stazione Idrovolanti. The commander was *Sottotenente di Vascello* Umberto Maddalena, with pilots STV Mario Baldini and Giovanni Paterniti, *Guardiamarina* Andrea Balzarotti and Luigi Ragazzi, *2° capo* Guglielmo Bocconi, Mario Conforti, Giorgio De Marchis, Amedeo Melis and *sottocapo* Ercole Bellingeri. The missions of this unit were ther escort of bomber and reconnaissance seaplanes and the protection of the Italian coastal cities from Austrian air raids.

At dawn of 9 June 1918 Austrian flying boats from Durazzo bombed Brindisi naval air station, causing many casualties. Baldini, after tending the wounded and the dying, took off and shot down K 387. On 7 July TV Francesco De Pinedo (later a world-famous pilot), Paterniti, Baldini and Conforti during an escort mission to Durazzo clashed with the Austrian naval fighters A 53, A 74 and A 84. The Italians claimed two enemy fighters driven down, but Austrian sources don't confirm their claims.

On 5 September Maddalena, Paterniti, Conforti and *2° capo* Giorgio De Marchi escorting bombers strafed enemy warships in the Adriatic. On the way back the motors of Maddalena and Paterniti failed, and the two pilots, separately, alighted. They were saved after a long and dramatic period at sea. No other particular eventis marked the wartime service of this unit.

## 263ª Squadriglia

The 263ª Squadriglia was formed on 10 November 1917 at Porto Corsini, near Ravenna, from the 2ª Squadriglia Idro of the Regio Esercito that used to be based at Grado, close to the frontline, and that was disbanded after the retreat of Caporetto. It was equipped with FBA bombers and a couple of new Macchi M 5 fighters.

On 21 November *sergente* Umberto Guarnieri flying for the first time a Macchi, serial 4868, shot down flying boat K 161. He returned to base and immediately took off on a FBA with squadron commander *cap.* Ferruccio Capuzzo to save the enemy airmen, Andreas Kaudek and *Fregattenleutnant* Alexander Lupinski. Alighting despite the heavy sea, they failed to find the Austrians, who most likely perished at sea.

On 1 January 1918 the pilots of the 263ª were *cap.* Capuzzo, *tenenti* Otello Cavara, Alberto De Angelis, Antonio Diana, *sottotenenti* Giacinto Baratozzi and Mario Sabatini, *sergenti* Guarnieri, Rapp and Luigi Spina, with the observers *cap.* Agostino Crosa di Vergani and Giuseppe Vezzoli, *ten.* Pierluigi Casagrande, Otello Civoli and Riccardo Danielli. The unit had 13 FBA and two M.5, serials 4875 and 7070.

In March the number of Macchis was six. In the summer the unit operated also against land objectives. On 17 July its aircraft took part in the first large scale day bombing of Pola naval base.

Meanwhile the Italian government had stipulated an agreement for turning over Porto Corsini to the U.S. Navy, for American pilots to operate there with Italian seaplanes. The 263ª Squadriglia

Macchi M.5 serial 13016, individual number 9, belonged to the 262a Squadriglia at Brindisi.

"Mutt", a very famous and highly decorated Macchi M.5 of the 263a Squadriglia, USN Naval Air Station, Porto Corsini.

The Macchi M.5s of the 264ª Squadriglia, Ancona.

left its planes and installations, and its pilots were scattered among other units. U.S. Naval Air Station Porto Corsini was activated on 24 July 1918. The commander was Leutenant Willis B. Haviland, a veteran of the Escadrille Lafayette, and the American pilots were trained at the Lake Bolsena flying school.

The first war mission was on 21 August, when four M.5 escorted Macchi M 8 19008 of Walther White and Albert P. Taliaferro on a leaflet mission to Pola. They were intercepted by four Phönix land fighters, that were faced by Ensign George Ludlow, on M.5 13015, Ensign Austin Parker and pilot Charles H. Hammann, while Ensign Dudley Voorhees had to get away, as his guns were jammed. Ludlow damaged a Phönix, then he was shot down by *Fregattenleutnant* Stephan Wolleman flying A 118. Ludlow brought his burning Macchi to alight in the open sea, and then Hammann alighted near him, picked him up and flew him to safety. Hamman, who died in a Macchi accident on 24 June 1919, received for his bravery the Medal of Honor, the first one ever assigned to an U.S. Naval aviator.

In the following months the American aviators repeatedly attacked Pola. At the end of the war, the Americans left Italy, and on 1 January 1919 the Naval Air Station was formally returned to the Italian Regia Marina.

## 264ª Squadriglia

The 264ª Squadriglia was formed on 19 November 1917 at Ancona with the men and equipment of the local Stazione Idrovolanti at the command of STV Eugenio Casagrande. On 19 January 1918 command passed to TV Aldo Pellegrini: the unit was then equipped with fifteen FBAs, flown by pilots STV Arrigo Cardori, Francesco Di Brocchetti, Giuseppe Dalle Ore and Mario Salvestroni, *Guardiamarina* Enrico Ulivi, *capo* Gaetano Cagnaroni and *2° capo* Tullio Crosio. As the installations of the base were very bad, the unit flew just a few exploration missions and gave most of its seaplanes to other units.

On 1 June 1918 the 264ª with commander Pellegrini had pilots *capo* Cagnaroni, *ten.* Giovanni Naccari, STV Alberto Briganti, GM Vittorio Centurione Scotto and Goffredo Cionni, sailors Burattini, Carlo Tonti and Arnaldo Vitali and seven observers. The unit then had eight FBA bombers and four Macchi M.5 fighters.

On 18 July Burattini, flying Macchi '8' shot down the Austrian seaplane K 186. Two FBAs then took off from Ancona and picked up its crew, *Fliegermeister* Egon Priboschitz and *Seefänrich* Lothar Tippner. On 5 September three Macchis pursued K 233 and K 234 that were reconnoitering Ancona. Over Pola three Phönix and two Ks raised up to meet them and in the ensuing combat Burattini claimed another victory. Austrian sources don't admit this loss, but report that K 167 was destroyed on alighting. This unit was disbanded soon after the end of the war.

# Biographies

## Michele Allasia

Michele Allasia was born on 20 June 1895 in Ferrara, the beautiful and historical city in the region of Emilia, just south of the delta of the Po River. He was a short and healthy young man, with brown hair and blue eyes, who by trade was a lathe operator. In November 1913 he joined the Battaglione Aviatori, the aviation department of the Italian Army, as a volunteer, and his mechanical training was probably the motivation for this choice.

He progressed from mechanics to pilot training, after the outbreak of the war, at the Busto Arsizio flying school and he was licensed as a military pilot on 22 May 1916 on Maurice Farman with Renault engine. He was immediately assigned to the 37° Squadriglia, a Farman equipped unit that had been established in April 1916 for the defense of Bergamo, a city about 100 km from the frontlines, but that could be the target of enemy raids. Only one *Sezione* (flight) was initially active, and a second one, 2ª *Sezione*, with Farmans 783, 1508 and 1511 was activated on June 6, when *sergente* Allasia arrived together with *tenente* Luigi Bourlot. For a strange coincidence, both Bourlot and Allasia were to die on the same type of airplane, on the same airfield, a few days apart, two years later. Allasia's period in the 37ª Squadriglia was uneventful. The unit, confined in a secondary sector, flew just 172 flights during the year, and the only noticeable event was a storm that destroyed its hangars and five airplanes in August. Allasia's logbook records 40 standing patrols, from 29 June to 10 November 1916.

Like every good pilot wanted to do, Allasia was transferred to fighter training in November 1916, at the Cascina Costa training field, where he became familiar with the Nieuport. Despite the winter period, his conversion was quick, and on 28 February 1917 he was assigned to the 80ª Squadriglia Caccia of the I Gruppo, a unit that was activated that same day on the airfield of S. Maria la Longa, at the center of the Isonzo front. Commanded by *capitano* Mario Gordesco, not an ace but a very respected and loved leader, the unit was then equipped with Nieuport 11 and, besides Allasia, had in its ranks another future ace, *serg.* Alvaro Leonardi. Allasia arrived flying his Nieuport serial 2195.

Allasia's first combat took place on 18 March, during an escort to a Voisin reconnaissance biplane. Two days later he met again the enemy, during an alarm flight. More combats followed on April 10, April 12, April 25, May 4 and May 13, each time in missions of escort to slow Farmans or Voisins.

His first Bronze Medal was awarded for his mission on 25 April 1917. On that occasion, together with *sergente maggiore* Amleto Degli Esposti, also of the 80ª Squadriglia, Allasia was escorting a Voisin of the 25ª Squadriglia, crewed by *serg.* Resch (a future fighter ace) and *cap.* Gabbin when he had a combat against two enemies over Dosso Faiti and drove them away. Maybe one of his adversaries was *Hptm.* Bruno Schonowsky of Flik 34 flying Brandenburg C.1 69.07, who reported a combat with four Nieuports that severely damaged his airplane.

On 30 April 1917 the 80ª Squadriglia was moved to nearby Aiello, closer to the Adriatic, in the sector of the 3rd Army, where it teamed up with the 77ª Squadriglia, that became its twin unit. The squadron was heavily engaged in the 10th Battle of the Isonzo that began on 10 May 1917.

Allasia's encounter with the enemy, on 17 May, was negative: his aircraft, Nieuport 11 2168 engaged in a Voisin escort mission was attained by ack-ack fire, the pilot was wounded, forced to land near Doberdò, and was hospitalized.

In July 1917 the 80ª Squadriglia started to be re-equipped with the Nieuport 17, or Nieuport 110 hp, as it was invariably called in Italy. Allasia returned from hospital and on 9 July first flew the new and better fighter. On the 13th he claimed his first victory, an enemy seaplane engaged during the escort of a Caudron that was reconnoitering no less than Trieste, the main enemy base. He was flying with *serg.* Alvaro Leonardi, of the 80ª: over Prosecco he spotted an enemy seaplane and fired a whole magazine from his Lewis at it, and saw it spiraling down toward the pier of Trieste:

> "I took off from Ajello airfield at 8,45 am to escort a twin engine Caudron tasked with a photographic reconnaissance, at 9 am and at 3700 meters of height over Prosecco I spotted an enemy seaplane that approached and attacked me. I engaged

Michele Allasia, right, in front of the wingless fuselage of a Farman MF.14 of the 37ª Squadriglia, an airplane of practically no value as a defense fighter.

Allasia standing next to his Macchi Nieuport Ni.17 of the 80ª Squadriglia Caccia, showing the five-pointed star superimposed on the fuselage roundel.

and I managed to get on its tail and I fired my entire gun magazine. The enemy aircraft was hit and went down to Trieste not far away from the pier spinning many times. I soon noticed a motorboat that moved quickly toward it, obviously for rescue operations. I couldn't watch longer because I had to remain with the Caudron that meanwhile had gone to take its photographs."

There is no confirmation from Austrian sources of this event. It is just reported that three seaplane fighters took to the air against an Italian raid and that ace Banfield, flying A24, fired at an enemy aircraft at a height of 3,000 meters.

In August 1917 the 80ª Squadriglia was heavily engaged in the huge 11th Battle of the Isonzo, the Italian effort to break through the Austrian defenses on the Bainsizza plateau, that came just short of achieving a major penetration, bringing instead the conquest of the plateau itself and little else. The unit flew 398 combat missions, engaged the enemy twenty times and scored three victories, one of them by Allasia on the first day of the offensive, 10 August 1917, when together with Sabelli of the 91ª Squadriglia he shot down an enemy over mount Stol. The Austrians that day mounted a large bombing action, with 16 two-seaters against the Italian positions in the evening, but some pilots of the 80ª and 91ª Squadriglia intercepted their formation, shooting down two, Br. C.1 229.25 victim of Allasia and Sabelli, and Br. C.1 69.92 victim of Ranza and Degli Esposti. This is Allasia's combat report:

> "I took off at 7.30 p.m. for an alarm over Dosso Fait, reaching the level of 4000 meters I saw five enemy aircraft that were flying to the Faiti. I engaged combat with one of them at about 7.55 p.m. near Dosso Faiti and Stol at a height of 4000 and after firing fifty rounds with my machine-gun I saw the aircraft burning and falling in the direction of Mount Stol. I had this combat together with two Spads but one of them didn't pursue the enemy aircraft, while the other one kept on flying toward the enemy. I followed the aircraft until I saw it on the ground near the Stol."

Right: A signed photo of Michele Allasia sitting in the cockpit of a Nieuport 17 of the 80ª Squadriglia. The belt-fed Vickers machine gun was an excellent weapon.

Allasia was a consistent flyer, performing every day all sorts of missions: patrol flights, escorts, and a large number of test flights. On 1 September 1917 Allasia received from the Duke of Aosta, commander of the 3rd Army, his first Silver Medal "in the field": The citation reads:

> "(A) Daring fighter pilot, he performed with faith and enthusiasm many fighter, escort, cruise and photographic reconnaissance flights. He brilliantly sustained many combats with enemy aircraft, showing his brilliant daring and spirit of sacrifice. Dauntless against anti-aircraft fire, he managed, on 1 September 1917, in a very difficult escort to two Voisins, to protect them from the attacks of three enemy airplanes and with exact and violent machine gun fire, he forced the enemy to give up the fight, allowing the Voisins to complete their missions and return safely to our lines."

It is to be noted that this mission was his second on that day, following another escort to Savoia Pomilio's, but nothing particular was reported in his flight log.

In the month of September alone, Allasia flew about 60 flights that included 15 escorts to Capronis, Pomilios and Voisins, 26 cruises, 13 flights on alarm, and some test flights. In October his activity didn't slacken, and it included a mission to Pordenone for night fighting training.

Allasia and his Spad 7 (an early model, without frontal radiator shutters) of the 77ª Squadriglia at Marcon airfield, near Venice.

An unknown pilot of the 77ª Squadriglia and his Spad 7. He is showing the typical apparel of an officer in aviation service: the officer's cap, nicknamend "the flowerpot", lined leather jacket, cavalry breeches and boots.

Reportedly, the personal Spad of Allasia, with an unusual variation to the Red heart insignia of the 77ª Squadriglia. Over the heart there is a ribbon with some letters, unfortunately unreadable.

In October 1917 Allasia was admitted to the course for *Aspiranti Ufficiali*, that is, for officer training, and on 1 March 1918 he was commissioned as a *Sottotenente*, 1st Lieutenant. While in the Austrian armed forces the distinction between officers and troops was very rigid, and it was exceptional for a petty officer to be promoted, the Italian army had many opportunities for making the most of its brightest pilots.

When the Austrians and Germans broke through the Italian lines at Caporetto, Allasia, like all other Italian airmen, multiplied his efforts. He flew three missions on 25 October, with one combat, three more on the 26th, two on the 27th, then he followed his unit as it retreated, from Aiello first to Comina, near Pordenone, then to Arcade and finally to Marcon, near Venice, that was to be its base for the rest of the war.

On 2 November 1917 he claimed an enemy airplane shot down near Codroipo, in the morning. Then on 7 November 1917 Allasia together with *sergente* Felice Avon shot down an enemy over the Livenza River, during an escort mission to a SAML two-seater in the early afternoon.

After the Caporetto retreat, Allasia thus described the situation in a letter to a friend, dated 18 November:

> "I am well, but we live in an incredible way, there is nothing to eat, and we get only cans and cracker, we sleep where it happens, and especially now that it is getting colder, it is not nice at all. But it's all right: the service is as usual, and on the 14th I shot down a German airplane, that was flown by two officers, who both died, and this is my fifth victory. A proposal was made for another Silver Medal, and I was praised by the Command of the Group. Now, where we are, we are not sure that we will remain, actually we have the order to be ready to move, and it is reported that we will go to Ferrara".

The last sentence meant that even a retreat across the Po River was contemplated, leaving all of the Veneto to the Austrians, before a line of resistance could be established along the Piave River.

On 10 November 1917, Allasia was shifted from the 80ª to the 77ª Squadriglia. On that day the two squadrons had reached their new base of Marcon. The 77ª Squadriglia during the retreat had burned six Spads, and now it had only three on strength.

The fact that Allasia was a reliable and determined pilot is proven by his flight log: despite the bad weather, in the second half of November and in December 1917 he flew no less than 59 combat missions, most of them escorts to two-seaters, during which he had three air combats, he strafed enemy troops three times and once he attacked a kite balloon.

He was assigned in December his second Silver Medal, for his deeds in 1917, that were described as four victories in 23 air combats, and two photographic reconnaissance missions. Some pilots, like Cerutti, were described as "naturals": each time they attacked, the enemy was shot down, other ones, like Allasia, were just professionals, doing their part, fighting the enemy, but not necessarily scoring each time they had to fight.

Allasia opened the New Year with two missions, a standing patrol of 100 minutes, followed by an escort flight to a SAML over Susegana lasting 75 minutes. He met the enemy on February 3rd,

A line-up od Spad 7s of the 77ª Squadriglia, an effective unit, that claimed 50 enemy aircraft shot down during the war.

Allasia, second from the left, lying in the grass in front of the SVAs of his unit, the 5ª Sezione SVA, at Fossalunga. The plane on the right has a personal decoration, most of its fuselage is painted in a undetermined very light color.

and again on March 9, but he never scored, while his strafing missions became more and more frequent.

On 11 March 1918 Allasia was sent on a reconnaissance mission over Pordenone, across the Piave River, and he was attacked by a patrol of eight enemy fighters. He managed to get out of that scrap, but his Spad was hit eighteen times by enemy bullets. This combat could easily correspond with the 11th victory claim by the Austrian ace Benno Fiala von Fernbrugg, who claimed a "Sopwith" over Spresiano, seen going down in a spin.

From March 12 to April 16, 1918 he did not fly, probably he was on leave, which was surely well deserved, considering the hectic and stressing pace of his flight activities.

Allasia seems to have been a pilot for all seasons: one day he was sent to escort two-seaters, the next day he flew fighter sweeps, then he was sent on photographic reconnaissance missions, using camera-equipped Spads, with at least one mission every day, all through the spring of 1918. His last flight with the 77ª Squadriglia, a photo mission over Conegliano and Vittorio, was on 6 June.

In 8 June 1918 Allasia was assigned to the 5ª Sezione SVA, that was formed on 14 May 1918, based at Fossalunga, and assigned as a strategic reconnaissance unit to the 3rd Army. These *sezioni* (flights) SVA reporting to the Army commands were élite units, to which initially only particularly reliable pilots were assigned. The SVA was considered a fast but heavy single-seater, unfit for dogfights, but the 5ª Sezione, heavily engaged in the frantic air combats during the Battle of the Piave, the mid-June final effort of the Austrian Army, scored several victories: two claims for *sergente* Giulio Bellaria, on 15 and 19 June and two claims for *s.ten.* Allasia, one on 15 June over Fontigo, and one on 23 June over Falzè di Piave.

For these actions, Allasia was assigned posthumously his third Silver Medal:

> "Wonderful fighter and reconnaissance pilot, always first in every daring deed, many times he flew at low level to strafe the enemy infantry and artillery. He attacked and shot down an enemy reconnaissance airplane, and forced two to land. Returning from a reconnaissance, he met two enemy fighters and attacked them with remarkable daring, shooting one down in flames and driving away the other one. - Sky of the Montello, 15-18-23 June 1918."

It is remarkable that the citation of the medal (our only source, as the Sezione papers are not preserved) reports one victory and two enemies forced to land and actually the Austrians in that area lost four airplanes, one shot down and three forced to land.

On 20 July, 1918 the 5ª Sezione SVA was shifted from Fossalunga to Marcon, the old airport of the 77ª Squadriglia near Venice, but on that very day Michele Allasia crashed to his death: his SVA 9 two-seater, serial 11712, went down out of control, hitting a house, killing also his passenger *capitano* Giuseppe Graglia. Despite the fact that all press reports at the time of his death assigned to him twelve air victories he was assigned only five confirmed victories in the official list of Italian aces issued in 1919.

## Michele Allasia
## 9 victory claims - 5 confirmed victories

| Claim | Vict. | Date | Sqn. | Aircraft | Hour | Place | Victim |
|---|---|---|---|---|---|---|---|
| 1 | - ? | 13.7.1917 | 80 | Ni 17? | 9h+ | Miramar-Trieste | Seaplane |
| 2 | 1 ? | 10.8.1917 | 80 | Ni 17? | 19h55 | Dosso Faiti | EA |
| *[with ten. Sabelli, 91ª Sq.] Zugsf. Hugo König WIA/Fähnr. i. d. R. Wolfgang Klemperer WIA, FlG. 1, Br. C.1 229.25* | | | | | | | |
| 3 | 2 ? | 2.11.1917 | 80 | Ni 17? | 9h30 | Codroipo | EA |
| 4 | 3? | 7.11.1917 | 80 | Ni 17? | 15h+ | Livenza | EA |
| *[with serg. Avon, 80ª Sq.] Maybe: Gefr. Wilhelm Appelt/Lt. Paul Wilkening, FA (A) 204 or: Zugsf. Friedrich Schieg/Oblt. Friedrich Wihrt, FA ?.* | | | | | | | |
| 5 | -? | 14.11.1917 | 77 | ? | - | ? | German a/c |
| 6 | 4 | 15.6.1918 | 5ª Sez. | SVA | - | Fontigo/Collalbrigo? | Two-seater |
| 7 | - | 15.6.1918 | 5ª Sez. | SVA | - | Fontigo/Collalbrigo? | Two-seater |
| 8 | - | 15.6.1918 | 5ª Sez. | SVA | - | Fontigo/Collalbrigo? | Two-seater |
| *Maybe: Fw. Rudolf Berger KIA/Oblt. Ernst Petric KIA, Flik 26/DS, Br. C.1 329.48 or: Stfw. Franz Koudela UNH/Oblt. Anton Chrzaszcz WIA, Flik 26/DS, Br. C.1 329.22 or: Unknown pilot/Lt. i. d. R. Johann Reinstaller WIA, Flik 52/D, Br. C.1 or: Unknown pilot/Lt. Ernst Winkle WIA, Flik 52/D, Br. C.1* | | | | | | | |
| 9 | 5 | 23.6.1918 | 5ª Sez. | SVA - | | Falzè di Piave | EA |

## Antonio Amantea

Antonio Amantea was born on 28 September 1894 in Lecce, a beautiful city in the Apulia region, the heel of Italy's boot, called "the Florence of the South" for its art and baroque architecture. Little is known of his early years, other than by trade he was an electrician. In September 1914 he was drafted and assigned to the 13th Field Artillery Regiment, automobile company, which proves that he had some sort of mechanical knowledge, so much that on 31 October he was assigned to the Comando Scuola Aviatori (command of airmen school) for pilot training. He got his pilot's brevet, on Bleriot monoplane, on 1 September 1915. The Battaglione Aviatori, the Italian aviation service, was manned with personnel drawn from all corps (cavalry, artillery, engineers, Carabinieri etc.), that maintained the assignation and the uniform of their original service.

Amantea reached the war zone on 15 November 1915, and at the same time he had the rank of *sergente*. He was assigned to the 3ª Squadriglia per l'Artiglieria, artillery-spotting squadron, equipped with Caudron G.3 and based at Medeuzza, on the Isonzo front. This squadron was a gallant unit that included among its pilots another future ace, Ferruccio Ranza. Its commander was *cap.* Felice Porro, who later wrote an excellent history of the Italian air force in World War I and who commanded the Italian 5th Air Force in North Africa in World War II.

Amantea flew his first reconnaissance mission on 24 November 1915, with *ten.* Fabio Jannoni Sebastianini as an observer. He flew three more missions until December 22, when the unit moved to nearby Gonars. Amantea flew four more missions, and many test flights, until 7 February 1916, when he went on leave. He returned to his unit in March, and from then on he flew a series of reconnaissance missions, all uneventful, despite the fact that those defenseless biplanes were flying over the most cruelly contested mountains of the Isonzo front, like the Sabotino and the San Michele.

In mid June 1916 the 43ª Squadriglia (as the 3ª Squadriglia per l'Artiglieria had been renamed since April 16) was hurriedly moved to the Asiago front, to supplement the Italian forces resisting the Austrian offensive southward called "Strafexpedition". The unit was based in Castel di Godego. Thus, Amantea first flew in the area where he was to score his air victories, the Altopiano di Asiago, a zone also called "gli altipiani", the plateaus, by the Italians, that faced the Alps and barred the way to the plain of Vicenza. On the new front, Amantea again flew several missions of spotting for the guns of the artillery. The Caudron G.3 despite its archaic look was much appreciated by Italian airmen, as it was reliable, slow but safe. Its defense armament, non-existent when the airplane was introduced, was always of limited value, on account of the frontal propeller and back tailbooms that limited its field of fire, yet many times the Caudrons survived encounters with Austrian fighters. Amantea, however, never met enemy fighters during his period on the Asiago front, and his main problems were bad weather and unreliable engines.

Antonio Amantea, in the parade uniform of a sergeant, with the winged propeller symbol of a pilot on its sleeves.

Right: Three outstanding pilots of the 71ª Squadriglia: Antonio Amantea, *ten.* Antonio Riva, later commander of the 78ª Squadriglia, and *sergente* Francesco Tola, in front of a Spad.

A famous image of the Austrian Oeffag Albatros 53.33 flying above the typical Alpine landscape of the Asiago front. This fighter was forced down by Amantea, Riva and Tola on 24 August 1917 with pilot Kowalczic, and it was shot down on 29 November 1917, by the Italian Panero.

In late July 1916 the unit moved back to the Isonzo front, based now at Sammardenchia. Amantea was active in the summer of 1916, flying 24 times in August alone, but again without meeting the enemy. Among the new pilots assigned to the 43ª Squadriglia now there were also Silvio Scaroni and Giorgio Michetti, two future aces too. Amantea's activity was outstanding, and he flew often with lieut. Pietro Pinna, the Sardinian who became the chief of the Italian Air Force in East Africa in World War II. His total log with the 43ª Squadriglia, from November 1915 to February 1917, stands at 173 flights, including 69 reconnaissance and artillery spotting missions.

On 20 February 1917 Antonio Amantea was posted to fighter training, and he quickly mastered his new mounts, because on 24 March 1917 he was assigned to the 71ª Squadriglia Caccia. Based at Villaverla and equipped with the Nieuport 11 80-hp fighter, this unit patrolled the front of Asiago (or South Tyrol, for the Austrians). His first fighter mission was a protective patrol over Bassano and Asiago on 4 April 1917.

On 16 April 1917 Amantea had his first combat: while cruising over Asiago, Bassano and the Pasubio mountain he spotted an enemy airplane and attacked it, together with another Nieuport, forcing the enemy to escape. After a few uneventful flights, another combat soon followed, on 28 April, when he met an enemy over Bassano, and attacked it, shooting 100 rounds in four bursts. The enemy was pursued back to its own lines. Amantea didn't claim a victory, and his possible opponent was a Brandenburg C.1 of Flik 21, that returned to its base with a dead observer, *Oblt.* Eduard Appel, while the pilot, *Zugsf.* Adolf Kind, was unharmed. The gallant Austrian pilot reported that he was attacked by four Nieuports, one of which he claimed shot down, and he had to fight his way out for 20 km before landing at Pergine, with his observer mortally wounded by two bullets in the chest. The aircraft was written off. There was only one other Italian Nieuport taking part in that fight, however, the one piloted by *serg.* Arrigo Rossi, who shot at the enemy from 100 meters and then had to drop off the combat, returning to base.

A similar combat took place on 15 May. Again, during a protective cruise between Cima 11 and Asiago, Amantea spotted an enemy over Spitz Tonezza and Verena, attacked it together with another Nieuport, shot 100 rounds (two drums of the Lewis gun) and saw it gliding down to its own lines. This time, enemy artillery took part in the action, and Amantea returned with a shattered windscreen. In this period Amantea also followed a course for promotion to officer.

In June, in that region there was the development of the tragic Ortigara offensive, an attack by Alpini troops to an Austrian-held mountain, the first large Italian thrust outside of the main Isonzo battlefront, that caused terrible losses, without any gain. Flying very often, Amantea had many opportunities to meet the enemy, and he

The mixed equipment of the 71ª Squadriglia: a brand new Nieuport 17, French built, serial 3123, and behind it a Spad 7.

had three duels, on June 2, 4 and 5. The first time he forced one enemy to escape, the second time he suffered a jammed gun, and in the inconclusive duel, his own Nieuport 11 was hit four times, the next day it was his enemy that had to retire.

On June 10 Amantea strafed enemy troops over the Val d'Astico. From then on, he started flying the new Nieuport 110 hp, or "15 Square meters", the Ni. 17. The new fighter didn't serve him on June 19 when he pursued an enemy that was just too fast for him.

In July sergeant Amantea flew almost every day, and had five more combats, on the 6, 9, 10, 13 and 29 July, all of them without victories, and two of them cut short by gun jams. On 2 August 1917 he scored his first confirmed victory: on an interception flight over Schio, he found an enemy and attacked it repeatedly. He saw the enemy observer slumping into the cockpit, and the Austrian aircraft started to fly irregularly. While he was about to finish it off, his engine cut abruptly, for a magneto failure. He had to interrupt the combat and crash-landed, wrecking his Nieuport. There is no confirmation of his victory, but there is an Austrian claim for a Nieuport, by the crew of *Korp.* Ludwig Lorenz and *Oblt.* Anton Zita of Flik 48 on that day, in that area. *Serg.* Cerutti reported a combat in that same area and with the same details, including the detail of the wounded observer, and it is possible that the two aces were in the same combat together.

On 6 August Amantea had a long duel over mount Toraro that lasted 30 minutes, against two enemies. Amantea fired 240 rounds, one of the enemies escaped, the other one glided back to its lines, visibly hit. His first victim was possibly a Brandenburg C.1, 129.54 of Flik 48, that escorted by Albatros 53.44 of *Zugsf.* Meissner (one of the few Alb. D III fighters assigned to the Tyrol front) had a combat with three Nieuports, and was brought to a forced landing, with the crew of *Korp.* Ludwig Lorenz and *Lt.* Franz Gróf unharmed. The Austrians claimed one Nieuport shot down. Another Austrian aircraft had an unlucky encounter with the enemy in that area, Brandenburg C.1 229.16 of Flik 17, whose pilot, *Korporal* Rudolf Horatschek was wounded at both arms but managed to bring back his airplane to Gardolo airfield, and this perhaps could be Amantea's victim. It is difficult to determine what exactly happened, because in that same morning two other pilots of the 71ª Squadriglia, *ten.* Antonio Riva and *serg.* Montalto Ercoli had another combat above Arsiero with two Austrians, one of which was claimed as shot down.

Three days later, on August 9, Amantea fought two enemies and forced them to retire. On August 18, in a single flight Amantea met two pairs of enemies, attacked all of them and forced the four Austrian airplanes to escape. That day, the Austrians claimed two Italian fighters shot down, a Spad on Mount Lisser, in that area, by *Lt.* Friedrich of Flik 24, and a Nieuport on Mount Pau, but only one fighter was damaged, the one flown by *serg.* Reali.

On August 23 over the Pasubio, Amantea engaged a duel with two enemies, one of which disappeared behind his own lines. The next day, August 24, flying together with *ten.* Antonio Riva and *serg.* Francesco Tola, he had a quick and furious combat against an enemy that was seen burning and going down toward Luserna. This victory was confirmed to Amantea, and his victim was *Feldwebel* Julius Kowalczik, an Austrian five-victories ace who force landed his burning Albatros D.II 53.33 of Flik 24 in the Astico Valley and got out unharmed.

September passed on at the same frantic pace: Amantea flew no less than 25 combat missions and had seven combats, on September 5, 9, 11, 15, 24, 28 and 30. In the first three occasions, his gun jammed and the enemy escaped. On the 15th of September, he saw the enemy gliding away, on the 24th, at 11:20 am, he scored against a pair of enemies, one of which was confirmed by the Italian Supreme Command over Mount Pau. This victim could be

Albatros D.III 53.51 of *Korporal* Julius Frühwirth of Flik 48/D which was damaged and made a crash landing at Terlan after a free hunt mission in the Sarca Valley and Pasubio area. The pilot was unharmed, but he suffered from a bad nervous shock, so he was granted a period of rest.

The two last duels of the month, Sept. 28 and 30, were inconclusive and the enemy escaped him.

In October 1917, Antonio Amantea flew 13 combat missions and had dogfights with the enemy four times, on October 13, 15, 21 and 26.

The same rate of action went on in November, and on the 18th Amantea claimed another victory. His logbook reports: "Pursuit and combat against two enemy aircraft, crashing one behind the Spitz of Tonezza". The victim is not identified, but the daily report of the Italian Aviation Command reports that both enemy airplanes could possibly have been shot down. In the following days, the 71ª Squadriglia, commanded by *cap.* Amerigo Notari, had to retreat from Villaverla, which was now under range of Austrian heavy guns, and was frequently shelled, and moved to Sovizzo.

In November Amantea flew 12 combat missions, and he converted to the Spad fighter, with which he flew his first combat mission on the 28th. Meanwhile, on 7 November, he had been promoted to "*Aspirante Ufficiale*". He received his commission as a *Sottotenente* on 6 February 1918.

In December Amantea flew exclusively the Spad, with which he performed 12 missions, most of them escorts to the SAMLs and SPs that were trying to counteract the enemy pressure, and he had only one combat. This took place on 7 December, a day of frantic air action in that area, with heavy Italian losses, but Amantea had to break off combat for the failure of a bracing wire in his plane. With those final combats, the last Austrian offensive in the area of Melette had ran out of steam.

Since April, in nine months of 1917, Amantea had flown 174 flights, more than half of them combat missions, battling with the enemy 29 times.

His first flight in 1918 took place on 15 January, after a brief period of leave: together with his squadron, he flew his Spad to Padua, and from there they escorted a large raid of Caproni bombers against Arsiè. The escort was so much effective and well coordinated that the pilots of the 71ª Squadriglia, including Amantea, were praised by the Aviation Command of the 1st Army.

Amantea flew 12 more times in January and on the 28th, during a long cruise over the Brenta valley, he scored again. He was flying Spad 1701 and he spotted an enemy airplane, attacked it firing 50 rounds and saw it diving away and disappearing in the fog in Val Frenzela. Despite the dubious circumstances, this combat was confirmed in the daily Bulletin, which, however, quoted it as only his fourth recognized victory. On that day, while Italian troops were in action at the Mount Valbella, Italian pilots according to the War Bulletin claimed ten victories, eight of them in that area, which don't find correspondence in known enemy documented losses, which amount to only one aircraft, a Brandenburg of Flik 21/D shot down behind Italian lines by Reali.

Antonio Amantea and his Spad. It was reported that it was decorated with a sentence in his native dialect of Apulia, but before his death the Ace wrote to an historian that he didn't have any photograph of it.

Amantea had 11 more combat flights in February, at the controls of Spads 1531 and 1669 together with many test flights and five more combat flights in March, until March 22. On that day, flying a protective patrol, he met an enemy aircraft and forced it to retire rapidly behind its own lines. He could not complete the pursuit because his engine was giving problems. Again, this combat was confirmed as a victory, despite the fact that Amantea himself, in his logbook, didn't claim it as such. On that day and in that area, the Aviatik D.1 38.16 of *Feldwebel* Franz Slanina of Flik 23 did have to force land, with a wounded pilot on board, but Austrian sources recognize Italian anti-aircraft fire, and not the bullets from the Spad, as the cause of his misfortune. It is to be remarked, however, that the area of the combat corresponds in both accounts, so possibly *Fw.* Slanina didn't even realize that a fighter had jumped him.

In April 1918 Amantea flew only a few times, ferrying fighters or training new pilots. On May 2, he took off for an alarm mission with an unusual mount, a Hanriot. The next day he was back on his Spad, to score his last victory. Flying as an escort for a SAML two-seater, he dueled against an enemy aircraft, fired 150 bullets and saw the enemy go down between Spitz di Tonezza and the Luserna fort. This victory was officially confirmed, and the pilot received the usual prize of Lire 500. It is to be remarked that Amantea's were the only victories scored by the 71ª Squadriglia in this period. The unit in mid May moved from Sovizzo, which was flooded, to nearby Castelgomberto.

He had seven more flights in May, three of them combat missions, and then on June 7 he was posted as an instructor to the Campo scuola (training field) of Busto Arsizio, near Milan, where fighter pilots were trained.

Meanwhile, he had gained a certain notoriety, and he was interviewed by the magazine *Cielo* that presented him as the only Ace of the 1st Army (Bedendo had not become an ace yet). Amantea spoke simply of his deeds:

Austrian troops gathered around the wreck of Aviatik Berg D.I 138.16 of *Feldwebel* Slanina, on 22 March 1918. According to the Austrians, it was forced down by groundfire, but it likely was a victim of Amantea. *(Photo Masajdek).*

> "I reached the front on 15 November 1915 with Caudron single engined airplanes. I flew 90 reconnaissance missions over the enemy in the Gorizia area. In February 1917, at last, I passed to fighters. I was assigned to the 1st Army. In the Trentino the quarry was scarce in those times. Only on 2 August I shot down my first enemy. On the 24th of the same month, I flamed one over Luserna. The third one, I shot down on 24 September at Verena. The other ones during the offensives of November and December. In the action of Valbella and of Col del Rosso in the days of January 27th, 28th, 29th I flew over five hours every day always strafing and escorting, and in the second day of the offensive I managed to shoot down an enemy.
>
> How do you shoot down an enemy aircraft? It is very easy; all you have to do is see it in your gunsight and shoot. If your machine-gun works, the enemy cannot escape. The problem is that often the gun jams and the enemy runs away. One day I had five combats and each time my gun failed me. Another time, Austrian fighters came to pay a visit to our airfield. I took off immediately with Capt. Barattini, commander of my Squadron. (*Actually, Barattini was a lieutenant, not the Squadron commander*) We followed the enemies to the airfield of Caldonazzo. I made contact with an enemy, I aimed it in my gunsight, the gun jammed and then I had to work to get free from the Austrian's attack. These are the risks of the fighter, but if the bullet starts, it reaches its target and then it's a piece of cake."

Antonio Amantea was awarded two Silver Medals, one in August 1918 for its two victories of August 2 and 24, 1917, and a second one for his victories of 24 September 1917 and 28 January 1918. A third Medaglia d'Argento al Valor Militare was awarded to him in 1922 for his victories of 22 March and 3 May 1918: thus, his official *palmarés* credits him with six victories.

After the war, Antonio Amantea was raised to the rank of *tenente* in May 1919 and then he was posted to a most dislikable destination, the life imprisonment military jail on Asinara Island. In October 1919 he left the Armed forces. In June 1922 he returned to the active service. It is reported that, in order to be accepted, he dressed in civilian clothes, but with all his medals, and he sat in front of the Ministry of Aviation, working as a shoeshine. This shocking form of protest worked immediately, he was called in, and admitted as an officer in the active service on that very day.

In August 1922 Amantea was posted to the Ghedi Flying School as Chief Instructor. When the Regia Aeronautica was formed as an independent air arm, on 28 March 1923, he left the Army and donned

A line-up of the Spads of the 71ª Squadriglia. This unit intitially didn't mark its aircraft, then it introduced individual numbers on the rudder, then nicknames or slogans were painted on the fuselages.

Amantea, in the center, in Libya, after having flown on a Meridionali Ro I, with a Group commander pennant affixed to its strut, two exceptional passengers: Crown Prince Umberto and princess Maria Josè, who stand at his sides for a picture.

the blue uniform of the air force. On 15 March 1925 he was posted to the 1° Stormo Caccia (Fighter Wing) and on 20 November 1925 to the new 2° Stormo. In 1926 he was posted to the Genio Aeronautico (air engineers) and he rose to the rank of captain.

From 1928 to 1932 he served in the Colonial Aviation, during the counter-insurgency operations in Libya, and became a Major. In 1933 he returned to Italy, posted again to the 1° Stormo Caccia, and then in November to the 4° Stormo, where he took command of the X Fighter Group. In December 1934 he also became judge in the Military Tribunal of Trieste, and in 1935 he was a Lieut. Colonel.

During the conquest of Ethiopia, 1935-36, he commanded the 44th Bomber Group, and was awarded two more Silver Medals and was knighted in the Military Order of Savoy. In this role, he is reported to have had the integrity to punish the commander of the 15ª Squadriglia, Galeazzo Ciano, who also happened to be the Minister of Culture and Mussolini's son in law and heir apparent. His honest, impulsive and daring character was reported by many anecdotes. In February 1937, when a terrorist attack by Ethiopian patriots wounded Gen. Graziani, the Viceroy of Ethiopia, Amantea who was standing next to him grabbed a carbine and shot at the attackers. In 1938, now a Colonel, he commanded the 50º Stormo Assalto (assault Wing) and then the Foggia flying school. In 1941 he was the commander of Castelvetrano air base, in Sicily. By the time of World War II he had reached the rank of General. After the Italian armistice, 8 September 1943, Amantea was in command of the Galatina air base, near his hometown of Lecce. While most Italian commands had melted, leaving the troops without orders, Amantea kept his post, faced the vengeful Germans, organizing a few armed airmen and forced them to abandon the area without sabotaging the structures, so that he could deliver them safely to the Allies. The airfields in the region of Apulia were to be much important in the war effort of the USAAF.

Amantea retired from the service in 1946, and spent the rest of his life in Lecce. He lived with his wife, Mrs. Fortunata Caramis, and with a colored girl that had followed him from Africa, and whom he had adopted. He died on 13 July 1983, forgotten by most, but not by the airmen who served under his orders, and who have petitioned the city of Lecce for naming a street after him.

Another important passenger for Antonio Amantera in Libya: standing next to him is General Rodolfo Graziani, the ruthless destroyer of the Libyan rebellion, then the hapless commander of Italian land forces in 1940.

Italian officers in Ethiopia, during the war of conquest, 1936. Second in the forefront is Farinacci, a fascist leader; next to him Colonel Magliocco, aeronautical advisor to commander in chief General Badoglio, then Antonio Amantea.

## Antonio Amantea
## 10 victory claims - 5 confirmed victories

| Claim | Vict. # | Date | Sqn. | Aircraft | Hour | Place | Victim |
|---|---|---|---|---|---|---|---|
| - | - | 28.4.1917 | 71 | Ni 11 | - | Bassano | EA |
| *Zugsf. Adolf Kind UNH/Oblt. Eduard Appel KIA, Flik 21, BR.. C.I* | | | | | | | |
| 1 | 1? | 2.8.1917 | 71 | Ni 17 | - | Arsiero | EA |
| 2 | - | 6. 8.1917 | 71 | Ni 17 | - | M.t Toraro | EA |
| *Maybe: Korp. Rudolf Horatschek WIA/unknown observer, Flik 17, Br. C.I 229.16 or: Korp. Ludwig Lorenz UNH/Lt. Franz Gróf UNH, Flik 48, Br. C.I 129.54* | | | | | | | |
| 4 | 2 | 24.8.1917 | 71 | Ni 17 | 10h | Luserna | EA |
| *[with ten. Riva and serg. Tola, 71ª Sq.] Stfw. Julius Kowalczik UNH, Flik 24, Alb. D.III 53.33* | | | | | | | |
| 5 | 3 | 24.9.1917 | 71 | Ni 17 | 11h20 | Mount Verena | EA |
| *Korp. Julius Frühwirth UNH, Flik 48/D, Alb. D.III 53.51* | | | | | | | |
| 6 | - | 18.11.1917 | 71 | Ni 17 | - | Mount Spitz | EA |
| 7 | - | 18.11.1917 | 71 | Ni 17 | - | Tonezza | EA |
| 8 | 4 | 28.1.1918 | 71 | S. 7 1701 | - | Bassano | EA |
| 9 | 5 | 22.3.1918 | 71 | Spad 7 | - | Vallarsa | EA |
| *Fw. Franz Slanina WIA, Flik 23, Aviatik D.I 38.16* | | | | | | | |
| 10 | 6 | 3.5.1918 | 71 | Spad 7 | - | Spitz- Tonezza | EA |

In the pilot's official papers and in his medals' citations Amantea seems ot have six confirmed victories while in the 1919 list of aces he is credited with five only. It is not known exactly which one was deleted, and why. Possibly it was the first one, of 2 August 1917, as Cerutti received no confirmation for this claim either.

# Giovanni Ancillotto

Giovanni Ancillotto, usually called Giannino or Nane, was one of the Italian World War I aces whose memory is best preserved: he was a sort of local hero, and his monument, a large cascade of marble, dominates his hometown of San Donà di Piave, a peaceful, elegant and wealthy town in the Southern part of Veneto, just east of the Piave river. Actually, it is so bulky that some townsfolk, in the current anti-militaristic mood, have proposed to dismantle it and move it somewhere else. Ancillotto was immensely popular for his most dashing deed, flying through a burning enemy kite balloon, that gained him the Medaglia d'Oro, the Gold medal for bravery, and also gained him the front page of the "Domenica del Corriere", Italy's most popular weekly magazine, and the visual image of a small fighter limping home carrying on its wings the shreds of the enemy balloon has remained in memory to this day. Ancillotto, however, was more than that: he was doubtlessly an outstanding pilot, the first Italian night fighter, and the first man in the world to shoot down two enemies in a single night engagement. He had a strong personality, that made him a dominating figure, despite his very young age.

The classical portrait of Giannino Ancillotto with his Spad, and on the breast the ribbons of all his decorations, including the Medaglia d'Oro.

Giannino Ancillotto was born on 18 December 1896 in a rich family of landowners and industrialists, in San Donà di Piave. His father, besides improving his domains, was the first person in Italy to extract and exploit methane gas. When war broke out Ancillotto was just eighteen, with a strong passion for mechanics and motoring sports, a student in the Politecnico university of Turin, Italy's most renowned school of engineering (also attended at the time by another future ace, Bartolomeo Costantini). He immediately volunteered, in November 1915, in the Battaglione Squadriglie Aviatori (the Italian Army air service) and was admitted as a student pilot at the Gabardini flying school in Cameri. He was licensed as a pilot in April 1916, graduating on the Gabardini 80 hp monoplane in a class of a dozen pilots that included two other future aces, Guglielmo Fornagiari and Antonio Chiri.

Giannino Ancillotto in the cockpit of Macchi Nieuport 17 serial 3664. This silver-painted fighter shows the emblem of the 80ª Squadriglia: a silver star in relief painted over the roundel. The five-pointed star, *stelletta* in Italian, worn on the lapels is the basic Italian military emblem.

In May 1916 he was posted to the Busto Arsizio flying school where he was licensed as a military pilot on the Farman on 14 June 1916.

He was assigned on 25 June 1916 to the 30ª Squadriglia, a reconnaissance unit based at Tombetta, near Verona, attached to the 1st Army, deployed along the northern front. Flying the obsolete Farman model 1914, the 30ª Squadriglia was a highly professional unit that gained fame as one of the best reconnaissance units of Italy. He flew on the northern front for less than a month, then on 21 July 1916 the 30ª Squadriglia was recalled to Chiasiellis, near Udine, as the danger of the enemy offensive on the northern front, the Strafexpedition, had been checked. Here the unit was assigned to the 2nd Army. Ancillotto flew his first orientation flight on the new front, on 24 July, with another pilot, *sergente* Arturo Dell'Oro. It was maybe an omen: also Dell'Oro, shifted to a fighter unit, was later to gain the Gold Medal for Bravery for a suicide ramming attack against an enemy airplane, in which he lost his life.

The seaplane bomber K366 floating damaged on Doberdò lake, a victim of Giannino Ancillotto on 27 October 1917.

All through the summer of 1916, Ancillotto flew photographic reconnaissance missions over the hottest part of the front, the Bainsizza plateau and the Chiapovano valley. On 31 August he was promoted to corporal.

On 10 October *cpr.* Ancillotto and *ten.* Sobrero attacked with their machine gun an enemy kite balloon that they spotted near Comen, forcing it to be pulled down. Strangely, this flight is reported in the squadron papers but not in his logbook. He flew a total of just 20 missions with the 30ª, but most of them were risky and eventful.

In mid October 1916, still a corporal, Ancillotto was shifted from the 30ª to the 27ª Squadriglia, based in Campoformido, also part of the II Group. On 18 October, flying with *s.ten.* De Vizzi, he was attacked by an enemy fighter, that hit their Farman but was driven back by their fire.

Ancillotto immediately proved his mettle: on 2 November he brought back his airplane severely damaged by enemy flak: his observer was again lieutenant De Vizzi. For this reason he was specially praised by the commander of the II Group, *magg.* Capuzzo:

> "I send my highest praise to observer s.Tenente mister Giuseppe de Vizzi and corporal pilot Ancillotto Giovanni of the 27° squadriglia for the serene behavior and the calm demonstrated in the reconnaissance of 2 November in which they managed to return to our lines despite having their aircraft damaged and the tank pierced by the fire of enemy artillery."

He flew again on 13 November, and again his airplane was hit by anti-aircraft fire. On 20 December Ancillotto with *ten.* Ambrogio Comi flew a reconnaissance mission at low level over the Vippacco River, and their airplane was wrecked by the enemy artillery, and glided back into the Italian lines with the engine off. For this action, he was officially commended in the Bulletin of the 2nd Army:

> "In an air reconnaissance done at very low level for the bad conditions of visibility, he went on performing his duty and contributing to the success of the operation, despite the strong fire of enemy anti-aircraft guns that hit his airplane in several vital parts – Vippacco Valley 20 December 1916."

In January 1917 Ancillotto, skipping the rank of sergeant, became an *Aspirante*, or cadet in the officers' corps. On 18 February 1917 he was posted back to the 30° Squadriglia. He had flown just nine flights with the 27ª, and he flew four more with the 30ª.

Such a dashing pilot could not endure for long the limitations of reconnaissance, and the role of being just a target for the enemy cannons. On 13 April 1917 Ancillotto managed to be recalled to the Aviation Command in Turin, for conversion on fighter aircraft. He

Right: Giannino Ancillotto standing in front of a Nieuport 27 armed with a Lewis on the top wing and Le Prieur rockets on the wing struts. This fighter, which apparently was never flown by Ancillotto, carries French roundels under the top wing and the red diamond over a circle insignia of the 83ª Squadriglia. This unit was based at Marcon for a brief period, from November 1917 to February 1918, assigned to the XIII Gruppo. *(Photo Perissinotto)*

was posted to Malpensa (where the present international airport of Milan is located) for training on the new Pomilio C fighter. This airplane, however, proved extremely dangerous to fly, and was assigned with some misgivings only to some reconnaissance units. At Malpensa Ancillotto was breveted on the Nieuport 80 hp fighter on 2 June 1917. On 11 June 1917 Ancillotto went to San Giusto training field, near Pisa, for the brief gunnery training on the Nieuport 80 hp le Rhone, and immediately he took his final tests on the type. All the events had shown that this intense young man, who had managed to be outstanding in two crack units, had all the marks of an ace.

On 14 June 1917, now a *sottotenente*, he was assigned to the 80ª Squadriglia Caccia, equipped with Nieuports and based at Aiello, and that very day he flew his first fighter patrol. He flew very much in the usual chores of patrols and alarms, and had combats with the enemy on June 21, 27 and 29, and July 14. In just 45 days, from mid June to July 1917, he flew 44 combat missions, in August he flew 45 more, having combats on August 18, 19, 23, and twice on the 28th, in two different missions. In September he flew 45 further missions, with combats on 7 September, (twice), and 23 September. Only victory was still eluding this determined fighter. He had a further combat on 2 October. On 3 October he flew for the first time an experimental night mission. On 7 and 19 October his Nieuport was hit during air combats. A remarkable mission was the one of 16 October, when he escorted Italian seaplanes over the enemy naval base of Pola. He had flown 18 missions by 24 October, when the Austro-German offensive in the Plezzo-Tolmino area broke through the Italian lines.

It looked as if the national tragedy had broken the spell of Ancillotto's fruitless fights. As soon as the weather cleared, on 25 October, he flew three times. The next day, 26 October 1917, flying with other pilots he met a formation of four enemy seaplane bombers escorted by two seaplane fighters, with two outstanding pilots, the ace Gottfried Banfield and Pichl, that were attacking Italian infantry. Well defended and excellently manned, the Austrian seaplanes were no easy prey, actually it was common knowledge among Italian airmen that the Austrian naval aviators were the most redoubtable enemies in the air. In the ensuing combat, Ancillotto shot down two enemies, the bomber K212, that force landed on Doberdò lake with a badly wounded pilot (a victory shared with Alvaro Leonardi and Ernesto Cabruna also of the 80ª Squadriglia and Francis Lombardi of the 77ª) and K367 that fell near the coast off Sistiana, and whose crew, *Frwg.* Eduard Tröber and *Lt.* Max Obendorfer was saved by an enemy torpedo-boat. On seeing the Italian attack to the seaplanes *Lt.* Franz Gräser of Flik 42/J came to the rescue and claimed a Nieuport shot down over Doberdò as his fifth victory. There are no known corresponding Italian losses in this combat, although surviving records of the 80ª Squadriglia are incomplete. That same day, Ancillotto flew four other combat missions.

The next day, 27 October, he was in the air again, and again he met a similar formation of four seaplane bombers and three Luftfahrtruppe landplanes, with the same escort of Banfield and

Giannino Ancillotto in front of a Nieuport 17, serial 3592, of the 80ª Squadriglia. *(Photo Lucchini)*

Pichl, repeating the attack on Italian infantries. Again Ancillotto fought them, and this time it was K366, that had escaped the previous day's action, that was shot down, crashing again in Doberdò Lake with the loss of its crew, Gröger and Mericka. This victory was probably shared again with Francis Lombardi and Sambonet of the 77ª Squadriglia also claimed that day in this area. In that hectic action, the defending Austrian fighters belonged to the redoubtable Flik 42. Aces Hefty and Teichmann claimed one Nieuport forced down, but Udvardy had to limp back to Prosecco airfield for a forced landing, maybe a victim of Francis Lombardi's fire.

According to a soldier's diary, as the Austrian seaplane was floating in Doberdò Lake, some Italian soldiers approached it. One of the Austrians drew a pistol and shot a Bersagliere, killing him. Then a Carabiniere leveled his rifle and shot, killing the Austrian. The other enemy airman, wounded, was taken out of the seaplane and carried to the Division command, where Italian doctors cared for him, while a cortege of soldiers followed him crying "Kill him, kill him!"

The 80ª Squadriglia then retreated, first to La Comina, and then to Arcade, still fighting. On 3 November 1917 Ancillotto took off on alarm, and over Oderzo, a town a few miles from his hometown,

Above: A well known image of Ancillotto with his Nieuport 11, as it came back after the flight through the burning kite balloon. Below: Another view of Ancillotto with his now famous old fighter. This Nieuport didn't carry any armament besides the Le Prieur rockets.

he shot down a German reconnaissance two-seater, whose crew, *Unteroffizier* Wilhelm Noss and *Leutnant* Richard Monninghoff, did not survive. This victory again was shared with Francis Lombardi.

On 10 November 1917 the 80ª Squadriglia settled in Marcon, near Venice, and on that same day Ancillotto was assigned to the twin unit, the 77ª Squadriglia. It may be possible that in those dramatic days, the most reliable pilots, such as Ancillotto and Allasia, were shifted to the unit that had the more effective fighters, the Spad 7s. Without even a conversion flight on the Spad, Ancillotto was back in action on 15 November 1917, escorting a Pomilio and protecting kite balloons. All through the month, he escorted the Italian two-seaters that were trying to map the new enemy positions, and he had combats against enemy aircraft on the 15th, 16th and 27th of the month.

On 15 November Ancillotto had a clash with the German fighters of Jasta 31 who on that day claimed two kite balloons flamed, by *Lt.* Thurm and *Lt.* Kosslick. He pursued a fighter, described as having a red fuselage band, but with no exit on account of a jammed gun:

> "Combat Report S.Ten. Giovanni Ancillotto: today I left at 1.45 p.m. to escort a P, I had a combat against 3 enemy aircraft that attacked a Kite balloon. The combat lasted about 10 minutes, but it was not successful because my gun would jam at every bullet. The fighter that flamed the kite balloon carried a red band around its fuselage, and I followed it across the Piave."

In these actions, he had understood the importance of captive observation balloons, that were the only spotting position on the totally flat plains of lower Veneto, and that were directing enemy cannon fire against the Italian lines. Ancillotto's determination to destroy those balloons was such, that he managed to fly on an obsolete Nieuport 11 fighter, only because it could be equipped with Le Prieur incendiary rockets, while the much better Spad was not yet equipped for them. Flying Nieuport 2265, on 30 November 1917 Giannino Ancillotto shot down an enemy kite balloon that fell down burning near Fossalta, just across the Piave River from his villa, that was on the enemy side of the lines. This was his fifth confirmed victory and it made him an ace.

On 3 December 1917 he flew three times, Twice the balloon was lowered before his attack, the third time Ancillotto scored, and flamed another kite balloon by S. Polo di Piave, a few miles upstream from Fossalta. The next day he attacked once more a balloon that was lowered.

On 5 December, he shot down his third kite balloon at Rustignè. a hamlet on the road from Oderzo to the Piave River. Three Spads from his squadron escorted his Nieuport and such was the dash of Ancillotto that he flew straight through the burning mass of the enemy airship. His Nieuport returned to base draped with the cloth of the enemy, that was recovered and used to tailor a raincoat offered to the Duke of Aosta, the commander of the 3rd Army, while other pieces were signed, framed and offered as souvenirs to friends, some of them having been preserved to this day. This action turned Ancillotto into a national hero. He had already been awarded two Medaglie d'Argento, one for his missions in the reconnaissance, a second one for his air victories of 26 October and 3 November. Now he received the Medaglia d'Oro, the Gold Medal for bravery that during the war was assigned to only 21 airmen, with this motivation:

> "Fighter pilot of admirable daring, from 30 November to 5 December 1917 in a series of daring attacks he fired three enemy balloons and forced others to stop their observations. In a special circumstance, he attacked his enemy with such dash, that he flew through the burning airship, bringing back on his severely damaged airplane parts of the ripped envelope."

The event is better commented by Francis Lombardi, who witnessed it:

> "How could that little Nieuport withstand the surely violent crash and then, with its power of just 80 hp, to keep on flying, overcoming the resistance of the heavy ripped cloth, nobody knows. Anybody, in Ancillotto's place, realizing to be still alive, after the crash that must have been scary, would have immediately landed in the first small field across the Piave. But not Ancillotto. With his iron willpower, and realizing that, unexplainably, he could still fly, with his extraordinary pilot ability, he wanted to return to his airfield, carrying the glorious shreds of the shot down balloon. We welcomed him in a triumph, astounded and moved."

Gabriele d'Annunzio, always a tinkerer with the Italian language, called him *"l'Ala incombustibile"*, the wing that doesn't burn, and coined a Latin motto for him: *"Perficitur Igne"*, through the fire he is made perfect.

Through the rest of the year, Ancillotto kept on flying, with a total of 21 missions in December 1917, but now photographic reconnaissance missions and ground strafing mixed with his usual balloon-busting flights. It is reported that among his targets was his own villa, that had been turned into an enemy command post, and against which he vented his rage. The Austrians were shooting back, however, and on both 28 and 30 December enemy artillery hit his aircraft.

Despite his wealth and his success, Ancillotto was modest and self-effacing. He was commonly called Nane, a dialect shortening of Giovanni. Despite his young age, he was much admired for his perfect flying skill, for his stamina and aggressive fighting spirit, and for the fact that he never boasted, or even commented about his exploits.

In January 1918 he flew fifteen war missions, many of them long recce missions, including one of two hours over his old airfield of Aiello, now in Austrian hands. In February he started night interception missions, against the German bombers, mostly AEG

Left: *Ten.* Carlo Alberto Conelli de Prosperi, of the 91ª Squadriglia, in front of his Spad 13 with individual number IX marked on the fuselage side and top. Right: *Capitano* Filippo Serafini, commander of the 77ª Squadriglia, who flew together with Ancillotto in the winning mission of 28 October 1918. Serafini, formerly a Caproni pilot, after the war became the founder of the Aero Club of Bologna, to which it passed the red heart insignia of the 77ª Squadriglia, that is still used today.

G.IV of Bogohl 4 that were effectively raiding Italian cities, airfields and railroads. The data in his logbook show his eagerness: on the night of 3 February he scouted the sky twice, first for 105 and then for 130 minutes. The next night he flew three times, for 70, 90 and 80 minutes, and once he met the enemy, firing 200 rounds. Overall, in February he flew 30 times, including 14 night missions. During the day, he strafed enemy troops, and on 25 February he was again attained by enemy groundfire.

It is incredible how he could sustain such a relentless rhythm of combat. In March again he flew seven times, with strafing missions, until 12 March, when, returning from an attack to a kite balloon, his engine failed and he had to force-land near Monastier. This crash has been connected to the claim made by the Austrian ace Gräser, of Flik 61J, who on that day claimed a SVA shot down by Monastier, as his 16th victory, flying the Albatros D.III 153.106. However, the Italian pilot's papers don't mention a combat between Ancillotto and enemy fighters.

At this time, somebody must have stepped in and forced him to go on leave, before he collapsed.

Ancillotto flew once more on 8 May 1918, taking off on alarm from Arcade, then he returned to the 77ª Squadriglia only on July 16. Thus, he missed the frantic days of the mid-June offensive on the Piave, and the chance of increasing his score - or of being shot down. He was immediately back in the thick of the action, with a combat on 17 July, and more missions, including his first leaflet dropping, until the fateful night of 24 July. The German heavy bombers had gone, but the Austrians still attacked Italian cities at night, particularly Treviso. Ancillotto took off with an Hanriot that was considered the best fighter for night flying, thanks to its excellent handling qualities. One after the other, he spotted two Brandenburg C.1 bombers, and shot both of them down. They belonged to Flik 101/G, and both crashed in Italian territory, one of the crews losing their lives, the other one being taken prisoner. This was his combat report:

"Pilot: tenente Ancillotto Giovanni
Time of departure: 0.35
Time of landing: 1.15
Shots fired: 100
Type of attacked enemy aircraft: Brandenburg (large bomber)
Time, height and place of combat: 0.55 800m. C. Marcon
Way of attack: from behind
Apparent result: shot down
Exact place of crash of the shot down aircraft: on Sile River (Trepalade)
Comments: I took off on alarm at the airfield at 0.35, then I cruised for some time above the airfield; I saw an enemy airplane that was in the beam of a searchlight. Immediately I flew toward the plane, as it was still visible. I reached it quickly and I shot a long burst and made it crash near Trepalade. I was returning to the field to land after the first victory when again I noticed shooting above Casale. So I went on with my cruise and suddenly I saw another enemy plane. I went after it and fired with my gun. So I managed to force down a second airplane near Sant'Elena sul Sile.

Pilot Ancillotto Giovanni"

The city of Treviso named its defender an Honorary citizen in 1924, and for this action, he received his third Silver Medal, assigned, however, only in 1925, after his death. At the time, however, the only result he got was an official reprimand for having picked up a twisted machine-gun from one of his victims as a souvenir, for which he had to write a letter of apology, explaining that picking souvenirs was "the general habit of all those who shot down enemy aircraft". That gun and other relics, including the tail of Br. C.1 169.121, are now preserved at Treviso military airfield. The next night he was back in the air.

Overall, in July Ancillotto flew 9 combat missions. In August he flew 28 more, that included the whole gamut: night interceptions (no less than five night missions in the single night of 25 August!) leaflet dropping, strafing, attack to captive balloons, photo-reconnaissance, and he scored one more victory: together with *tenente* Carlo Alberto Conelli de Prosperi, on 21 August, he shot down an enemy Brandenburg between Ponte di Piave and Salgareda, again a few miles from his San Donà. No identification of this claim is available among Austrian sources: "Having seen an airplane targeted by anti-aircraft batteries above Salgareda Bocca Collalta we attacked it and followed it down from 4,500 to 100 meters from the ground, firing short bursts. The enemy observer returned fire aggressively."

On 29 August, he twice attacked kite balloons and forced them to be pulled down. Despite the stream of bullets from his gun, 370 in the first action, 150 in the second one, the balloons did not catch fire. Attacks on "drakens", as they were called, were very dangerous on account of the many guns that protected them and fighter pilots of all nations disliked this kind of mission. Ancillotto flew five more missions in the first four days of September, then he was assigned on 6 September 1918 to the Commissariato Generale di Aeronautica. His experience was required for some staff work, possibly test flights, but when the final Italian offensive was launched on 24 October, exactly one year after the Caporetto retreat, Ancillotto managed to be sent back to his old Squadron, the 77ª.

In the final days of the war, his activity was frantic: between 27 and 31 October he flew no less than 14 combat missions, strafing and bombing enemy troops. On 28 October together with the Squadron commander, *cap.* Filippo Serafini, he went of a free hunt mission over San Fior airfield. There they saw a Sopwith Camel attacked by an enemy fighter. Ancillotto engaged the enemy, but it was too late: Camel E 1579 of No.66 Squadron was shot down in flames, with the death of Lt. G.A. Goodman. Ancillotto hit the Austrian, and saw it falling down vertically, following it down to 200 meters. Serafini tackled another enemy, wrongly identified as a Pfalz, and shot it down. Two more Camels and three Austrian fighters joined the combat, and soon the Italians saw two more fighters burning on the ground, although they couldn't identify them. Their allies were pilots of No.66 Sqn., among them Lt. A. Paget flying Camel E 7215 claimed two "Alb DV" burned east of Godega, for his 5th and 6th victories. The British pilot also was hit after an attack on S. Giacomo di Veglia airport, and made an emergency landing at Treviso.

> "Pilots. Cap.no Serafini Filippo=Ten. Ancillotto Giovanni
> Time of departure: 8.25
> Time of landing: 9.25
> Shots fired: 500
> Type of attacked enemy aircraft: Pfaltz fighter
> Time, height and place of combat:
> Way of attack: time 9 – 1,000 meters – S. Fior airfield, from behind and below Cap. Serafini, from behind and above Ten. Ancillotto. The one of Capitano Serafini crashed . The one of Ten. Ancillotto followed in a vertical dive down to 200 meters from the ground.
>
> Exact place of crash of the shot-down aircraft: near the airfield.

San Fior airfield in an Italian reconnaissance photo. The flat fields of Veneto, usually seeded with corn or wheat, were easily converted to airfields, and the common presence of large brick farms or landowners mansions provided good accommodation for squadron personnel and offices.

> Other airplanes that took part in this combat: two Englishmen who engaged combat with three more enemy fighters.
>
> Additional notes: we took off for a free hunt flight and flew to S. Fior airfield. We saw an enemy attacking a British fighter. Immediately Ten. Ancillotto rushed to defend it but he failed in saving it, as it caught fire. Then he attacked the enemy and saw it diving vertically while being hit and followed it down to 200 meters. He considers it shot down. Capitano Serafini meanwhile attacked a Pfaltz at a height of 800 meters and this was shot down and crashed, as witnessed also by Ten. Ancillotto. Meanwhile two more British fighters and three enemy fighters arrived and engaged combat. Two aircraft went down in flames but we couldn't identify their nationality.
>
> The pilots
> Ten. Ancillotto Giovanni
> Capitano Serafini Filippo
> 27 October 1918"

After 31 October, Ancillotto did not fly any longer in the war. Possibly, during those final days he rushed to his home, across the Piave River, to see what had remained of it after the Austrian occupation and his own attacks.

His career of fighter pilot (not counting the period as a two-seater pilot) can be resumed in a total of 315 operational flights with 24 combats, 11 kite balloon attacks, 14 strafing missions and at least 27 night flights.

After the war, the ace felt restless, like so many other teenagers who had become men on the battlefield, and now were unaccustomed to the silence of the guns. Probably he couldn't stand the sight of the empty airfields, and of the warplanes left to rot. In December 1918 he got to be sent to Rome, to the Nucleo Comunicazioni Aeree, a new unit that, under the leadership of Mario Mercanti, was experimenting air mail services from Rome to the main Italian cities, and long distance raids, partly to keep communications going in an Italy troubled by never-ending strikes and agitation, and

Ancillotto in Somalia after the war, standing with local personalities in front of a SAML two-seater.

In elegant civil clothes, and looking more mature than his age, Ancillotto in Peru in front of an Ansaldo Balilla.

The monument to Ancillotto, in the center of his hometown, San Donà di Piave. The bronze bas relief above the marble eagle heads depicts his flight through the balloon and his portrait.

The birthplace of the ace, the large villa in the outskirts of San Donà, now belongs to a circus family, as intimated by the elephants on the gate posts.

partly to lay the groundwork for the future airline services. On 11 March 1919 he flew from Rome to Trieste, the newly liberated town, a flight of 800 km in 4 hours 40 minutes with an Ansaldo SVA. On 11 September 1919 Ancillotto, together with *ten.* Danilo Mainardi (who later crashed to his death in Latvia, while instructing the local pilots on the A.1 Balilla) and Polish Lt. Zawisza flew from Rome's Centocelle airport to Warsaw, as an ambassador of goodwill to the new Republic of Poland and a salesman for Ansaldo aircraft. The Polish government actually bought large quantities of Ansaldo A.1, SVA and A.300.

Returning to Italy, Ancillotto joined Gabriele d'Annunzio in the occupation of Fiume, the border town between Italy and Yugoslavia that the Allies refused to deliver to Italy, and that was turned by the poet-adventurer into a small revolutionary Republic, animated by patriotism and by radical ideologies.

Ancillotto remained in Fiume for a year organizing the tiny "Legionnaire air force" of that "corsair" state, and taking part in cloak-and-dagger missions, until the city fell to the Italian armed forces that finally stepped in to bring back legality. While d'Annunzio was moved to his gilded cage, the villa on Garda Lake that was called the Vittoriale, and where the SVA that flew d'Annunzio to Vienna is still preserved, Ancillotto moved to Peru, in one of the many missions that the large Ansaldo group was organizing in order to find new markets for its airplanes. Ancillotto organized a daring raid from Lima, the capital, to Cerro de Pasco, a short raid, but one that required flying above the Andes, more than 6,000 meters, and landing on a makeshift airfield in the farm of two Italians, at a height of 4,800 meters. The raid was successfully accomplished on 2 May 1921 with an Ansaldo A.1 Balilla that returned to Lima safely. He received what can be rightly called a hero's welcome. Peru did not forget him, and when he died that nation built a monument to him in Lima (just as the Italians had built in Domodossola a monument to Geo Chavez, a Peruvian, the first man ever to fly over the Alps on 23 September 1910) and they contributed to the erection of the large monument in his hometown.

In a few years, Ancillotto had had more adventures than most men can dream in a lifetime, but it was not enough. He moved to Africa, Somalia, the long forgotten Italian colony, where he performed many exploration flights, and even took part in some local action against native rebels. When he returned to Italy, all roads were open to him, especially a political career, in the Fascist party, that had taken power and that attracted the interest of most veterans and aviators. Ancillotto however refused any political involvement, feeling unprepared on account of his young age. He was interested in starting a career with the aviation industry, and he talked to his friends about performing some outstanding record flight, but on 17 October 1924, as he was driving his fast car from Turin, the industrial center of Italy, back to his town, San Donà di Piave, to participate in a patriotic ceremony, he had an accident, and crashed to his death by a village called Caravaggio. His monument was erected in 1931, as eternal memory to a young man who had all the tracts of the winning fighter pilot, the ace. His villa still remains: as his family died out, it was sold to the Nones Orfei family, who run the most important Italian circus, and it is used as a winter stop for the circus people, caravans and animals.

## Giovanni Ancillotto
## 11 confirmed victories

| Claim | Vict. # | Date | Sqn. | Aircraft | Hour | Place | Victim |
|---|---|---|---|---|---|---|---|
| 1 | 1 | 26.10.1917 | 80 | Ni 17? | - | Brestovica | Seaplane |
| *Frwg. Eduard Tröber UNH/Lt. Max Obendorfer UNH, Seeflugstation Trieste, K 367.* | | | | | | | |
| 2 | 2 | 26.10.1917 | 80 | Ni 17? | - | Doberdò Lake | Seaplane |
| *[with serg. Leonardi & m.llo Cabruna, 80ª Sq., ten. Lombardi, 77ª Sq.] Frg.Lt.Maximilian Kramer Edler von Drauberg WIA, POW/Matr. Marcello Anasipoli POW, Seeflugstation Trieste, K212.* | | | | | | | |
| 3 | 3 | 27.10.1917 | 80 | Ni 17? | - | Doberdò | Seaplane |
| *Fähnr. Richard Gröger POW/E. Frwg. Alois Mericka POW, Seeflugstation Trieste, K366* | | | | | | | |
| 4 | 4 | 3.11.1917 | 80 | Ni 17? | - | Ravarè | EA |
| *[with ten. Lombardi, 77ª Sq.] Unteroffz. Wilhelm Noss KIA/Lt. d. Res. Richard Monninghoff KIA, FA 14, DFW C.V* | | | | | | | |
| 5 | 5 | 30.11.1917 | 77 | Ni 11 2265 | - | Fossalta | Kite balloon |
| *Lt. i. d. R. Franz Bibica KIA, BK.13* | | | | | | | |
| 6 | 6 | 3.12.1917 | 77 | Ni 11 | - | S.Polo | Kite balloon |
| *Lt. i. d. R. Max Riegert KIA, BK.10* | | | | | | | |
| 7 | 7 | 5.12.1917 | 77 | Ni 11 2265 | - | Rustignè | Kite balloon |
| *Lt. i. d. R. Hans Bratina, KIA, BK.2* | | | | | | | |
| 8 | 8 | 24.7.1918 | 77 | Hanriot | 0h55 | Treviso | Brandenburg |
| *Zugsf. Josef Hartmann, KIA/Oblt. i. d. R. Wilhelm Fidrmuc, KIA, Flik 101/G, Br. C.1 129.38* | | | | | | | |
| 9 | 9 | 24.7.1918 | 77 | Hanriot | 1+ h | Treviso | Brandenburg |
| *Lt. i. d. R. Franz Löffler WIA, POW/Oblt. i. d. R. Otto Halir WIA, POW, Flik 101/G, Br. C.1 169.121* | | | | | | | |
| 10 | 10 | 21.8.1918 | 77 | Spad 7? | - | Ponte di Piave | Brandenburg |
| *[with ten. Conelli 77ª Sq.]* | | | | | | | |
| 11 | 11 | 28.10.1918 | 77 | Spad | 9 h | San Fior | Pfalz (?) |
| *[with RAF pilots, 66 Sqn.]* | | | | | | | |

# Flaminio Avet

Flaminio Avet came from a distinguished family that traced its origins in Savoy, and that gave to Italy and France important jurists, such as his great-grandfather Giacinto Fedele, who as Minister of Justice signed the first Italian Constitution with king Carlo Alberto in 1848, and valiant soldiers. He was born at Bendejun, near Nice, on 3 August 1890.

After attending high school, the young count, tall and elegant, with a handsome face, followed the family tradition and in 1914 he graduated with top grades in Law School at the University of Rome. The outbreak of war forced him to drop the lawyer's robe and get in an army uniform, with the red collar of the Lancieri di Firenze cavalry regiment, as a *sottotenente di complemento* (reserve 1st lieutenant). Modern war, with trenches, barbed wire and machine guns, left no room for the romanticism of a cavalry charge, and Avet, like many other officers, passed to another arm. In August 1915 he was in force with the 37° Reggimento Artiglieria da Campagna (37th Field Artillery Rgt.) on the Carso. Very soon he could prove his valor during the bloody 4th battle of the Isonzo against Mount San Michele. On 25 November 1915 he commanded a battery engaged against an enemy attack. Despite the Austrian reaction with strong and accurate counter-battery fire, that destroyed one of his cannons, Avet with cool determination directed fire until the enemy threat was checked, earning his first Medaglia d'Argento.

Despite being a good artillery officer, Avet wanted something more, so he volunteered for aviation service, he had his medical check on 21 May, two days later he was admitted as a student at the Venaria Reale flight school, near Turin, and on the 25th he had his first ground lesson on an Anzani powered Bleriot. Avet got his first brevet on 15 September, and then he moved to Mirafiori, also near Turin, for his 2nd brevet on Aviatik, and then to Cascina Costa, near Milan, where he got his military brevet on 15 November.

In early 1917 Avet was assigned to the 73ª Squadriglia SAML that he reached on 3 February, after a short leave. That unit was based at Verona, tasked with defense and reconnaissance, and there Avet found *sottotenente* Leopoldo Eleuteri, an old friend from Cascina Costa, who also became an ace and whose career was closely intertwined with the one of Avet.

After a series of familiarization and test flights, Flaminio Avet flew his first war mission on 20 April 1917, a patrol flight from Col Santo to Rovereto at the controls of SAML S.1 1587. Five days later, 25 May, coming back from a leaflet-dropping mission, Avet discovered shrapnel fragments in the spar of his right wing. Up to the start of the summer 1917 Avet flew several visual and photo reconnaissance missions and even some Caproni bomber escort flights, such as the mission of 19 June, with SAML 1590. The use of this slows two-seater as a fighter was justified, as the Austrians had only a handful of Brandenburg KD in that sector of the front.

On 31 August Avet was assigned back to the Battaglione Scuole (current name of the schools command) for fighter conversion. First he practiced gunnery at Furbara, and then he was posted to Malpensa for conversion on the new Ansaldo SVA, a type of aircraft then reserved only for the best pilots. There he was graded as "very good" in target practice. His instruction was completed on 27 October, just in time to be rushed back to the front, as the Austrians had broken through at Caporetto and the Italian commands were struggling to close the breach sending men and equipment.

Flaminio Avet in the elegant uniform of a lancer. Before World War I, few could predict that the days of cavalry were coming to an end.

In a letter to his relatives he wrote: "... the past doesn't matter any longer. Let's place our trust in the future...". On 1 November he was posted to the 82ª Squadriglia Hanriot that on 10 November moved to Istrana. This squadron, formed by very good pilots, still was considered the "Cinderella" of the X Gruppo that had much more accomplished units. Thus in the mess jokes the 91° was the "Squadriglia of aces", the 70ª was the one of "semi- aces" and the poor 82ª the squadron of "pigeons".

Avet met the enemy for the first time on 27 November, while escorting a SP in Val d'Assa, but it got away. On 13 December, cruising over the Asiago plateau with *sergente* Ferdinando Nelli, of the 70ª Squadriglia, Avet near Col dell'Orso met a patrol of enemy aircraft and engaged combat, an Austrian got on his tail but Nelli got him out of harm's way.

In the morning of 26 December Avet was on alarm duty together with Benvenuti and Teobaldi on Istrana airfield. When approaching enemy aircraft were spotted, Avet told his two NCOs to take off, then rushed to squadron commander *tenente* Antonio Fochessati, for instructions. The general alarm was sounded, Avet got in his

Hanriot, but it as it was taxiing it was riddled with bullets. Then the pilot got into the last available Hd, his motorist Solenghi launched the propeller and he took off as bombs were falling on the airfield. He found a German two-seater north east of Camalò, already under attack, Avet got into attack position, fired and the bomber crash-landed in a small field, turning over. Only one man got out of the wreck, with burning trousers, who was immediately helped and then captured by Italian soldiers. The ace Silvio Scaroni, Michetti and a British pilot, among others, took part in this action. It is not clear, however, if Avet claimed a participation in this victory.

Avet became acting commander of the 70ª Squadriglia and then, on 18 January 1918, effective commander, replacing *tenente* Rino Corso Fougier, also acting commander in the place of *capitano* Ferdinando Martinengo, sick. The unit had suffered, changing five commanders in a few weeks, and it was now Avet's task to turn it into a well-knit and motivated fighting outfit.

On 25 January Avet together with *ten.* Alessandro Resch had an inconclusive combat over the plateau. On 1 February he yielded command to *maggiore* Vincenzo Lombard, then he went on a two weeks furlough. On 18 March, anti-aircraft guns targeted his fighter cruising between Oderzo and Nervesa, and broke a rib. On 3 April he protected Resch and Eleuteri as they attacked an Austrian kite balloon at Conegliano.

On 17 April 1918 Avet, flying Hd 523, lead a flight of Hanriots cruising from Ponte alla Priula to Quero, when they found a two-seater of Flik 52/D spotting for the artillery with the escort of a large formation of Albatros D.III of Flik 42/J, six or nine of them, according to reports. The Italians, identified by the Austrians as "Sopwith", got in a favorable position and started their attack. The first to fall was Albatros 153.152 that shed its wings in its dive, killing *Oberleutnant* Michael Gassner-Nordnon von Laudon. Then, despite the frantic intervention of the Albatros, among them the ace Ferdinand Udvardy, the Brandenburg C.1 169.35 was also shot down, falling behind Austrian lines, with the death of pilot *Korporal* Viktor Stibor and observer *Leutnant* Ladislaus Rajcsics. According to Italian documents Avet "shot down a two-seater and took part in the destruction of two fighters". The second one could be the mount of *Oberleutnant* Richard Fitz, who, according to Austrian sources, after the combat in which two aircraft were shot down was attacked again and repeatedly hit but not, apparently, shot down nor forced to land. The Austrians claimed two victories, but all the Italian fighters returned safely to base. British pilots also claimed heavily that day on the Piave front, but place and details of this combat correspond much better with the Italian claims.

On 25 April Avet and his patrol claimed two victories in a combat above Conegliano, during a leaflet-dropping mission to Susegana. His victim could possibly be Brandenburg C.1 169.82 of Flik 4/D, forced to land near Ramera with its two aviators unharmed. The place corresponds, but the Italian reports state that the victims were two fighters. Actually two Albatros fighters of Flik 68/J were

Flaminio Avet during pilot training, in front of Bleriot XI serial 304. The obsolete Bleriot, unfit for combat duties, was an excellent trainer and was operated at Mirafiori flying school well into 1918.

lost with their pilots on that day in that area, but Austrian sources state that it happened due to a collision before landing and don't report a combat, so the question remains open. Both claims received no confirmation.

On Friday 3 May 1918 Avet lead a flight of fighters escorting a SAML reconnaissance to Grossaga. The Italians clashed with a flight of fighters of Flik 63/J coming from Motta di Livenza airfield, lead by the commander *Oberleutnant* Josef Lederer, flying Phönix D.1 228.42. The Austrian unit, just recently assigned to the front, paid dearly for the inexperience of its new pilots. The Austrian squadron commander was shot down at 12 a.m. between Zenson and S. Andrea di Barbarano, behind the Italian lines. The fighter was totally destroyed; the Italians recovered the body of the "very young" pilot. Avet, for this victory behind own lines, got the usual prize of 1000 Lire. The other components of the Flik 63/J flight didn't fare any better, soon after they were bounced by pilots of RAF No. 66 Squadron, who claimed three fighters shot down and one "out of control". *Feldwebel* Ladislaus Tomana was killed, shot down in flames at Isola di Sopra while *Zugsführer* Rudolf Nemetz managed a forced landing on Mansuè airfield. Despite the overclaims of the British pilots, it was really a black day for the Austrian squadron. Teobaldi, 82° Squadriglia, was credited with an active participation in the victory with the British pilots, and an Italian anti-aircraft battery claimed it too.

In the morning of 17 May Avet took part in the combat that cost the Austrian ace Gräser his life. Initially several Italian fighters from the 91ª, 79ª and 78ª Squadriglia attacked an Austrian formation and scattered it: when Avet joined the fray, he saw a single two-seater with an escorting fighter crossing the Piave near Lovadina. As Avet attacked, the fighter fled leaving the slower airplane to be shot down near Maserada, behind the Italian lines. Avet saw the two enemies leaving the wreck, one of them supporting the other. It was Brandenburg C.1 229.30 of Flik 12/Rb, and the captured crew was formed by observer *Leutnant* Viktor Herzog, unharmed, and pilot *Feldwebel* Johann Valenta, wounded. Two pilots of the 78ª Squadriglia had a share in this victory, *sergente* Cosimo Rennella and *tenente* Mario Fucini. The latter in his memories wrote that, after a few rounds, his gun jammed and as he was trying to fix it a fighter with two diagonal bands on its top wing had stepped in, evidently Hanriot 11411 of Avet. Confirmation of the victories of 17 May was controversial; it seems that in the end the Brandenburg was credited to Avet alone, to the regret of Rennella and Fucini.

Two months passed before Avet scored again. In the meantime the commander of the 70ª Squadriglia, *capitano* Ernesto Sequi, who had replaced Lombard on 1 February, was wounded in air combat on 17 June, and Avet replaced him as acting commander. On 25 June he had a combat over Codignano with two fighters called *"type Halberstad"*, one of which was seen to spin down to earth, the other diving but recovering near the ground.

On 15 July 1918 Avet lead a patrol in a cruise from Sernaglia to Conegliano and fought against a large group of Austrian fighters

A portrait of Flaminio Avet early in his carreer. His officer's bonnet is decorated with two crossed cavalry lances and an airman's engine and propeller, below the royal crown.

Right: Three aces and friends of the 70ª Squadriglia: *serg.* Aldo Bocchese, Flaminio Avet and *ten.* Leopoldo Eleuteri in front of a Spad.

Flaminio Avet in the cockpit of his Hanriot, serial 11411. Over the top wing one can see little roundels that patch bullet holes.

A collective portrait of the pilots of the X Gruppo, 70ª and 82ª Squadriglia Caccia. Avet is sitting in the center, at his right sits *magg.* Ercole, commander of the group, standing behind him is Alessandro Resch and sitting at his feet Leopoldo Eleuteri. The officer in leather jacket on the right is the not-yet-famous Arturo Ferrarin.

escorting two reconnaissance airplanes. At the end of this combat the Italian pilots were credited with two victories, that the 17° Sezione Aerostatica (balloon unit) saw falling down, one between Sernaglia and Fontigo, the other one near Pieve di Soligo. Avet shared these victories with his good friend Eleuteri and the gallant *sergente* Aldo Bocchese. This was his combat report:

> "We attacked the strong enemy patrol (2 reconnaissance airplanes and about ten fighters) frontally from above, while they tried to cross the lines, having seen us. The attack of our patrol was resolute and many dogfights followed. So while I was engaging the first one, Tenente Eleuteri, near me, attacked another one and Sergente Bocchese went after a third one. Then I lost sight of my plane because there were many other aircraft close to me."

The remains of a Phönix D.I fighter serial 228.42 shot down by Flaminio Avet on 3 May 1918: its pilot *Oblt.* Josef Lederer was killed. This fighter belonged to Flik 63/J.

There are no known enemy losses, except for a Flik 74/J pilot who force landed at Pianzano airfield due to a wing failure.

In the morning of 29 July Avet flew with the escort of the Capronis that were out to bomb the la Comina airport, the important base now in enemy hands. South of Santa Lucia he stepped in to drive an enemy away from a Caproni, it was a fighter that he described as, "with rounded wings like a Lloyd and tail like a D.III – type never seen before". It was a hard combat, Avet got a bullet in his lower wing and Eleuteri was hit eleven times, his wing caught fire but he managed to return to base. Their enemies were probably the fighters of Flik 56/J, a fighter, possibly Eleuteri, was claimed as shot down by *Oberleutnant* Vuk Edler von Hreljanovic.

On 16 August Avet went on leave again, until 1 September. Back to the front, he claimed another victory on 3 October, during a cruise along the Piave River, between Pederobba and Grave di Papadopoli.

His last victory in the conflict was shared once more with his usual companions, Eleuteri and Bocchese, on 28 October, during the final battle of Vittorio Veneto. The three Italian pilots all had jammed guns, but they forced with their presence alone an Austrian airman to give himself up and land at Arcade. This victim has not been identified, but the Black Cross from this plane was picked up by Avet and survives to this day, framed. In the final days of the war Avet, again as commander of the 70ª Squadriglia, made an effort to fulfill his role despite a very bad health, probably affected by the "Spanish" flu.

Avet had claimed 13 victories, but he was officially credited with 8 confirmed victories in the list of the Supreme Command in 1919. He had flown about 198 combat missions, 30 of them with two-seaters. He was decorated with two Silver Medals, both issued after the war, the first one with a citation referring to five victories between 17 April and 17 May 1918, the second one to his two victories on 15 July and to the ones of 3 and 28 October.

After the armistice Avet was first assigned to the Comando Scuole Aviatori (schools command) in Rome, then he was a member of a mission to the French experimental establishment at Villacoublay, finally he collaborated with *colonnello* Pier Ruggero Piccio in the International Committee for Reparations in Paris in 1920, surely exploiting his law degree and his perfect knowledge of the French language.

Avet in the cockpit of a rare camouflaged Spad 7 of the 70ª Squadriglia.

A general view of all the fighters of the 70ª Squadriglia lined up on Gazzo airfield: from the left one can see an Ansaldo Balilla, six Hanriots, four Spads and then more Hanriots.

It was now time for Avet to leave the uniform, but his return to civil life was not a happy one. His health was compromised by the efforts of war, so he was forced to long periods of rest in different clinics. On 5 February 1925 he married in the church of San Marco in Rome Maria Luisa Nicolis of the Counts di Robilant, a young noblewoman who had earned two War Crosses for her work as a volunteer nurse. The couple lived in the Palace del Grillo, a patrician palace near the Coliseum overlooking the Roman Forum, where Vivaldi first performed his music.

On 19 January 1925, the death in a flying accident of Leopoldo Eleuteri, his companion in so many war flights, was a hard blow for Avet. He was asked to give the farewell speech at the funeral, but his illness was so bad that all he could do was to write a letter.

The birth of a daughter was a happy interlude, but Avet's health went from bad to worse, despite the loving care that surrounded him. The Italian ace finally died on 21 August 1928 in the house of Bendejun where he was born. His funeral was followed by a large number of French and Italian civil and military officials. His body now rests in the family chapel in the cemetery of Verano, in Rome.

A portrait of Flaminio Avet in front of a Spad 7 of the 70ª Squadriglia. The unit number is repeated over the top wing, center.

A view of the palace of the Avet family in Rome, next to the Forum of Julius Caesar.

## Flaminio Avet
## 12/13 victory claims – 8 confirmed victories

| Claim | Vict. # | Date | Sqn. | Aircraft | Hour | Place | Victim |
|---|---|---|---|---|---|---|---|
| - | - | 26.12.1917 | 82 | Hd | 9h | Camalò | Two-seater |
| *[with s.ten. Scaroni, Michetti & Fucini, 76ª Sq., cap. Riva, 78ª Sq., serg. D'Urso, Teobaldi, Benvenuti, ten. Comandone?, 82ª Sq., 2nd Lts. Jarvis, Mulholland & Frayne, No.28 Sqn., RFC]<br>Lt. Kessler WIA, POW/Lt. d. Res. Edebohls KIA, FA 2, DFW CV or:<br>Uffz. Willi Petzold KIA/Lt. d. Res. Wilhelm Voigt KIA,, FA 2, DFW CV or:<br>Vzfw. Karl Uecker/Lt. d. Res. Heinrich Pfeiffer DOW, FA 2, DFW CV or:<br>Vzfw. Pohlmann WIA, POW/Lt. Schlamm ,WIA, POW, FA(A) 219, DFW CV* | | | | | | | |
| 1 | 1 | 17.4.1918 | 70 | Hd 523 | - | Valdobbiadene | Two-seater |
| *[with ten. Eleuteri, Resch and serg. Bocchese, 70ª Sq.] Korp. Viktor Stibor KIA/Lt. Ladislaus Rajcsics KIA, Flik 52/D, Br. C.I 169.35.* | | | | | | | |
| 2 | 2 | 17.4.1918 | 70 | Hd 523 | - | Valdobbiadene | Fighter |
| *[with ten. Eleuteri, Resch, serg. Bocchese, 70ª Sq. & serg. Rennella, 78ª Sq.] Maybe: Oblt. Michael Gassner-Nordon von Laudon KIA, Flik 42/J, Alb. D.III. 153.152* | | | | | | | |
| 3 | 3 | 17.4.1918 | 70 | Hd 523 | - | Valdobbiadene | Fighter |
| *[with ten. Eleuteri, Resch, serg. Bocchese, 70ª Sq. & serg. Rennella, 78ª Sq.] Maybe: Oblt. Richard Fitz,UNH, Flik 42/J, Alb. D.III, returns to own airfield with damages.* | | | | | | | |
| 4 | - | 25.4.1918 | 70 | Hd | - | Conegliano | Fighter |
| *[with ten. Eleuteri, Guglielmotti & Croci, 70ª Sq.] Maybe: Korp. Rudolf Kusyn UNH/Lt. Julius Knüppel UNH, Flik 4/D, Br.C.I 169.82* | | | | | | | |
| 5 | - | 25.4.1918 | 70 | Hd | - | Conegliano | Fighter |
| *[with ten. Eleuteri, Guglielmotti & Croci, 70ª Sq.]* | | | | | | | |
| 6 | 4 | 3.5.1918 | 70 | Hd | 12h | Zenson | Fighter |
| *Oblt. Josef Lederer KIA, Flik 63/J, Phönix D.I 228.42.* | | | | | | | |
| 7 | 5 | 17.5.1918 | 70 | Hd 11411 | - | Maserada | Fighter |
| *[with ten. Fucini & serg. Rennella, 78ª Sq.] Fw. Johann Valenta WIA, POW/Lt.i.d.R. Viktor Herzog POW, Flik12/Rb, Br. C.I 229.30* | | | | | | | |
| 8 | 6 | 15.7.1918 | 70 | Hd | 10h30 | Sernaglia/Conegliano | Fighter |
| *[With ten. Eleuteri and serg. Bocchese, 70ª Sq.]* | | | | | | | |
| 9 | 7 | 15.7.1918 | 70 | Hd | 10h30 | Sernaglia/Conegliano | Fighter |
| *[with ten. Eleuteri and serg. Bocchese, 70ª Sq.]* | | | | | | | |
| 10 | - | 3.10.1918 | 70 | Spad 7 | - | Moriago | Fighter "D.5" |
| *[with ten. Eleuteri and Guglielmotti, 70ª Sq.]* | | | | | | | |
| 11 | - | 8.10.1918 | 70 | ? | - | Oderzo | Fighter |
| *[with ten. Eleuteri, 70ª Sq, who gets confirmation]* | | | | | | | |
| 12 | 8 | 28.10.1918 | 70 | ? | - | Arcade | Two-seater |
| *[with ten. Eleuteri and serg. Bocchese, 70ª Sq., cpr. Ramberti 76ª Sq.] Maybe: Phönix C.I 121.68, Flik 62/K, missing in action* | | | | | | | |

# Francesco Baracca

Francesco Baracca, eighty years after his death, is still Italy's best known aviator, his name celebrated by streets in all Italian cities, and his memory preserved by his insignia, a prancing black horse, carried by a fighter Group of the Italian Air Force and by the Ferrari racing cars. Francesco Baracca was the perfect hero: young, handsome, compassionate, and blessed by a romantic death, in the thick of battle, carried away by a cruel destiny, but not bested by the enemy. The character and the thoughts of Baracca are well known, thanks particularly to the long and detailed letters that he wrote to his mother, and that portray him as a serious young man, deeply engaged in war but never boastful, always respectful for the humble infantrymen stuck in the trenches, and for the enemy airmen. Many aces, of all nations, that were hailed as heroes in their time, have later been criticized for different reasons: for unproven and exaggerated claims, for hate and cruelty against the enemy, for selfishness in using their wingmen in order to beef up their scores. No criticism can be leveled at Baracca: his claims are always perfectly accurate, actually current research demonstrates that he underclaimed, and his actual score could be 36 victories, not 34 as credited to him. He always defended his adversaries, the Austrian fliers, that he considered his fellow gentlemen, and that he kindly visited in hospital or in captivity. He was considerate with his subordinates, allowing them to grow into accomplished fighter pilots. His figure is well known: he is not a two-dimensional icon, but a full-fleshed man. Francesco Baracca was something more than a hero: he was a man.

Francesco Baracca was born on 9 May 1888 in Lugo di Romagna, a small, wealthy agricultural town in the heart of Romagna, the flat region south of the Po river that produces a special breed of Italians: warm, generous, hot-blooded and hot-headed, always burning with the three local passions: music, politics and racing engines. The Romagna is a region of genial eccentrics.

Francesco Baracca before the war, at the controls of a Macchi-Nieuport Monoplane.

Baracca was the only child of a well-to-do family: his father, Enrico was a wealthy landowner and businessman, his mother Paolina Biancoli was a countess. Francesco could easily have settled for a comfortable and rewarding life attending the family properties. He was a young man of culture, he played the cello and his letters show a good literary talent, he had all the qualities of a country gentleman. Instead, after having studied at home and in Florence, in 1907, at the age of 19, he joined the Officers School of the Scuola Militare at Modena. His lifetime choice of a military career was not a makeshift, but a willful choice. His course at Modena was very successful, he was graded as an excellent cadet, then in the Autumn of 1909 he progressed from Modena to the School of Cavalry (Scuola di Applicazione di Cavalleria) at Pinerolo, a town north of Turin, at the feet of the Alps. He was by then a *Sottotenente* (Sub-Lieutenant). In his selected course of 50 cavalrymen there were many names that were to become important in Italian military history: Raffaele Cadorna, son of General Luigi Cadorna, chief of the Italian Liberation Army in World War II, and five future aviators, de Bratti, Lombardi, Avogadro, Ripamonti and Pacinotti. In December 1910

Baracca, left, and *cap.* Tacchini, center, in front of Nieuport 10 serial 1037 and, behind, 1035, that the two pilots flew on 19 November 1915 when they successfully intercepted Austrian aircraft raiding Udine.

Baracca was assigned to the Reggimento Piemonte Reale, a highly prestigious regiment, then part of the garrison of Rome. Baracca's life was rigorous, studies were hard (despite the current joke that cavalry officers were supposed to be *"bel, blond e bestia"*, handsome, blond and dumb). But riding activities were dominant, and Baracca always scored top grades in horsemanship. After a period of garrison in Rieti, in November 1911 Baracca was again based in Rome, and he lived fully in the careless atmosphere of high society. His looks and his qualities made him an instant success with the ladies of Roman nobility, who were willing to explain to the handsome officer, hero of the steeplechase field, why the Belle Époque was so belle. Baracca also became a Freemason, which in Italy in those times meant having secular, progressive and patriotic ideals, sharing the values of the "Risorgimento", but it also meant accepting the harshest condemnation by the Catholic Church.

In 1912 he put forth a request to be assigned to aviation, the new arm that was showing its potential in the Italo-Turkish war then being fought in Libya. Baracca commented in a letter the exploits of Palma di Cesnola, a Lieutenant of the Piemonte Reale who had become a pilot and was then flying in Libya. He was accepted, and in the spring of 1912 he was part of a group of four Italian officers sent to the flying schools at Bétheny, near Reims, France. His companions were *tenenti* Domenico Bolognesi, Arturo Oddo and Cesare Buzzi Gradenigo. He admired the development of French aviation, the number of airfields and the quantity and quality of flying activities. The military school at Reims had at that time more than 50 French student pilots and sixty hangars with different types of airplanes. His type was the Hanriot monoplane, and his instructor was Mr. Hanriot, the 19 years old son of the constructor. His first flight as a passenger was on 4 May 1912, and soon flying lessons began. The summer of 1912 in France must have been a glorious time, that Baracca spent flying in the Champagne, riding horses and visiting Paris. He received his pilot's license, #1037 in France, and Italy's sport license #171, on 9 July 1912. In the summer he returned to Italy, and, now assigned to the Battaglione Aviatori, the air service, he was posted to Cascina Malpensa airfield. His first exploit was an exercise on 12 November, a mass flight of eight airplanes in front of an American mission. On 8 December 1912 he received the Italian military pilot license. In 1913 he was assigned to Mirafiori airport, near Turin, then in July he was shifted to Busto Arsizio, in the 6ª Squadriglia, equipped with Nieuport monoplanes. In August he took his father for a flight and in September took part in the cavalry exercise in which for the first time four complete air squadrons participated acting as reconnaissance for the Red and Blue parties.

In 1914, still in Busto Arsizio, he was a pilot in the 5ª Squadriglia, commanded by *cap.* Pier Ruggero Piccio and consisting of pilots Guido Tacchini, Giulio Palma di Cesnola, Filippo Valdimiro and himself. Piccio became an ace, and the chief of Italian fighters, Guido Tacchini had an uneventful life as commander of the 70ª Squadriglia and later a General, Palma di Cesnola became the leader of a SVA squadron and later a General, Valdimiro piloting a Caproni was shot down and killed by Hungarian ace Joszef Kiss on 17 September 1916. Baracca meanwhile was gaining a reputation as a high-class pilot. As World War I broke out, Italy started moving its armed forces East, expecting that its neutrality would be followed by a conflict with the Double Monarchy, and at the end of the year the 5ª Squadriglia was moved to Pordenone, in the Friuli region, near the border with Austria.

In the last days of May 1915, before the Italian declaration of war (Italy had on 26 April 1915 signed a secret pact in London pledging to denounce its alliance with Austria and Germany, and to join the war on the Allied side) Baracca was sent to Malpensa to test fly the Macchi Parasol, an artillery spotting airplane that he found satisfactory, but that in operations was dangerously unstable and was quickly retired. Baracca was spared the life, usually brief, of a recce pilot, however, because he was chosen to return to France to pick up Nieuport biplanes, armed fighters that France had promised to Italy. He reported the street demonstrations in which excited crowds were clamoring for war, but without adding any jingoistic

*Tenente* Baracca next to his Nieuport 11 serial 1451 right after his first official air victory, 7 April 1916.

A famous portrait of *ten.* Francesco Baracca in the cockpit of his Nieuport 11 clearly showing details of the attchment of the Lewis gun over the top wing.

comment of his own. Like all military professionals, Baracca accepted war, but was not excited about it. He arrived in Paris on 23 May, the day before the declaration of war. He visited the airport of Le Bourget, where he learnt all about the new characteristics of the war: air fighting. He commented in a letter to his mother:

> "The airplane that I am learning is a small Nieuport biplane with a speed of 140 km per hour, and is used exclusively as a scout and fighter: it is beautiful, and both the pilot and the passenger can fire with a machine-gun and automatic rifle and its great speed gives it superiority over the enemy airplane. The pilots tell that the best satisfaction is to shoot down a Taube with the machine-gun, and I also hope to experience this satisfaction."

Baracca, left, and Ruffo di Calabria discuss their experiences of fighter pilots with King Victor Emmanuel III (second from right) at Santa Caterina.

In June, Baracca wrote:

> "Our training unfortunately is slow, and we were hampered by bad weather and strong winds. We must almost start from scratch, because the old Nieuport that I used to pilot had totally different controls than the ones now used on all airplanes: while I moved the rudder with my hands, now you move it with your feet, so we started with a small 35 hp airplane, rolling on the ground for some ten days. When I learn the new

Right: A bad quality but puzzling picture: Baracca surrounded by well-wishers after one of his early victories. The man in the center with a beard is often erroneusly identified as Keller. The Nieuport 11, however, seems to be equipped with a different, apparently belt-fed machine gun, and not with the usual Lewis.

Below: Baracca, fourth from the left, at Santa Caterina airfield with a group of unknown officers and a chaplain. The airplane in the left background is his Nieuport 11, serial 1451, showing a polished metal cowling.

Brandenburg C.I 61.57 shot down by Baracca on 7 April 1916 drew quite a lot of attention. From a technical point of view it was more advanced than any Italian airplane at the time. It was then used for propaganda in-flight photographs and then, painted with Italian markings, it remained flyable until October 1917, when it was destroyed.

controls I'll be able to pilot any type of aircraft. Our Nieuport reaches 145 to 150 kph, it is difficult and you must progress with care. I wonder if it will be much used in our terrain, because it needs a large airfield for take-off and landings, but now even better airplanes are produced, and that can be arranged… Here in France they fly with too much carelessness and no prudence: each day we see serious falls always caused by the pilot, never by the airplane. In this respect, in Italy we had a much better school and a flight discipline that here is totally lacking."

Baracca had to wait in France until the end of July, training at Le Bourget. Then, as France had agreed to send a Nieuport squadron for the defense of Venice, he was assigned to that unit, that needed an Italian officer for coordination. This project didn't appeal to him, who remarked that as aircraft raiding Venice would approach from the sea, they would be spotted only when over the city, that is to say, too late for interception. Instead, he was posted to a new fighter unit, the 8ª Squadriglia equipped with Nieuport biplanes, which was established at Mirafiori on 26 July 1915 and moved immediately to Aviano.

The 8ª Squadriglia Nieuport Biplani was formed by commander *cap.* Tacchini, and pilots Captains Oreste Salomone and Maffeo Scarpis and Lieutenants Domenico Bolognesi and Francesco Baracca. It was equipped with Nieuport 10 serials 384, 387, 388, 595 and 596. The unit spent the month of August fitting machine guns to its aircraft and testing them. On 21 August it moved two airplanes to Campoformido for the defense of Udine, the city where the Italian Supreme Command was located, after the first air raid by the Austrians, who dropped bombs that caused damages and casualties. On 25 August Baracca, still with a Nieuport without machine-gun, and *cap.* Scarpis flew the first combat mission, an interception. Baracca carried a gunner, soldier Vittone.

On 9 September the whole unit moved to Santa Caterina, near Udine. Salomone had left the unit, going to the Caproni bombers. On 18 February 1916 he gained the first "Medaglia d'Oro", the gold medal that was Italy's highest award for bravery, bringing back a Caproni maimed by enemy fighters, with all the other crewmen dead. Salomone died in an accident on the night of 3 February 1918. On 7 September 1915 there was the first air combat. *Cap.* Bolognesi with Nieuport 387 and Baracca with 596 intercepted an enemy aircraft raiding Udine. Baracca fired at close range, but his gun jammed. Bolognesi fired 100 rounds at the enemy, and finally also *cap.* Tacchini joined the fray, pursuing the enemy. The airplane of *ten.* Baracca was hit in three places:

"Finally the meeting arrived, at 10 in the morning. 1,200 meters high above Udine, I see sharpnels over Palmanova, then I see the Austrian airplane coming toward Udine. That's it. What is going to happen? It won't escape me, it is very high, and I pull the stick and climb to Palmanova. It is coming closer quickly: I meet it at 2,200 meters, he is 300 meters above me, and seems not to have seen me. I shoot two, three rounds; the machine gun is not working properly. A steep turn, I reach it again but it is still too high and the gun doesn't work. I shoot, again two or three rounds go, then the gun jams. I pull the breech, shoot, it jams again. How maddening! And the guy goes on to Udine, then turns back. I am in his sights, but he doesn't shoot, maybe I still have not been seen, I turn too and get below him. I shoot two, three rounds more, then the gun jams definitively. I glide and follow it to Gradisca. I want to see if it goes down, I may have wounded the pilot, nothing happens. I return to my field quite angry. But the emotion of the fight was sublime."

On 16 May 1916 the Austrians again attacked Udine with a formation of a dozen bombers. Baracca intercepted a straggler on its way to the city, attacked it at close range and saw it going down over Gorizia. This victory was confirmed only many months later, from information gained from prisoners. His second victim was Lohner B.VII 17.42 of Flik 12 and its crew of *Korporal* Lajos Knelly and *Fähnrich* Ladislaus Hórvath-Tima was killed. The pilot had been mortally wounded; the observer tried to get the plane down but died in the crash landing on the roof of a house.

Also his third official victory shared with *ten.* Ruffo on 23 August 1916, over a Br. C.1, 61.61 of Flik 19, piloted by *Zugsführer* Anton Himmer, unharmed, was not confirmed until eight months later. In August 1916 Baracca was deservedly promoted to the rank of Captain.

On 13 September 1916 Baracca scored a victory that is not recognized in the official list nor is carved in the marble of his monument, probably because it is not quoted in his letters. It is, however, recorded by the Squadron diary:

> "In the morning Cap. Baracca sees an enemy airplane above Monfalcone, and approaches west of that town, at 3,500 meters. The enemy turns against our airplane and shoots with the M/G above the wing, maneuvering so as to prevent Cap. Baracca to get behind it. Finally getting next to it, Cap. Baracca fired his gun. Only 10 or 12 bullets were fired, as one bullet jammed, but the fire was enough to cause the enemy to dive quickly toward Comen. Reloading his gun he fired once more against the airplane at a distance of 200 meters. Our airplane was hit four times by the enemy fire, one hit the lower spar close to the seat, an exploding bullet ripped the cloth."

The enemy was a Brandenburg C.1 of Flik 28, 64.07 that returned home with a wounded pilot, *Feldwebel* Josef Koschiczek and a dying observer, *Lt.* Ottokar Hauser. The same day, Baracca lead his squadron in an escort flight for the Caproni bombers that attacked Trieste harbor.

Three days later, 16 September, cruising with Ruffo and Olivari over the Caporetto area, where the land battle was raging, Baracca spotted an enemy two-seater. Attacked by the three best pilots of Italy, the Austrian had no chances:

> "He went down for 500 meters trailing smoke, and we were behind him. Its fall was impressive: we saw the airplane out of control, but with the engine running, spinning by itself, and standing in the air with the big black crosses standing out on the yellow wings. I thought it would crash but maybe the observer, although wounded, could grasp the rudder cables and straightened it before touching the ground. It crashed its wings against the trees of a wood on the slopes of Mount Stol, turned over and burned. We landed in Udine, got a car and drove to Potocki, a Slavic village on the mountain where the plane had crashed. Pilot and observer in the fall were thrown out of their seats and didn't burn with the plane. The pilot – a corporal with two silver medals for valor –was dead with two bullets in his head, he was still grasping his Mauser pistol. The observer, a 2nd lieutenant in mountain artillery, a very distinguished officer, seriously wounded in the guts, was operated, and is not likely to survive. (*He died three days later*) They came from Villach airfield. Their plane was a Löhner with 160 hp engine. We got a few bullets in our planes."

French-built Nieuport 17s of the 70ª Squadriglia: first is N 3140 of *serg.* Poli, then N 3139 of Ruffo di Calabria and then N 2614 of Baracca showing his black prancing horse.

The Spad 7 of Francesco Baracca, with his famous black prancing horse. *(Photo Pierluigi Moncalvo).*

The victim was not a Löhner, but Lloyd C.III (WKF) 43.74 of Flik 16, then based at Seebach bei Villach, and the casualties were *Zugsführer* Franz Morozko and *Lt.* Anton von Csáby. In his pocket was the letter of a lady "the beautiful unknown one", who invited him for a date in Villach that very night.

On 25 November 1916 Baracca scored what he claimed as his fourth victory (actually the sixth). By this time, free from the orders that limited him to the Udine area, he was roaming far away, and on the Carnia Mountains he spotted an Austrian formation bombing Tolmezzo. He attacked what he called an Albatros, actually a Brandenburg C.1, and shot it down with 30 rounds. Later, he reached his victim by car, and found the pilot dying, and the observer wounded: "I talked with him a long time in the hospital and I learnt a lot. They were coming from Villach; we found on the airplane a machine-gun with a belt of 1,000 rounds and two Mauser pistols. I was alone in the combat".

His victim was Brandenburg 68.03 of Flik 16, with the crew of *Korporal* Fritz Fuchs and *Lt.* Kálmán Sárkozy, but the victory was assigned also to *serg.* Gorini of the 70ª Squadriglia.

No more victories came during the year 1916. Baracca remarked: *"I started the New Year flying with a magnificent weather."* On this day, 1 January 1917, there is the legend of the encounter between Baracca and the Austrian ace Gottfried von Banfield, with the two heroes saluting each other in the sky, but this event never happened. Baracca met another enemy, though: at 4 p.m. he attacked a two-seater that gave him a tough fight, maneuvering smartly and leading him over the Austrian flak, until it got away. Baracca claimed a probable: "The artillery observers saw it going down, but could not determine where it touched down because of the hills. But I believe I shot it down." His likely victim was a Brandenburg of Flik 12 with the skillful crew of *Korporal* Franz Neuwirth and *Lt.* Viktor Huber. The pilot returned to his base, but the observer was killed in the combat. Baracca received official confirmation for this victory only later on. That same day *ten.* Ruffo forced down another airplane with a dead observer, that also could not be confirmed from the ground. *Korporal* Neuwirth unfortunately was killed by Italian artillery exactly two months later on another Brandenburg of Flik 12.

On 11 February 1917 Baracca, again defending Udine, scored what was then his 5th official victory, and received the public glory of an ace. His combat took place watched by all the population of Udine and by the King himself. Baracca reached a lone Austrian flying on a recce mission over Udine at 4,000 meters:

> "The enemy, with great courage, accepted the combat, trusting that he could repeal us all. A Nieuport attacked, but I saw it spinning down and disappear, so I thought it was hit (*it was serg. Caselli, who landed safely out of gas*). Then I attacked, and I realized immediately that I faced one of the best Austrian pilots, but I was sure he was not going to return home. We started a close dogfight with tight turns and loops, so I could fire just short bursts, but he also couldn't aim well, and I was not hit. After a few moments other Nieuports arrived, but they also could shoot occasionally on account of the great speed and agility of the enemy airplane. The worst danger, besides the bullets, was to crash against each other as I, during the dogfight, was looking just at the enemy, and couldn't see the other Nieuports. I got as close as 20 or 30 meter many times, but neither of us could fire. So I gave up a tail attack, and aimed straight at its side while he was steeply banking and I managed to fire some 30 rounds as I had closed down the distance to less than 20 meters, then I kicked my rudder and passed above it. The other ones were shooting too. The enemy crash-landed near Remanzacco hitting a tree.... The enemy airmen were

taken away: the observer is a Polish cavalry lieutenant of a noble family. The airplane was all spotted with blood."

The enemy that Baracca described so chivalrously was the crew of *Korp.* Ludwig Fleck and *Lt.* Wilhelm Graf Siemienski of Flik 35, flying Brandenburg C.1 27.74. Two months later, Baracca wrote:

> "We went to visit the Austrian lieutenant wounded over Udine. He is suffering, his wound is still open but he will make it. I brought him the photographs of the airplane, I promised him a piece of the propeller as a souvenir and books to read. His mother wants some news through Switzerland: we learnt that he is her third son, only survivor of the war, and her family keeps one hundred Italian prisoners who work on their fields and are treated very well."

No other major Allied ace had Baracca's consideration and humanity for his victims, although Austrian aces showed similar consideration for their victims. The airplane was exhibited in Udine for the benefit of the Red Cross.

Many widespread texts assign another victory to Baracca on 23 March 1917, at Merna. No such event happened, however, and the Squadron diary confirms that on that day nobody flew, on account of the snow.

On 19 March 1917, Baracca first flew the new fighter, the Spad 7, two of which had just been delivered. He flew Spad 7 serial 4688.

At the end of April, with the new camera-equipped Spads, the pilots of the squadron started to fly strategic reconnaissance missions. Baracca flew to Bruneck and Brixen (now Brunico and Bressanone) on 25 April. Baracca returned to victory on 26 April 1917, although flying a Nieuport 17, for what he claimed as his 8th success. It was another dramatic combat, against a very determined and skillful Austrian crew that dueled for a long time against Baracca, Olivari, (who had to quit with a jammed gun), and Gorini. Although Baracca doesn't report it, another pilot took part in this fight, *serg.* Imolesi of the 78ª Squadriglia, whose role is reported in the Squadron Diary. The victims were *Zugsf.* Josef Majsai and *Lt.* Emmerich Treer of Flik 35, on Brandenburg 129.17.

On 1 May 1917, Baracca's squadron, 70ª Squadriglia, gave birth to a new unit, the 91ª Squadriglia, the "Squadron of the Aces". *Cap.* Tacchini, a good but lackluster officer commanded it, but Baracca was its biggest star, followed by Ruffo, Olivari and Ranza. Actually, Baracca was formally assigned to the Supreme Command, and simply attached to the unit, and this is the reason why he was not yet its commander.

The ace celebrated the birthday of the 91ª chasing an enemy airplane over Monfalcone. He believed that the observer was wounded, and that the airplane force landed by Nabresina, as was confirmed by the observers, but his enemy was *"abilissimo"*, he didn't run but stood his ground to dogfight and put a few bullets into Baracca's aircraft. No wonder: the "very able" pilot was Frank Linke Crawford, flying with his observer *Oblt.* Johann Albert Graf Herberstein Freiherr zu Neuberg und Gutenhag in Brandenburg 229.08 of Flik 12. One month later Baracca commented: "I learnt from Peppino that my airplane of May 1st force landed at Nabresina on account of my fire, I believe it will be counted as one of my victories." Actually, it was not.

Baracca and Ruffo di Calabria, both showing wear and fatigue on their faces, in front of one of their victims, Brandenburg C.I 27.74.

As the great 10th Battle of the Isonzo began, with massive air support by the Corpo Aeronautico Militare, duels raged in the air. 10 May 1917 was a day of countless combats and Baracca scored his first victory on a Spad, became officially a double ace, destroyed his first fighter and unknowingly met his Austrian counterpart. Near Gorizia, he jumped two Austrian fighters: one got his first burst, and crashed behind its lines, the second one counterattacked immediately, put one bullet into the Spad and saved the life of his wingman. He was Godwin Brumowski, the top Austrian ace, while his crashed wingman was *Zugsführer* Rudolf Stöhr of Flik 41, flying KD 28.17, whose carreer ended with this flight. This is the original combat report:

> "Taking off for a fighter flight I cruised between Plava and Vippacco. At about 9:50 I saw two enemy airplanes, that then I recognized as very fast fighters, entering our lines one behind the other at a distance of 300 meters. Flying about 500 meters above them (I was at 4,200) I waited for the right moment and I dove on the first one, that was turning from Vertoiba to Gorizia, getting 30 meters behind it and firing a burst of 60 rounds; the enemy surprised didn't have time to escape my attack and after a few disordered movements it crashed; while the enemy started to fall I heard a quick burst close to me that I avoided with a sharp turn and I saw the other fighter passing 200 meters over my left wing going down. He enemy had fired attacking me frontally, but for the very direction of the attack he had the time to fire just twenty rounds. His fire hit my airplane and one bullet crossed the stabilizer and fuselage, just missing the rudder cable. Because of the attack of this fighter, I couldn't see the fall of the other one. I had descended to 3,200 meters and then I saw three reconnaissance airplanes passing by very high going toward our lines. I had the time to go and attack one of them at 4,000 meters, but it was already returning to the lines and it dove firing from 300 meters. I didn't

return fire from that distance. At 10:30 being over the Sabotino Mountain I heard fire behind me, that I avoided with a sharp move, seeing an enemy fighter that attacked me from a distance of 100 meters. The enemy immediately returned to his lines and avoided my counter attack. The shot down airplane was a single seat fighter of the latest type, with crossed struts, very fast (about 175 kph) and the other two that attacked me were similar. I landed at Borgnano airfield as I was short of fuel."

Baracca must have guessed something about his rival, because one week later he wrote:

"I beware of enemy fighters and it is unlikely that they will jump me. A few days ago I jumped one of them who had just shot down one of our two-seaters, but I was a few seconds late, and it got away. Maybe he was that Austrian pilot, Count X, who has also shot down ten of our airplanes, as you can read in the Austrian communiqué."

Despite all the combats, the Italians on that day lost just two aircraft, and claimed just two. All the pilots flew many missions those days, while their names became known nationwide: the Supreme Command had lifted its ban on naming winners of air duels, and the popular press built up the fame and reputation of the Aces, like in the other warring nations.

Scaroni commented the events of 10 May 15 years later:

"That day was almost fatal to Baracca: from 9:50 am to 10:30 am he was jumped twice, and his good luck had saved him. After so much experience, after so many examples, there are still people who discuss the possibility of surprise in air combat. The German ace of aces Richthofen, who had destroyed eighty allied airplanes, was jumped and shot down, without his squadron pilots realizing it. Dorme, excellent French ace, was shot down and his wingman Deullin, another splendid fighter, didn't realize it. Lieut. Bozon-Verduras, wingman of the great Guynemer, didn't even realize that a German fighter had jumped and shot him down. This, just to name the great aces. Everybody must understand that there is nothing easier than surprise in the air, despite the lack of ditches, woods, walls and other obstacles to prepare a trap. Surprise in the air is due mainly to the extraordinary speed of the airplane, and to certain conditions of light. An airplane in the sun cannot be seen even from a short distance. An airplane hidden by distance gets on your tail in a few seconds. Those sinister black spots that move up there in a boundless scenery, turn into a stream of bullets in the small span of time that is necessary to check an

*Lt. i.d.R.* Wilhelm Graf Siemienski, of Flik 35, shot down by Francesco Baracca with *ten.* Ruffo di Calabria on 11 February 1917. *(Photo Karl Meindl).*

Luigi Olivari guides Baracca in discovering the details of the SVA at Santa Caterina, August 1917. The large and heavy airplane was not found fit for use as a fighter. *(Photo Casirati).*

Captain Baracca talks with captain d'Annunzio, surrounded by a crowd of fans. The down-to-earth no-nonsense style of Baracca was the opposite to the flamboyance of the poet.

Oeffag Albatros D.III 153.54 in which the Austrian ace *Oblt.* Szepessy Sokoll found his death, shot down by Baracca on 6 November 1917.

instrument or the ground, or while you are engaged in determining the nationality of an airplane flying close to you. All pilots are subject to such surprises, independently of their ability as fighters."

On 13 May, Baracca flew with Ruffo and attacked a trio of Austrian planes, the duel was short, and the pilot was killed immediately, the observer fell out of the aircraft, which crashed in the Italian lines. *Fw.* Julius Busa and *Oblt.* Hermann Grössler were the crew of Brandenburg 129.20 of FlugGeschwader 1.

One week later, 20 May, the Ace shot down what he believed was his 12th victory. As the battle raged on Monte Santo, he reached Piccio that had just shot down an enemy, and attacked a Brandenburg above Plava, flaming it with 50 incendiary bullets: "They use incendiary bullets, because a few days ago they burned the plane of Ruffo, and now we use them too with good results." This victim was possibly Brandenburg 229.10 of Flik 12, the crew of *Zugsf.* Ludwig Ferschl and *Lt.* Eugen Csutka was killed, but another possibility is another Brandenburg of the same unit lost in this action, C.1 129.52. This is Baracca's Report:

"Taking off for a cruise at 9 a.m., I reached our lines over the Vodice at 2,400 meters and cruised at that level between Gorizia and the Sabotino, as I couldn't go higher on account of the clouds. At 9:30 as I followed our lines from Monte Santo to Plava I saw two Spads pursuing two Albatros at about 1,500 meters: I also saw an enemy fighter and other airplanes that I couldn't recognize over the Bainsizza. I saw one of the Albatros falling down, attacked at close range by a Spad but I didn't follow it because in front of me an enemy far away was flying from the Jelineck to the mount Corada. I attacked it near Globna at a level of 2,200 meters, I emptied my gun with incendiary bullets in its right wing and firing I got to 30 meters, when I saw flames developing in two points, behind the back of the pilot and next to the observer who was still firing, and one second later the Albatros crashed in flames leaving a pillar of smoke. It fell on the peak of Hill 363 east of Plava and I saw a large flame and a column of black smoke that went on for 15 minutes."

The next day he had a long duel with three enemy fighters, two "star-strutters" and another single-seater, without results. On 23 May he flew escort to a large Caproni attack on the Isonzo front and took part in the battle that developed with defending Austrian fighters (that claimed two Capronis, and actually one was forced to land and one was damaged) but he missed the opportunity to shoot down a fighter because his gun jammed. Even an ace like Baracca sometimes goofed. Here is his combat report for 25 May:

"I took off at 7.15 to Gorizia; at 7:45 I saw at 4,000 meters two enemy Albatros that were climbing toward our lines. One of them entered from the Vodice to the Sabotino at 3,400 meters. I spiraled down onto it attacking it alone and diving, but on account of my high speed and having misjudged the distance I could fire only 20 rounds, after that I had to stop firing and turn to one side, not to crash into the enemy, that I missed for

Right: This is the only known photo of Baracca in flight, protecting with his Spad the Pomilio of *ten.* Augusto Gauthier, 117ª Squadriglia, early December 1917.

20 meters. I saw the gun and the observer, who was surprised and didn't fire. The Albatros got away diving toward its lines. At 8:30 from 4,000 meter I saw a Nieuport 80 hp attacking an Albatros at 3,000 meters and forcing it down with a burst; I didn't have time to join the combat, I followed my Albatros as it escaped toward Aidussina".

What Baracca described was the 4th victory of Baracchini, against Brandenburg 129.38 of Flg.G.1.

One 3 June, there was another victory in the same circumstances, a lone attack against a formation of Br. C.1 bombers from Fl.G.1. Baracca's technique was now well defined: he zoomed at very close distance from the enemy, 20 meters, and dispatched it with a single burst. He realized that the enemy was dead (*Zugsf.* Johann Rotter and *Oblt.* Max Bednarzik, on Br.C.1 129.51) but he couldn't spare much sorrow, as he watched from above the carnage of Italian infantry on the Carso rockbeds.

In the first week of June, Baracca moved with the 91ª Squadriglia, of which he assumed command, to Istrana to support the ill-conceived and bloody offensive of the Ortigara, in the Asiago sector. He remained one whole month in the Trentino sector, but scored no victory, just damaged an enemy on 8 June.

On 7 July Baracca was back in Udine, and during his first mission he scored again. Late in the evening, he spotted two enemy two-seaters above the Faiti Mountain; he attacked them at close range and sent one spinning at full speed, until its wings broke off in the air. The victim was Brandenburg 129.68 of Flik 46 crewed by *Fw.* Sigmund Duschek and *Lt.i.d.Res.* Andor Bózo von Dinyeberki et Agod. Back to base, Baracca assigned his Spad to his mechanic, but a spark fired the gas fumes in the engine, and the plane burned on the ground.

Baracca had by now served two years at the frontline. His 15th victory, on 31 July, was once more against an aircraft of Flik 46, Brandenburg C.1 69.93, crewed by *Zugsf.* Julius Klausz and *Oblt.* Erich Bock. The airplane caught fire in the air, and the observer jumped from the cockpit:

"I arrived on the spot of the fall one hour later, among a big crowd of officers and soldiers. The lieutenant observer was found one kilometer away from the plane. They had dropped bombs on a camp of grenadiers; one hit the car of a colonel who was passing nearby, without wounding them. The tracer bullets that we use should not set the gasoline on fire, however this happens rather often, and it is very impressing to see an airplane burning at 3,000 meters and the airmen who jump, as it always happens, and I think that I won't use them, because this is already the third one that I set on fire. But those bullets are very useful to correct the aim, because you can see where you shoot, and the Austrians use them too. That plane was one of 20 that dropped bombs on our lines."

This is the same event in the combat report:

A signed photograph of *cap.* Baracca, with a dedication to his wingman Giorgio Pessi, dated 15 March 1918. *(Photo Casirati).*

"Taking off on alarm for an enemy raid from the Corada I got over Gorizia when an enemy squadron had already returned to its lines. Cruising at 3,000 meters I started to go down and at 2,000 meters I spotted anti-aircraft bursts over the Vodice and an enemy much higher that was entering our lines flying to S. Martino. I climbed quickly, never losing it from sight and turning behind it. The enemy turned southeast flying to S. Floriano. A Nieuport 110 HP was following it far away, during the pursuit, reaching 3,600 meters, the level of the enemy, who was starting to turn and go down at full throttle, I passed the Nieuport by 20 meters also going down at full throttle and leaving the Italian fighter behind, as the Albatros was 500 to 600 meters away. As I got closer to the enemy it started firing, and twice it dove at full throttle to gain speed, and then leveled again, firing, but I avoided his move the second time, putting my Spad in an almost vertical dive and getting below and behind the enemy, so I got 30 meters from the tail of the Albatros, slightly going down and almost a still target. I immediately aimed correctly and after six rounds that I saw getting into the fuselage with their tracer trail, I saw smoke and flames getting out and I stopped firing. The enemy started a steep dive that turned into a crash as flames engulfed it, and it crashed northeast of Peuma. The attack took place above Oslava at 6.30. I flew low over the spot of the fall, then I landed at Borgnano and went to the spot of the crash."

On 3 August 1917 Baracca was flying over Tolmino, in the morning, when he saw a Brandenburg pursued by *ten.* Baracchini, of the 76ª Squadriglia, flying another Spad. They attacked the enemy deep behind enemy lines in the valley of the Sava River, and saw it going down toward Wocheiner Feistritz. Baracca's Spad was hit in this engagement. This victory is not confirmed by Austrian sources, but it was reported that Baracchini later on took photographs of the downed airplane.

On 6 August Baracca attacked two enemy Brandenburgs above Tolmino, but they evaded him with quick and skillful maneuvering, not before putting a bullet through his Spad's elevator.

In mid August 1917 the 11th Battle of the Isonzo took place. The Italian effort was huge, as many as 200 airplanes flew missions on a single day, the Bainsizza plateau was conquered, but the offensive went no further, the Austrian defenses were drawn to the limit, but miraculously they held. The Italians had total mastery of the air, and the pilots of the 91ª could indulge in strafing. On 19 August Baracca met a two-seater escorted by a fighter and shot it down with his usual tactics:

> "Taking off for a Caproni escort I passed the lines at 2,400 meters among intense flak fire. After the bombing action I went down toward Castagnevizza at 500 meters above the ground shooting several times with the machine gun at the lines east of Castagnevizza and east of Selo. Climbing up I spotted at 2,500 meters an enemy fighter and an Albatros. The Albatros from Duino was going to the Hermada and to our lines between Oppacchiasella and Castagnevizza. I crossed him at 2,400 meters at 7.25 above Selo. With a quick maneuver to the right I glided getting behind it. The enemy fired quickly, turned east and then north. At 50 meters climbing steeply, I fired a burst from behind and after about 10 rounds I saw flames getting out of the fuselage. The Albatros went down in flames, at 200 meters below me its wings broke apart. The burning fuselage fell 1 km north of Selo."

And to his mother he wrote: "I jumped on it, I mislead it with my maneuver, I shot, it defended itself, but after as few as 15 bullets it caught fire, fell for 200 meters then its wings broke and splintered, and the burning fuselage with the airmen fell near Selo, just behind our lines. If the offensive goes on, we will find it. Also one of ours fell, put on fire by an enemy fighter." His victim was once more a Brandenburg of Flik 35, 229.23 piloted by *Korp.* Engelbert Gerlich and with as observer *Lt.* Raimund Wania.

On 20 August Baracca escorted again a Caproni raid, then attacked a lone Brandenburg above Voiscizza:

> "I turned left attacking it on the left side with a first burst. The enemy turned steeply many times, firing occasionally. I fired again, but it was difficult to aim, as the Albatros dove and turned all the time. For a second I was below and behind its fuselage but I couldn't fire, as I would have crashed into it. I passed it 20 meters to the side, I saw the pilot but didn't see the observer, who must have been wounded, the machine gun was abandoned and a long cartridge belt hang from the fuselage. We were at 1,200 meters, the enemy dove and I couldn't attack it, it went down and flew away toward Reifenberg."

The most famous photograph of Francesco Baracca, now *maggiore*, in front of his Spad 7 with the prancing horse. Original copies of this picture show that there is a difference in tone between the horse and its tongue, proving that the horse was actually black and the tongue was red.

On 28 August 1917 Baracca again escorted a raid of Caproni against gun batteries at Panovizza. He spotted a group of fighters far away and then he saw them attacking and shooting down an S.P. The ace dove toward them, but it was too late. Baracca had lost his chance to attack again Brumowski who, flying KD 28.69, has just shot down Savoia Pomilio S.P.3 4651 of 22ª Squadriglia, killing *serg.* Ernesti and *ten.* Piacentino: "The enemy squadron had gone away immediately, and I saw it far away toward Ternova. The last group of Capronis arrived and I saw many bombs exploding in the middle of the Aisovizza airfield."

On 1 September 1917, while the battle was still raging on the San Gabriele mountain, in the final phase of the 11th Offensive, Baracca was cruising with his whole squadron when a formation of eight Brandenburg bombers of Fl.G 1 was bombing the Bainsizza plateau. As the Austrians got away, Baracca spotted a straggler, left the formation and bounced it:

> "One of them turned toward Gargaro, next to it, higher, a fighter took the direction of Bainsizza. I attacked the first of them that looked like a Brandeburgo (*by now the correct name*

*was sometimes used for the Brandenburg Br.C.1, usually called Albatross*) at 2,400 meters. Time was 19:40 over Gargaro. I fired a first burst of thirty rounds from 100 meters on its right side. I found myself behind it, in a dive and I went down as the enemy was firing, I fired again at 100 meters and soon later the enemy started to go down, I kept on firing getting to 50 meters behind until it disappeared, diving, severely hit under the wings by my fire. I saw it again, rapidly going down in tight spirals, I went down to 1,200 meters and I followed its fall almost down to earth, losing it below my wings, in the woods of Hill 664 north-east of Zagorie, where the Albatros must have fallen. As it was late, visibility was bad, I believe that the enemy fighter close by has not seen my attack. I believe our fighters cruising over Gorizia have not seen the enemy squadron because they were too far south and too high. In fact it is difficult to see enemy airplanes at dawn and sunset from high above."

His victim was either Brandenburg 69.10, (the observer *Leut.* Odilo von Khuepach zu Ried, Zimmerlehen und Haslburg was killed, the pilot *Fw.* Rudolf Nölscher was wounded but force-landed at Haidenschaft) or Brandenburg 229.43, that was shot down by a Spad in the area of Britof-Gargaro, the pilot, *Zugsf.* Johann Trnka and the unknown observer both unharmed.

In his letters the ace remarked:

"The Austrians are not much active in the air, they received new fighters (*probably the Albatros D.III*) that we have met and repelled. Our victory turns out to be a very big one, it was a real defeat for the Austrians, but we cannot advance further in the center because the wings in Tolmino and the mountains above Gorizia hold. We made a huge booty of provisions and equipment. Still, the war will not end now, we generally believe that it will last long. I am always very busy and oppressed by hard work, I wish for an early winter, to have a bit of rest."

Early in the morning of 6 September, Baracca had a combat over the Sabotino Mountain that he described as beautiful. He and *ten.* Sabelli attacked a Brandenburg bomber:

"It dove for 500 meters, then it recovered, passed its lines and landed a bit further. The enemy with the back gun put up a devilish fire with tracing lighted bullets, without hitting us. He maneuvered very well, but we were attacking with such élan at a distance of 30 meters that we disturbed each other, for the fear of colliding, and I proposed a silver medal for Sabelli. Thus I have reached my 19th victory, and I look forward to the 20th one, with a bit of luck".

The victim was possibly Br.C.1 129.50 of Flik 34, the pilot *Zugsf.* Stefan Morth was wounded, the observer *Oblt.* Béla Gerey died. In the combat report, Baracca added: "The Albatros was dark yellow, a color that blended with the ground, and its tail was painted red."

On 15 September Baracca had a duel with a fighter but his gun jammed. In this period, he tested the SVA, two of which had been delivered to the squadron. He flew five times on this plane, and he produced a very detailed technical report, dated Sept. 16th, in which he criticized some design features of this airplane and particularly its lack of agility and the incorrect position of the guns, excluding that it could be used as a fighter.

On 23 September he attacked single-handedly a flight of four fighters over Lokve, but his initial quarry escaped his attack, and then the fighters attacked him. Baracca easily got away, but this event could correspond to a claim over a Spad by Austrian aces Udvardy and Strohschneider of Flik 42/J in that area (although the two Austrian aces state that their victim had French roundels).

One more duel with a fighter took place on 29 September:

"Flying on a fighter sweep at 17:15 over the San Marco at 4,000 meters I spotted enemy airplanes, I got down to 3,000 and attacked over the San Gabriele an enemy fighter firing at its side from the left. The enemy dove in a spiral giving me the time to shoot another 20 rounds from 50 meters and I went down toward other enemies that I saw over Bainsizza, but immediately I was jumped by the same fighter from above. I went down in a spiral a few hundred meters; the enemy passed quite close to me shooting and went below my airplane. I attacked it again from above and the enemy went down toward Ternova attacked by another Spad and by a Nieuport 80 HP far away. I lost from sight the enemy toward Ternova below 1,000 meters."

By this time, Baracca, who had been promoted to Major for his war merits, was showing some worry, compounded by the sorrow for Olivari's death:

"I am considering coming for a couple of days to Lugo, but first I need my 20th one, otherwise I'll stay here. The Austrians received new fighters, they also fly in big formations and now it is no longer easy to attack them. Their fighter pilots are getting bolder, but they don't accept combat with us, and they just attack those of our airplanes that are not so well armed."

After a busy September, in October flying activities were mostly hampered by bad weather, which protected from Italian reconnaissance the build-up of Austrian forces and the arrival of German divisions in preparation for an offensive of the Central Empires. Then, all of a sudden, Baracca met the Germans in the air. The day was 22 October, but it is quoted as 21 October in the Ace's notes, which were used as a source by all later accounts. However, the Squadron diary demonstrates that no flying activity took place on the 21st, and that it was the next day that the combats took place. Baracca, flying for the first time a 200 hp Spad 13 with two guns, was on a free hunt mission over Mount Nero when he was bounced by five enemy fighters, Albatross D.III with the leader's painted in red; he got away from this attack spinning his Spad. Continuing his flight, he spotted two fast new two-seaters over the Bainsizza, he attacked and one immediately went down, falling behind Italian lines. The second two-seater identified by Baracca as an Aviatik, turned on him and attacked, but it also was quickly dispatched, it crashed in the Italian lines and burned:

> "The two airmen were killed. The two other airmen of the first airplane, two German Lieutenants, were also dead, the pilot with 11 bullet wounds, I got the rudder of that plane (dark green color), its photograph was lost in the retreat, and remained behind the Tagliamento river; it was a superb victory and a triumph for me."

The two victims can be identified as DFW C.5s of Flieger Abteilung 14, with crews of Zimmermann and Fischer, Guttler and Hansberg.

At dawn on 24 October 1917 the Germans spearheaded an attack at Caporetto, in the sector of the long-suffering 2nd Army. The initial shelling cut connections, Italian artillery didn't intervene until too late, and the Army, whose commander had disregarded orders to prepare for the defensive, collapsed, jeopardizing the whole front. The gravity of the situation was not perceived until the evening, when it was too late. The Austro-German offensive was so quick that the whole Italian front collapsed, leaving behind first the Isonzo then the Tagliamento rivers.

Weather was horrible on the 24th but the next day, as the sky cleared; the whole of Italian aviation was thrown into the furnace to stop the enemy advance. The 91ª Squadriglia had five combats with the Germans and Austrians, shooting down six of them, and suffering two pilots killed, lieuts. Sabelli and Ferreri. Baracca shot down an Austrian "Albatros" (that is to say, a Brandenburg) over San Marco at 11.10 in the morning, together with Piccio. This victory is obscure: it was identified as 29.63 of Flik 19, the killed airmen being *Fw.* Anton Lambert and *Lt.* Jan Pappius. However the times don't correspond, as 29.63 was shot down at 14:30, so it is much more likely to be Ruffo's victim, in a later combat. To further complicate matters, the 91ª's diary reports this victim as an "Aviatik". Baracca, however, was sure that his adversary was Austrian, not German. In a later combat, Baracca attacked a German two-seater that got away.

On the 26th:

> "Another day of triumphs: I took off on a patrol with ten. Parvis, who covered my back. At 11.15 I attacked a German 'Aviatik' north of Cividale, and after a tough fight the enemy fell flaming at S. Lucia di Tolmino; at noon another combat north of Cividale. The German maneuvered very skillfully and its fire was well aimed. I felt twice my Spad getting hits, I was many times in a bad situation, but in the last attack, at 3,000 meters I saw flames on board, the German started to lose height burning and after a few seconds the Aviatik fell engulfed in an impressive red flame and crashed within our lines, by Matajur that was still ours on the 26th. That is my 24th victory. A few minutes later, my engine stopped working at all and I had to land in a field near Cividale. A strut of my Spad was broken, the fuel tank was pierced from end to end by a bullet, luckily not an incendiary one, the engine hit by a bullet and wrecked, the propeller hit, the wings also. I left the plane and walked to Cividale, where I had to witness the tragic retreat of our troops. German airplanes were flying over the city every half-hour; at 1 p.m. I watched the air combat between a Spad of my squadron and a German, the German was shot down near Cividale, later I learnt that the winner was Costantini of my squadron (2nd victory)."

This time, Baracca was again flying a single-gun 140 HP Spad 7.

In the afternoon of 27 October the 91ª Squadriglia received the order to abandon its airfield, Santa Caterina. The Austrians were getting close to Udine, base of the Supreme Headquarters. The Spads took off on the morning of the next day, under a pouring rain. Baracca remained alone: seeing a squadron of cavalry that was protecting the airfield, he was tempted to join them, for a final charge against the enemy:

The last victim of Baracca: Albatros D.III 153.266 shot down on 15 June 1918; its pilot *Lt.* Sigmund von Josipovich was captured.

The men who possibly shot down Baracca: *Oblt.* Barwig and *Zugsf.* Kauer in front of their Phönix C.I 121.17 of Flik 28. *(Photo Selinger).*

> "In the morning of October 28th, the Austrians were at the outskirts of Udine. I let all my squadron take off, in the midst of a thunderstorm with the clouds down to 100 meters and a furious wind. All the airports of Campoformido were burning, dark pillars of smoke, the warehouses of Udine were burning, and everything was in flames. I had a squadron of Genoa cavalry that was defending the airfield: I remained, and departed the last, uncertain whether to abandon my airplane and mount on a horse to charge the Austrians."

Instead, he set to fire the hangars and the airplanes that could not be savaged: four Spad 7s, two SVAs, one Nieuport 17 and the Brandenburg 61.57, his victim of 7 April 1916. Then he took off, bitter and sad.

The squadron moved to La Comina airfield, Pordenone, then, on 31 October, to Arcade, near Treviso, behind the Piave River that became the line of resistance of the Italian Army, after the Tagliamento river could not be held. The next day the squadron moved to Padua, the beautiful city of the Veneto that became the new base of Italian high commands. Baracca, busy organizing the reformation of his unit, that received many new pilots, including the future aces Novelli and Magistrini, did not have combats until 6 November.

That day, according to the squadron diary, Baracca with *ten.* Parvis met two fighters, erroneously identified as Berg, over Portogruaro and after a long combat shot one down near Fossalta (one of six villages with the same name in the region, this one next to Portogruaro). Baracca's report is more detailed:

> "After having rested a few days, rested so to say, meaning I didn't fly, I took off in the morning of the 6th for the lines with ten. Parvis behind me. The weather was fair, I was in the mood for sweeping the sky: we found two enemy fighters above Portogruaro at 10, I attacked one, Parvis the other one, the one of Parvis after a brief dogfight escapes toward Latisana, mine, impressed by my attack, gets lower and lower in spirals. Firing occasionally, we get down to 50 meters over the trees, I am always behind him harassing him, until the enemy hits the ground smashing the airplane. It was one of my best dogfights, the Austrian fighter pilot didn't get out of the plane, so he must have been killed or badly hurt."

Actually, Baracca's victim was *Oberleutnant* Rudolf Szepessy-Sokoll von Négyes et Renó, of Flik 41/J, a 5-victories Austrian ace, who was mortally wounded and nursed his Albatros D.III (Oef) 153.46 to force-land in no man's land, that was soon conquered by the advancing troops. Baracca's second opponent, the one who escaped, was *Feldwebel* Radames Iskra, also of Flik 41, flying Alb. 153.08, who somehow landed behind Italian lines, despite the fact that Parvis admitted that he hadn't fired, and was taken prisoner. This victory is erroneously dated Nov. 4th on the marble basement of the monument to Baracca in Lugo.

Iskra, who reported that he had flown his first and last combat flight over Italian territory that day (actually he was flying in Brumowski's Flik 41/J since June and claimed a victory over a Spad on 17 September) presented himself as a defector and gave a lot of information to his interrogators, including some sour, slandering and not necessarily reliable, comment on the "professional jealousy" among Austrian aces.

Half an hour later, Baracca and Parvis met two German two-seaters above Conegliano, escorted by two fighters that stood out of harm's way. The first one returned fire and escaped, the second one was attacked at close range and shot down: "I land a Arcade, and soon I get there with a car, and find the airplane, brand new, I take two guns and other objects; the airmen are from Berlin, one dead, the other one seriously wounded, now he should be dead. The 'Aviatik' is set on fire by the cavalry, because there are no means to transport it."

It was a DFW from Fl.Abt.(A) 219, and its crew was Binder and Schroeder. The two fighter pilots spent the night in Arcade, and the next day, 7 November, they took off, as the battle was raging. First they attacked three fighters that were dogfighting with an Italian Nieuport, and dispersed them, and then they spotted a lone German flying high above Conegliano. The German two-seater fought valiantly:

> "He spots our attack, turns back and aims at me, with a quick maneuver I attack him from the left, he defends himself very well, it is a German Aviatik, 170 km of speed and very well armed. Parvis attacks him, then I again from 100 meters. After 150 rounds it is hit and goes down, I follow it to the ground shooting, then I land at Arcade. Back to the car, and we go to recover the plane. It was near Orsago. The Austrians were 1 km away, we want to get there at any cost, we hear machine-guns, squadrons of cavalry were galloping by. We find the beautiful Aviatik, the two Germans dead, we take the guns, the rudder, and the black crosses, and then we set the gasoline on fire and go. The airplane is destroyed with its Benz 220 hp engine. So I shot down my 27th, and Parvis his 7th."

According to a report of the Red Cross, the aircraft was DFW C.5 3955/17 and its crew was *Gefreiter* Wilhelm Appelt, wounded, and *Lt.* Paul Wilkening, killed in action.

Operations were curtailed by bad weather the following days. On 15 November Baracca had what he described as a "*tragic and terrifying combat.*" He was in patrol with *ten.* Parvis, flying a Spad 13, when he spotted a German two-seater high above Istrana, and attacked it at close range:

> "It defended itself well, but after 120 rounds I saw flames on board and it started to go down, I saw the airplane engulfed by flames at 4,000 meters and the two airmen jumped out, and the Aviatik fell in a field near Istrana I landed immediately and a few minutes later I was there. The German airmen were two lieutenants of very distinguished aspect; one of them had a gold ring, a wedding ring and a nice portrait of a lady in a leather case. He wore the Iron Cross, a German military award. I keep from that plane the guns, the tubes of the burned out rudder, the cameras."

The victims were *Lt.* Alfred Müller and *Lt.* Erich Peucer of Fl.Abt. 14, but strangely the Red Cross reports a third casualty of that crash, *Feldwebel* Wilhelm Noss. Baracca further commented:

> "There is little activity in these days. I often take off leading flights of 4 or 5 pilots and we go over the lines: woe to whom meets us. On the other side there is the flight of captain Brumowski who has shot down 22 of our airplanes, he flies an Albatros fighter with two guns, all red, but we never met."

On 23 November the 91ª Squadriglia had several combats, lost two Spads, too much damaged to return to flight, with no harm to their pilots, Parvis and Piccio, and claimed three victories. One of them belonged to Baracca, who in the afternoon shot down a fighter near Falzè di Piave:

> "The dogfight was exciting, from 3,500 meters it went down to 500 meters in a sequence of fire and maneuvers to escape the fire, finally the enemy glided on the Piave, on a stretch of dry riverbed near our side, and it overturned. I followed it down to 100 meters, then I climbed because I heard machine guns firing from the opposite bank, the German pilot was wearing a black silk scarf, maybe wounded, he put his plane on fire and then he managed to escape and hide. The plane was an Albatros D.III fighter with two guns, the best German fighter. It was my 29th and Novelli's 3rd."

The squadron diary reports that the German pilot was found dead in the fuselage: he was *Feldwebel* Überschär of Jasta 39. This was the last of eight German airplanes shot down by Baracca.

On 30 November Baracca together with Piccio attacked a kite balloon floating near Oderzo, in the southern part of the front. Despite repeated attacks with incendiary bullets, the balloon refused to catch fire, and it was hoisted down, as the two observers jumped with their parachutes. No claim was made.

The Piave front was now stabilized, the Austro-German offensive had captured a large amount of Italian territory, had taken many prisoners and a considerable booty, but had not put Italy out of the war. The Italian Army stood fast on the Piave. The last Austrian attack took place at the beginning of December 1917 on the Asiago plateau, on the northern front, but it was stopped after some initial success. On 7 December, the aircraft of the 91ª moved from Padua to Nove di Bassano, to support the Asiago front, and flew three patrols. During the second one, Baracca was flying with *ten.* Novelli and *aspirante* Olivero. He saw some AA fire south of Asiago, and spotted an Austrian two-seater:

> "I discovered it only when I was above it, at 400 meters. Its wings were painted black, and only part of the fuselage was yellow, so it mixed with the color of the woods below. It was an Austrian Albatros. I engaged combat at a level of 2,600 meters, just 600 meters above the ground, after a sharp exchange of fire the observer must have been hit, because he

The preserved Spad 7 of Baracca, the airplane flown when he scored his victory at Kaberlaba on 7 December 1917, exhibited in rather bad conditions during an air race in 1920.

stopped firing, and then with three more well aimed rounds I set the airplane on fire. It was a really lugubrious vision to see that airplanc all black with the yellow fuselage and trailing fire, fall in the wood behind mount Kaberlaba."

His victim was Brandenburg C.1 29.20 of Flik 45, crew of *Fw.* Hermann Schütze and *Lt.i.d.R.* Hermann Mehrfurt. In the afternoon Baracca found another two-seater, pursued it over the Grappa Mountain but couldn't reach it:

"French and English fighters are also wonderful: three of them, captains, have fallen in combat. To my order are coming English pilots who will form another fighter squadron together with mine, with their fast and well armed aircraft: I will ask for one. I am also waiting from France the 200 hp Spads with two machine guns. We see many American officers who are full of enthusiasm and can hardly wait to enter into the battle and are ready to make war for years and years, trusting in victory. We have again full air supremacy and each enemy airplane that shows over the lines is shot down by us or by the Allies. Even if the Germans could manage to advance, the land will be contested inch by inch and we will not retreat to the Adige River."

On 13 December, an unlucky number, Baracca flew two missions, in both of them he met an enemy airplane and both times it escaped him. At this time, in December he was awarded the Medaglia d'Oro al Valor Militare, Italy's highest award for valor, with an exalting citation:

"First fighter pilot of Italy, undisputed champion of ability and courage, sublime confirmation of the Italian virtues of élan and boldness, tempered in sixty-three combats, he has already shot down thirty enemy airplanes, eleven of them during the latest operations. In the latest combats he returned twice with his airplane severely hit and damaged by machine gun bullets."

As the front settled, and winter fell in, Baracca finally had a period of leave. On Christmas day, 1917 he was sent to Turin, together with *t.col.* Piccio and *cap.* Ruffo, to test new fighter models, leaving the command of the unit to *cap.* Costantini. He visited the aviation factories of Turin, and tested the new Ansaldo Balilla fighter, finding it less than satisfactory.

They returned on 24 January 1918. Baracca, who had not flown in combat since 7 December, returned in action on 1 February, when he attacked an enemy fighter above Asiago:

"Flying in free hunt and being south of Astigo I spotted an enemy at 4,400 meters and I tried to cut his escape going to Valstagna, and I reached the enemy over Cima Ecker. It was a fighter that defended itself with maneuvers, turning quickly and escaping my attack, going down to Marcesina. I couldn't pursue it because my throttle jammed."

Two days later, he had another duel in the same area, which he had to interrupt as his gun jammed immediately. On 6 February the kings of Italy and Belgium visited the base, and the valiant King Albert proudly decorated Baracca with the Croix de Chévalier de la Couronne. He didn't fly much in February nor in March, activity was limited and the weather was bad most of the days, while in the night Padua and its airfield were raided by the German bombers that were remarkably effective. Moreover, the unit was having trouble with the Italian-built Hispano Suiza engines of its Spads. On 19 March he left his unit again, going to Milan. He was solemnly awarded the Gold Medal in the famous Theatre della Scala by Special Commissioner for aviation Eugenio Chiesa, the respected politician who managed Italian aircraft production. He returned to the 91ª Squadriglia, that meanwhile on 1 April had moved from Padua to Quinto di Treviso.

On 3 May 1918 Baracca was back in action. Flying with *serg.* Nardini, he attacked two separate patrols of six enemy fighters, possibly scoring hits on one, then he spotted a lone Brandenburg that was reconnoitering the Italian lines along the Piave, *serg.* Nardini attacked it but had to desist with a jammed gun, then Baracca got in. It was a tough adversary, and Baracca had to attack three times before shooting it down in flames on the banks of the Piave. His victim was Br.C.1 369.28 of Flik 19/D and two aviators lost their lives, *Zugsf.* Josef Friedrich and *Oblt.* Karl Rosenbaum.

This is Baracca's version of the combat:

> "Serg. Nardini attacked, as he was closer, but his gun jammed. Then I stepped in and fired a burst, but the 'Albatros' defended itself very well, it went lower and then waited, accepting combat over the Grave di Papadopoli. I came in at full speed attacking it on the side, and the 'Albatros' quickly turned around, but the pilots must have been hit and at my last attack at 1,500 meters it caught fire in front of our lines. It was impossible to recover it because the ground was under artillery fire."

In these days of limited enemy activity, Baracca took part in the humble but useful missions of propaganda leaflet launches over the enemy. The Italians had organized an efficient psychological warfare office, which prepared flyers in the different languages of the Austrian Empire, fomenting dissension among the nationalities.

On 19 May the Squadron had eight pilots in the air, escorting a Caproni raid in the Feltre area, when it met a similar Austrian squadron of nine fighters. A huge dogfight developed, with Baracca being the first one to attack the Austrian leader, but at the end of it the Italians could claim only two damaged, with two of their own also having received some hits. At the same time, and in the same area, the British pilots of Sopwith Camel squadrons No. 28 and 66 claimed several enemy fighters shot down, and in fact the Austrians lost two Phönix D.1 of Flik 14/J.

Baracca scored again on 22 May. Flying together with *serg.* D'Urso, he met an enemy fighter at high level above Cimadolmo, and quickly sent it crashing down, south of Borgo Malanotte. The victim was *Fähnrich* Ernst Freiherr von Pontalti, of Flik 51/J, who was piloting Albatros D.III 153.155:"Yesterday at 9.30 a.m. I shot down my 32nd enemy airplane, a fighter that was part of a flight of 6 airplanes. It crashed in flames at Borgo Malanotte, beyond the Piave, toward Grave di Papadopoli. I cannot write long."

This is the combat report:

> "I saw a reconnaissance plane escorted by 6 fighters going toward Cimadolmo at a height of 4,000 to 4,500 meters parallel to the Piave. I crossed the enemy formation flying higher and I aimed at the fighter at the extreme left and higher than the other ones. The enemy planes were going toward Tezze di Cimadolmo; the enemy fighter had not seen me while I believe I was spotted by the two-seater that turned inland. I dove under the fighter that was flying at 4,400 meters and then climbing I fired a first burst below and behind from 50 meters, I saw its left lower wing folding without breaking, I stopped firing, the fighter dove, I fired again from above at 30 meters of distance, I saw the left wing breaking off from the fuselage and the plane crashed breaking apart. I saw it falling and burning and it crashed south of Borgo Malanotte. During the attack my back was covered by sergeant D'Urso who followed me close and gave me the security from the attack of other enemies."

Baracca's character, earnest, modest and serious, is expressed in anecdotes. An old peasant from Quinto, who loved him, told him to stop flying, as he had done his duty well enough. Baracca replied: In this job, if you do your duty you are dead in six months. Me, I am still alive.

At this time, Austrian aviation had reduced its activity, but the enemy was mastering its forces for the final offensive. The bad weather that protected the preparation was of no avail: this time, the Italians expected the enemy attack, their lines of resistance were well placed and well manned. The Austrian attack, the so-called "offensive of hunger", unleashed not on a single point but all over the front, from east to west all along the Piave, and from north to south by the Grappa mountain, was doomed from the start. On 15 June the Flieger-Kompanien were all in the air, but Italian fighters jumped on them with a vengeance. The Luftfahrtruppe lost dozens of airplanes, while Italian losses were minimal.

The 91ª Squadriglia was in action from dawn to sunset, and shot down five enemies, Baracca accounting for two: a two-seater shot down at low level at Saletto, on the right bank of the Piave River, and a few minutes later a fighter, shot down into the Italian lines. His first victim has been possibly identified as Brandenburg C.1 369.116 of Flik 32/D, whose crew of *Oblt.* Albin Tomasic and *Leut.* Arthur Fellner died, but there can be no certainty, among the several Austrian losses. The second one was *Lt.* Sigmund von Josipovich of Flik 51/J, wounded and taken prisoner, whose Alb. (Oef.) D.III 153.266 marked with a six-pointed star was taken in good condition. A famous Italian aviation historian of the past, who painted beautiful aviation paintings, refused to paint this duel because of his rabid anti-Semitism. Actually, the six-pointed star often painted on Austrian airplanes had not to do with the Jewish faith, but it was the distinctive mark of Austrian soldiers, like a five-pointed star marked Italian military men.

"Taking off at 11 a.m., on a patrol with sergeant Aliperta, I cruised over the Piave. At 12.30 west of Saletto I saw an enemy "Albatros" recce, I attacked it resolutely and having fired 120 rounds the airplane took fire turning over and breaking apart before reaching the ground. I had a gun jam. Having shot down the first one that was bombing our troops, I saw another airplane and attacked it, but both guns jammed and I didn't have time to clear them because the enemy dove away over the Piave and was soon at a level of 300 meters. At the same time sergeant Aliperta attacked another enemy. At 12.45 I spotted an "I DV" fighter plane flying toward San Biagio. I attacked it at a height of 800 meters, from the side and tail. After a burst of about 100 rounds it fell within our lines turning over and remaining intact north of San Biagio di Collalto. During the combat sergeant Aliperta cooperated with me cutting the escape of the enemy. The airplane was part of a big flight of 25 enemy fighters that were flying at 1,200 meters; they were pursuing our airplanes that returned to their lines. Many guns were firing at me from the ground, and a bullet struck my rudder."

The events surrounding the death of Baracca, on 19 June 1918, like Manfred von Richthofen's, are controversial. The accepted version is that Baracca, who was flying Spad 7 with 180-hp engine serial S.5382, was hit by some unknown Austrian infantryman during a strafing mission on the Montello, but at the time of his death, the press reported that he had shot himself in order to prevent capture. *Cielo*, the main aviation magazine of the period, reported:

"Francesco Baracca falls gloriously fighting from above against enemy infantry. Francesco Baracca has died strafing the Austrians from low level, as he was fighting like a soldier among the soldiers who were breaking the enemy attack on the Montello. No German or Austrian airman will be able to boast of having shot him down."

But another article in the same issue added:

"Rather than fall in the hands of the enemy, I will kill myself had declared Baracca. Our glorious soldier has kept his promise. The body of major Baracca was recovered on the Montello next to his burned airplane. Lieuts. Ranza and Osnaghi of the squadron of the aces and our (journalist) Raffaele Garinei ventured on the harsh slopes of the Montello looking for the body of the gallant airman; after having checked all the terrain, at about 3 p.m. beyond the first ridge, exactly near a trench between Collesel della Zorla and Sorgente dei Frati they found the body of the hero. On its right temple a hole produced by a pistol bullet makes one believe that Baracca falling with his burning airplane had the willpower to keep the promise that he had often repeated – if the Austrians flame my airplane, rather than falling alive in the hands of the enemy, I'll kill myself. The right hand of Baracca, in fact, was frozen near the temple and his revolver was picked up not far from the corpse. Wrapped in a gray blanket, posed on our car, the body of Baracca was carried to the airfield of the aces, where Piccio, Ruffo, Novelli, Costantini, Keller, Olivero, Bacula, Nardini, Aliperta, Magistrini, D'Urso, Decorato, Dronetti were waiting in sorrow to know the lot of their great teacher. To the recovered body of Baracca great honors will be offered."

Recently, a modern author has offered the unsubstantiated hypothesis that Baracca had shot himself to avoid the horrible results of an airplane fire. Finally, a common and repeated theory is that the Austrian Phönix C.1 121.17 piloted by *Zugsf.* Kauer and *Oblt.* Barwig of Flik 28 shot down Francesco Baracca. This opinion was immediately disproved by the Italian Supreme Command, that issued a communiqué:

"The Austro-Hungarian war bulletin announced as proven that lieutenant Barrug (*sic*) shot down on 19 June the famous Italian airman Baracca. This news is groundless, and one must condemn this attempt, new in the history of war, of attributing false laurels to whomever for the death of a valiant soldier. The shooting down of major Baracca is the obscure claim of an unknown enemy infantryman, who shot to death with his machine-gun the heroic airman, while from very low level he was gallantly strafing units moving on gangplanks over the Piave. In that moment no enemy airman dared to fly over the Montello. On that day the enemy aviation, already trounced by ours, was absent from the sky of the battle."

In the author's opinion, the first theory (suicide) seems impossible, and the second one (an Austrian air victory) unproven. The time between the battle damage and the crash of a low-flying fighter is so short that no man, struggling to control his airplane, unsure of its actual conditions, could have the time to ponder, decide, get out his gun and shoot. If the Spad was on the ground, Baracca could get out of it, and avoid getting burned. As for captivity, besides the fact that it would not have been dishonorable, Baracca fell in a contested area, where the Italians were advancing: too many doubts to justify a suicide.

Also the testimonial of *tenente* Osnago, who was flying behind Baracca, reports of a sudden burst of flame beneath the Spad, where a bullet struck the fuel tank, and an immediate crash, confirming the theory that a single bullet had penetrated, hitting Baracca's head. Osnago reported that they were alone in the sky when Baracca fell, disproving the theory that Kauer and Barwig had shot him down. However, he may have missed seeing the Austrian airplane, or else his declaration may have been doctored by Italian authorities, who wanted to preserve the myth of the undefeated ace. Arnold Barwig, who survived the war, in later years discussed in good faith his role in the crash, explaining that the enemy Spad has suddenly appeared, coming out from a cloud, and he had been quick to fire his machine-gun at it. The accepted fact that the fuel tank burst in flames would exclude the effect of a simple

## Francesco Baracca
## 34 confirmed victories

| Claim | Vict. # | Date | Sqn. | Aircraft | Hour | Place | Victim |
|---|---|---|---|---|---|---|---|
| - | - | 19.11.1915 | 1 | Ni 10 1035 | 8h | Udine | - |
| *[with cap. Bolognesi, 1ª Sq.] Korp. Adolf Veselic UNH/Lt.i.d.R. Ludwig Riesenecker WIA, Flik 4, Alb. B.1 24.17* | | | | | | | |
| 1 | 1 | 7.4.1916 | 1 | Ni 11 1451 | 7h | Medeuzza | Albatros |
| *Fw. Adolf Ott POW/Oblt. Franz Lenarcic KIA, Flik 19, Br. C.1 61.57* | | | | | | | |
| 2 | 2 | 16.5.1916 | 70 | Ni 11 | - | Gorizia | Two-seater |
| *Korp. Lajos Knelly KIA/Fähnr. Ladislaus Hórvath-Tima KIA, Flik 12, Lohner 17.42* | | | | | | | |
| 3 | 3 | 23.8.1916 | 70 | Ni 11 | - | Merna | EA |
| *[with ten. Ruffo, 70ª Sq.] Zugsf. Anton Himmer UNH/unknown observer UNH, Flik 19, Br. C.1 61.61* | | | | | | | |
| 4 | 4 | 13. 9.1916 | 70 | Ni 11 | morning | Monfalcone | EA |
| *Fw. Josef Koschiczek WIA/Lt.i.d.R. Ottokar Hauser KIA, Flik 28, Br. C.1 64.07* | | | | | | | |
| 5 | 5 | 16.9.1916 | 70 | Ni 11 | - | Caporetto | Lohner |
| *[with ten. Ruffo and sold. Olivari, 70ª Sq.] Zugsf. Franz Morozko KIA/Lt. Anton von Csáby KIA, Flik 16, Lloyd C .III 43.74* | | | | | | | |
| 6 | 6 | 25.11.1916 | 70 | Nieuport | 11h | Tolmezzo | Albatros |
| *[with serg. Gorini, 70ª Sq.?] Korp. Fritz Fuchs KIA/Lt. i. d. R. Kälmän Sarközy POW, Flik 16, Br. C.1 68.03* | | | | | | | |
| 7 | 7 | 1.1.1917 | 70 | Ni 2614 | - | Castagnevizza | Albatros |
| *Korp. Franz Neuwirth UNH/Lt. i. d. R. Viktor Huber KIA, Flik 12, Br. C.1* | | | | | | | |
| 8 | 8 | 11.2.1917 | 70 | Ni 2614 | - | Ozzano | EA |
| *[with ten. Ruffo and serg. Gorini, 70ª Sq.] Korp. Ludwig Fleck POW/Lt. i. d. R. Wilhelm Graf Siemienski POW, Flik 35, Br. C.1 27.74* | | | | | | | |
| 9 | 9 | 26.4.1917 | 70 | Ni 17 | 11h10 | Gradisca | Two-seater |
| *[with serg. Gorini 70ª Sq. and serg. Imolesi 79ª Sq.] Zugsf. Josef Majsai KIA/Lt. Emmerich Treer KIA, Flik 35, Br. C.1 129.17* | | | | | | | |
| - | - | 1.5.1917 | 91 | ? | 10h | Latisana | Albatros |
| *Oblt. Frank Linke Crawford UNH/Oblt. Johann Graf Herberstein UNH, Flik 12, Br. C.1 229.08* | | | | | | | |
| 10 | 10 | 10.5.1917 | 91 | Spad 7 | 10h | Vertoiba | Fighter |
| *Fw. Rudolf Stöhr WIA, Flik 41, KD 28.17* | | | | | | | |
| 11 | 11 | 13.5.1917 | 91 | Spad 7 | 10h15 | Mount Korada | Albatros |
| *Fw. Julius Búsa KIA/Oblt. Hermann Grössler KIA, FlG.1, Br. C.1 129.20* | | | | | | | |
| 12 | 12 | 20.5.1917 | 91 | Spad 7 | 9h30 | Plava | Two-seater |
| *Zugsf. Ludwig Ferschl KIA/Lt. i. d. R. Eugen Csutka KIA, Flik 12, Br. C.1 229.10, or: Korp. Michael Scholz, KIA/Oblt. Ludwig Varga, KIA, Flik 12, Br. C.1 129.52* | | | | | | | |
| 13 | 13 | 3.6.1917 | 91 | Spad 7 | 9h30 | N.E of Plava | Albatros |
| *Zugsf. Johann Rotter KIA/Oblt. Max Bednarzik KIA, FlG.1, Br. C.1 129.51* | | | | | | | |
| 14 | 14 | 7.7.1917 | 91 | Spad 4702 | evening | Carso | Albatros |
| *Fw. Sigmund Duschek KIA/Lt.i.d.R. Andor Bózo de Dinyeberki et Agod KIA, Flik 46, Br. C.1 129.68* | | | | | | | |
| 15 | 15 | 31.7.1917 | 91 | Spad 5418 | 6h30 | Podgora | Albatros |
| *[with ten. Buzio, 81ª Sq.] Zugsf. Julius Klausz KIA/Oblt. Erich Bock KIA, Flik 46, Br. C.1 69.93* | | | | | | | |
| 16 | 16 | 3.8.1917 | 91 | Spad 7 | 9h50 | Valley of the Sava | Brandenburg |
| *[with ten. Baracchini, 76ª Sq.]* | | | | | | | |

| Claim | Vict. # | Date | Sqn. | Aircraft | Hour | Place | Victim |
|---|---|---|---|---|---|---|---|
| 17 | 17 | 19.8.1917 | 91 | Spad 7 | 7h25 | NE of Selo | Albatros |
| *Korp. Engelbert Gerlich KIA/Lt. Raimund Wania KIA, Flik 35, Br. C.1 229.23 or: Zugsf. Artur Zeides KIA/Oblt. Karl Balzar KIA, Flik 28, Br. C.1* | | | | | | | |
| 18 | 18 | 1.9.1917 | 91 | Spad 7 | 19h40 | Zagorje | Albatros |
| *Fw. Rudolf Nölscher WIA/Lt. Odilo von Khuepach KIA, FG 101, Br. C.1 69.10, or: Zugsf. Trnka UNH/ unknown observer UNH, FG 101, Br. C.1 229.43* | | | | | | | |
| 19 | 19 | 6.9.1917 | 91 | Spad 7 | 18h30 | S. Gabriele | EA |
| *[with ten. Sabelli, 91ª Sq.] Zugsf. Stefan Morth WIA/Oblt. Béla Gerey KIA, Flik 34, Br. C.1 129.50* | | | | | | | |
| 20 | 20 | 22.10.1917 | 91 | Spad 13 | 12h | Sabotino | Aviatik |
| *Lt. Werner Zimmermann KIA/Lt. Arthur Fischer KIA, FA 14, DFW C.5* | | | | | | | |
| 21 | 21 | 22.10.1917 | 91 | Spad 13 | 12+h | Volnik | Aviatik |
| *Lt. Wolfgang Güttler KIA/Lt. Heinrich Hansberger KIA, DFW C.5* | | | | | | | |
| 22 | 22 | 25.10.1917 | 91 | Spad 7 | 11h | Bainsizza | Aviatik or Br.C.1? |
| *[with mag. Piccio, 91ª Sq.]* | | | | | | | |
| 23 | 23 | 26.10.1917 | 91 | Spad 7 | 11h15 | S. Lucia Tolmino | Aviatik |
| 24 | 24 | 26.10.1917 | 91 | Spad 7 | 12h | Matajur | EA |
| *[with ten. Parvis, 91ª Sq.] two out of the following German crews:*<br>*Vzfw Bruno Fuchs KIA/Lt. Hans Joachim Schlüter KIA, FA 17*<br>*Untoff. Eduard Fricke KIA/unknown observer*<br>*Vzfw. Kurt Berroth KIA/Lt. d. Res. Kurt Frommhold KIA, FA 39*<br>*Unknown pilot/Lt. d. Res. Gustav Krannig,*<br>*Unknown pilot/Lt. d. Res. Erich Lucas KIA, FA 19*<br>*Lt. d. Res. Erich Lucas/Lt. d. Res. Werner Schulze KIA, FA 39* | | | | | | | |
| 25 | 25 | 6.11.1917 | 91 | Spad 7 | 10h | Fossalta | Berg |
| *[with ten. Parvis, 91ª Sq.] Oblt. Rudolf Szepessy Sokoll KIA, Flik 41/J, Alb. D.III 153.54* | | | | | | | |
| 26 | 26 | 6.11.1917 | 91 | Spad 7 | 10h30 | Godega | Aviatik |
| *[with ten. Parvis, 91ª Sq.] Vzfw. Walter Schroeder KIA/Lt.d.Res. Albrecht Binder KIA, FA (A) 219, DFW C.5* | | | | | | | |
| 27 | 27 | 7.11.1917 | 91 | Spad 7 | 14h55 | Orsago | Aviatik |
| *[with ten. Parvis, 91ª Sq.] Gefr. Wilhelm Appelt KIA/Lt. Paul Wilkening KIA, FA (A) 204, DFW C.5 3955/17* | | | | | | | |
| 28 | 28 | 15.11.1917 | 91 | Spad 13 | 12h30 | Istrana | Aviatik |
| *[with ten. Parvis, 91° Sq.] Lt. Alfred Müller KIA/Lt. Erich Peucer KIA, FA 14, DFW C.5* | | | | | | | |
| 29 | 29 | 23.11.1917 | 91 | Spad 13 | 15h05 | Falzè | Fighter |
| *[with ten. Novelli, 91ª Sq.] Fw. Karl Überschär KIA, Jasta 39, Alb. D.V or D.III* | | | | | | | |
| 30 | 30 | 7.12.1917 | 91 | Spad 7 | - | Kaberlaba | Two-seater |
| *[with ten. Novelli, 91ª Sq.] Fw. Hermann Schütze KIA/Lt. Hermann Mehrfurt KIA, Flik 19, Br. C.1 29.20* | | | | | | | |
| 31 | 31 | 3.5.1918 | 91 | Spad | 11h | Salettuol | Two-seater |
| *[with serg. Nardini, 91ª Sq.] Zugsf. Josef Friedrich KIA/Oblt. Karl Rosenbaum KIA, Flik 19, Br. C.1 369.28* | | | | | | | |
| 32 | 32 | 22.5.1918 | 91 | Spad 7 | 9h | Cimadolmo | Fighter |
| *[with serg. D'Urso, 91ª Sq.] Fähnr. Ernst Pontalti KIA, Flik 51, Alb. D.III 153.155* | | | | | | | |
| 33 | 33 | 15.6.1918 | 91 | Spad 13 | 12h30 | Saletto | Two-seater |
| *Oblt. Albin Tomasic KIA/Lt. Arthur Fellner KIA, Flik 32/D, Br. C.1 369.116* | | | | | | | |
| 34 | 34 | 15.6.1918 | 91 | Spad 13 | 12h45 | San Biagio | Fighter |
| *[with serg. Aliperta, 91ª Sq.] Lt. Sigmund von Josipovich POW, Flik 51/J, Alb. D.III 153.266* | | | | | | | |

## Flavio Torello Baracchini

Flavio Torello Baracchini was probably Italy's most effective ace, and possibly one of aviation's most deadly fighters of all times, and yet he was an eagle with clipped wings, hardly remembered outside his hometown, and remarkable mainly because his name, in Italian, sounds like "little Baracca". And yet Baracchini enjoyed great popularity in the summer of 1918, when, after Baracca's death, he was Italy's leading ace, twice honored by a specific mention in the daily War Bulletin, on 24 June, when the Bulletin announced that he had scored his 29th victory, and two days later, when he scored his 31st. But when the final score of Italian fighter pilots was officially issued and distributed to the press by the Supreme Command, in February 1919, Baracchini's accomplishments had shrunk from 31 to 21 victories, placing him as Italy's fourth ace, instead of second, as he was considered during the war. While very many inaccuracies and omissions abound in that official document, none is so flagrant as Baracchini's, and his relatives had their own explanation, that *ten. col.* Piccio, then chief of fighter forces, condoned it, as it meant his rising to 3rd place among aces, with 24 victories. On the other hand, research seems to prove that probably there were no such intentions and the reduction of official victories for some pilots seems to be, at least in part, fully justified.

But not only officialdom clipped Baracchini's wings: also the fortunes of war did it. He was in action in the cockpit of a fighter for only two seasons, the spring of 1917 and the spring of 1918, and each time a serious wound put him out of action, depriving him of the chances of fighting against the enemy in the decisive battles of both years' autumns. One wonders what havoc could he have caused in the enemy ranks if he had been able to fight more than the six months in which he built up his fantastic score.

Flavio Torello Baracchini was born on 28 July 1895 in Villafranca Lunigiana, in the beautiful and history-rich Lunigiana, an area of wooded valleys overlooking the Thyrrenian Sea, where three regions, Tuscany, Emilia and Liguria meet. His family was wealthy, owning a sweets and pastry factory. Young Flavio Torello (torello in Italian means little bull) was an easygoing, friendly young man, short and apparently of small build, but actually possessing an exceptional physical strength. After completing technical studies, he volunteered in the Army, getting assigned to the 3rd Engineers Regiment, in December 1914 but he immediately volunteered for pilot training. He was sent to the flying school of Venaria Reale, near Turin, where he was trained on the Bleriot XI getting his brevet in 31 March 1915, and then to Mirafiori for training on the Voisin pusher.

In 1916 Baracchini, with the rank of *Aspirante* (officer cadet) was assigned to the front, reaching on 28 February the 7ª Squadriglia Ricognizione e Combattimento (7th reconnaissance and combat Squadron) a Voisin- equipped unit based at S. Maria la Longa, facing the Carso front. His first flight with that unit took place on 14 March, Baracchini flying with a Major Revelli for tests of a new machine gun: possibly the SIA-Revelli two-barreled light gun.

On 31 March, in a recce mission, his second combat flight, flying with the observer *ten.* de Bratti, Baracchini met enemy fighters, but the two Italians managed to drive them away, firing as many as five machine-gun magazines, and got back to base with their airplane shot up.

A few days later, on 9 April, Baracchini's engine stopped working, but the pilot managed to glide back for 18 km (12 miles) under constant firing from the enemy trenches, getting his Voisin and the passenger, observer *ten.* Pacinotti, safely on the ground in Lucinico, for which feat he received an official praise from the command of the I Gruppo. While the enemy fire finished wrecking his airplane, Baracchini shook his fist at the sky and cried to the Austrians: *"You'll pay for this!"*

Flying training with Bleriots at Venaria Reale, Turin. Baracchini is the first on the left, the second one is Alfredo Barbati, who landed in Switzerland and was interned, the third one is Antonio Fochessati, later commander of the 82ª Squadriglia and of the 5ª Sezione Difesa.

On 15 April 1916 the 7ª Squadriglia Voisin changed its name, becoming 26ª Squadriglia. Engaged on the most dangerous Isonzo front, the 26ª Squadriglia had frequent engagements with the enemy, that didn't cause any losses on account of the moderate aggressiveness of the Austro-Hungarians and the incredible strength and resilience of the Voisin. On 19 May, flying in a protection cruise with gunner soldier Icardi, Baracchini had another duel with an enemy aircraft.

On 29 June Baracchini flew with the observer *ten.* Massimo Vitale on a bombing mission to Duino and Jamiano. Twice the *bora*, the cold wind that blows over Trieste, pushed the slow Voisin back, on the third attempt a Fokker monoplane attacked them above Cormons. Baracchini was already showing the mettle of a fighter: on 14 July he spotted two seaplanes on the Gulf of Trieste and attacked them, driving them away with his fire. On 6 August, this time with *s.ten.*Caldera, he attacked another enemy aircraft over San Marco.

He had less luck on 22 September, when the enemy anti-aircraft fire hit his Voisin (the observer was *s.ten.* Lioy, future director of the Italian Air Force's historical office) and put it out of use. Baracchini managed to land on Aiello airfield. His final flight – the 58th – with the 26ª Squadriglia took place on 20 November 1916.

In December, Baracchini moved to the fighter school of Cascina Malpensa, near Milan, where he trained on the Nieuport 10 two-seater and the Nieuport 11 Bébé, both with 80 hp engines, remaining there until March 1917. In April 1917 Baracchini, now a fighter pilot and a *Sottotenente*, was assigned to the 81ª Squadriglia Nieuport, a new unit that was formed in Arcade in March 1917 and that on 20 April moved to its battle position on Borgnano airfield, near the contested city of Gorizia. This unit had as commander *cap.* Salvatore Calori and was equipped with a dozen Nieuport 11s.

Baracchini's eagerness was not matched by his skill, because on his second combat flight, 24 April 1917, he met two Austrian aircraft, attacked, found himself with a jammed gun and barely escaped with his fighter shot up. His only comment was, *"I think I've still got something to learn"*.

On 1 May the 81ª Squadriglia got its first victory, when *serg.* Sorrentino shot down an enemy Brandenburg, but Baracchini was not part of this combat, as he had crashed his Nieuport on take-off. He took off again, reached the combat and shot 50 rounds from his Lewis at other aircraft.

His rate of activity was phenomenal: in the first two weeks of May he flew no less than 27 times, had five combats, all inconclusive, and had to force-land with a stopped engine twice.

Finally, on 15 May 1917, Baracchini claimed his first victory, an enemy two-seater shot down in flames behind San Marco, that was not confirmed to him. This aircraft is possibly identified with a Brandenburg C.1 of Flik 23 that was damaged in combat.

Baracchini applied tactics that he had long studied, remembering that the Austrian fighters who used to shot at his Voisin from long range were ineffective: spotting the enemy, he first maneuvered to place himself in the sun, then he held his fire until he was less than 200 feet from the enemy. Those were the hectic days of the 10th Battle of the Isonzo, and in the next five days Baracchini flew 11 missions, fighting enemy aircraft four times. On 20 May Baracchini claimed another Brandenburg shot down in the area of Vodice, shared with *magg.* Pier Ruggero Piccio, of the 91ª Squadriglia. The enemy crew of Flik 12 had no chances to survive. Their Brandenburg crashed in flames near Madoni and both were killed.

His rise to acedom was meteoric: he scored victories again on 23 and 25 May, and on June 3, 6, 18, 19 and 22, for a total of nine

Baracchini sitting in the nacelle of a Canton-Unné powered Voisin of the 26ª Squadriglia.

Right: Baracchini posing for the photographer in his Voisin, serial V659. These aircraft came from the factory with the word Squadriglia painted on the nacelle, to which each unit added its own number in different styles.

After a success, Baracchini stands behind the rudder of his Nieuport 11 serial 2179, surrounded by visiting officers. His rather worn fighter shows a personal insignia in an unusual position, on top of the fuselage, and a little shield roughly painted below the serial number.

Baracchini's Nieuport 17 with black shield insignia.

victories in 36 days, which was quite an accomplishment. The victories of June 19 and 22 were scored by Baracchini flying a 110 hp Nieuport together with *ten.* Gastone Novelli flying an 80 hp Nieuport. Their victims on 19 June were a very good enemy crew: the pilot was *Zugsf.* Stefan Fejes, the observer was *Lt.* Josef Purer both listed as aces, who optimistically claimed a Nieuport shot down in the combat. Fejes force landed his aircraft after being hit in the engine. The airplane he shot down on 25 May, Brandenburg C.1 129.38 of Fluggeschwader 1, was repaired and put back in service, and then was shot down again, with fatal results, on 24 July 1918, a victim of Giannino Ancillotto.

The press began to pay attention to Baracchini, but he was extremely unrewarding to interviewers, as he came out as a shy and reserved young man, of few words and given to understatements. Described by his fellow pilots as good hearted, reserved and *"with the soul of a teen-ager"* who met visitors dressed in a pajama before his daily evening "free hunt missions" he had in himself, however, the ruthlessness and daring of the natural ace, fuelled by adrenaline, as turns out from his few recorded words: "How do I shoot down airplanes? As the other ones do. In those moments I get excellent eyesight, I feel as if I am becoming almost reckless. It is stronger than me."

This is how he described his 20 minutes long combat of 18 June, for which he received confirmation for a victory: the enemy aircraft was possibly Brandenburg C.1 129.14 of Flik 32, that force landed in its own airfield, Aisovizza, with the observer *Leut.* Demeter Birth badly wounded while pilot *Korporal* Alois Loos escaped damage. "I got them with a ruse. At 4,000 meters, when I saw the enemy right below me, shooting, I spun my fighter and got down to 1000 meters. The other guy, still shooting, thought that I was hit and got down also. Then I jumped at it and shot it down." For his quick maneuvering and for his penchant for very close attacks, the press nicknamed him "the d'Artagnan of the air". It was in this period probably that he painted on the fuselage of his Nieuport the black shield of the French swordsman.

On 15 July 1917 Baracchini was shifted from the 81ª Squadriglia to the 76ª, also located on Borgnano airfield, and part of the VI Group together with the 78ª, 81ª, 83ª and four reconnaissance squadrons, supporting the central 2nd Army in the region of Gorizia, ready to launch the 11th Battle of the Isonzo. The next day, 16 July 1917, Baracchini had first one combat flying a Nieuport, and he drove off an enemy airplane. Later on the same day, he flew his first combat mission on the Spad 7, a couple of which were assigned to the 76ª next to the Nieuports.

On July 17 he got his first victory on the Spad, shooting down the Br. C.1 129.57 of Flik 12. The Austrians had dueled for over 15 minutes, scoring hits on his aircraft, and they also used Baracchini's own trick, of spinning in order to pretend to be out of control, but the Italian ace had not fallen for it. Anyway they had impressed him for their cold courage and ability, so much that he wished he could shoot down the airplane without wounding them. Of the two crewman the observer, *Lt. i.d.R.* Ernst von Nagy was killed during the fight, while the pilot, *Zugsf.* Kanler, badly wounded, was forced to land his aircraft behind the Austrian lines wrecking it in the landing.

His next victory arrived on the 29th, a Brandenburg forced down near Kneza: this is his description of the event:

> "As soon as I got in the Tolmino area, I saw one enemy aircraft. I saw one that was trying to attack one of our planes, already engaged by a black cross plane. I rushed to it firing freely but all of a sudden my gun jammed. I had to get down in a spin and started to fly again lower than my contender. Because of my attempts and the effort to pull the breech you can see my fingers, that are bruised and still marked with blood. I cleared the machine gun. I tested it successfully. I climbed back

Baracchini in front of his Nieuport Ni.17 bis showing his personal insignia, a Black shield. No detail is known on the supply of the Nieuport Ni.17bis to Italy.

> up aiming from below the enemy aircraft that tried to escape. Quickly I was below it; I pulled the nose up at the cost of stalling, and rat-a-tat I placed a whole magazine of rounds into it. The airplane crashed down to the right of Tolmino. It went well."

The mention of "a magazine of rounds" suggests that he was flying a Nieuport 17 equipped with a Vickers and a Lewis on the top plane instead of the Spad mentioned in his flight logbook. In this period his documents mention some flights with a "140 hp Nieuport" and there is photographic evidence of this.

On 3 August Baracchini had a most remarkable victory, shared with the ace of aces, Francesco Baracca: the two of them pursued an Austrian airplane from Caporetto deep in enemy territory, in the valley of the Sava river, and shot it down by the Wocheiner stream. Soon after, Baracchini flew back over the scene of the combat and brought back photographs of the downed enemy.

The events of this combat are not clear, nor does Baracca's combat report clarify them:

> "Alerted by the signals on Mount Korada I flew toward Tolmino where a few minutes later I saw an airplane coming from the Korada and closely pursued by another one. Seen from the front I confused it with a Spad and it passed me 200 m. above, I was at about 3800 meters. The Albatros, as I recognized it too late, crossed its lines and kept flying East, while the plane that followed it was a Spad marked "76" and attacked it first. The enemy defended itself quite well with quick banks and with its fire: I attacked it twice but I couldn't aim and fire. There was a very strong wind that moved the combat away from our lines toward the Lake of Wochein, I got close to the enemy at 100 meters and found the right moment to fire about 20 rounds that I judged well aimed: I believe that the Albatros was repeatedly hit from behind. Also the other Spad attacked many times. The combat was down to 3000 meters. As my engine was losing power I turned back to our lines, that were 15 km away, but then the engine came back so I turned once more to the Austrian that was now far away and was going down following the valley of the Sava of Wochein river. I saw again the other Spad. I lost sight of the aircraft near the village of Wocheiner Freistritz. This combat took place around 9.50 a.m. The last attack was east of Cima Novi. As I was coming back I was targeted by strong artillery fire."

On 7 August he flew four missions, for a total of 6 and a half hours in the air, his usual rate, in the last one of which he attacked three enemies, drove them away and then shot down something described as "*a small round enemy balloon*".

The next day, however, his luck ran out: he had a long duel with an enemy, a highly combative and well manned Brandenburg, forced it down in the Tolmein region but was wounded. He landed back in his base, got out of the fighter but blood was seeping from the scarf around his face: an enemy bullet had shattered his jaw.

A proud Baracchini standing in front of his Nieuport 17bis. This aircraft carries a double armament, a synchronized Vickers gun over the cowling and a Lewis gun over the top wing.

This according to the official version: there are photos, however, showing Italian soldiers getting a wounded pilot out of a Nieuport crashed in a wood and then carrying him on a stretcher, who could be Baracchini. The wound turned out to be serious; Baracchini was hospitalized in Udine and was visited by Gen. Maggiorotti, chief of the Italian Air Service. On 15 August, the king of Italy, Victor Emmanuel III visited him, together with his guest, the president of France Poincarè, and the two eminent visitors awarded him on the spot the Medaglia d'Argento and France's Croix de Guerre with Palms. But, meanwhile, he had also been assigned Italy's highest award, the Medaglia d'Oro, (the fifth one awarded to aviators up to that time) after his wound of 8 August with this motivation: "Very able and daring fighter pilot, with serene disregard of danger and dauntless courage, in 39 days of service at the front, he faced brilliantly and victoriously 35 air combats, shooting down nine enemy aircraft – Sky of the lower and middle Isonzo, 15 May, 22 June 1917."

The wounded hero spent a long time recuperating, but the cares he received couldn't hide that his chin was disfigured. Adding insult to injury, as he was traveling to the Pontremoli hospital (near his home) for further treatment, passing through Bologna, somebody stole his kit, containing all his war memories and decorations. In November he was back home in Lunigiana, where he participated in patriotic events, that he much disliked: "My good people here wanted to give me big honors, but the racket of their celebrations has just annoyed me. I did appreciate the generosity of my townspeople, but I asked them to give that amount to the refugees. This forced idleness, this boring quietness are wearing me down and disheartening me."

During his leave, while he traveled every day to a dental surgeon for the reconstruction of his jaw, he wore a perfect black uniform with shiny booths and a black cape. Once in the train two officers from the front in combat uniforms started making remarks about the dandy shirkers in fancy uniforms: Baracchini just stood up dropping his mantle, showed his scars and his Gold Medal ribbon, and introduced himself: the two officers apologized and embraced him.

He was not deemed able to return to flying until March 1918, when he returned to the 76ª Squadriglia, that meanwhile had become a crack top scoring unit, and that was located in Casoni, near Bassano, in the crucial area of the Grappa mountain, the bastion of Italian defense. In the meantime there had been many changes at the front, many new pilots had made themselves aces and one of the best was in his own Squadriglia, *s.ten.* Silvio Scaroni who at the time had reached 17 victory claims.

Immediately Baracchini was back in action with his new mount, the Hanriot Hd.1, on which he painted the four aces as a personal marking: his intention of becoming the ace of aces of Italy could not be more evident. On 3 April 1918 he scored his first victory for 1918, a kite balloon shot down together with Scaroni and Giorgio Michetti, also of the 76ª Squadriglia (a victory not acknowledged in the 1919 victories list). However this was the only success obtained together with Scaroni, their characters being too much different: they admired each other, but used different tactics to score their successes.

One of the victims of Baracchini: Brandenburg C.I serial 69.09 of FlugGeschwader I, forced down as it was trying to intercept Caproni bombers on 23 May 1917. *(Photo Masajdek).*

Baracchini flew many free-hunting missions that failed to bring him more victories, only on 1 May he had a combat, but his gun jammed. Then on 2 May he tackled with a squadron of eight enemy fighters and shot one down by Cimadolmo. His victim however was surely an Aviatik C.1 two seater of Flik 50 that for its small dimensions could easily be mistaken for a fighter. It crashed near Tezze and the two crewmen met their death.

The next day another fighter fell under his guns near Sernaglia, a victory shared again with Michetti, but confirmed only to Baracchini, who received a prize of 500 Liras. On 11 May he fired at a kite balloon, the observer took to his parachute and the enemy

The 76ª Squadriglia received in mid 1917 a few Spads, but most of these fighters went to the 91ª and then to the 71ª. Baracchini flew this aircraft, and Baracca noted down the number 76 on its fuselage in the report on the combat of 3 August 1917.

Left: Very thin and emaciated, Baracchini recuperating from his bad chin wound at the seaside in Tuscany with two friends. His uniform, including sabre, is remarkably sophisticated, down to riding breeches and boots. The badges on his sleeves represent commission to officer for war merit. Right: Back to action after his long period of healing, Baracchini returned to the 76ª Squadriglia but he stayed there briefly. Here he is standing in front of Hanriots of that unit.

balloon was pulled down, damaged. On 13 May he attacked on the Montello a patrol of three Albastros D.III of Flik 68/J that were coming back from an escort flight. The consequences of the fight were devastating for the enemies. Their leader *Offstv.* Emanuel Stumpa, a very experienced pilot credited with two victories, fell in flames behind the Italian lines after a fierce fight, another pilot was able to cross the Piave river and to force land his aircraft behind his own lines while only the third one reached by luck his own airfield. From the wrecked fighter Baracchini took some souvenirs: a wheel, a magneto and a piece of propeller, for which action he was denounced by the commander of the Aquila Brigade.

Five days later, he shot down another kite balloon, shared with Di Loreto of the 76ª. The balloon was only deflated and it was not acknowledged in 1919. On 20 May one more fighter, likely to be Albatros 153.163 of Flik 42, was shot, victim of Baracchini, Di Loreto and Baggini, the enemy pilot, *Korp.* Sandor Szjarto, survived. However Austrian data are not clear, one source reporting 153.163's loss as due to anti-aircraft fire, but another source reports also of a combat in which the pilots of Flik 42/J claimed a "Sopwith" shot down. The place where Szijarto force landed however corresponds perfectly with Baracchini's claim.

On 26 May 1918 Baracchini was involved in a big combat in which six Albatros D.III of Flik 42/J in an escort mission suffered the attack behind Italian lines of several Italian and British fighters, all mounted by experienced pilots. There were three Spads of the 91ª Squadriglia, two Hanriots of the 78ª, and three Camels of No. 28 Squadron other than Baracchini flying in that area. In the big combat the enemy formation was put to rout and the Spads of 91ª Squadriglia shot down an Albatros and a two-seater behind Italian lines, Mecozzi and Capparucci of the 78ª claimed a fighter and Baracchini claimed another one behind enemy lines. The British pilots also claimed three enemy aircraft shot down, in effect the same ones of the Italians. It is not easy to discern who shot down whom, but Baracchini's victim was possibly Austrian ace *Zugsf.* Udvardy, forced to land behind his own lines.

This was the last victory for Baracchini at the 76ª Squadriglia: on 31 May 1918 he was assigned again to the 81° Squadriglia, still the twin unit of the 76ª in the VI Gruppo, that had mixed equipment of Hanriots and Nieuport 27s. It is reported that he did not go along well with Silvio Scaroni, the rising ace of the 76ª Squadriglia, who considered him conceited and selfish, and this could be the reason for his passage to the twin squadron. Here he changed again the badge on his aircraft resuming his old black shield that so much successes had given him one year earlier.

Another victory came on 2 June, when at 5:45 a.m. he shot down a two-seater at Susegana after a long duel above the Montello, so in exactly one month, Baracchini had claimed again nine victories, bettering his record on one year before, when he scored nine victories in 37 days.

On 15 June, as the Austrians launched their final bloody offensive, trying to breach the Italian lines all along the Piave, combats reached their peak. Baracchini claimed three victories on the 15th. On 18 June he took part in a big attack by Italian two-seaters and fighters on the Montello, the little hill overlooking the Piave River that had become the keystone to the battle, and where Baracca was killed the next day.

The day after he applied to a fighter some rockets, or "Turpilles" as he called them in his logbook. He tried to shoot down the kite balloon at Cimadolmo, but three out of four rockets didn't ignite. Again on the 20th he tried to shoot down another balloon, this time at Susegana, but he failed, and returned with several hits in his petrol tank and oil reservoir.

A brand new French built Hanriot, serial 515, for Baracchini, who emblazoned it with the four Aces. His surviving Hanriot is now preserved at the Italian Air Force Museum in Vigna di Valle, painted in these colors.

When Baracchini moved to the 81ª Squadriglia, he left his Hanriot that was passed on to *ten.* Ludovico Censi, who here appears sitting in the cockpit. The pilot standing in front is Renato Donati, later holder of the absolute height record. Behind the tail stands the tall *cap.* Giulio Lega.

A line-up of the Hanriots of the 81ª Squadriglia at Casoni in June 1918, as a Caproni lands in the background. The first fighter is the one of Baracchini, showing his black shield, the second one, painted in white and red stripes, is the mount of *cap.* Zoboli, the squadron commander.

*Sergente* Palpacelli flew this highly decorated Nieuport 27 of the 81ª Squadriglia, serial 11339. Later this airplane was fitted with Le Prieur rockets and it may have been used by Baracchini.

His wish to shoot down a kite balloon came true on the 21 June, when he destroyed the "sausage" floating above S. Pietro. The enemy observer *Oblt.* Tomenko of BK 4 saved his life with a parachute. Austrian sources described the attacking aircraft as *"a French fighter equipped with rockets"*.

On 22 June while flying over the Montello he met the Nieuport 27 of another ace, *serg.* Reali of the 79ª Squadriglia and together with him attacked a big enemy patrol claiming one shot down in flames and two out of control. The enemies once more were probably the fighters of Flik 42/J that had a pilot forced to land on the S. Vendemmiano airfield and suffering nervous breakdown. One of his companions claimed a Nieuport shot down, although both Italians reached their own airfields undamaged. Both Baracchini and Reali received no confirmation for these three claims.

All this frenetic activity of fighting aircraft and balloons didn't prevent him from participating in the daily raids in which his squadron was involved to strafe the now retreating Austro-Hungarian troops, sometimes making even six flights in a day. One gets the impression, however, that Baracchini in those days was going beyond the limits to increase his scoreboard. Baracca had died on 19 June, so Baracchini had a good chance to become the top Italian ace. Records for this period are not foolproof, and it is not easy to match precisely Italian claims with the relevant Austrian losses.

On 25 June Baracchini and *sergente* Corti attacked six enemy fighters between Oderzo and Motta di Livenza at sunset. While the sergeant had his gun jammed, Baracchini pursued an enemy down to low level and shot it down, but as he flew back above the trenches he was wounded by ground fire, a bullet passing right through his abdomen. He managed to return to base, and was rushed to a field ambulance. Luckily, the bullet had burned its way through the intestines, and no infection developed. But the young airman was out of the war; another slow and painful recovery was expecting him.

Reporting his fall, the official communiqué of the Italian Supreme Command credited him with 31 victories (thereby indirectly confirming the claims on June 22nd and a final claim on June 25th).

In August he was promoted to lieutenant in permanent service. For a certain time, he was hailed as Italy's leading ace, after the death of Baracca, but fame was no consolation for him, unable to

Fighters of the 81ª Squadriglia. The first one is the Hanriot of Baracchini, with two guns and a black shield. The third one is the Nieuport 27 of Palpacelli, with Le Prieur rockets.

Above: Like all aces, Baracchini received the gift of a surplus fighter after the end of the war. He kept it near his home in Villafranca and after his death his family gave it to the Museo del Risorgimento in Florence, where it was pictured in the late Thirties. When the Museum was dismissed, the Hanriot luckily returned to the Italian Air Force that preserved it. Right: The large and aesthetically debatable monument to Baracchini in his hometown, beautiful Villafranca Lunigiana.

participate in the final victory. He was awarded the title of Knight of the Order of Saints Maurizio and Lazzaro, the youngest recipient in history.

But surely he must have felt very bitter for the unexpected and unexplainable degradation from 31 to 21 victories, despite the fact that his victories were officially announced in the bulletins. The explanation of this event exists: the 5ª Sezione of the Comando Generale d'Aeronautica, charged with establishing the official list of aces, assigned to Baracchini victories on 20 May 1917, 23 May 1917, 25 May 1917, 3 June 1917, 6 June 1917, 18 June 1917, 19 June 1917, 22 June 1917, 17 July 1917, 29 July 1917, and 3 May 1918, 13 May 1918, 20 May 1918, 25 May 1918, 15 June 1918 (two), 21 June 1918, 25 June 1918, for a total of 18, disclaimed the victories of 3 May 1918 (at Sernaglia) and 13 May 1918 (across the Piave), 15 June 1918 (on the Montello) and 22 June 1918 (two at Falzè) and ignored the claims of 15 May 1917, 3 and 8 August 1917, 3 April 1918, 2 May 1918, 11 May 1918, 18 May 1918, 26 May 1918, 2 June 1918, 22 June 1918 at Mansue, 25 June 1918. The official list was surely prepared in good faith, but too quickly. Without analyzing Austrian records, the 5ª Sezione sent out its conclusions to all pilots (all who could be reached, that is) in mid January 1919, allowed a few days to the pilots to write in their comments and additions, and then hurriedly in February 1919 issued its final decisions, that have not been challenged ever since.

The wreck of one of Baracchini's victims, surely the Albatros D.III 153.189 of *Offstv.* Stumpa.

Despite his feats, this ace received only one other Italian decoration, besides the Gold Medal: the original Medaglia d'Argento issued after his face wound on 8 August 1917. It was really embarassing to give more awards to a man who already sported the Gold. After the war, the British gave him a Distinguished Service Cross.

In 1919 Baracchini was part of the delegation of surviving Gold Medal recipients who carried the casket of the Unknown Soldier in the National Shrine in Rome, but photographs show an ailing man, with companions ready to take his place, in case he failed under the weight.

After the war, Baracchini, like all top aces, received as a gift from the government a fighter, a Hanriot that was first based at Luni airport, near his hometown, which he used to buzz his town. In the autumn 1919 its magnet was seized by military authorities, lest Baracchini defected to Fiume, to join the volunteers of d'Annunzio who occupied that contested city in defiance of the Italian government. The Hanriot then moved to the Museum of the Risorgimento in Florence. Later recovered by the Air Force, it is now exhibited in the Museo Storico dell'Aeronautica Militare in Vigna di Valle, near Rome, painted in a spurious finish and with the insignia of the four aces briefly used by Baracchini.

Returning to civil life, Baracchini organized the expedition of twenty surplus aircraft to Turkey, then at war with Greece, and smuggled them on a small ship, challenging the Greek blockade, delivering them to Kemal Ataturk. Back in Italy he married Maria Novarese in Rome on 20 July 1921. He remained close to the air force, and he created a company, Stabilimento Pirotecnico in Rome, for the production of signal rockets of his own design, that were dropped from an aircraft and fired by a small parachute. On 29 July 1928 a fire broke out in his factory on the Via Aurelia. It was a Sunday, but Baracchini was there with two workers, Luigi Simonetti and Walter Simoncelli testing a new explosive mixture. Suddenly there was an explosion and flames engulfed the men. Baracchini was badly wounded in the head. He was carried first to a hospital, then to the private clinic of Professor Bastianelli, one of the most famous surgeons and a pilot himself. He remained in agony for one month, in the Clinica Bastianelli, in Rome, and died on 18 August 1928, a victim of his valor. His body was brought to the Basilica of Santa Maria degli Angeli, in Rome, the church reserved for state funerals, on a gun carriage. He was buried in the Verano Cemetery in Rome.

# Flavio Torello Baracchini
## 31 victory claims - 21 confirmed victories

| Claim | Vict. # | Date | Sqn. | Aircraft | Hour | Place | Victim |
|---|---|---|---|---|---|---|---|
| 1 | - | 15.5.1917 | 81 | Ni 11 | - | Aisovizza | EA |
| *Unknown crew UNH, Flik 23, Br.C..1 129.46* | | | | | | | |
| 2 | 1 | 20.5.1917 | 81 | Ni 11 | 9h30 | Vodice-E. di Aisovizza | EA |
| *[with magg. Piccio, 91ª Sq.] Zugsf. Ludwig Ferschl KIA/Lt. i. d. R.. Eugen Csutka KIA, Flik 12, Br. C.1 229.10* | | | | | | | |
| 3 | 2 | 23.5.1917 | 81 | Ni | - | S. Marco-Vertoiba | EA |
| *Korp. Johann Trnka UNH/Lt. i. d. R. Wenzel Vrtovec WIA, FlG. 1, Br. C.1 69.09 FTL* | | | | | | | |
| 4 | 3 | 25.5.1917 | 81 | Ni | - | Aisovizza | EA |
| *Korp. Otto Spenn UNH/Lt. i.d.R. Fritz Raab DOW, FlG. 1, Br. C.1 129.38.* | | | | | | | |
| 5 | 4 | 3.6.1917 | 81 | Ni | - | Sober | EA |
| *Maybe: Zgsf. Anton Hoffmann KIA/Oblt. Friedrich Wowy KIA, Flik 4, Br. C.1129.41 or: unknown crew, Flik 4, Br. C.1 129.02 FTL* | | | | | | | |
| 6 | 5 | 6.6.1917 | 81 | Ni | 9h30 | Vodice | EA |
| *[with ten. Olivari, 91ª Sq.] Korp. Alexander Veszprémy KIA/Lt. i. d. R. Ernst Pirnos KIA, FlG. 1, Br. C.1 229.19* | | | | | | | |
| 7 | 6 | 18.6.1917 | 81 | Ni | - | Aisovizza | EA |
| *Korp. Alois Loos UNH/Lt. i. d. R. Demeter Birt WIA , Flik 32, Br. C.1 129.14 FTL* | | | | | | | |
| 8 | 7 | 19.6.1917 | 81 | Ni | 8h45 | Aisovizza | EA |
| *[with ten. Novelli, 81ª Sq.] Zugsf. Stefan Fejes UNH/Lt. i. d. R. Josef Pürer UNH, Flik 19, Br. C.1 29.63 FTL* | | | | | | | |
| 9 | 8 | 22.6.1917 | 81 | Ni | - | S. Marco | EA |
| *[with ten. Novelli, 81ª Sq. & serg. Piaggio, 77ª Sq.] Korp. Stephan Hegedüs DOW/Oblt. Johann Miklosi KIA, Flik 35, Br. C.1 229.05* | | | | | | | |
| 10 | 9 | 17.7.1917 | 76 | Spad 7 | - | Lom-Tolmino | EA |
| *Zugsf. Karl Kanler WIA/Lt. i. d. R. Ernst Von Nagy KIA, Flik 12, Br. C.1 129.57* | | | | | | | |
| 11 | 10 | 29.7.1917 | 76 | Spad 7 | - | Tolmino-Val | Ladio |
| *Fp. Oblt. Fritz Losert UNH/August Raft Edler von Marwil WIA, Flik 2, Br. C.1 FTL* | | | | | | | |
| 12 | 11 | 3.8.1917 | 76 | Spad 7 | 9h50 | Woch. Feistritz | Albatros |
| *[with cap. Francesco Baracca, 91ª Sq.]* | | | | | | | |
| - | - | 7.8.1917 | 76 | Spad 7 | - | Tolmino? | Round balloon |
| 13 | 12 | 8.8.1917 | 76 | Spad 7 ? | - | Tolmino S Lucia | Albatros |
| 14 | - | 3.4.1918 | 76 | Hd | - | Premaor | Kite balloon |
| *[with ten. Scaroni & Nannini, 76ª Sq.] Kite balloon, BK.17* | | | | | | | |
| 15 | 13 | 2.5.1918 | 76 | Hd | - | Cimadolmo | EA |
| *Maybe: Korp. Leopold Höllriegel KIA/Lt. i. d. R. Franz Steininger KIA, Flik 50/D, Av. C.1 137.06* | | | | | | | |
| 16 | 14? | 3.5.1918 | 76 | Hd | - | Sernaglia | Fighter |
| *[with ten. Michetti, 76ª Sq., confirmed originally only to Baracchini]* | | | | | | | |
| 17 | - | 11.5.1918 | 76 | Hd | - | Canove | Kite balloon |
| 18 | 15 | 13.5.1918 | 76 | Hd | - | Arcade | Fighter |
| *Offstv. Emanuel Stumpa, KIA, Flik 68/J, Alb. D.III 153.189* | | | | | | | |

| Claim | Vict. # | Date | Sqn. | Aircraft | Hour | Place | Victim |
|---|---|---|---|---|---|---|---|
| 19 | - | 13.5.1918 | 76 | Hd | - | across Piave | Fighter |
| *Fw. Jacob Vesely,UNH, Flik 68/J, Alb. D.III 153.196 FTL* | | | | | | | |
| 20 | - | 18.5.1918 | 76 | Hd 6272 | - | Collabrigo | Kite balloon |
| *[with asp. Di Loreto, 76ª Sq] Confirmed by 18ª Sez. Aerostieri. Victory later disclaimed* | | | | | | | |
| 21 | 16 | 20.5.1918 | 76 | Hd | - | NW Sernaglia | Fighter |
| *[with asp. Di Loreto and cpr. Baggini, 76ª Sq.] Maybe: Korp. Alexander Szijarto, Flik 42/J, Alb. D.III 153.163 FTL* | | | | | | | |
| 22 | 17 | 26.5.1918 | 76 | Hd | - | Cornadellu | Fighter |
| *Maybe: Zugsf. Ferdinand Udvardy UNH, Flik 42/J, Alb.D.III FTL* | | | | | | | |
| 23 | - | 2.6.1918 | 81 | Hd. | 5h45 | Susegana | Two-seater |
| 24 | 18 | 15.6.1918 | 81 | Hd | 8h | Near Moriago | Two-seater |
| *Maybe: Zugsf. Alexander Sandor MIA/Lt. i. d. R. Julius Kardhordo MIA, Flik 38/D, Br. C.I 369.113* | | | | | | | |
| 25 | 19 | 15.6.1918 | 81 | Hd | 9h15 | Piave/Montello | EA |
| *[with cap. Mazzucco, 81ª Sq.]* | | | | | | | |
| 26 | - | 15.6.1918 | 81 | Hd | - | Piave | Fighter? |
| *Possible victims:* | | | | | | | |
| *Fw. Rudolf Berger KIA/Oblt. Ernst Petric KIA, Flik 26/DS, Br. C.I 329.48,* | | | | | | | |
| *Stfw. Franz Koudela/Oblt. Anton Chrzascz WIA, Flik 26/DS, Br. C.I 329.22* | | | | | | | |
| *Unknown pilot/Lt. i. d. R. Johann Reinstaller WIA, Flik 52/D, Br. C.I* | | | | | | | |
| *Unknown pilot/Lt. Ernst Winkler WIA, Flik 52/D, Br. C.I* | | | | | | | |
| 27 | 20 | 21.6.1918 | 81 | Ni.27? | 11h30 | S. Pietro/Col S. Martino | Kite balloon |
| *Oblt. i. d. R. Theodor Tomenko UNH, Draken BK 4* | | | | | | | |
| 28 | - | 22.6.1918 | 81 | Hd 11383 | 8/9h35 | Montello | Fighter |
| *[with serg. Reali, 79ª Sq.] Maybe: Fw. Hubert Graf ,UNH, Flik 42/J, Alb. D.III , S. Vendemmiano* | | | | | | | |
| 29 | - | 22.6.1918 | 81 | Hd 11383 | 8/9h35 | Montello | Fighter |
| *[with serg. Reali, 79ª Sq.]* | | | | | | | |
| 30 | - | 22.6.1918 | 81 | Hd 11383 | 8/9h35 | Montello | Fighter |
| 31 | 21 | 25.6.1918 | 81 | Hd | 8h15 | Motta di Livenza | Fighter |
| *[with serg. Corti, 81ª Sq.] Maybe: Pfeiffer, KIA, Flik 43/J, Ph D.IIa 222.02* | | | | | | | |

# Sebastiano Bedendo

Sebastiano Bedendo, normally called Nino, was born on 18 July 1895 in Rovigo, the small and quiet capital of the Polesine, the region at the delta of the Po River. He came from a family of the middle class. He was drafted as a soldier on 1 June 1915, one week after Italy's entry in the war, and he was assigned to the 8th Field Artillery Regiment in Verona. On 4 July 1915 he was commissioned as a *sottotenente*, and assigned to the 5th Regiment of Fortress Artillery. Somehow he immediately got a connection with aviation, because in September he was posted as an observer to the 7th Aerostatic Fortress Section, operating captive observation balloons. The lure of flying must have gotten him, because on 6 April 1916 Nino Bedendo was assigned to the Battaglione Scuole Aviatori (Aviation training battalion) and he got his flying brevet in a remarkably short time: on 20 April he had his first flying lesson, and on 1 July 1916 he was licensed on a Caudron G.3, probably at the Mirafiori training field, near Turin. His second-degree brevet as military pilot followed, on Farmans on 1 September 1916 at the Busto Arsizio training field. On 26 August he was posted to Mirafiori, Turin, to train on the Caudron G.4 twin engine reconnaissance two-seater, and in a very short time, on 7 October 1916, Bedendo was assigned to the 48ª Squadriglia based at San Pietro al Campo, near Belluno.

This marked the beginning of a permanent connection of Bedendo with the Asiago front, the Alps and the plateau between Austrian-held Trento and Italian-held Vicenza that remained all through the war secondary compared to the main Eastern front, which was the one along the Isonzo river up to late 1917, along the Piave river afterward. The 48ª Squadriglia, the only one assigned to the 4ª Armata, was a remarkable unit, equipped with Caudron G.3 and Caudron G.4 reconnaissance airplanes, commanded by Felice Porro (who was the chief of the Italian Air Force in North Africa in World War II and the author of a semi-official history of the Italian air force in World War I). The 48ª Squadriglia, engaged in a very difficult mountain theater, for which it was equipped with the safe and powerful G.4, produced many outstanding pilots, like Natale Palli, later the pilot of d'Annunzio during his flight to Vienna, and Aldo Finzi, the Jew who became Mussolini's Secretary for Aviation in 1923 and who was shot by the Germans in 1944.

No combat flights with the 48ª Squadriglia are on record for Bedendo, who on 24 November 1916 returned to Mirafiori. In December 1916, he was assigned to the 42ª Squadriglia of the V Gruppo, another Caudron G.3 equipped unit, employed for artillery observation, and based in Gonars, in the southern part of the Isonzo front. His logbook reports only two combat missions in February 1917, in one of which enemy ground fire hit his airplane, with shrapnel damaging fuselage and wings. He fell ill and was hospitalized in Palmanova on March 13, and for a long time he did not fly.

No more flights are recorded for Bedendo until June 22, 1917 when he flew a photographic reconnaissance over Gradiscutta, followed by one mission dropping leaflets on 10 July, and three ground attack bombing missions on 19, 21 and 23 August 1917, during the

Sebastiano Bedendo in the nacelle of a Farman with Aldo Finzi, a man who had a complex and dramatic life.

11th Battle of the Isonzo, when all the Italian Air Force was employed in mass attacks against the Austrian positions for the bloody conquest of the Bainsizza plateau.

Probably his war flights were many more than the ones recorded in what official papers survived the passing of time and the Caporetto retreat.

In the lull that followed the battle, Bedendo moved to fighter training, and he was licensed as Nieuport pilot, at the Malpensa school, in late autumn 1917. On 1 November he was assigned to the 70ª Squadriglia Nieuport, and on 17 November he was moved to the 72ª Squadriglia. This unit commanded by *ten.* Giuseppe Rigoni, was created on 21 October 1917 and based in Castenedolo, in the most western sector of the front, for the protection of Brescia, a city important for its production of weapons, and that was uncomfortably close to the westernmost extension of the front line.

Bedendo spent his first days in the new squadron training on the Hanriot, and flying the fighter serial 6178. His first recorded fighter mission was an escort to a SP two-seater on 27 November that lasted 2 hours and 22 minutes. His combat missions were scarce in this period, due to bad weather on the Asiago front. Two escort missions to SAMLs on 26 and 27 December with Hanriot 6178 represent his final flights for 1917. The whole 72ª Squadriglia flew only 58 missions in 1917, without losses or claims. Austrian air activities were close to nil, but the memory of the air bombings of Brescia in 1916 still kept the Italians wary.

1918 began for Nino Bedendo with another SAML escort flight on 4 January, and then he was shifted to the *Sezione Hanriot* (Hanriot flight) of the 72ª Squadriglia assigned to the 71ª Squadriglia. As the 72ª didn't have that much to do, it was called to reinforce the 71ª, a Spad equipped unit then based in Sovizzo, near Vicenza, that was heavily engaged supporting the Italian troops that were fighting the Austrians over the Asiago plateau, the area where, according to the constant thought of the Austrian supreme commander Field Marshal Conrad, the main attack to crush Italy should be exerted. The main adversaries of the 71ª Squadriglia were the Luftfahrtruppe squadrons based around Trento, like the redoubtable Flik 24 reconnaissance unit and the fighter Flik 55/J of Pergine (that, however,

The 71ª Squadriglia had its share of personal markings in 1918, but surviving photos are very scarce. Here a bottle of wine christens the Spad 7 named Enrica of *ten.* Consigli. Later in the war that pilot flew a Spad 13 named Enrica III°.

lost much of its sting since January, when its ace Kiss was wounded and Arigi, another outstanding ace, was shifted), while in June Flik 3/J of commander Navratil arrived, based first at Gardolo and then at Romagnano.

Bedendo began his activities with an escort mission for Capronis on 12 January 1918, which ended for him with a forced landing near Bassano due to an engine failure. On his next flight, on 29 January, Bedendo had his first combat. Cruising over Val Gamarana and Val Brenta, he met two enemy airplanes, against which he fired several bursts, and saw them fly back to their own lines. He flew one more escort on the 30th, and a SAML escort on the 31st, that ended with another forced landing at Villaverla airport for failure of the magneto shaft of his Hanriot.

He flew five more uneventful missions with the 71ª Squadriglia in the first ten days of February 1918 flying French-built Hanriot 520 before returning to the 72ª, as the fighting on the ground had receded. Back in Castenedolo, Nino Bedendo flew five more long and uneventful cruises above the Alps north of Brescia, in the second half of February. On 2 March 1918 he was assigned to the 71ª Squadriglia, where he was entrusted with several missions of escorts to Pomilio two-seaters and cruises. In his first four missions, he flew Hanriot serial 6307, and then he passed on the Spad S.7. On the fifth such mission, on 19 March, escorting a photo-equipped Spad and a SVA on reconnaissance, while flying Spad S.7 serial 6344, Bedendo had a brief duel at 11.10 a.m. with an enemy Albatros D.III that he saw diving down toward his own lines, an enemy surely put to flight but not necessarily destroyed. Unknowingly, Bedendo has messed with a special occasion: Emperor Karl of Austria was on a visit to Pergine airfield, home of Flik 55/J, and the Austrian fighter was probably in the air to protect the august visitor from intruders.

From March to June 1918 Bedendo flew a total of 38 combat missions, most of them escorts to SAMLs and Capronis, using Spads 6344 and 6360. In one of these missions, on June 7, an interception sortie, flying together with a British Sopwith Camel, he met two enemy aircraft, dueled with them and forced them to retreat behind their own lines. His companion most likely was Lt. G. McIntyre, of Squadron No. 45, who was flying Camel B 2443 "K" and who claimed a DFW out of control over Folgaria.

The day was hectic on that section of the front: his squadron commander *ten.* Aldo Anesini claimed a possible victory, *serg.* Mario Pereno attacked an enemy Brandenburg, 369.52 of Flik 17 that returned to its base damaged, while the observer *Leut.* Gustav Zechenter was killed, and the Royal Air Force's Squadron No. 28 claimed three victories, one of them identified as Brandenburg 369.20 of Flik 8D, with pilot *Zugsführer* Adolf Gnamusch prisoner, and observer *Lt.* Rudolf Huss killed in action.

In the first week of July Bedendo flew seven cruise missions, in only one of which he met the enemy. Then he had a couple of weeks of furlough. He flew again on 24, 27 and 28 July and finally things got moving for him. On 29 July 1918 in the late morning he spotted in the area between Mount Spitz and Tonezza enemy airplanes that had attacked a Pomilio PE of the 136ª Squadriglia. He recognized a fighter and a two-seater. He shot at the latter, that dove and escaped, and then his gun jammed. He avoided the attacks of the Austrian fighter from below, disengaged his gun, shot from above at the fighter, which went into a straight dive in Val Terragnolo. This was his first confirmed victory. No correspondence with Austrian data is available.

The next day, 30 July 1918, Bedendo went again cruising over the Asiago plateau and the Lagarina valley, and he met four enemy fighters. He fired two bursts at one of them from a distance of 100 meters, and saw it falling down over Monte Cadria. He then discovered that his gun was jammed, and he escaped the attack of the surviving three enemy fighters spinning down out of harm's way. The only corresponding Austrian data for the day is an escort flight of *FP Zugsf.* Richard Bruckner, of Flik 54/D, who had a combat flying Phönix 228.35 but who returned safely to his base, Campo Maggiore.

The next they again, 31 July, Bedendo claimed his third victory in three days. Flying to escort a Pomilio over Rovereto, he spotted an enemy aircraft, approached it and fired seventy rounds from a distance of 200 meters. Then his gun jammed, so Bedendo broke off the dogfight, managed to repair his gun and attacked again, firing 200 more rounds at his enemy, until his gun jammed for a second time. His wingman, *capitano* Breglia, then entered the fight, shooting more bullets into the Austrian machine that was seen falling vertically north of Calliano. The same day later on, Bedendo flew one further long cruise in the same area.

In August 1918 Bedendo flew 20 combat missions with the 71ª Squadriglia. On 6 August, while flying in patrol with *sergente* Giampiero Vecco of the same unit, he spotted two enemy aircraft, a fighter and a two-seater, in the sky above Tonezza. They immediately engaged combat and after some burst they saw the two-seater diving away steeply toward Lastebasse, and the fighter falling vertically, leaving behind a long trail of yellow smoke. This victory was assigned to Nino Bedendo. On 9 August his plane was hit by a fragment from an enemy shell, and lost its gasoline, so that he had to land in Verona. On 10 August he again met the enemy. He was flying as an escort to a Pomilio SP.3 of the 31ª Squadriglia, crewed by *ten.* Agostino Gostoli and *ten.* Mario Rondi, in the area of Mount Pasubio and Val Terragnolo when he met five enemy fighters. He engaged combat and claimed one of them, marked by a black and white checkerboard on its fuselage. His own Spad was hit by the enemy and returned with holes in the top wing and struts. White/black checkers probably indicated an aircraft of Flik 3/J, but this Austrian fighter unit didn't suffer losses on this day.

*Sergente* Giampiero Vecco, here with a plain Nieuport 17 showing no serial number but with an unusual wing roundel, shared with Bedendo his victory of 6 August 1918.

Two days later, on 12 August, Bedendo had one further combat. He was flying above the Pasubio and the Lagarina valley together with *serg.* Augusto Levrero when they spotted two enemy aircraft. The Austrian airplane that he targeted was seen diving steeply toward Folgaria, but Bedendo didn't claim it as shot down.

On 22 August Bedendo had one more combat, again during an escort to a Pomilio two-seater. This time he met two enemy fighters, had a duel with them, shooting 150 rounds, and saw one of them fall toward Noriglio trailing a long plume of smoke. According to squadron papers, but not to his logbook, in this duel *cap.* Breglia, flying another Spad, accompanied him. Again a possible opponent could have been *FP Zugsf.* Richard Bruckner, who reported a duel while flying Phönix 228.55, and who again returned safely to his base.

In September the main activity of Bedendo was photographic reconnaissance, using a camera equipped Spad. Although his logbook has no more entries after 7 September, he is known to have

A Spad of the 71ª Squadriglia in 1918 with individual number 7 and personal marking "Ocio!!!" (Watch out) at Sovizzo aifield. In the background there is a SVA of the 1ª Sezione.

17° Sezione Aerostatica (balloon flight) confirmed that two Austrian aircraft were shot down, one at Sernaglia and one at Pieve di Soligo. In that area an enemy pilot of Flik 74/J force landed his Aviatik D.1 on Pianzano airfield after suffering "wing failure", but at present it is impossible to know if this event can be connected with that combat.

The young pilot had earned himself a good reputation, as proven by the fact that he was often the leader of patrols, even including officer pilots. The last victory of the three friends arrived in the final days of the war, 28 October 1918. With their presence alone, despite jammed guns, they forced an enemy airplane to land at the airfield of Arcade, with the pilot giving himself up. The joy for this capture is evident in the excited combat report written on arrival:

> "The patrol as soon as it spotted the enemy aircraft started going after it, reached it and surrounded it, and although it did barrel rolls – spins – dives – and although it flew into a cloud we didn't lose contact and it considered itself lost and glided in the direction of Arcade – I followed it down to the ground and I saw it landing. The other two pilots did the same."

For his wartime service Bocchese earned only one Medaglia di Bronzo, with a citation referring to one victory on 17 April 1918 and to his strafing flights during the battle of mid June.

Bocchese and his personal mount, Hanriot 19209 with individual number 17. The 70ª Squadriglia had both 13 and 17, a peculiar fact, because squadrons usually avoided these unlucky numbers.

On 4 January 1919 Bocchese was posted to the 91ª Squadriglia Baracca, now based at Zaule airfield, near Trieste, that was now an Italian city, and he remained there until September, when he was demobilized and returned to civilian life. In the mid Thirties he still belonged to the discharged reserve of the Italian Air Force, but in the *Ruolo Servizi Categoria Governo*, that is to say non-flying support personnel.

Bocchese, left, and his unknown *tenente pilota* friend show details of the checquered Hanriot, with individual number 1.

A nice portrait of Aldo Bocchese with signature and dedication to Cesare Magistrini. Behind him is a Spad 7 with striped tail and a white patch where the insignia of the 91ª Squadriglia is going to be painted.

Another view of the same aircraft, Spad S.2471 at Zaule, Trieste, right after the end of the war. The white background for the squadron insignia has already been applied.

He was occasionally recalled to service for brief periods, but always with bureaucratic tasks. One wonders if Bocchese, now facing paperwork on his desk, ever recalled when, twenty years before, he had given up paperwork to become a pilot. Now the air force was much different from what it used to be, the Fascist regime used it as a propaganda tool, his good friends were gone, Eleuteri had died in 1926, Avet two years later. Young pilots didn't pay much attention to the campaign ribbons of an old sergeant major who used to fly when they were not even born yet. Bocchese recalled the old times with his friend Leonida Schiona, a former fighter pilot with the 78ª and 84ª Squadriglia, with two confirmed victories, who had become an important commander with Ala Littoria, the national airline.

In May 1940, when Italy was about to enter the tragedy of the Second World War, Bocchese was dismissed from the air force for permanent inability to military service. He lived in Rome, then he moved to a cottage in Lastra a Signa, just outside Florence, until his death, on 15 March 1976. He was buried, as he had specifically required, wearing his war decorations.

Aldo Bocchese next to a Spad of the 91ª Squadriglia after the Armistice. Still no medal ribbon on his jacket.

Three old friends at a spa, drinking mineral water and talking about the old times: from the left, Leonida Schiona, Aldo Bocchese and Cesare Magistrini

## Aldo Bocchese
## 6 confirmed victories

| Claim | Vict. # | Date | Sqn. | Aircraft | Hour | Place | Victim |
|---|---|---|---|---|---|---|---|
| 1 | 1 | 17.4.1918 | 70 | Hd 6219 | - | Valdobbiadene | Two-seater |
| *[with ten. Avet, Eleuteri and Resch, 70ª Sq.] Korp. Viktor Stibor KIA/Lt. Ladislaus Rajcsics KIA, Flik 52/D, Br. C.I 169.35* | | | | | | | |
| 2 | 2 | 17.4.1918 | 70 | Hd 6219 | - | Valdobbiadene | Fighter |
| *[with ten. Avet, Eleuteri and Resch, 70ª Sq.] Maybe: Oblt. Michael Gassner-Nordon von Laudon KIA, Flik 42/J, Alb. D.III* | | | | | | | |
| 3 | 3 | 17.4.1918 | 70 | Hd 6219 | - | Valdobbiadene | Fighter |
| *[with ten. Avet, Eleuteri and Resch, 70ª Sq.] Maybe: Oblt. Richard Fitz UNH Flik 42/J, Alb. D.III. damaged* | | | | | | | |
| 4 | 4 | 15.7.1918 | 70 | Hd | 10h30 | Sernaglia/Conegliano | Fighter |
| *[with ten. Avet and Eleuteri, 70ª Sq.]* | | | | | | | |
| 5 | 5 | 15.7.1918 | 70 | Hd | 10h30 | Sernaglia/Conegliano | Fighter |
| *[with ten. Avet and Eleuteri, 70ª Sq.]* | | | | | | | |
| 6 | 6 | 28.10.1918 | 70 | Hd | - | Arcade | Two-seater |
| *[With ten. Avet & Eleuteri, 70ª Sq. and cpr. Ramberti, 76ª Sq.] Maybe: Phönix C.I 121.68, Flik 62/K MIA* | | | | | | | |

# Alessandro Buzio

Alessandro Buzio is a difficult personality to track: the not so many memory books of Italian World War I pilots, who flew side by side with him, tell us lots of details about other pilots, but ignore completely Buzio. Although he had one son, it is likely that he died before his father, because Buzio had no heir, and he could have died alone, were it not for the constant friendship of a younger friend, with whom however he never talked about his aviation experiences. No paper from him has surfaced, not one diary nor a letter nor a photograph. Even worse, his wartime service, that should be preserved in official documents, is lacking in documentation, because the *Diario Storico,* the squadron journal of the 76ª Squadriglia, in which he served a long time, is very much deficient.

Alessandro Buzio was born in Pavia, a beautiful city in Northern Italy, on 13 January 1893 from a wealthy family, that gave him the means to get a good education, and he got a diploma as an accountant. Like many young men at the start of the 20th Century, Alessandro was an enthusiastic sportsman, an excellent swimmer and bicycle racer, and he rode a motorcycle. This meant an obvious fondness for mechanics, and Buzio after his conscription was placed on leave, to be assigned to the course for Allievi Ufficiali di Complemento (reserve officer training) after which he was posted to the 5º Reggimento del Genio (5th Engineers regiment). He asked to be assigned to the Direzione Tecnica dell'Aviazione (Technical directorate of aviation) and on 9 June 1915, Italy having just entered the European War, he was admitted as a pupil in the Battaglione Scuole Aviatori. By the end of the year he had got his flying licenses, he served briefly as an instructor on the Macchi Parasol high-wing two-seater, and then on 24 April 1916 he was posted to the 75ª Squadriglia. This newly created unit, equipped with Nieuport fighters, set up its base at Tombette airfield, Verona. This squadron was part of the III Gruppo and its main task was the protection of Verona, close to the border and the base for many high commands and military warehouses.

On 16 May 1916 a cannon fired, as an air raid alarm, and Buzio climbed over the historical city together with *sergenti* Guido Nardini and Guido Consonni to confront the raiders. The Italian pilots attacked with such élan that the Austrian two-seater while running away dropped a camera, which was recovered in a field. The commander of the stronghold of Verona commended the three fighters.

The first victory was not late in coming. On 27 June Buzio, Consonni, Nardini and *ten.* Mario De Bernardi, who will become probably Italy's greatest test pilot, winner of the 1926 Schneider Trophy Race at NAS Hampton Roads, Virginia, went up for another interception mission. They attacked above Verona an airplane of Flik 21 engaged in a reconnaissance flight, Brandenburg C.1 26.11 with pilot *Zugsf.* Josef Holub and observer *Fähnrich* Friedrich Edler von Langer that landed at Arzignano. The two slightly wounded Austrian airmen were captured.

This victory raised a sensation, and honors were showered on the winners. A group of businessmen of Verona offered a considerable sum that the four airmen turned over to a relief fund for disabled veterans. Also the Fiat and Michelin companies gave money prizes and finally Buzio, by now holding the rank of tenente, together with his three companions, was awarded the Medaglia di Bronzo al Valor Militare on 19 November, pinned on his breast by General Bompiani, commander of the Military Division of Verona.

His later activities with the 75ª Squadriglia were unremarkable, until 27 April 1917, when Buzio was assigned to the new 81ª Squadriglia on the airfield of Borgnano, replacing the unlucky *sergente* Riccardo Ciotti, killed on 23 April in a flying accident.

In the new unit, based close to the most contested part of the front, Buzio flew many missions of protective cruise and Caproni escort, without particular events beside the fact that he twice over-

The winners of the combat of 27 June 1916 receive their decorations: from the left, Guido Nardini, Guido Consonni, Mario De Bernardi and Alessandro Buzio.

turned his fighter on landing, on 16 and 20 May 1917. He had some brief clashes against enemy aircraft in late June and July, without scoring victories. On 31 July he attacked a reconnaissance two-seater above Mount Sabotino. After his initial fire the enemy tried to escape in a dive. Buzio followed it, but a faster fighter joined in, the Spad of Francesco Baracca, who firing at pointblank hit the plane, Brandenburg C.1 69.93 of Flik 46 which crashed against Mount Podgora, killing its occupants, *Korporal* Julius Klausz and *Oberleutnant* Erich Bock.

In the morning of 7 September Buzio met a reconnaissance airplane flying with four fighters from Ternova. He attacked the two-seater, but the escort fighters drove him off. On 26 September he flew an escort mission to the Capronis bombing Grahovo and he witnessed the fall of Ca 4066 of the 14ª Squadriglia, hit by an enemy anti-aircraft shell, that crashed in flames near Tolmino with the death of the whole crew. On 1 October Buzio left the front, on furlough.

On 27 October after the enemy breakthrough the 81ª Squadriglia moved back from Borgnano to Aviano, but Buzio was forced down by bad weather near Spilimbergo. On 31 October, as the 81ª moved further back to Arcade, Buzio was shifted to its twin unit, the 76ª Squadriglia. On 3 November he flew his first mission with this squadron, an offensive patrol with Mario Fucini. One week later the 76ª moved to Istrana.

On 30 November Buzio flew together with *serg.* Luigi Fanti and *cpr.* Paolo Avenati an escort mission for a SP.3 of the 22ª Squadriglia. Two enemy fighters attacked him, but his two wingmen got him out of troubles. The Squadron Diary reports his mission of 19 December with Avenati and *asp.* Mario Donati: "They found an enemy reconnaissance airplane coming from the Asiago plateau. They attacked it and shot it down at the start of the Val Seren. The airplane of ten. Buzio was hit in its top right wing."

A smiling Mario De Bernardi driving a forlorn-looking Austrian airman captured after being shot down by him and by Alessandro Buzio on 27 June 1916.

On 26 December there is the so-called Battle of Istrana, and Buzio participates in the dramatic interception of the German bombers, engaging the enemy twice but with no results. The war journal of the 76ª Squadriglia for January 1918 has not survived, while in February the only important event for this pilot was a strafing attack against barracks in Val Cismon on the 26th together with *ten.* Ludovico Censi. Meanwhile on 2 February the 76ª had moved from Istrana, constantly under attack from enemy night bombers, first to Isola di Carturo and then, on the 17th, to Casoni.

On 1 May Buzio dropped messages over the enemy airfield at Feltre, probably giving information about Austrian casualties or asking for news about Italian ones, as was customary among the chivalrous airmen on that front. On 3 May Buzio in a patrol in the area of Mount Grappa with four squadron mates, *ten.* Censi, *s.ten.* Donati and Nannini, *serg.* Ticconi, attacked a two-seater flying with two fighters over the channel of the Brenta, but the reaction of the escort fighters stopped him from following the dive of the Brandenburg. All five Italian pilots were credited with a shared victory.

This pretty Nieuport 11, marked with a visible tricolor band across its fuselage and with its serial number Ni.1763 written in unusual characters, was reportedly the personal airplane of Alessandro Buzio.

When, mid 1916, and where, Verona, far away from the frontline, shooting down an aircraft was an unusual event, the decoration of the victorious pilots with the Medaglia di Bronzo was an important affair. Buzio is the first one toward the camera.

The commander of the VI Gruppo, *magg.* Lombard, with his personal Hanriot, showing a dark front fuselage, probably red, and a tricolor shield with the baton of command.

Pilots of the VI Gruppo at ease. From the left, Alessandro Buzio, Michele Borea D'Olmo, Giuseppe Retinò, Silvio Scaroni. Standing: *cap.* Barbanti, medical officer, and Giulio Lega.

A group of pilots of the 76ª Squadriglia. Second from left, Baracchini, third Alessandro Buzio, seventh Giorgio Michetti, eighth Silvio Scaroni.

On 21 June Buzio flew with Scaroni in offensive patrol from the Montello to Susegana. "They spotted at about 7.30 p.m. an enemy patrol of six fighters returning to their lines. At 7.55 p.m. the enemy fighters crossed the lines. They attacked one of them and shot it down after intensive fire near C. Ancillotto." Their victim was *Feldwebel* Jacob Vesely of Flik 68/J, who was flying a standing patrol over the bridges on the Piave, and who lost his life in the fall of Aviatik D.1 115.31. Curiously, in his memoirs Scaroni wrote that he had lost his companion in the clouds, and doesn't mention him when describing his lone attack on the Austrian flight!

On 15 July Buzio had to abort a standing patrol over the Grappa and return to Casoni with engine troubles. As he was rolling on the runway and returning to the parking area of the 76ª Squadriglia, a British R.E.8 (serial D4823 flown by pilot Lieutenant J.W. Davies) was taking off, and crashed into his Hanriot, destroying it.

On 29 July, an anti-aircraft battery located between Col della Chiesa and Coll dell'Orso, in the Grappa area, proved particularly bothersome, and Buzio did not hesitate to go down and strafe it, firing 300 rounds.

In September, after two years of uninterrupted service, Buzio was called back from the frontline and briefly sent to the flying school at Busto Arsizio, then to the aerial gunnery school at Furbara, where he passed along his combat experience to the fledgling pilots. Then the war ended, and on Christmas Eve he was honorably discharged.

He was decorated with two Silver Medals. The first one was issued in the field for his part in the air victory of 31 July 1917, the second one, also in the field, read in its citation:

> "Quiet and courageous fighter pilot, with constant example of boldness and enthusiasm he flew many fighter, patrol and escort flights. In daring combats he shared effectively in the destruction of two enemy airplanes. Sky of the Isonzo, 22 September 1917 – Sky of the Piave and Brenta, 3 May 1918 – Val di Seren, 19 December 1917 – Mount Asolone, 3 May 1918."

After the war Buzio married a teacher, and he worked in the field of aviation, first as station manager and then in 1940 as Direttore dell'Esercizio (operations manager) of the airline Aviolinee Italiane. He rose in rank in the reserve of the Air Force, but he was removed from the list of flight personnel, and he was finally retired on 31 July 1958.

Without close relatives and now a widower, Alessandro Buzio spent the last years of his life in Milan, enjoying the close friendship of Francesco Magni, a former pupil of his wife, who shared a common experience in aviation. It was Magni who assisted him at his death, on 1 October 1972, and who according to his wish gave his medals to the Museó del Risorgimento of Pavia, his native city.

Hanriots of the 76° Squadriglia late in the war, three of them sharing a cartoon character on the rudder. The first aircraft has a black circle with a white swastika and was the airplane of *serg.* Pietro Mondini; the second one, with a white owl, was the plane of *serg.* Antonio Pasta; the third one, with a red cross on white shield, was the plane of *ten.* Camillo Sivori; the last one, with checquered tail, was the plane of *asp.* Raimondo Di Loreto.

## Alessandro Buzio
## 5 confirmed victories

| Claim | Vict. # | Date | Sqn. | Aircraft | Hour | Place | Victim |
|---|---|---|---|---|---|---|---|
| 1 | 1 | 27.6.1916 | 75 | Ni 11 | - | Arzignano | Two-seater |
| *[with s.ten. De Bernardi, serg. Nardini & cpr.Consonni, 75ª Sq.] Zugsf. Josef Holub POW/Fähnr. Friedrich Edler von Langer POW, Flik 21, Br. C.I 26.11.* | | | | | | | |
| 2 | 2 | 31.7.1917 | 81 | - | 6h35 | Sabotino/Podgora | Two-seater |
| *[with cap. Baracca, 91ª Sq.] Korp. Julius Klausz KIA/Oblt. Erich Bock KIA, Flik 46, Br. C.I 69.93.* | | | | | | | |
| 3 | 3 | 19.12.1917 | 76 | Hd? | - | Val Seren | Two-seater |
| *[with cpr. Avenati & serg. Donati, 76ª Sq.].* | | | | | | | |
| 4 | 4 | 3.5.1918 | 76 | Hd | - | M. Asolone | Two-seater |
| *[with serg. Ticconi, ten. Censi, s.ten. Donati & Nannini, 76ª Sq.] Korp. Leo Sigl UNH/Oblt. i.d.R. Franz Weintritt WIA, Flik 16/D, Br. C.I 169.78* | | | | | | | |
| 5 | 5 | 21.6.1918 | 76 | Hd | 20h | Sud di Mandre | Fighter |
| *[with s.ten. Scaroni, 76ª Sq.] Fw. Jakob Vesely KIA, Flik 68/J, Aviatik D.I (Lo) 115.31* | | | | | | | |

# Ernesto Cabruna

The Arma dei Carabinieri, arm of military police, is one of the most reputed bodies of the Italian army and of Italian society, and also the official list of Aces includes one pilot belonging to the Carabinieri. Although he was not one of the top aces, this personality is still well remembered, his feats brought him on the front covers of the magazines during the war, and earned him the Medaglia d'Oro al Valor Militare, an award seldom assigned to living combatants.

Ernesto Cabruna was born in Tortona, province of Alessandria, the rich, agricultural southern part of the Piedmont region, on 2 June 1889. His was a family of small shopkeepers and Cabruna, after technical junior high school, moved to Alessandria to attend technical high school. Only one event from his youth is known: when he was twelve years old he ran away from home and walked all the way to Liguria "to look at the sea" as he explained to his parents when he was picked up. He was described as a lively young boy, with an all-pervading sense of justice.

At the age of 18 he decided to enlist and he was accepted as *Allievo carabiniere* (Carabinere trainee) on 18 October 1907. He was eager for action, and he hoped to be assigned to Southern Italy, where in the past the Carabinieri had operated against the brigands. One year later he distinguished himself during the rescue operations after the terrible earthquake of Messina (the tragedy that gave Field Marshal Conrad the idea of launching a surprise attack against Italy, then an ally of Austria. Luckily Emperor Franz Josef gave no heeding to this bizarre suggestion.)

Cabruna on 30 September 1911 was promoted to the rank of vice brigadiere, sergeant, and probably he was already attracted by aviation. On 12 October 1910 he presented the request for a patent for an airplane with a very large lifting surface; three days later he lodged another invention of his, described as "a new propeller, or more properly a new mean of propulsion that can be applied to any system of locomotion in the air and in the water." Not much more is known about these supposed inventions that prove the enterprise and eagerness of the young Carabiniere. His father presented a request to attach his son to the Battaglione Specialisti del Genio, (the fledgling air arm of Italy) so that he could work on his ideas, but it was denied.

In 1911, when the war between Italy and Turkey broke out, Cabruna volunteered. Just before departure he received news of the death of his mother. He returned home for the funeral, but then he immediately went back to his unit, although he was allowed to stay home. He sailed for Libya from Naples on 28 April 1912 and remained there until May 1913. Then he took part in the occupation of the Dodecannese islands. Cabruna took part in the combats following the landing on Kalimnos island, that ended only when the Turkish garrison was surrounded in the center of the island and besieged for three days in a fort. *Vice Brigadiere* Cabruna at the time was chief of the police attached to the 4th Bersaglieri Battalion.

Then he was appointed chief of the garrison on the little island of Kos, and he performed his duty rigorously, showing his character of a rebel against any injustice. Despite being a military man

Ernesto Cabruna in a retouched portrait in front of a brand new Nieuport 17.

When flying with the 80ª Squadriglia Cabruna visited Cavazzo Carnico, the airfield of his former unit, the 29ª Squadriglia, and he is ready to take off in a Nieuport 17 showing the star insignia of his unit.

living for duty and discipline, he didn't stand for the abuses of power of some officers against the local population and never accept to ignore what he felt was unfair.

He returned to Italy in 1914, in the meantime his father also had died, and he served in the Legione Carabinieri of Turin, and was promoted brigadiere on 31 January 1915.

When the war with Austria-Hungary broke out he was commander of a small Carabinieri station at Salbertrand, near the border with France. He was impatient with the war news coming from all of Europe. A letter to his sister Fillide explains his feelings at the time: "My only worry is simply to have to live here without danger, without anxieties, without satisfaction, unable to follow the reckless impulse of my heart. All of my feelings and also all of my duties are humiliated and destroyed. If you really love me, don't ask me to get away, wish me instead to have my real deep wishes come true." A few days later he volunteered and reached the front with the 10th Reali Carabinieri mobilized company.

He didn't manage to serve in the frontline. We don't know of his activities until May 1916, when the Austrians launched an offensive in the plateau of the seven towns. During the enemy advance onto Asiago he distinguished himself in the relief operations for the civilians of that town. Cabruna, amidst falling enemy shells, operated with such courage helping the wounded that he was awarded a Medaglia di Bronzo.

And yet he absolutely wanted to try something new, daring and romantic, fit for a dreamer as he. Again he asked to be assigned to the air service, and this time he was accepted. On 12 July 1916 he was posted to Turin for pilot training. Three months later, on 16 October, he got his pilot's wings flying on Farman #1435 at Malpensa airfield, near Milan.

War in the air had fascinated him. Above all, Cabruna was struck by the personality of the warrior poet, Gabriele d'Annunzio, a man that one could either hate or love, no middle ground. His words and deeds were a source of inspiration for Cabruna: "We fly with the intention of fighting with death and of winning always, even when defeated by death. Thus death brings forth life, that's what d'Annunzio told us airmen."

After a brief period as an instructor he was posted on 10 November 1916 to the 29ª Squadriglia da ricognizione, based at Cavazzo Carnico. This unit had the difficult task of providing reconnaissance and surveillance in the Carnia area, a very large section of the front, where the two armies faced each other across high mountains, in a most difficult terrain. Despite the freezing cold (not even the engine heat could help, because in the Farman the engine was behind the pilots) and the risk of failure of those underpowered engines, with no open space for emergency landings, the unit performed well, under commander *cap.* Venanzi first, *cap.* Ferrero later. There had been occasional clashes with enemy aircraft, particularly the ones of Flik 16 based in the Gail Valley that performed the same missions of their Italian rivals flying from Villach.

Just a few days before Cabruna's arrival a crew had a combat with an enemy airplane and returned badly hit; on 25 November another crew had better luck during a mission over the Gail Valley, having three combats with enemy planes, driving them away and possibly shooting down the fighter of *Hptm.* Stojsavljevic, commander of Flik 16, who had claimed, somewhat optimistically, several victories over the Farmans of 29ª Squadriglia.

Cabruna had no combat and no remarkable adventures during his stay with this unit. He flew his first combat mission on 2 January 1917, and until 8 April he flew 19 reconnaissance missions. On

Fighters of the 80ª Squadriglia, Ni.17 and Ni.11 with the red star. *(Photo Moncalvo)*.

A postcard of Cabruna with his Spad 7 serial 1420, showing the red heart insignia of the 77ª Squadriglia and the prancing lion shield of his city Tortona.

that day he was posted again to Malpensa for fighter training. He got his fighter conversion on 14 June flying Nieuport 11 serial 2127.

He was promoted to the rank of *maresciallo* and on 15 June he went to S. Giusto, Pisa, for the air gunnery school, then on 21 June he finally joined a fighter squadron, the 84ª Squadriglia, then under formation at Arcade.

All the pilots of this unit were inexperienced as fighters, and they had to fly many training flights before reaching the frontline. Cabruna logged 21 flights until 3 July on Nieuport 2301 then the whole squadron moved to S. Maria la Longa.

The Nieuport 11 at that time was obsolete, and the 84ª was employed mainly for protection and escort to the slow SP and Caudrons of the artillery squadrons and to the Voisins of the outstanding 25ª Squadriglia. His first emotion came when he overturned on 27 July, his fighter was completely wrecked but the pilot was unharmed. In August the Italians launched the big offensive that brought the conquest of the Bainsizza plateau, and air activity became frantic. On 6 August three Austrian airplanes attacked, but Cabruna managed to protect the two-seater he was escorting and to escape. On 28 August he returned with his plane hit in five places.

On 5 September he escorted a Caproni bomber mission, and heavy anti-aircraft fire hit both his fighter and a Caproni nearby. On the way back many airplanes tried to intercept the Italians. Cabruna, having seen that the Caproni was safely behind the Italian lines, turned back and faced the five Austrians. He managed to escape them, but he was hit also by ground fire and force landed, his Nieuport was badly damaged but he was unharmed. From 20 September to 26 October he flew as many as 32 combat mission. In this period he was shifted to the 80ª Squadriglia, equipped with the more modern Nieuport 17 fighters that carried the unit insignia of a five-pointed star.

His first victory came in the days of the Caporetto breakthrough. Air activity was intense all along the front and Austrian naval aviation supported land operations, sending its flying boats to bomb Italian troops at the mouth of the Isonzo River. On 26 October Cabruna shared in the destruction of enemy seaplane K212 at Lake Doberdò, behind the Italian lines, together with other fighters of the 77ª and 80ª Squadriglia, who also were to become aces: Lombardi, Leonardi and Ancillotto. The two crewmembers were captured.

That same day the squadron was ordered to retreat, destroying its inefficient aircraft. After a first stop at La Comina airport, where all retreating Italian squadrons were concentrated, on 1 November the 80ª moved to Arcade, across the Piave River, then on 10 November it reached its final base, Marcon, where it remained for the rest of the war with its twin unit within the XIII Gruppo, the 77ª Squadriglia.

Badly lacking in equipment but highly combative, the two units fought hard against enemy aircraft on the southern sector of the Piave front. Cabruna shared in this activity flying almost every day. On 14 November during an escort flight he faced five enemy fighters and brought back the two-seater into Italian lines. On 24 November he made a forced landing, but it cannot be determined whether due to engine failure or to combat. *Lt.* Bertelsmeier of Jasta 39 claimed on that day a "Sopwith" shot down at Montebelluna, which is rather north of the usual area of operations of the 80ª

Personnel of the 77ª squadriglia: in the center, in front of the propeller, stands commander Serafini and to his right is Cabruna, in a dark leather jacket.

Fighters of the 77ª Squadriglia at Aiello, mid 1917. The Spad was a very compact fighter, and its dimensions can be judged against the man standing behind it.

Cabruna and his good friend Alvaro Leonardi leaning on the fuselage of the Spad 7 of the ace. The little black crosses marking victories along the spine of the fuselage don't appear in this wartime photograph.

Squadriglia; still his victim may possibly have been Cabruna. He had another combat on 27 November and another engine failure on 3 December, resulting in a forced landing.

Cabruna scored his second victory on 5 December. Details of this combat are lacking; probably he fought together with *serg.* Carabelli of the 77ª and *serg.* Degli Esposti of the 83ª against a Brandenburg of Flik 32/D. This second victory brought him a Medaglia d'Argento.

Winter brought heavy snow, but that didn't stop him. His activity from mid November to mid January 1918 is resumed in the number of combat missions, 56, almost an average of one a day, in a period when most days were unfit for flying. He didn't meet enemy aircraft, but still had his shares of engine problems, force landing on 30 December and 15 January.

On 26 January 1918 Cabruna was shifted to the 77ª Squadriglia. That meant changing fighter, because the 77ª used the Spad 7. He remained in the same airfield, with his friends of the 80ª, particularly his good friend *sergente* Leonardi. In his first day with the new unit he had a combat, attacking a balloon without results.

Until 20 February he flew 24 combat missions, mostly escort and alarms, and on 4 February again he attacked a balloon, with no visible result. The next day enemy ground fire hit his Spad. He even refused the winter leave in order to stay at the front, his relatives were worried but he explained his life to them: "At the airfield times passes merrily and very quickly, always in the hope of having an air combat and of showing to the enemy what we Italians can do. You live in the nightmare of war, we live in the enthusiasm of victory."

In early March he strafed enemy troops three days in a row. Finally on 12 March victory returned to Cabruna, who during a cruise attacked and shot down an enemy plane that was seen crashing near Case Bartolotto, south of S. Donà, 500 meters from the Austrian bank of the river.

After other inconclusive flights, 29 March was a memorable day that brought fame to this pilot. This is how Cabruna himself described the events in his Combat Report:

> "Time 11.00: patrolling between Ponte di Piave and Grisolera I left the flight and attacked 10 airplanes (3 of them red) escorting a two-seater.
>
> Either unseen or confused for one of them, I followed them to hinder their intentions, then as I saw them aiming for Ponte di Piave-Treviso with the 3 red ones leading, over the bridge I jumped the two planes that were following the leader and I confused them. Then I attacked the leading red plane that was startled so I could get close and shoot. I kept following it as it escaped to its lines and I fired several bursts from close range on account of which I saw it banking badly, gliding and then diving in an excessive dive without pulling back. Due to the thick fog and mainly for the effort to avoid the menace of the remaining ones, which were above me, I failed to observe where the leader fell. The confusion among them, also for the loss of their leader, was such that they gave up returning across the lines, and one by one they aimed inland. Three of them stayed, however, but they also dove and flew to their territory as soon as my flight reached me.
>
> Mar. Ernesto Cabruna."

Left: A postcard celebrating the feat of 29 March 1918 which brought the Medaglia d'Oro to Ernesto Cabruna. The ace signed this postcard. Right: *Sergente* Nicola Malaspina reached the 77ª Squadriglia too late to take part effectively in the war, but in this signed photo he stands in front of Cabruna's Spad, which shows the individual number XIII over the top wing and no black crosses on the fuselage spine.

This event, as described by its protagonist, may seem minor, but this lone attack against such a strong enemy fighter squadron became legendary. The front cover of the "Domenica del Corriere", Italy's most famous weekly, carried a painting by artist Beltrame showing the feat of Cabruna. Actually the pilot couldn't determine if the red Albatros was actually shot down, nor could the ground observers provide confirmation, because of the thick fog, but the enemy fighter diving vertically had witnesses, soldiers of the 23° Rgt. Infantry and of the gun battery at Mogliano, so this victory was confirmed to Cabruna. Austrian sources don't report losses of fighters in that date and place. An all-red fighter could suggest that Cabruna might have attacked the famous ace Godwin Brumowski, who however didn't fly on that day.

Beyond the actual result of this combat, Cabruna on that occasion showed the courage of a hero. When he landed all his companions circled him to cheer and embrace him, but he, with his usual modesty, just asked for a cigarette.

The good weather returned, flying intensified, but the only event was one more forced landing due to engine failure. Activity increased in June, in the expectation of a new Austrian offensive. The Italian air force attacked ground targets, and on 13 June the fighter of Cabruna was damaged.

In the first day of the attack, 15 June, he was immediately in the air. His first combat was inconclusive, but then he shot down an enemy aircraft near Tezze, for his fifth victory. When he couldn't find airplanes, he went down to strafe enemy troops across the Piave, and on 20 June, flying with *sergente* Piermattei and *caporale* Sottani he found an enemy plane and after a series of furious attacks shot it down over Musetta (unknown locality: possibly Contrada Fossetta, west of S. Donà di Piave, at that time still held by the Austrians). The enemy crew got out of it unscathed, the unknown pilot and the observer, *Lt.i.d.R.* Franz Sedlmaier, of Flik 22/D reporting a combat with four Spads.

The next day together with *s.ten.* Pietro Molino, Cabruna attacked and shot down in flames a kite balloon at Ceggia, using 380 rounds. This was one of three balloons shot down on June 21, *tenente* Baracchini and the only French ace on the Italian front, sergeant Andrè Robert Levy of Escadrille N 561 claiming the other two. Cabruna however didn't get confirmation of this victory:

> "21 June 1918
> Pilot. Maresciallo Cabruna Ernesto
> Time of departure: 6.29
> Time of landing: 8.50
> Type of enemy aircraft attacked: Draken
> Time, height and locality where the combat began; 7.30-1800 m. Ceggia.
> Way of attack: diving
> Result: I forced it to go down to very low level where I left it still burning.
> Other airplanes that took part in the combat: another Spad of S.Tenente Molino.
> Additional notes: having backed the other Spad that aimed at the balloon first, but that had to give up with its gun jammed, I got very close and shot a long burst following it in its quick descent.
>
> The pilot
> Ernesto Cabruna"

In the following days he flew strafing missions every day, getting hit again on 26 June. The Austrians had retreated to their original positions all along the Piave except for a small marshy area where the Italians used strong pressure to eliminate that last bridgehead. The 77ª Squadriglia operated in this area until 14 July.

When the land battle was over also air activity decreased in the lower Piave area. Cabruna had to wait until the end of the war to meet enemy aircraft again. On 28 July 1918 he was commissioned to the rank of *sottotenente* "in the field" for war merit. On 26 September he had a bad accident flying the new and somewhat maligned new Italian fighter, the Ansaldo A.1 Balilla, serial 16548. These new airplanes had just reached the front, and were assigned in small numbers to all squadrons for operative evaluation, but as opinions were mixed, the Comando d'Aeronautica on 18 September 1918 issued an order to deliver all Balillas to the 74ª Squadriglia based at Castenedolo. According to the biography of Cabruna written in the early Seventies by his friend Tomaso Cartosio, also an airman and a d'Annunzio enthusiast, he had flown a long reconnaissance mission from Marcon, near Venice, all along the frontline up to the Trentino, landing at the opposite part of the front, at Castenedolo. Actually it had been just a normal ferry flight, although it may have been that Cabruna, instead of flying the shortest route, may have gone take a look at the enemy, still it surely was not a strategic reconnaissance. Landing at Castenedolo he realized that both the tank and the oil pressure hand pump were not working. The scalding hot oil started flying everywhere hitting him in the face. Cabruna, blinded, lost control of his aircraft and landed "roughly", as the papers report, breaking his right shoulder bone and suffering shock and a bruised hand and chin.

He was hospitalized, but when one month later the final offensive began, despite being on sick leave he forced his return to the squadron, with an arm still in plaster. He returned to his 77ª Squadriglia on 27 October, two days later he flew a training flight, and on 30 October he returned to action, flying escort, strafing and cruise missions. His final combat mission was on 2 November. During a visual reconnaissance flight he spotted six airplanes taking off from Aiello, attacked them and destroyed two of them. The citation of his Medaglia d'Oro, issued many years later, wrongly dates this action 25 October, when he was still in hospital.

In fact, his commanders advanced a request for Italy's highest decoration, but the Medaglia d'Oro was assigned only in 1924, because of political reasons. Its citation reported a constant show of daring, his final combat, with two aircraft flamed while taking off, a total of 900 combat flight hours and his risky one-against many attack, reporting however no less than thirty airplanes as attacked single-handedly.

Cabruna remained in the service, on 5 May 1919 the Duke of Aosta, commander of the 3ª Armata, ordered him to fly his fighter to Genoa for an exhibition of war material. In his message to the city, the commander described Cabruna as "the most glorious among my war wings". In a few years, thanks to his valor, Cabruna rose in the ranks of the Carabinieri in a remarkable way but, according to his character and his beliefs, he threw away all that he had earned, in order to follow what he believed was a just cause. He was, as we said, a great admirer of Gabriele d'Annunzio. After the victory celebrations, Italy had to face the problems of its new borders. The peninsula of Istria, to the east, had an Italian majority, but also a sizeable Slovene and Croat population. With the birth of the new nation of Yugoslavia, both this kingdom, supported by England, France and the United States, and Italy claimed possession of certain parts of the Istrian coastline and of Dalmatia. The fact that all these parts had been promised to Italy in the secret "Pact of London", April 1915, had become irrelevant according to the principles fostered by President Wilson. England and even more France were not happy at the idea of an Adriatic Sea controlled by Italy.

Cabruna in Fiume during the rebellion lead by Gabriele d'Annunzio. The pilots with him are, second from left Vittorio Pomilio, then Arcidiacono, Granzarolo, Cabruna and Savino. The other two are unknown. The airplane is a Macchi M.5, serial 13021, formerly of the US Naval Air Station Porto Corsini, still carrying the nickname "Jeff" painted by its American owners.

One of the thorniest knots was the city of Fiume (now Rijeka, Croatia) that the Pact of London didn't assign to Italy, but that was now claimed, according to a request of its population. The Italian government of that time did not please a large part of public opinion and particularly veterans, looking too weak in the confrontation with its allies, who seemed to deny the rightful spoils of victory. Thus was born the legend of the "mutilated victory", as d'Annunzio pointedly christened the situation.

The French had set up a base in Fiume to contrast Italy, and in a climate of excitation Italian and French soldiers clashed, so both contingents were pulled back and a British force replaced them. Politicians however had discounted the combativeness of veterans. On 12 September 1919 the poet Gabriele d'Annunzio lead a force of *Granatieri* and *Arditi* élite troops who disobeyed orders, marched on Fiume and occupied it, among cheering crowds. Besides the contemporary judgement of history, that event marked a collision between traditional power and the new generation formed in the crucible of war, impatient of injustices, proud, combative and reluctant to return to the normalcy of peacetime. Cabruna was one of the many officers (meanwhile he had become *tenente*), who by instinct left their posts and joined the poet in his mad adventure. He was the first pilot to land at Fiume. Such a gesture meant giving up everything, deserting and throwing to the wind all that he had accomplished. He did all that with no remorse, and became one of the closest collaborators of the "*Comandante*", as d'Annunzio was called by his followers, better, his disciples. He was his liaison officer, charged with "secret affairs". He was to remain in Fiume a long time, among exaltation, deprivations and disappointments. He was ready to follow his chief but at the same time he criticized him when he thought the poet was wrong. He had to remind him not to waste his time with futility, particularly with women, to concentrate on their cause and their ideals.

As he had relinquished his post, Cabruna mailed to the Ministry of war his resignation from the commission of Carabinieri officer, "as I no longer trust the Monarchy and the Institutions".

At Fiume he started suffering with the disease that was to kill him, and he had to rest in a Sanatorium. The diagnosis was "cardio aortic syndrome producing rheumatic polyarthritis and malaria, due to the exceptional service activity that he performed along the Lower Piave during the war."

Together with another famous airman, Guido Keller also formerly of the 91ª, an artist and a dreamer, the most eccentric pilot of all times, he organized raids to supply the besieged city with food and provisions. He never forgot, however, that he was a Carabiniere, and very proud of it. In the midst of the bedlam that was Fiume, he found the time to challenge to a duel Mario Carli, commander of the "iron heads" legionnaires, who had made snide remarks about the Carabinieri. It was a pistol duel, and Cabruna was the loser, getting a bullet in his chest, that put an end to the challenge.

The crucible of Fiume, declared a "free end independent state", went on until d'Annunzio, under the threat of the Army and of a naval bombardment of the city, went away with his followers, on 18 January 1921. Those who returned to their posts were pardoned of their rebellion, but not Cabruna, who secretly remained in Fiume as the poet had ordered him, holding the task of commander of the Gendarmeria, that is the local militia. Politics in the city went on with clashes between the opposing factions: Cabruna, fearing irregularities, organized an attack against the courthouse where the election ballots were checked, and two days later, as president of the Military Committee, he lead an armed occupation by legionnaires and fascists of City Hall, forcing the resignation of Riccardo Zanella, chief of the pro-Yugoslav autonomist party. Later on, when Mussolini was Prime Minister, the treaty of Rome between Italy and Yugoslavia signed in January 1924 assigned Fiume to the former. The city was returned to the Slavs in 1945.

The Spad of Cabruna exhibited in not too good conditions ath the Milan Aviation Show, 1934.The Spad behind it was exhibited as the one of Novelli.

Cabruna, when he felt his mission was completed, left Fiume in April 1922, full of honors and of love from the local population, but with no money, no officer commission and no pilot's job. The fascist party did not attract him. He wrote to a friend: "..what our Commander wrote to Giacomo Treves should apply to you and to all of us: the so-called new Italy (the Fascists) imitates emphatically my gestures and my speech, but they naturally hate my spirit. Very soon they will banish me, or else I will go in voluntary exile."

He went to visit d'Annunzio in Gardone, at the posh villa, the "Vittoriale" that the government had given to the poet to keep him out of troubles: he was kept waiting three days before being let in for a talk, because the poet was correcting some drafts. D'Annunzio then welcomed him and offered financial help, but Cabruna proudly refused. He had come to report that his mission was accomplished, not to ask for help.

He was so depressed that he considered suicide. Once for a change, the government came to his help. Although a rebel, Cabruna still was a famous veteran, and the man in power knew how to exploit idealists. In 1923 his resignation was declared void, and he was readmitted in his commission, entering on 17 February 1924 in the ranks of the Regia Aeronautica.

Meanwhile his hometown Tortona had honored him so much that the Duke of Aosta, who obviously liked Cabruna very much, gifted the city with the Spad 7 of the ace that was still in Genoa.

The biggest honor, anyway, was the granting of the Medaglia d'Oro al Valor Militare that so far had been mothballed on account of his rebellion. This is its citation:

> "Magnificent fighter ace of aviation, perfectly executing orders or in daring initiative, in winning combats with incredible courage, even alone against an overwhelming number of dangerous enemies, often with failing health, in every circumstance of war he offered his untiring activity, with the mettle of Roman heroism. In the final large offensive he wanted to participate at any cost, he got out of a hospital where he was cured for a wound, and although his right arm was still immobilized and hurting, and therefore in conditions of absolute inferiority, with incomparable willpower and daring, he attacked many enemy aircraft ready for take-off on a far away aviation field and he set two of them on fire. In another event he attacked a group of thirty enemy aircraft, shot one of them down and drove the other ones away from their target, as for him the superiority of the enemy was an incentive to engage combat. In various and difficult circumstances, from the beginning to the end of the war, he flew over 900 hours and did not hesitate facing the most daring deeds, and giving great and important service to his country. Aiello, October 1917, Sky of the Piave, June, July, November 1918."

Captain Cabruna at Ciampino airport, Rome, in 1927 with Ansaldo AC.3 metal monoplane fighters of the 91ª Squadriglia.

Below: Cabruna, third from the left with a walking stick, in Libya with officers of the 37ª Squadriglia, a unit based at Apollonia and equipped with the SVA 10 two-seater, one of which is in the background.

After this decoration Cabruna the rebel, the radical for country, nationalism and action, slowly faded away. In 1925 he rose to the rank of *capitano* and General Piccio, Chief of Staff of the air force, appointed him aide-de-camp. Still, Cabruna always remained on the fringe of the fascist regime. Occasionally he served as duty officer for "his" Commander d'Annunzio, whom the dictatorship had curiously appointed General of the Arma Aeronautica "ad honorem".

In May 1927 Cabruna was assigned to the Aerocentro Specialità Idrovolanti, (seaplanes conversion unit) then to the VII Gruppo Autonomo of the 8° Stormo Misto. There he commanded an experimental squadron at Ciampino airport, Rome, that tested the new Ansaldo AC 2 fighters, Italian version of the Dewoitine metal monoplanes. A weakness in the forging of the strut attachment points made this fighter insecure. There were many fatal accidents and Cabruna decided to stop flights with this model, and he privately advised Minister Italo Balbo and the young Amedeo d'Aosta, Duke of the Puglie, (son of the Duke of Aosta) against that airplane. Luckily the future Viceroy of Ethiopia heeded to his advice, because the fighter readied for him crashed with another pilot. The top brass however did not appreciate this fact, and he was condemned to three months of hard confinement because, "...talking with the Duke of the Puglie after an accident he made inappropriate comments."

Mussolini pardoned him, and then Cabruna was assigned to North Africa. There he took part in the operations for the reconquest of Cirenaica as commander of the 23ª Squadriglia, a SVA-equipped unit based at Apollonia, Libya. He remained in Libya just eight months, because his disease was getting worse, and in June 1929 he returned to Italy. In October he was deemed medically unfit for service, was placed on leave of absence and in June 1932 he left the armed forces.

In 1930 he made a long trip to Latin America on behalf of the Società Dante Alighieri, (sponsor of the Italian language) then when he returned to Italy he left any public position and concentrated in his studies, working on social doctrines with inspiration from Greek philosophers, fathers of the Church, Mazzini and many more. Strangely, there were no quotations of d'Annunzio. His political thought returned to the years of Fiume, when he had theorized a pro-Bolshevist state, planning even the formation of Soviets in Fiume and in Italy.

In 1935 he visited Russia and was much impressed. He also traveled much in Greece and Turkey. He gave his Russian travel notes to Mussolini, but the dictatorship censored them and never allowed their circulation.

By this time much of his life was spent in hospitals all over Italy. He much regretted that some fascist chieftains claimed merit for the military actions in Fiume that belonged to him. The regime didn't pay attention, his protestations were ignored. This growing delusion pushed Cabruna to an ever more radical opposition to fascism. His closeness to anti-fascist movements is reported in documents preserved at the Public Records Office in the United Kingdom, published in the 1980s by Professor Alberto Santoni. There he is openly quoted as a sympathizer for "free Italy" as *"Gold Medal Cabruna"*, a.k.a. "X 19".

The grave of Cabruna in the Vittoriale, the beautiful villa of Gabriele d'Annunzio at Gardone, overlooking the Lake of Garda.

After World War II, coming out of a long period in a Sanatorium, Cabruna strove to bring Italy out of the ashes of defeat. In 1944 he accepted the position of Commissioner for the Association of disabled war veterans. He managed this association his own way, with strict rules, forbidding the use of cars for himself and for the other managers, giving no wage to those who had other incomes, and gaining new enemies. In 1945 the similar French association of veterans invited him to Paris, and unexpectedly he was welcomed as a brother. Then he went to Yugoslavia to get repatriation of Italian prisoners.

His final years were shrouded in discretion. He isolated himself, lead a simple life and had an attitude of opposition, due to the wrongs that he had suffered and also to his bad health. What he wrote to a friend shortly before his death revealed his attitude: "...now I have nothing ugly to throw away, and nothing beautiful or useful to give..."

Cabruna died on 9 January 1960 in the Villa Azzurra hotel of Rapallo, assisted by a woman friend from Fiume and by his always-loving sister Fillide.

Three years later his body was moved to the Vittoriale, the outlandish villa of d'Annunzio on the Lake Garda, buried on the "Hill of the Arks". The most famous and loved legionnaire of Fiume rests next to his Commander. There are twelve arks there, three holding the remains of airmen: Adriano Bacula, of the 91ª Squadriglia and then SIAI test pilot, Antonio Locatelli and Guido Keller.

An important memory of the pilot Carabiniere remains, his beautiful Spad 7 with the insignia of the 77ª Squadriglia and of his town, Tortona, exhibited in the Aeronautica Museum at Vigna di Valle, near Rome. The Arma dei Carabinieri preserves another Spad with the same insignia at Pratica di Mare air force base.

A view of the enbankment at the Miraglia Naval Air Station with the Macchi M.5s of the 260ª Squadriglia and, closer to the camera, the single Hanriot floatplane tested at Venice.

M.5 alighted in the lagoon of Alberoni. While the other pilots flew overhead keeping at bay the Austrian soldiers, Ravazzoni alighted in the swamp next to his friend, who was punching holes through the hull of his Macchi, to sink it. Having thrown out all useless weight, Ravazzoni asked Calvello to take the controls and he sat on the fuselage holding to the engine struts. After a long run the flying boat took off from the muddy waters and got away, with the Austrians shooting their rifles at it.

In the final days of the war, Calvello strafed the bridge at Latisana and enemy convoys in the canal of the Marano lagoon. His final war mission, the last one of 183, was a reconnaissance over Trieste, by now ready for the arrival of the Italians, on 2 November 1918. The formation of the 260ª Squadriglia led by group commander *Tenente di Vascello* Pierozzi included Calvello, Dri, Ravazzoni, Bartolozzi and Pagliacci. Seaman Pagliacci broke ranks and alighted his M.5 in the harbor of Trieste, the first Italian to enter the liberated city.

After the armistice, Calvello flew occasionally in mail service between Venice and Pola. On 1 December 1918 he was assigned to the new Trieste Naval Air Station, where he remained until 1 June 1919, then returning to Venice. He was awarded two Medaglie d'Argento al V.M., the first one in March 1918, the second one in May 1919.

One of the victims of Umberto Calvello: the Austrian flying boat R1, that he shot down on 22 April 1918. *(Photo Csizek).*

He died on 10 August 1919, while test flying a SIAI S.9 flying boat that he was supposed to exhibit at the ELTA air show in Amsterdam. In the morning, he first flew this new aircraft, together with Umberto Maddalena (who later became one of the greatest Italian pilots) who had ferried the S.9 from Sesto Calende. In the afternoon, he took off alone. While flying in his typical steep bank, its engine faltered, and the heavy aircraft crashed in the lagoon of Venice, killing him instantly. He was just 22 years old.

Pilots of the Miraglia Naval Air Station: from left *STV* Ivo Ravazzoni, *STV* Umberto Calvello, then commander *Capitano di corvetta* Roberti di Castelvero and *TV* Orazio Pierozzi.

Above: The Macchi M.5 serial 7282 of *STV* Ivo Ravazzoni, who saved Calvello when he was forced to alight on 31 october 1918. The eagle's head is the emblem of the 260ª Squadriglia.

Right: A Macchi M.5 fighter seaplane taking off from the channel if S. Andrea island. Blurred in the background there is the hangar of the 261ª Squadriglia and some Macchi M.5s of the 260ª, with their fuselages showing green and red stripes.

## Umberto Calvello
## 3 confirmed victories

| Claim | Vict. # | Date | Sqn. | Aircraft | Hour | Place | Victim |
|---|---|---|---|---|---|---|---|
| 1 | 1 | 22.4.1918 | 260 | M.5 7103 | 16h | Fasana | R1 |
| *[with 2º capo Rivieri & marò Pagliacci, 260° Sq.] Maschinenmaat Stefan Bauer UNH /Seefähnrich Alfred Edler von Herbetz UNH , R1* | | | | | | | |
| 2 | 2 | 4.5.1918 | 260 | M.5 | - | Trieste | A91 |
| *[with TV Martinengo, 2° capo Jannello, Rivieri, marò Pagliacci, 260ª & 261ª Sq.] Fähnr. i. d. R. Josef Niedermayer POW, A91* | | | | | | | |
| 3 | 3 | 4.5.1918 | 260 | M.5 | - | Trieste | A78 |
| *[with TV Martinengo, 2° capo Jannello, Rivieri, marò Pagliacci, 260ª & 261ª Sq.] Flgst. Franz Boros POW, A78* | | | | | | | |

Right: Calvello's Macchi M.5 in flight, showing that the Happy Hooligan insignia was symmetrical on both sides.

# Marziale Cerutti

MIR, *Marziale Imperatore Romano*, "Martial, the Roman emperor" was the mocking war insignia that adorned the Nieuport 27 of Marziale Cerutti, and that showed the character of this young pilot: carefree, light-hearted, but determined and deadly in flight, at the controls of his fighter. His many photos reveal his character: always smiling, always surrounded by friends, like sergeant Antonio Reali or sergeant Attilio Imolesi, always active, in groups or on a motorcycle, or proud of his quirks, like a woman' stocking used as a cap. If an actor were to have played Cerutti on the screen, it could only have been Steve McQueen.

Marziale Cerutti was born in Brescia on 10 March 1895, but he lived in the nearby town of Lonato.

In his little town, where the Alps meet with the plain of the Po River, he was considered a reckless young man. Once he had climbed the facade of the large baroque cathedral of Lonato, settling in a niche prepared for a statue. As a boy, he survived pneumonia and typhus, a remarkable feat at the beginning of the 20th century, showing a tough body. He retained his prankish character all through his life. Once he invited his friends for a trip to the nearby Garda Lake for a swim. Driving an open car, he just raced over the beach, and he dived into the water, with all of his friends and the car. He had a rich and religious aunt, who when she died willed all her properties to other people, and not to her mischievous nephew.

In January 1915 he was conscripted, as a soldier in the 3rd Regiment of Railroad Engineers, but he was quickly dismissed for medical reasons. By that time, he had attended Liceo (high school) and was registered as a student. In July 1915 he returned as a volunteer, assigned to the 25th Field Artillery Regiment as a car driver. Although, as a student, he could be admitted to officer training, he didn't want to wait for the paperwork, and accepted the rank of soldier in order to speed up his draft. He was probably employed as a car driver. It was reported that Oreste Salomone (the future winner of the first Gold Medal) carried him for a joyride on a Caproni bomber and this stimulated his interest in aviation. As he was a passionate motorcycle rider, for this technical prowess (which was a rare quality in Italy at those times) he was admitted on 30 October 1915 to the Battaglione Aviatori, the Italian Air Service, and he was sent one month later, November 1915 to the flying school at San Giusto, Pisa, and he was licensed as a pilot on 15 April 1916, on a Bleriot.

On 26 June 1916 he was sent to Busto Arsizio training field to get his military brevet on Farman pusher biplanes, then he was briefly assigned to Malpensa to train on Caproni bombers, on 11 September, and then on 16 September (the day in which his license as a military pilot was issued) he got to be moved to Cascina Costa, for fighter training. His training had lasted almost one year, while Cerutti was impatient. Only two months later, on 26 November 1916, he was assigned to the 79ª Squadriglia Caccia with the rank of corporal. He remained at this unit all through the war.

The 79ª Squadriglia, based at Istrana, was being formed in those days, commanded by *cap.* Francesco Chimirri and equipped with Macchi Nieuport 11. Its task was to defend the area of the Asiago plateau, a front that was secondary compared to the Isonzo, but particularly difficult for its mountain terrain. Among the pilots who became friends of Cerutti there were three future aces, Attilio Imolesi, Giovanni Nicelli and Antonio Reali, all sergeants, only one of whom survived the war.

Cerutti's first war mission was a patrol flight above Asiago on 21 January 1917, soon followed by many more. On 3 March he dropped patriotic leaflets for war bonds over Treviso. As not much action was going on in that front, on 28 March a flight of three

Marziale Cerutti in training, standing in the cockpit of Bleriot XI serial 35.

Another aspect of flying training in World War 1, with an obsolete but still fascinating Farman model 1912.

fighters was detached for the defense of Padua, and Cerutti was one of them.

Later he returned to Istrana, and on 3 June 1917 he had his first combat: flying together with Reali, they met an Austrian two-seater at Borgo Roncegno. Although repeatedly hit by the fire of the two Italians, and with the observer visibly wounded, the enemy aircraft was not confirmed as shot down. On 7 June Cerutti met again an enemy airplane between Gallio and Cengio, had a combat with it but the enemy escaped him, because the gun of his Nieuport 2124 jammed.

On 14 June Cerutti gained his first victory that was also the first confirmed victory of the 79ª Squadriglia. He took off with Nieuport 2269 for a patrol over the "altipiani" with *ten.* Moresco flying Nieuport 2266 and at about 7 a.m. they spotted an Austrian airplane near Monte Vezzena and attacked it. After a few rounds, the Lewis of the officer jammed, and the prey was left to Cerutti, who shot it down. Observers on the ground confirmed the success of this attack. The victim of the Italian pilot was Hansa-Brandenburg D.I 28.35. This was his combat report:

"About 8 a.m. cruising the line Caldiero and Asiago near m. Verena I saw an enemy airplane type Albatros that was flying at my same level, 3,800 m. The airplane came resolutely toward me attacking, it opened fire first at a distance of about 300 m. Maneuvering in order to attack it profitably, my airplane banked about 50 m. As soon as I recovered I saw the enemy airplane very near, at about 50 m., I managed to get on its tail and started firing. The enemy airplane turned trying to get into its own territory. I followed it shooting with successive bursts. After some fifty rounds I saw the airplane banking on its right wing in a quick descent. I kept following it unloading the rest of the bullets in my twin magazine down to a level of 3,000 m. The enemy kept falling down to earth. It must have fallen near Vezzana."

The next day, 16 June 1917, Cerutti was promoted to *sergente.*

His second victory arrived a few days later, on 22 June, against a two-seater that, under his fire, remained with a stopped propeller and then disappeared gliding into the clouds over the Val d'Assa.

This is his combat report:

> "At about 6:10 I saw an enemy airplane, Albatros type that was coming from the Asiago plateau going to Val d'Assa at an altitude of about 3,500 meters. I went down and attacked it from behind shooting 8 or 9 rounds, and right away I saw its propeller stopped. I followed it as it went down quickly and I fired at intervals another twenty rounds. I believe I hit it again because then its glide was not at all regular. Suddenly it disappeared into a cloud and I didn't want to follow it as I had gone down to an altitude of 2,500 m and over enemy land. The enemy airplane must certainly have crashed against the banks of Mount Verena as it didn't have enough height (most reasonably) to get beyond, with a still engine."

During the summer of 1917, Cerutti kept on flying patrols looking for the enemy that was staying away, suffering from the heavy losses suffered in June. On 2 August 1917 as he was flying above Asiago, he spotted an enemy, attacked it at very close quarters and thought he saw the enemy observer hit, but the enemy managed to escape him hiding into a cloud. During this long duel, Cerutti's weapon jammed ten times. The next day, August 3, he saw an enemy two-seater escorted by a fighter. This time, discretion was the best part of valor, Cerutti attacked from the safe distance of 150 meters, and the two Austrians escaped him.

In the autumn 1917 the situation changed, and with the support of the Germans the Austrians inflicted to the Italians the defeat of Caporetto. On 26 October 1917 Cerutti, flying a new Nieuport

Cerutti in leather jacket and smiling, stands in front of a Nieuport 11 of the 79ª Squadriglia. This airplane is marked with a diagonal white band, showing that it is assigned to the 1st Flight.

Cerutti showing off, riding a motorcycle and wearing a lady's stocking as a hat, while Attilio Imolesi leans down to pet a dog.

Ni.17, was escorting a Caproni when, near Primaccò, he spotted another Nieuport attacked by an enemy fighter, and rushed to help it. At his first bursts, the enemy disengaged going into a spin. Cerutti followed it, and as soon as the fighter, described as *"looking very much like our SPAD and painted dark green"*, got back flying straight and level, he fired again. This time the enemy could not get out of the dive, and crashed near Volzano.

This is his combat report:

> "I was cruising at an altitude of 3,600 meters escorting Capronis, when I saw an enemy airplane fighting one of our Nieuports. I went toward them and I saw the Italian fighter spinning as the enemy attacked it. Then the enemy abandoned it and went to its lines. Then I attacked it frontally placing 4 or 5 rounds into its engine. When it got out of its spin I clearly saw that its propeller was still. I attacked it from close firing at intervals some 30 rounds. Then it spun again but I saw it fall near Volzano."

In that area, near Caporetto, in that action German fighters shot down five Caproni bombers. Cerutti's victim was possibly an Albatros of Jasta 31, that unit having two wounded pilots that day, and the attacked Nieuport probably was the one of *serg.* Poli of 70ª Squadriglia.

This was at the time his first official victory, while the two ones in June were confirmed later on thanks to papers found on an enemy crewman shot down in January 1918 by *sold.* Panero.

Meanwhile the 79ª Squadriglia moved from Istrana to Nove di Bassano. Cerutti was not always the hunter, sometimes he was the prey, as on 30 October, when, flying with Nieuport 17 3657 as an escort to Italian Pomilio two-seaters above Monfalcone, not a fighter attacked him, but a bold German two-seater that forced him on the defensive, riddled his airplane with bullets and forced him to retreat to his base. Before that, however, he had attacked an Austrian seaplane that, according to the Pomilio crew, alighted on the Isonzo River:

> "I took off at 2 p.m. from Comina to escort Pomilio airplanes. I was near Monfalcone at an altitude of about 1,300 m. when I saw an enemy seaplane near us. I attacked it and followed it down to 700 m. Suddenly I heard a machine gun behind me and turning I saw a German reconnaissance airplane that was attacking me. I banked and got below its tail. As I was short of ammunition I kept a completely defensive tactic and after a few maneuvers I got free and returned to base. In this second combat an enemy bullet hit my top left wing. The Pomilio crew that I was escorting say that the enemy seaplane was forced to alight on the Isonzo."

Two Austrian seaplanes were lost in that area, but the dates, as reported by Austrian publications, don't correspond: K225 was shot down by a Pomilio on the 29th, K368 by the Italian anti-aircraft artillery on the 31st.

Again on 10 November his gun jammed during a duel with two Austrian airplanes near Folgaria, and it was the enemy who hit his airplane. On the 18th he protected a Farman from the attack of four fighters over Roana. As he dove to get out of harm's way, this event could explain the victory claim over a Nieuport by Austrian aces Kenzian, Stumpa and Lahner in that area.

On 24 November 1917, as the front had now settled, Cerutti escorted a SAML reconnaissance airplane, and then he decided to remain in the air. It was a good choice, because soon after he spotted two Austrian two-seaters escorted by a fighter, and attacked them. With a few rounds he flamed the gravity tank of the first Brandenburg, and as it crashed burning, he turned to the second one, claiming it shot down also.

For this daring action, Marziale Cerutti was assigned the Medaglia d'Argento *"sul campo"*, the Silver Medal assigned on the field, with this citation:

> "Airplane pilot, having seen two enemy airplanes, he reached them and with admirable resolution he attacked them. Defying the fire of their machine-guns, with skillful offensive

maneuvers he took up the combat with both and, despite a motor failure, he insisted until he shot both of them down, one of them falling in flames. Monte Grappa, 24 November 1917."

His fame was growing among airmen, and Baracca quoted him in a letter to his mother: "The day before yesterday a gallant sergeant, who is very promising, and is a lost soul, alone shot down two enemies beyond the Monte Grappa. His name is Cerutti, and this is his fourth victory. Many of ours also have gone down in these days, but the Germans are losing more."

A "lost soul", a reckless daredevil: the judgment of Baracca was very fitting. At the time, however, Cerutti's claims stood at six, not four. One more victory came on 27 November, when he shot down an "Albatros", that is to say a Brandenburg two-seater at Alano di Piave. Actually it was the enemy that had bounced him, but Cerutti, feeling the bullets, managed to climb and get on his attacker's tail, piercing its fuel tank. His victim may have been a German DFW C.5, as Feldfliegerabteilung 39 reported the loss of one of its aircraft forced to land near Mosnigo, a few miles downstream from Alano, on that day, and it was typical of German two-seater pilots to engage a fighter without hesitation.

Then, for a long time, Cerutti fell ill and didn't fly. In January and February 1918 the 79ª Squadriglia was engaged in the support of Italian operations for the conquest of the Tre Monti in the sector of the Altipiani, which had become the keystone of the Italian defense. On 5 January, escorting a SP in Val Brenta, Cerutti discovered from the explosions of gunfire the passage of an enemy fighter and a two-seater. He attacked and shot at both of them, but then his gun jammed. Later, he saw the two-seater getting back to its own lines, but not the fighter, so he presumed it shot down.

On 10 January Cerutti and Reali escorted a SAML over Primolano and they engaged combat with five Austrian fighters. Reali claimed one of them and then he had a dogfight with the other four, helped by Cerutti who could do very little, as his gun had jammed. Back to base, Reali counted twenty holes of incendiary bullets in his airplane. This event could be the basis for the 15th victory claim by the Austrian ace Linke Crawford, who claimed a Nieuport over Monte Asolone, south of Primolano.

Trying to look intense, but still smiling, Marziale Cerutti in the cockpit of Nieuport 11 serial 2126.

Another image of Cerutti with his Nieuport 11 2126 showing many details of this much successful fighter.

A Nieuport 11 of the 79ª Squadriglia is drawn out of its hangar.

On 28 January 1918, as the Italian troops attacked for the successful conquest of Col del Rosso and Cima Valbella, their first offensive return, there was strong air activity. In the morning Cerutti had a combat between Asiago and Meletta, at the end of which the ground observers saw an enemy going down. Cerutti was escorting a SP two-seater when he saw seven enemies, and attacked them alone, getting at a two-seater quickly dispatched over Fontanelle. Later on, in the early afternoon, there was a series of dogfights of Reali and Cerutti that, according to the observers, caused the fall of a fighter at Val Melago and another one at Valle Ronchi. For these events, he got his second Silver Medal "on the field" with this citation:

> "Bold fighter pilot who fought many combats with exceptional élan, courage and ability. In a long escort flight, he shot down an enemy fighter, saving our reconnaissance airplane. Taking off again on a patrol, together with another pilot he attacked a big enemy patrol and after a long dogfight he shot down a fighter, getting his eighth air victory. Sky of the Asiago plateau, 28 January 1918."

In February and March Cerutti claimed four victories, using the qualities of the new Nieuport 27 fighter, called in Italy Super Nieuport, fast and agile, but very difficult and dangerous.

On 5 February, together with *serg.* Giovanni Nicelli, he saw over Sasso Rosso an enemy two-seater spotting for the artillery, escorted by two fighters. The enemy single-seaters fled, and the Brandenburg became an easy prey for the two Italians. Trying to escape, it flew lower and lower until it was forced down near Case Girardi hitting the ground and crashing. The two Austrians got out of it unscathed and were taken prisoners. This airplane was described as painted dark green above, gray below.

On 11 February, during a patrol flight on the Altipiano, he spotted over Bertiaga three fighters, wrongly identified as Germans, escorting a two-seater. He attacked them and he shot down a fighter described as completely black that trailed a plume of white smoke and then crashed near Asiago. Actually, the strange looking fighter can possibly be identified with a Phönix D.1 of Flik 24, a new model at the front that may have confused the limited spotting capacity of the Italian pilots.

On 24 February, again cruising over Asiago, Cerutti saw a two-seater escorted by three fighters that were flying around Crespano. He attacked them and one fighter was shot down. This time, his victim was identified as a Berg, but again it may have been a Phönix of Flik 39/D.

On 25 March Cerutti was flying in the area of Montebelluna. Reaching the Montello, he spotted eight enemy airplanes. He attacked one that ran away. He followed it attacking and saw it crashing near Moriago-S. Pietro di Feletto. The squadron diary and the daily Aviation bulletin credit him with two victories, but there were many claims for other fighters in that spot and time, by the pilots of the 76ª Squadriglia and by the French, in a very likely case of over claiming.

Cerutti talking in front of an early model Spad 7. There is photographic evidence of many pilots of the 79ª Squadriglia with Spads, but no exact data has been found on the supply of one or more of these fighters to that unit.

Cerutti, still wearing his lady stocking, and Attilio Imolesi with a Nieuport 17 of the 79ª Squadriglia. The fighter carries no marking and its Vickers machine gun is not fitted.

Cerutti reached fame and glory. The magazine "Nel Cielo" placed him on the cover page on 10 February 1918 and again on 1 June. Its editor, Tullo Morgagni, fought a long battle against the opposition of the Commands, to spread the fame of the best pilots, and he was one of the creators of the concept of "ace" itself. His magazine held a "Premio Cacciatori del Cielo" (Sky fighters prize) in two editions, the first one closing on 31 October 1917. The second edition, sponsored by the Pirelli tire company, offered a prize of 20,000 lire for the pilot with more victories from 1 January to 30 April 1918. The list was updated with the data of the Comando dei Servizi Aeronautici of the Supreme Command, and the results were confirmed in August. Scaroni was the winner, followed by Reali and then by Cerutti, with five victories. Then in September the Commissariato Generale d'Aeronautica changed the confirmations, and thus also the results of the prize, so Cerutti and Scaroni were first, with the same number of victories. The two aces, both from Brescia, had another cover story on "Nel Cielo", one next to the other.

The magazine published in its 10 March 1918 issue a colorful report by Cerutti:

"On 28 January I shot down two, while on the ground there was the action for the conquest of Col del Rosso and Cima Valbella, one after the other. I found the first one near Fontanelle: it was a reconnaissance plane. With a few rounds I had the chance of shooting it down. With the other one, an Austrian single-seat fighter, I had a duel that was the most exciting ever. I found it alone cruising above Sasso Rosso. I followed it and attacked it and I understood immediately that he was a worthy enemy, although his plane was not as good as mine. It was a never-ending competition of turns, always trying to get the advantage of height. But I didn't want to let him get on top of me, at any cost. Then, as I tried to make a feint to get on his tail, I turned too harshly, I fell in a spin at full throttle and necessarily when I compensated for my mistake I was some 100 feet lower than my adversary.

He tried to use the advantage to hit me but I tried to regain the lost height and soon, with a hefty pull on the joystick, I was up again…and the dogfight started again. I tried to get him in my gun sight, he kept on evading me twisting like an eel… and this went on for 25 minutes. Finally I pretended to give up the combat and I let him get away for a while, then I pounced on his tail and shot some thirty rounds. Very soon he went into a spin and crashed to the ground.

On 5 February I was cruising with sergente Nicelli. We spotted an enemy two-seater over Sasso Rosso that was spotting for the artillery with an escort of three fighters. We both attacked forcing it to dive quickly leaving the escort fighters that ran away and disappeared. Nicelli on one side, me on the other side harried it, firing occasionally and forcing it to go lower and lower. It flew toward the valley of Case Girardi and that spelled its ruin. It was by now a few feet from the ground, surrounded by mountains on all sides. Nicelli and I got even closer, shooting all the time. It tried to get even lower to avoid us, but by now it was just two meters from the ground, and its landing gear struck a rock, broke off, the airplane lost its balance and fell near the Case Girardi. The two airmen were captured. They were unhurt, despite the bullets that passed between their seats and their rather rough landing.

My eighth victory was the one of 11 February. I found it in the sky of Bertiaga. It was a German fighter completely painted dark blue. With three other fighters it was escorting a reconnaissance "Albatros". This type of fighter was much better than all other types in climb, speed and above all agility. I attacked it resolutely, separating it from the other ones. It defended itself with very tight turns and banks. But I could aim it very well and after my second machine gun burst I saw a long trail of white smoke coming out of its fuselage. I thought that I had flames it so I stopped following it too close, expecting to see it crash in flames. Then I understood I had just pierced its water tank that was emitting a white plume of steam. I reached it again; we both were at low level over the Altipiano. I attacked it from the back shooting a full burst, very soon it banked, then it went down in a dive and crashed into the ground.."

In March the 79ª Squadriglia moved to the airfield of San Luca di Treviso. The allies knew Cerutti's exploits also, and in that month he was awarded the French *Croix de Guerre* and the "Star of Karageorge with swords (Silver)", a Serbian decoration.

By the end of spring, Cerutti was exhausted and waiting for a furlough. In mid May the municipality of Lonato invited him for a patriotic celebration, but he needed a special license form the command to participate. Finally, he got a furlough, but he returned to his squadron as soon as he knew of the expected Austrian offensive of mid June, so he was able to take part in the violent air combats that followed. Doctors had forbidden him to fly, for his exhaustion, but Cerutti didn't want to remain on the ground. In the evening of 21 June Reali and *s.ten.* Toffoletti took off for a patrol, and Cerutti joined them. They met over the Montello a flight of two enemy two-seaters returning to their lines, after, it was told, having dropped food to their troops; they attacked them and surely shot down one in flames, and possibly the other was also dispatched. Returning home, they spotted over the lines two more enemy two-seaters. They attacked them at very low level and forced one down. They all fell in the area between S. Lucia, Susegana and Ponte alla Priula. Actually, the daily Bulletin credits the three Italian pilots with three victories, over two-seaters that were dropping food to their troops, but the press reported that out of the three enemies, that were strafing, and not dropping food, only one was shot down, and the other two got back to their lines.

Three days later, on 24 June, again in the evening, Cerutti was flying with *cpr.* Edmondo Lucentini and *sold.* Carlo Rossi when they met another Brandenburg. "Spotted an enemy airplane in the direction of Susegana we attacked it closely, machine gunning it and forcing it to land north east of Susegana," said his concise combat report.

On 4 July his new victim was an observation balloon.

"Taking off on alarm over Cimadolmo, I flew toward one of the three drakens floating in the area between Oderzo and Fontanelle and precisely against the middle one and I attacked it twice. At my second burst I saw that a few meters above the ground it caught fire falling, and leaving on the ground a dense column of smoke. The other two drakens were immediately lowered."

Hansa Brandenburg C.I 29.05, of Flik 24/F, one of the victims of Cerutti, shot down on 24 November 1917. *(Photo Höfling)*.

On 10 August another enemy airplane was maybe shot down over the Montello, according to his flight log, but this event is not confirmed by the squadron papers, so it could be a wrong note pertaining to his victory of the next day. On 11 August in fact he attacked an enemy aircraft together with *s.ten.* Federici, but lost it from sight and couldn't get an exact confirmation. The Austrian two-seater dove from 2,000 to 150 meters before escaping Cerutti's vision:

"I was cruising above Moser when I saw an enemy airplane at 2,000 meters above Montebelluna. I reached it and fired a burst of 100 rounds. The enemy started a descent that to me seemed more than precipitous. I followed it shooting intermittently but it managed to get more and more distant. At 150 meters from the ground my gun jammed and I lost it from sight. I looked for it on the ground and in the air but I couldn't find it. It is unlikely that it fell nearby."

On 11 September 1918 Cerutti claimed one more victory; an enemy two-seater shot down at Ponte della Priula. At this time of the conflict, Austrian aviation was flying with much less intensity. Then he had to stop for medical reasons, and from mid September to mid October he didn't fly.

On 24 October 1918, exactly one year after the enemy breakthrough at Caporetto, the Italians unleashed their final offensive that broke the enemy front at Vittorio Veneto. For the aviators, the last days of the war brought a very intense activity, with many air combats and strafing attacks.

On 26 October Cerutti flew as an escort to the large Caproni 4 triplane bombers, and he met enemy fighters:

> "Returning from an escort to Caproni triplanes, I saw between Mandre and Malanotte three enemy airplanes lower than me. I reached them and aimed well at the last one, firing about sixty rounds. Then I saw a large burst of smoke and flame, and the enemy airplane falling. I attacked the other two, but as my gun jammed, I had to leave the combat. Against the latter one I fired about 150 rounds."

Cerutti sitting on the wheel of his most famous fighter, the Nieuport 27 with his flamboyant insignia, the letters MIR and the large Ace of Clubs card.

His victim could be identified as Aviatik D.1 338.26 of corporal Milan Soviansky, wounded in combat.

Cerutti claimed the last victory for the 79ª Squadriglia, just as he had claimed the first one. He got it in cooperation with *serg.* Chiri of the 78ª Squadriglia, in the early afternoon of the 27th, near Cittadella di Piave. The enemy fighter was seen to touch the ground and turn over its back. The Nieuport 27 of Cerutti, serial 19730, was adorned on its fuselage by the acrostic "MIR", Marziale Imperatore Romano, and by a big "ace of clubs" playing card behind the cockpit. Cerutti had a dogfight with a fighter:

> "I was cruising above Susegana when I saw the alarm direction pointing at Oderzo. I flew in that direction and I met another Hd – *that Cerutti describes as an Hd of the 78ª Squadriglia with its tail painted black, the one of Chiri, who on that day claimed three victories* – and we joined forces. Then we spotted two enemy fighters on our left. We attacked one of them, after a long pursuit I saw it diving steeply and for a moment I lost it from sight. Then I saw it again close to the ground. I don't know if it was trying to land, but it somersaulted and crashed. I fired about 200 rounds at it. Then I attacked the other one, fired 30 rounds and then I left the combat on account of a gun jam."

In little over two weeks, the final part of the war, he flew 22 war missions that included many dangerous strafing missions. Since June, when theoretically he was considered unfit for flying, he did not go away from the front and he didn't stop fighting, with 68 combat missions in four months. His palmarés consisted of three Silver Medals and the "exceptional commission as *Sottotenente* for war merit" in March 1919. His third "Argento", awarded in June 1921, had this citation:

> "Fighter pilot of untiring combative spirit, after having proved many times his exceptional valor, he yet managed to put on fire an enemy balloon and to shoot down three enemy

Cerutti with another Niueport 27 of the 79ª Squadriglia with attractive markings. We don't know, however, if he actually flew this particular aiplane, French serial 5907, Italian serial 11358.

aircraft. Sky of the Montello, Medium and Lower Piave, July-November 1918."

Cerutti showed all the qualities that make a true ace: master of his airplane, good acrobatic pilot and good shooter, but above all master of the tactical situation, able to impose his initiative and his will for combat. It is remarkable that so many of his victories were claimed against fighters, and that Cerutti was constantly able to impose and determine the combat, even when facing much larger formations.

The official victories of Cerutti are 17, placing him sixth in the roll of Italian Aces, right after Baracca, Scaroni, Piccio, Baracchini and Ruffo di Calabria. Cerutti himself claimed 19, but the total of his claims, as reported by his flight logbook and the Squadron papers, could be as high as 22. One has the impression that confirmation of claims came easier in the 79ª than in other units, but on the other hand in the mountain territory where the 79ª Squadriglia long operated, with scarce ground observations posts, and much of the vision obscured by high mountains, it was more difficult to get confirmations than in the plains full of trenches and observatories of the Isonzo and Piave fronts. Whatever his actual score, Cerutti was an exceptional fighter pilot and a natural-born ace.

In January 1919 he was briefly admitted to hospital for nervous exhaustion, a not uncommon occurrence to pilots, as the tension of the war years disappeared. His medical reports lists: fatigue, vertigo, malnutrition, tremors. Combat flying, especially over high mountains, was not easy, but Cerutti had held until the last day of the war.

In the immediate post-war period, as air squadrons were disbanded and pilots left the service, Cerutti didn't demobilize. We don't know why he chose to remain a military pilot. Surely, with his fame, his character and his qualities he would not have had problems readjusting to civilian life. The destiny of many of his colleagues, sergeants who were aces during the war, was not easy. After the war heroes are useless, the new young pilot Lieutenants didn't care much for old sergeants with their stories of forgotten battles. These old experts, often with little schooling, were left to the humble role of instructors or test pilots.

The career of Marziale Cerutti, instead, was totally different. The promotion to officer in 1919 for his war merit started an impressive rise in the ranks. In September 1919 he was posted to the famous 91ª Squadriglia "Baracca", now based in Trieste, and then in 1920 to the Mixed Squadron based in the Dalmatian City of Zara. Now a Lieutenant, in September 1921 he took part in the air races of Brescia, flying a beautiful bright red Hanriot. Other aces took part in that event: Mario Stoppani, now a test pilot for Ansaldo, flew an A.300C and Alvaro Leonardi, an instructor at Gabardini, flew a Gabardini monoplane. In October 1921 he lead a patrol of three Hanriots, with *serg.* Pereno and *serg.* Sartori (future test pilot at Macchi) in the First National Air Meeting at Florence. In October 1923 he passed in the ranks of the new Regia Aeronautica, the independent air force that was created on 28 March 1923. In 1924

At the Air Races of Brescia, September 1921, Marziale Cerutti flew a beautiful all red Hanriot with racing number 10. The Spad number 29 was flown by *cap.* Giuseppe Barattini, the Hanriot 31 by *serg.m.* Ennio Sorrentino, a four-victories fighter of the 81ª Squadriglia, the Breguet 14 I-BACF "25" by Giulio Baldi, later test pilot with Piaggio, the Caproni "23" by *t.col.* Armani and *serg.* Antonini, later test pilot with Caproni; and the Ansaldo SVA Brescia had as its pilot Battista Bottalla.

he was assigned to an experimental unit with the new Dewoitine monoplane metal fighters. In 1925 he was posted to the 1° Stormo Caccia, where he commanded the 84ª Squadriglia of the VII Gruppo, and then the 91ª.

From 1927 to 1929 he was at Cinisello, near Milan, with the 2° Stormo Caccia, then with the VII Gruppo of Rome, rising to the rank of Captain. In October 1927 he was punished with one month of arrest, for refusal to obey orders of a superior in rank. Meanwhile, in 1928, he married Gladys Wolter, a German girl, who was born in Korea and who gave him two sons. In April 1931 Cerutti became a Major, and from July 1931 to July 1933 he commanded the VI Gruppo Caccia of the 1° Stormo at Campoformido. Lieutenant Colonel in April 1934, Colonel in September 1935, he became commander of the 15° Stormo Bombardamento (bomber wing) at Ferrara and the next year, 1936, he commanded the 11° Stormo, also at Ferrara, in the period when it was re-equipping with the S.81 bomber. He was also appointed as judge in the Military Tribunal.

In 1938 he became vice chief of staff of the 3ª Zona Aerea Territoriale, (air region) and commander of Pantelleria air base. In June 1940, as Italy entered World War II, Cerutti was assigned to the Superior Command in North Africa, having reached the rank of General.

On August 3 1940, driving from El Feteja to Tobruk, his driver fainted and the car crashed at full speed against a cliff. Cerutti had a broken leg, and had to be sent to the famous Rizzoli hospital in Bologna for a long recovery. In April 1941 he had another quirk accident: he had to jump down to avoid a Caproni 164 trainer rolling along Gorizia airport with nobody on board, and this gave Cerutti another two months of disability.

In August 1941 he was appointed chief of staff of the Comando Servizi Aerei Speciali, that is to say the air transport component of the Regia Aeronautica, formed by the militarized airliners of the air transport companies. At the end of 1941 he was posted to the Air Force Staff (Stato Maggiore della R.A.) in Roma, and finally in February 1943 he became the commander of the Comando Caccia e Intercettori "Rex" of the 3ª Squadra Aerea, the group of units charged with the defense of Northern Italian skies from the raids of Allied heavy bombers.

Cerutti in a unarmed and highly polished Hanriot, racing in the Baracca Cup, 1922.

In the days of confusion after the armistice, 8 September 1943, Cerutti joined the Aeronautica Nazionale Repubblicana, the air force of Mussolini's republic, in the German-held northern part of Italy. He probably believed that remaining in the service was his duty. But somehow, he refused the spirit or the political dealings of the republic of Mussolini, because he was quickly dismissed, and he started to cooperate secretly with the Resistance. But he was not able in the field of politics. Other airmen who had served in the ANR, much more compromised than him, were accepted with all honors in the ranks of the winners. Cerutti was denied that. He was epurated and lost his rank, in the troubled days after the end of the war. Honor, justice and merit demanded a different end, but Cerutti didn't have the time to obtain justice. After the war, he was still in his village, showing his talent at the billiard table, but avoiding any memory or boast of his time as a fighter pilot. An enthusiastic motor man, he always drove at full speed, with whatever mechanical vehicle. On 26 May 1946, as he was riding his motorcycle, a concoction that he had assembled with pieces of different vehicles,

An unusual airplane for an uncommon pilot: Marziale Cerutti, right, at Aviano with the Travel Air Mystery Ship of the Italian Air Force.

very near his home on Lonato, in a road that is now named after him, he had an accident, trying to avoid a woman, Giuseppina Cozzaglio, standing in the middle of the road, and was killed.

The aeronautical presidium in nearby Desenzano did not come to his funeral, because he had been disgraced, but the whole town followed his last rites.

He received justice too late: he was exonerated from any accusation and reinstalled in his rank posthumously. Three years after his death, on 26 May 1949, the municipality of Lonato placed a plaque in the arcade of the city palace, in a public ceremony. It reads: "To Marziale Cerutti, General, decorated ace of aviation in the 1915-1918 war, with unconquered wings he marked in the sky of the Fatherland trails of glory. The Lonatese in memory and example, in the 3rd anniversary of his death. 26 May 1949."

General Rino Corso Fougier, left, commander of the Italian Air Force in 1942-43, talks with General Marziale Cerutti.

## Marziale Cerutti
## 22 victory claims - 17 confirmed victories

| Claim | Vict. # | Date | Sqn. | Aircraft | Hour | Place | Victim |
|---|---|---|---|---|---|---|---|
| 1 | 1 | 14.6.1917 | 79 | Ni.11 2269 | 8h | Mount Verena | Albatros |
| *Fw. Richard Schuster, WIA, Flik 21, KD 28.35* | | | | | | | |
| 2 | 2 | 22.6.1917 | 79 | Ni.11 | 6h10 | Mount Verena | Albatros |
| - | - | 2.8.1917 | 79 | Ni | 11h | Val d'Assa | EA |
| *[with serg. Imolesi, 79ª sq.?]* | | | | | | | |
| 3 | 3 | 26.10.1917 | 79 | Ni.17 | - | Volzano | Fighter |
| *[with serg. Poli, 70ª Sq?] Maybe: Addix WIA , Jasta 31, Alb. D.III or: Vzfw. Fritz Jacobsen WIA, Jasta 31, Alb. D.III* | | | | | | | |
| 4 | - | 30.10.1917 | 79 | Ni.17 3657 | 14h | Isonzo | Seaplane |
| 5 | 4 | 24.11.1917 | 79 | Ni.17? | - | Fonzaso | Two-seater |
| *Korp. Wilhelm Hanke KIA/Oblt. i. d. R. Robert Tárkányi KIA, Flik 24/F, Br. C.I 29.05* | | | | | | | |
| 6 | 5 | 24.11.1917 | 79 | Ni.17? | - | Mount Grappa | Two-seater |
| *Zugsf. Otto Kleber WIA/Lt. i.d.R. Karl Gsöllpointner WIA, Flik 17/D, Br. C.I 229.16* | | | | | | | |
| 7 | 6 | 27.11.1917 | 79 | Ni | - | Alano di Piave | Albatros |
| *Possibly: unknown crew, FA 39, DFW C.5 FTL* | | | | | | | |
| - | - | 5.1.1918 | 79 | Ni | - | Monte Corno | Fighter |
| 8 | 7 | 28.1.1918 | 79 | Ni | 11h15 | Asiago-Meletta | Two-seater |
| 9 | 8 | 28.1.1918 | 79 | Ni | 14h10 | Val Ronchi | Fighter |
| *[with serg. Reali, 79ª Sq.]* | | | | | | | |
| 10 | - | 5.2.1918 | 79 | Ni.27 | 9h | Case Girardi | Two-seater |
| *[with serg. Nicelli, 79ª Sq.] Korp. Karl Müller POW/Lt. i. d. R. Hugo Welser POW, Flik 45/D, Br. C.I 29.16* | | | | | | | |
| 11 | 9 | 11.2.1918 | 79 | Ni.27 | 8h10 | Asiago-Bertiaga | "German fighter" |
| *Oblt. Johann Kosch, Flik 24/F, Ph. D.I 328.06 FTL* | | | | | | | |

| Claim | Vict. # | Date | Sqn. | Aircraft | Hour | Place | Victim |
|---|---|---|---|---|---|---|---|
| 12 | 10 | 24.2.1918 | 79 | Ni.27 | 9h | Crespano | "Berg" fighter |
| *Maybe: Korp. Josef Santler KIA, Flik 39/D, Ph D.I 128.09* | | | | | | | |
| 13 | - | 25.3.1918 | 79 | Ni.27 | - | Moriago-S.Pietro di Feletto | EA |
| *[maybe with serg. Ciotti, 79ª Sq.]* | | | | | | | |
| 14 | 11 | 21.6.1918 | 79 | Ni.27 | 19h | S.Lucia-Susegana | Two-seater |
| *[with serg. Reali & s.ten. Toffoletti. 79ª Sq.] Korp. Raimund Praschl WIA/Oblt. Jakob von Jaskiewicz WIA, Flik 65/DS, Br. C.I 269.22* | | | | | | | |
| 15 | 12 | 21.6.1918 | 79 | Ni.27 | 19h+ | S.Lucia-Susegana | Two-seater |
| *[with serg. Reali & s.ten. Toffoletti. 79ª Sq.]* | | | | | | | |
| 16 | 13 | 21.6.1918 | 79 | Ni.27 | 19h+ | S.Lucia-Susegana | Two-seater |
| *[with serg. Reali & s.ten. Toffoletti. 79ª Sq.]* | | | | | | | |
| 17 | 14 | 24.6.1918 | 79 | Ni.27 | 19h30 | Ponte della Priula | EA |
| *[with cpr. Lucentini & sold. Rossi, 79ª Sq.] Maybe: Zugsf. Rudolf Herdin/Rittm. Karl Rossak, Flik 4/D, Br. C.I 369.67 FTL at Gajarine* | | | | | | | |
| 18 | 15 | 4.7.1918 | 79 | Ni.27 | 7h45 | Oderzo-Fontanelle | Kite balloon |
| *[with serg. Soliani, 79ª Sq]* | | | | | | | |
| 19 | 16 | 11.8.1918 | 79 | Ni.27 | 6h-6h50 | Moriago | Two-seater |
| *[with s.ten. Federici, 79ª Sq.]* | | | | | | | |
| 20 | 17 | 11.9.1918 | 79 | Ni.27 | 7h45 | East of Ponte della Priula | EA |
| 21 | - | 26.10.1918 | 79 | Ni.27 | 15h25 | Mandre-Borgo Malanotte | Fighter |
| *Maybe: Korp. Milan Soviansky WIA ,Flik 74/J, Av. D.I 338.26* | | | | | | | |
| 22 | - | 27.10.1918 | 79 | Ni.27 | - | Cittadella di Piave | Fighter |
| *[with serg. Chiri, 78ª Sq.] Maybe: Korp. Georg Hauger,Flik 74/J, FTL* | | | | | | | |

Marziale Cerutti with his colorful Nieuport 27, autumn 1918.

# Antonio Chiri

Close to the Gran Paradiso national park, in the region of Piedmont, the village of Locana lies below the ruins of a fortress of the Templar Knights. There Maria, wife of miller Matteo Battista Chiri, on 26 August 1894 gave birth to a child christened Antonio Quinto. Matteo Chiri immigrated to America, but when his wife died he returned to Italy, to take care of his family. He also died very soon, and twelve years old Antonio emigrated to France, reaching a cousin in Albertville, and learnt the trade of coppersmith. He spoke French well, but kept his Italian accent, so he decided to move to Normandy, where he could pass himself for a Frenchman from the south of France, avoiding the discrimination against Italians. He was a hard worker, and he became manager of production in a distillery, until he was called for his draft visit, which took place in the Italian Consulate in Paris, in early 1914. On 9 September the young man appeared in front of the Military District of Ivrea and a few days later he was assigned to the 17° Reggimento Artiglieria da Campagna (field artillery).

Maybe he had a penchant for mechanics, born from the gears of the mill, maybe he just loved flying. Antonio Chiri volunteered for aviation service and on 18 April 1915 he was accepted at the flying school of San Giusto. Nine days later he had his first lesson, a ground run on a Bleriot with Anzani 25-hp engine, but less than one month later he was excluded from pilot training. He remained at the airfield, with the less than exalted job of tinsmith. Chiri however did not give up, and he was readmitted to the school on 26 November.

Having now the rank of *caporale*, Chiri obtained his pilot's brevet, on a Bleriot, on 16 May 1916, and he flew four times on a Caproni at Malpensa. On 7 June he was shifted to Cascina Costa for conversion on the Nieuport fighter. This came very quickly, and on 8 June he was ordered to the war zone, assigned to the 77ª Squadriglia, a new unit that had just been formed at Istrana. The young pilot flew many training and familiarization flights, and on 17 August he was ready for his first combat mission from Cascina Farello, the new base of the 77ª.

On 11 October Chiri was transferred to the 78ª Squadriglia, and thus returned to Istrana, while at the end of the month he rose in rank to sergeant. On 31 October 1916 he was posted to S. Caterina airfield, base of the 70ª Squadriglia, as his squadron was often called to supply pilots and aircraft to reinforce the more active Isonzo front. By mid-November, he was back at Istrana. After a certain number of uneventful missions, on 24 November a shrapnel ball hit the Nieuport 11 of Chiri during a combat mission. The pilot brought down his fighter safely at Nove airfield, deserving a special commendation of the Supreme Command.

On 29 November Chiri had his first clash with enemy aircraft. Flying with *sottotenente* Giacomo Brenta above the plateau, they met an airplane and drove it off toward Pergine. From September to the end of 1916 the young pilot flew 13 combat missions and several test and exercise flights. In December he returned twice to S. Caterina. On 5 February 1917 he moved to Padua, and he remained there until the 28th, flying many uneventful standing patrol flights.

Sergente Antonio Chiri in training, with a Farman model 1914 serial number 432.

Nothing remarkable happened until 19 March 1917, when Chiri flying Nieuport 11 serial 1664 in an air patrol above the plateau attacked a Brandenburg of Flik 21 and scored his first victory. He attacked it five times at very close range hitting it in the fuel tank, radiator and engine, forcing Brandenburg C.1 27.55 to land at Rotzo, near Gallio, within the Italian lines. Its pilot *Korporal* Emanuel Mattl and its observer *Oberleutnant* Franz Resch, became prisoners of war. Chiri proudly had his picture taken in front of his victim. This victory was also reported in the War Bulletin the following day, and Baracca quoted it in his diary: "Yesterday a sergeant shot down another airplane at Bassano. It came down with a pierced tank, while of the airmen shot down on the 17th, the officer is dead, the other one is badly wounded. The Austrians dropped today a letter from the formers wife, asking for news, and we answered." (He is referring to Heintschel and Alapy, shot down by Olivari, and showing the general respect of Italian pilots for the enemy airmen).

One month and one day later, 20 April, the Piedmontese pilot was detached to Campoformido with three companions to escort a group of Caproni bombers. During this mission Chiri had five clashes with an Austrian fighter that was threatening one of the Capronis, and drove it away. *Capitano* Bolognesi, commander of the 78ª Squadriglia, remarked his performance and reported it to the command of the X Gruppo.

On 25 April 1917 Chiri was shifted once more from Istrana to S. Caterina. On 3 May he had a combat against seven enemy airplanes over Gorizia, shooting one of them down and, as he marked on his logbook, forcing two more to land, and yet claiming only one victory. Not much is known about this combat, as the 78ª back in Istrana did not report properly the sorties of its pilots flying with the new 91ª in S. Caterina. Two Brandenburgs of Flik 4 and Flik 34 were forced to land behind Austrian lines and there were no others Italian claims that day. It is possible that Chiri attacked the the first one together with Ruffo, the only other Italian to report a fight in this zone of the front: he claimed no victory but described a Nieuport as having attacked the enemy first, and maybe it was Chiri. One of the enemies was *Hptm.* Schonowsky, commander of Flik 34; all four enemy airmen escaped unhurt from the relics of their aircraft.

On 7 May he was posted to an advanced airfield, Bolzano (a village east of Gorizia, not to be confused with the city of Bolzano, Bozen) where in landing, possibly a Nieuport 17, he had on 11 May a spectacular accident. From Bolzano he flew daily combat missions, that included a duel with an enemy airplane that he drove away, on 21 May and another one four days later, when he engaged an enemy over Duino and pursued it all the way to Prosecco. On 2 June he flew his fighter back to S. Caterina and then to Istrana.

Back on the Asiago front, Antonio Chiri was still constantly engaged in combat missions: one or more often two each day. On 19 June he flew as escort to Caproni bombers, in the occasion in which his squadron mate *serg.* Fornagiari scored his first victory, and he also tangled with an enemy and drove it away. In July and August 1917, things didn't change: many transfers from Istrana to the Isonzo front and back, and duels with enemy airplanes on 1 and 10 July (on the northern front) and on 28 July, 17, 21, 23 and 25 August on the Isonzo front.

Finally Chiri had another victory on 26 August. On that day he defended from a fighter a Savoia Pomilio over Loque, and shot down the attacker. The crew of the reconnaissance airplane witnessed this victory, which was confirmed by the Italian command, although no corresponding Austrian data was found.

By this time the 78ª Squadriglia had moved to Borgnano, on the Isonzo front, and was receiving the first French-built Hanriot Hd.1 fighters. Besides several other combats, Chiri also had quite a lot of accidents, from which her always nursed his airplane back. An unusual one happened on 4 September, when he had to abort a mission because a part had detached from his machine gun damaging the propeller. Then again on 28 September he had a long duel with an enemy airplane, fired 300 rounds some of which, however, holed his own propeller. On 20 September commander Bolognesi issued a request for the concession of a Medaglia d'Argento al Valor Militare for Chiri, then on 1 October the gallant sergeant was rewarded with an eight days leave. He was back in action on 12 October and he even logged a flight, a standing patrol, on 24 October, the day of the Caporetto surprise, when most Italian air units were grounded by bad weather.

Twelve excellent brand new Hanriots of the 78ª were torched on Borgnano airfield in the rapid withdrawal after the Caporetto breakthrough. The unit then retreated and returned to Istrana, which was now a frontline airfield.

On 13 November Chiri, flying a Hanriot over the Piave, together with Guido Masiero, also of the 78ª, and other pilots, shot down a two-seater that crashed in flames near Arcade, behind the Italian lines. Fucini, who took part in this combat, wrote in his memories that the observer had jumped, to avoid being burned alive. This victim of Italian fighters was most likely a German aircraft of Flieger Abteilung (A) 232.

Another victory following a combat over Moriago on 20 November was not confirmed, but it generated another request for the concession of the Medaglia d'Argento to Chiri. This combat is very difficult to pinpoint: according to his logbook he claimed an airplane shot down over Moriago together with another Hanriot on the 19th, but according to squadron papers, this combat took place on the 20th, and it could be the same victory also claimed by Masiero, although the "Official List" of victories makes them appear as two separate victories on the 20th.

On 10 December Chiri joined Scaroni and Masiero for an attack to an Aviatik C.1 of Flik 58 that crashed in flames near villa Ancillotto with the death of its two occupants. The fifth confirmed victory arrived for Chiri in the hectic "battle of Istrana", 26 December 1917, when together with another Hanriot he shot down between Montebelluna and Biadene one of the German bombers participating in that mismanaged attack. His level of activity was amazing: during 1917 Chiri flew 249 flights, 190 of them combat missions, he had 22 air combats and claimed seven victories.

The New Year brought no respite to Chiri, who had his first combat on 12 January 1918 during a Caproni escort. On 24 January

A landing not perfectly executed, and a Nieuport 17 upside down. This photograph probably depicts an accident of Antonio Chiri at Bolzano on 11 May 1917.

One of the victims of Antonio Chiri, Brandenburg C.1 27.55 of Flik 21 shot down on 19 March 1917. The Austrian crew was captured.

1918 the 78ª Squadriglia left the VI Gruppo and joined the newly formed XV Gruppo. As night bombing raids went on, the Italians decided to get some squadrons away from Istrana, so the 78ª moved to Nove di Bassano together with the other Squadriglia in its Gruppo, the 79ª, and from its new base its flew its Hanriots supporting the ground action at Monte Valbella.

In the morning of 21 February 1918 Chiri, together with *sergente* Fornagiari and *soldato* Celio, both of the 78ª, escorted an SP.3 of the 33ª Squadriglia flown by *capitano* Fusco with the observer *tenente* Calì. A group of enemy aircraft attacked the Italian formation in the area of Gallio, and in the ensuing combat the Italians claimed two victories, one between Buso and Frenzela and another one near Gallio. Chiri later wrote a note of complaint asking to be assigned a victory, but failing, because it was not reported in the official list of victories issued after the war. His regret was well founded, because actually the Austrian ace Franz Lahner of Flik 55/J was wounded as he attacked a SP and was forced to land near Rodighieri, north of Asiago. His Albatros D.III 153.158 hit the ground near the lines and was destroyed by Italian gunfire. While Chiri got no confirmation, Lahner was credited with the non-existent destruction of the reconnaissance two-seater.

Chiri flew almost every day in February, and then on 3 March he went on leave, and returned to action on 25 March. On 3 April he attacked a kite balloon neat the Piave and then went down to strafe the trenches at Susegana. On 4 April he claimed a victory on the Montello, but this claim was explicitly dismissed after the end of the war. As spring came, Chiri kept up his level of activity that now included also the dropping of leaflets. At an unknown date, maybe 15 March, he raised in rank to *sergente maggiore*.

On 17 May Chiri participated in the combat that brought the demise of the Austrian ace Gräser, shot down behind Italian lines. Not only Chiri but also two pilots of the 79ª Squadriglia, Reali and Lucentini, took part in that combat and in 1919 were denied confirmation of the claim, that was assigned only to three pilots of the 91° Squadriglia: Nardini, Novelli and Magistrini.

Brandenburg C.I 27.55 dismantled and stored in an Italian hangar.

This is how Chiri described the tragedy of the enemy pilot in his combat report:

> "I took off for an alarm flight at 9.57 a.m. and I climbed toward Treviso, I was still lower and below an enemy patrol formed by six fighters and two recce planes. I could reach them only over the Piave, at a height of 3000 meters. The six fighters having escorted the two-seaters returned to their lines. By this time I was 200 meters above them and I attacked them. Meanwhile an Italian patrol formed by two Spads and one Super NI arrived. After a brief dogfight an enemy fighter fell down in flames near Pero, while the other ones got away. I shot 60 rounds."

A hand-written note in an Italian document mailed to the Red Cross gives the sad epitaph to the young Austrian pilot: "The re-

Antonio Chiri, third from left, with one of his Hanriots with a black tail. This one has serial 614. The airplane behind is the one of *cap.* Antonio Riva, his squadron commander.

Chiri with another of his Hanriots with a black tail. The object on the fuselage, partly covering the serial number 11457 (?), of course is just a bag, not a parachute.

mains were found in the relics of the fighter. His burial place is unknown."

Strangely, he didn't fly in the first days of the Austrian offensive that began on 15 June, but on the 19th he flew three sorties, two on the 21st and four on the 22nd. Another claim came after a combat on 23 June over S. Michele. Chiri was flying with *ten.* Fucini and *cpr.* Venier, and after the combat he went after five enemy fighters, that evaded him, and finally he returned with his fighter badly hit by A-A fire. This victory also failed to get its confirmation.

On 12 August 1918, Chiri, flying on patrol to Susegana, claimed an enemy aircraft shot down at Nervesa. On 20 August put up another claim for an enemy two-seater shot down near Moriago. This claim may correspond to the events of a Brandenburg of Flik 19/D, that reported a combat against two "Sopwiths" near Villanova, one of the dozens Italian villages with that name being next to Moriago. His many combat missions brought fruits, and in the early afternoon of 31 August Chiri claimed and finally received credit for another victory, an enemy airplane that was seen falling vertically near Mandre, shared with two other aces of 78ª Squadriglia . "As I was flying in patrol with Tenente Fucini and Serg Renella (sic) I saw a reconnaissance aircraft near Arcade. My patrol leader attacked first, I immediately moved to his side to help the patrol leader, I fired a long burst following it up to 700 meters beyond the Piave, there I lost it, because I had to avoid crashing into Serg Renella." The fight took place late in the evening, between 19.55 and 20.37 and the two-seater was maybe an Ufag C 1 of Flik 22/D that reported a fight at evening against...a Bristol fighter! The observer, *Lt.* Bauer, was badly wounded and died later on.

September and October went on as usual, with plenty of flights and missions for Chiri, doing all a fighter pilot could do, including, on 29 September, a standing patrol to protect a military parade from unwanted aerial visitors. Chiri claimed his last victories on 27 October, flying one of his Hanriots with a black tail:

> "I attack the reconnaissance that was bombing S.Polo, upward from below following it and shooting about 50 rounds. I saw it breaking up in flight and falling near the river of Monticano. About 5 minutes later together with sergente Cerutti of the 79ª Squad. I attacked 2 enemy fighters and after a sharp dogfight I shot one of them down near Cittadella and Codogne and forced the other one to land."

Again these two claims were not confirmed as victories to Chiri after the war, but possibly those two fighters belonged to Flik 74/J, which on that day reported the loss of two of its fighters, one missing and one forced to land. One must comment that the powers that be were particularly strict in confirming the victories of this gallant and long-fighting sergeant, who maybe deserved a higher place in the list of aces. In 1918 alone he claimed 11 victories but received credit for only one.

Furthermore, while the papers of the so-called Commissione Bongiovanni listed for him six confirmed victory (19.3.1917, 26.8.1917, 13.11.1917, 10.12.1917, 26.12.1917 and 31.8.1918) and one not confirmed (4.4.1918) without mentioning all the others, in the list of Aces issued in 1919 by the same Comando Generale d'Aeronautica he is credited with only five! The reasons why another one was deleted will probably never be known, it was possibly an error, which proves that the criticism of that list was not unfounded. While most of the victories acknowledged to famous aces of all nations generally suffer of a number of overclaims, re-

Above: The 78ª Squadriglia was surely the most colorful Italian unit. This Hanriot is decorated with the flags of the Allied nations: on the right side Italy, Belgium and the United States (an American soldier appreciates). On the other side there were probably France, Great Britain and Japan but we don't know in which order.

Below: A martial looking *serg.* Chiri with his black tailed Hanriot. The fighter behind, with the insignia of two crossed swords inside a circle, is the one of *serg.* Oreste Codeghini.

search in the Italian archives clearly proves that Chiri was not one of them and could have received more credit than the few victories listed above.

During the course of 1918 Antonio Chiri flew 184 flights, including 156 combat missions. He was awarded three Medaglia d'Argento al V.M., all of them in the field, a noticeable accomplishment for a "non-com", and contrasting with the small number of his confirmed victories. The first "Silver" was assigned for his victory of 19 March 1917, the second one for the combat of 26 August 1917 and the third one reported in its citation: "A fighter pilot, he showed the best military qualities. In 14 months he flew over 250 missions, engaging several combats and shooting down three enemy aircraft. Sky of the Piave, 27 August 1917 – 3 November 1918."

Peace came, but the year 1918 ended badly for Antonio Chiri, who on 21 December had a bad accident flying Hanriot # 815 from the airfield of S. Giacomo di Veglia. Just after takeoff, at a height of 200 meters, the controls of his fighter failed to respond, possibly because the plane had been parked in the open, with curious bystanders messing up with it. Despite the frantic efforts of the pilot, the aircraft crashed killing some spectators who were watching its takeoff. Chiri was badly wounded and was in coma; he was brought to an American hospital and his family received news of his death. Luckily, or thanks to good American medicine, the young pilot sur-

Right: Antonio Chiri in later years, in officer uniform, proudly showing his three Medaglie d'Argento.

Below: Instructor at the seaplane flying school at Como, Chiri in the cockpit of a Caproni 100 floatplane primary trainer in the late 1930s.

vived, and came out of his coma on 6 January 1919, a date that Chiri then considered a sort of second birthday. He was then posted to the Deposito Aviatori of Turin. On 11 November 1919 Chiri was sent on permanent leave, he was nominally assigned to the 17° Reggimento Artiglieria da Campagna in July 1920, but this was only a formality, because since 16 September 1920 he left the armed forces. On 8 August 1921 King Victor Emmanuel visited the College for the Rehabilitation of Mutilates of Turin, Chiri was among the guests and the king stopped to talk with him.

Despite a permanent leg disability, Antonio Chiri continued his military career in the reserve. In 1934 he was transferred from the army to the air force, with the rank of *sottotenente*, he was periodically called back to service flying the required number of training flights. In 1936 he was an instructor at the seaplane flying school of Como. His private life went on smoothly, he worked as an employee of the Post Office of Turin, had a family and retired with the rank of provincial director of the mail. The destiny of Antonio Chiri had been similar to the lot of so many other pilots of World War I: a brief period of fame, followed by an obscure life of work and family. He died on 6 January 1971 in Turin, where he is buried in the Monumental Cemetery.

## Antonio Chiri
## 18/20 victory claims – 5 (or 6) confirmed victories

| Claim | Vict. # | Date | Sqn. | Aircraft | Hour | Place | Victim |
|---|---|---|---|---|---|---|---|
| 1 | 1 | 19.3.1917 | 78 | Ni 11 1664 | - | Rotzo-Gallio | EA |
| *Korp. Emanuel Mattl POW/Oblt. Franz Resch POW, Flik 21, Br. C.1 27.55* | | | | | | | |
| 2 | - | 3.5.1917 | 78 | Ni | - | Gorizia | EA |
| - | - | 3.5.1917 | 78 | Ni | - | Gorizia | EA |
| - | - | 3.5.1917 | 78 | Ni | - | Gorizia | EA |
| *Maybe: Hptm. Bruno Schnowski UNH/Fähnr. i. d. R. Hugo Hierse UNH, Flik 34, Br. C.1 129.40 or: Fw. Josef Franke UNH/Lt. Arpad Pindter Von Pindtershofen UNH, FLIK 4, Br. C.1 129.41 FTL* | | | | | | | |
| 3 | 2 | 26.8.1917 | 78 | Ni? | - | Loque | Fighter |
| 4 | 3 | 13.11.1917 | 78 | Hd? | 15h45 | Arcade | German two-seater |
| *[with ten. Masiero, 78ª Sq., ten. Fucini & serg. Razzi 76ª Sq.] Maybe: Lt. Eduard Schulze Watzum KIA /Lt. Kamillo Ehrler KIA, FA( A) 232* | | | | | | | |
| 5 | - | 20.11.1917 | 78 | Hd? | - | Moriago | EA |
| *[Maybe with ten. Masiero, 78ª Sq.] Maybe: Offstv. Sokolowski/Lt. Fritz Kretschmer, FA(A) 219 FTL or: unknown pilot/Lt. d. Res. Binder KIA, FA (A) 219* | | | | | | | |
| 6 | 4 | 10.12.1917 | 78 | Hd | - | Villa Ancillotto | Two-seater |
| *[with s.ten. Scaroni, 76ª Sq. & ten. Masiero, 78ª Sq.] Maybe: Korp. Josef Bantz KIA/Oblt. Felix Kanitz KIA, Flik 58, Av. C..I 37.24* | | | | | | | |
| 7 | 5 | 26.12.1917 | 78 | Hd | - | Montebelluna/Biadene | Aviatik |
| *[with other pilots]Possible victims:<br>Lt. Kessler WIA/Lt. d. Res. Edebohls KIA, FA 2, DFW C.V<br>Uffz. Willi Petzold KIA/Lt. d. Res. Wilhelm Voig KIA, FA 2, DFW C.V<br>Vzfw. Karl Uecker KIA/Lt. d. Res. Heinrich Pfeiffer DOW, FA 2, DFW C.V<br>Vzfw. Pohlmann POW/Lt. Schlamm POW, FA (A) 219, DFW C.V.* | | | | | | | |
| 8 | - | 21.2.1918 | 78 | Hd | 10h55 | Buso/Frenzela | Fighter |
| *[with serg. Fornagiari, 78ª Sq.] Korp. Franz Lahner WIA, Flik 55/J, Alb. D.III 153.158.* | | | | | | | |
| 9 | - | 21.2.1918 | 78 | Hd | 10h55 | Gallio | Fighter |
| 10 | - | 4.4.1918 | 78 | Hd | - | Montello/Valdobbiadene | EA |
| 11 | - | 17.5.1918 | 78 | Hd | - | Pero | Fighter |
| *[Confirmed only to serg. Magistrini, Nardini & ten. Novelli, 91ª Sq.] Lt. Franz Gräser KIA, Flik 61/J, Alb. D.III 153.221.* | | | | | | | |
| 12 | - | 23.6.1918 | 78 | Hd | - | S. Michele | EA |
| *[with ten. Fucini & cpr. Venier, 78ª Sq.]* | | | | | | | |

| Claim | Vict. # | Date | Sqn. | Aircraft | Hour | Place | Victim |
|---|---|---|---|---|---|---|---|
| 13 | - | 12.8.1918 | 78 | Hd | - | Nervesa | EA |
| 14 | - | 20.8.1918 | 78 | Hd | - | Moriago | Two-seater |
| *Maybe: unknown pilot UNH/Oblt. i. d. R. Emil Kowarsch WIA, Flik 19/D, Br. C.I 369.143 or: Unknown pilot UNH/Lt. i. d. R. Ludwig Eisner WIA, Flik 5/F, Ufag C.I 161.105* | | | | | | | |
| 15 | 6 | 31.8.1918 | 78 | Hd | 13h30 | Mandre | Two-seater |
| *[with ten. Fucini & serg. Rennella, 78ª Sq.] Maybe: Stfw. Franz Prochàzka UNH/Lt. i. d. R. Emil Bauer DOW, Flik 22/D, Ufag C.I 161.10.* | | | | | | | |
| 16 | - | 27.10.1918 | 78 | Hd | 14h40 | Vazzola/Montegano | Two-seater |
| *[maybe with serg. Rennella, 78ª Sq.]* | | | | | | | |
| 17 | - | 27.10.1918 | 78 | Hd | - | Cittadella/Codogne | Fighter |
| *[with serg. Cerutti, 79ª Sq.] Maybe: Korp. Adolf Wissinger MIA, Flik 74/J* | | | | | | | |
| 18 | - | 27.10.1918 | 78 | Hd | - | Cittadella/Codogne | Fighter |
| *[with serg. Cerutti, 79ª Sq.] Maybe: Korp. Georg Hauger, Flik 74/J, FTL or: Stfw. Ferdinand Udvardy, Flik 42/J, FTL* | | | | | | | |

## Bartolomeo Costantini

Vittorio Veneto is the town in the Veneto region, from which the final battle of World War I is named, in which the Austrian army was routed. For most Italians this name still stands for the Great War and victory, and it was much used: a battleship in World War II, and a cruiser today use this name. At the turn of the century the town was called just Vittorio, and it got its name from King Victor Emmanuel II, as it was founded in 1866. It then had about 2,000 inhabitants and it was a crossroad toward the Alpine passes. Today very little remains of that small town: it was heavily bombed in World War II, it was rebuilt and it had an economic boom in the Sixties, with considerable development. The visitor can still find the living memories of the Great War: local libraries are stocked with books, from military texts to the minor but interesting personal memories. In the center of town there is the Museum of the Battle, that preserves many exhibits of World War I, including the remains of a German airplane shot down by Bartolomeo Costantini, also known as Bortolo or Meo. Vittorio in fact was his birthplace, on 14 February 1889 in via del Boschetto, son of Carlo, an engineer, and Eugenia Gei. He had an older sister, Marianna, and a younger brother, Severino.

His mother died when he was eight years old and he was raised according to the strict ways of middle-class of those years. He attended the high school of Belluno and then he went to the Polytechnic University of Turin, developing a strong interest for mechanics. He hadn't completed his courses when, on 30 September 1909 he volunteered in the Brigata Ferrovieri del Genio, (railroad brigade of the Army engineers) for the officer-training course. One year later he was dismissed but he was recalled to arms at the time of the war of Libya, 1911, and was assigned to the 5° Reggimento Genio. He was commissioned as a *sottotenente* on 22 January 1912 and then he managed to be assigned to the Battaglione Aviatori. He got license of military pilot n. 177 on 13 September at Aviano airport. On 3 March 1913 his father died.

During his period in Turin Costantini had made many acquaintances among enthusiasts of the new world of automobiles. He was a close friend of *ingegner* Giovanni Marmaglia, the son of a rich banker who had bought the Aquila Italiana car factory in Turin, and who gave the opportunity to Costantini to race his cars in 1914. He couldn't race very much because war was looming, so he was recalled to service as a military pilot on the Bleriot monoplane in September 1914. He briefly served with different Bleriot- equipped units, the 3ª, 13ª and then 14ª Squadriglia. At the outbreak of war he belonged to the 3° Reggimento Fanteria da Fortezza and served with the 13ª Squadriglia Bleriot, which operated from Pozzuolo del Friuli, before being disbanded in July 1915.

On 16 February 1916 he was promoted to *tenente*, then on 15 July 1916 he was posted to the Cascina Costa pilot school for training on the *"Nieuportino"*. On 12 August 1916 he was sent to the front, assigned to the recently formed 78ª Squadriglia Caccia.

The first base of the 78ª Squadriglia was the large La Comina air base near Pordenone. All its personnel under commander *cap.* Bolognesi was new, but among them there were many pilots who were to become top fighter pilots, like Brenta, Chiri, Magistrini and Di Rudinì. The latter, son of a former prime minister and future commander of the 76ª Squadriglia, wore spectacles and gained the reputation of an excellent fighter instructor.

Costantini, first from left, next to a Bleriot XI, serial 297, of the 13° Squadriglia. The other pilots are Lombard, Tacchini and Carnevale.

Costantini, sitting above the cockpit of a Nieuport, sports a romantic scarf in this pre-war photo

When Costantini reached his unit combat missions had not yet started. It took some time for the 78ª to become operative with its new Nieuport 11s. The squadron was moved to the front of Asiago, and set up its base at Istrana, then a new and unknown field, that became later on one of the most important air bases of Italy, and is still one of the main bases of the Aeronautica Militare.

The northern front facing the Trentino area did not buzz with air activity: there were a few Austrian units mostly based at Pergine airfield, at that time equipped only with two-seaters; there were no fighters facing the Italian Nieuports. Service mainly consisted of dull standing patrols along the frontline to preclude enemy reconnaissance flights, escorts to slow two-seaters or to the Capronis that occasionally bombed the enemy rear. The rare combats always ended with the enemy two-seaters escaping. During the whole period when Costantini was with the 78ª only one airplane was shot down, by *sergente* Chiri. It was an uneventful period, but probably useful for practice. In early 1917 Costantini, still assigned to the 78ª Squadriglia was detached to the Isonzo front, leading a flight that was sent as a reinforcement for the coming offensive, and that was attached to the Caudron-equipped 43ª Squadriglia. At the same time these pilots, coming from different units, were trained on the Spad 7, a new fighter recently arrived from France, with different characteristics from the Nieuports then equipping Italian fighter units.

The big change came when Costantini was assigned to the 91ª Squadriglia, formed just one month before. This unit had the best pilots, most of them officers, the best fighter, and an exceptional leader, *capitano* Baracca. The exact day of the arrival of Costantini is not known, it was end of May - early June 1917: his name is reported for the first time on the squadron diary on 6 June. That day the 91ª moved from the Isonzo to the Asiago front, to support the coming offensive of the Ortigara. So Costantini returned to Istrana, his original base. But with a fighter like the Spad 7 it was much easier to intercept enemy aircraft than with the obsolete and slow Nieuport 11.

His squadron mates had combats most every day, but Costantini met the enemy only once, on 26 June. He was flying alone in the early morning in a cruise above the Pasubio mountain when he spotted two aircraft, a Brandenburg escorted by a fighter, coming from Val Lagarina. The fighter ran away and Costantini was free to attack the two-seater, but his machine gun jammed after 15 bullets and the enemy could get away unscathed toward Trento.

This part of the front seemed to be jinxed for Costantini, as was proven a few days later, before the 91ª returned to the Isonzo. On 1 July he spotted above Tresche an "Albatros" enemy two-seater, that is to say a Brandenburg, and he dove to attack. When he tried to pull the throttle his engine failed and the Spad fell in a spin. He recovered, attacked and fired fifty rounds, but water temperature had risen above 90 deg and he had to quit, landing on the nearest airfield. When on the ground he discovered that the enemy observer was a good shooter, and had pierced his radiator and right wing.

The 91ª Squadriglia returned to Santa Caterina airfield and was immediately back in the thick of action. Costantini performed well in many escorts to Capronis and photographic reconnaissance flights, exploiting the high speed of the Spad. Sometimes, despite his efforts, his sortie was in vain, like on 8 July, when he shot 12 glass plates but development failed. Other times bad weather frustrated his missions.

On 14 July 1917 he witnessed the combat between the Spad of *tenente* Ruffo and an enemy fighter, but before he could join, the latter was dispatched. He noticed with horror a human body jumping from the burning aircraft at 4,000 meters; he was *Feldwebel* Karl Sailer of Flik 35.

During the great August offensive, the 11th Battle, Costantini was always in action, flying mainly as escort to the slow Capronis that bombed Austrian positions on the Bainsizza plateau, then, his job completed, he often went down to strafe the belabored Austrian troops. Still, no victories came.

Wearing a leather flying suit, Costantini, an inveterate smoker, doesn't seem concerned by the fact that the Nieuport 17 of the 78ª Squadriglia behind him is being refueled.

On 14 August 1917, a day of intense activity, as the Austrian launched a mass attack of Navy seaplanes and Army bombers against Venice, the Italian Capronis raided Assling. The 91ª Squadriglia sent three escorting fighters, including Costantini, but the squadron diary reported only a combat for Di Rudinì with no claims. Yet, the Silver Medal later assigned to Costantini mentioned in its citation his actions on that day. One Austrian fighter, Brandenburg D.I 28.66 of *Oblt.* Benno Fiala, did get shot up in the radiator and had to force land at Haidenschaft.

On 10 September he had an animated combat together with Olivari against a two-seater escorted by two fighters. Costantini engaged the fighters and Olivari the other plane. Having driven off his two enemies, he managed to shoot briefly against the two-seater, without observable result. Actually Brandenburg 329.12 of Flik 35/D was so badly damaged that it turned over and was wrecked on landing at Wippach airfield. Its observer, *Oblt.* Rudolf Hiller broke his safety belts in the crash and was projected out of the fuselage, suffering very serious wounds. The two Austrians claimed one Spad shot down, maybe the one of Olivari, who had had to break off attack with a jammed gun, but both Italian fighters returned safely from this combat. Thus, the one who had really shot down an enemy didn't claim it, while the ones who hadn't hit anything got the confirmation of a non-existent victory. It must be remarked that many victory claims by Austrian reconnaissance airplanes don't find any correspondence in the data of Italian losses.

The terrible days of the Caporetto battle brought his first confirmed victory to Costantini. Those were sad days for the 91ª Squadriglia, that in the first day of combat lost two of its pilots shot down by German fighters, but by the evening of 25 October the unit had shot down six enemy aircraft. One of them was a German "Aviatik" flamed over Tolmino, the victim of Costantini, recently promoted capitano, together with Ruffo.

Costantini in the cockpit of his Spad 7 of the 91ª Squadriglia, with his personal insignia of a black pennant. As usual, he sports a cigarette in his mouth.

The next day, while the gravity of the situation was becoming apparent, Costantini took off alone and very soon met another "Aviatik", as the Italians called the DFW C.V. After a brief duel the two-seater crashed over Castelmonte, behind the Italian lines, its two occupants killed. *Serg.* Fanti, of the 76ª Squadriglia took part in this combat.

On 27 October the order came to leave S. Caterina and to retreat. All that could not be carried was burned, the squadron moved under a pouring rain to la Comina, then on 31 October further back to Arcade, behind the defense line of the Piave. The next day it moved to Padua. It had been a tough period, but despite the dramatic retreat the 91ª Squadriglia shot down 13 enemy aircraft. Reinforced with eight new pilots it was ready to give battle to the enemy if it tried to cross the Piave.

On 5 November Costantini had a combat over Belluno in which his rudder was hit. His third victory came on 23 November. He was

Costantini, still a lieutenant, at Santa Caterina in front of the Spads of the 91ª and the Nieuport 11s of the 82ª Squadriglia.

A complete view of the Spad 7 of Costantini showing the black pennant and lack of roundels on the top wing.

flying with one of the new pilots, *sergente* Magistrini. Over Cornuda they spotted a German two-seater and shot it down behind the Italian lines after a brief combat. Its occupants, *Vzfw.* Krauser and *Lt.* Strathmann of FA (A) 232, were found dead in the wreck. One week later, on the 30th, he repeated himself shooting down one more German airplane behind the Italian lines with the help of two wingmen, *tenente* Adriano Bacula and the Argentinean aspirante Edoardo Olivero. Observation post 28 confirmed this victory: "One enemy airplane burned with machine gun by one of our daring fighters crashed at 11.30 near Rivasecca."

In the following four months there was a lull in operations, and not just Costantini but his whole squadron didn't repeat such successes. The unit was based at Padua, to keep its role of protection for the Supreme Command, now based at Abano, and to be able to act both on the Piave and on the plateau fronts, but it was rather far from the frontline, and that airfield was also often covered with fog. The new Italian built Hispano engines of the Spads were troublesome and unreliable, while the French couldn't supply more original engines. Moreover, German bombers attacked the airfield and in the night of 21 February they put out of action all the fighters of the squadron.

At the end of December 1917 Baracca, Ruffo and Piccio were invited to Turin to evaluate the new fighters of Italian design and for a few weeks Costantini became acting commander, and its many duties kept him away from flying. The situation didn't improve, from 20 January 1918 to 19 February he flew only seven times, and this situation went on until 15 April, when he was finally sent on "winter leave", even if winter was long gone.

We don't know where he went on leave, his hometown Vittorio was now in Austrian hands. When he returned, on 8 May, the situation had improved, the 91ª Squadriglia was now based at Quinto di Treviso, closer to the frontline and equipped with a good number of Spad 13 that had joined the older Spad 7.

On 19 May after almost five months Costantini met the enemy again. That day his squadron was escorting a formation of Caproni bombers in action over Feltre. Eight Spads lead by *magg.* Baracca met a gaggle of nine enemy fighters and attacked. The dogfight was very violent, the Austrian fighters were kept away from the Capronis, and two of them possibly were damaged by the Italians.

When the Austrian Kaiser's army launched its offensive, on 15 June, Costantini and his squadron were ready to meet that challenge, but for the captain never-ending gun jams and engine trouble always precluded him victory. For instance on 1 July he was forced to drop out of his formation twice for failures. When everything worked properly, then enemy aircraft ran away from a combat. On 5 July Costantini had a tough dogfight with four fighters during an escort flight for Capronis. He had to wait until 12 August 1918 to score another victory, a "type D.III" fighter that he met during a lonely cruise in the morning and shot down, with the collaboration of *cap.* Riva, commander of the 78ª Squadriglia, the enemy falling vertically over S. Lucia.

On 22 August he scored his sixth and final victory. Once more, he was flying alone over the lines when he intercepted a patrol of enemy aircraft. The squadron diary describes that unusual event with a sort of amazement:

> "Capitano Costantini, cruising from 9.50 to 11.50 attacked an enemy reconnaissance two-seater escorted by three fighters and shot it down in flames near Marano di Piave; the observer saved his life jumping with a parachute from the plane: it is the first time that such an event happens and that a parachute is actually used on an airplane."

Actually the airplane he shot down was a fighter from Flik 42/J of the pilot Friedrich Hefty, an ace, and it was the first parachute rescue on the Italian front. Possibly the sight of a parachute convinced Costantini that he had hit a two-seater, because so far parachutes were too big to be used on single-seaters. The Germans however had developed a new model that could fit into smaller cockpits and the Austro-Hungarians had received a few of them.

The enemy pilot reported his own version of that event:

> "We were flying in a patrol of three, with the engines at low speed, along the Piave at a height of 4,800 meters. As I was arranging the belts of my parachute to lock them across my breast I hear behind me the rattle of machine gun fire, so I gave full throttle. Right away however my aircraft began to emit smoke, the firewall broke and a big fire erupted from the fuel tank. I understood immediately that I could do nothing but jump. I fixed the joystick for horizontal flight, I unfastened my seatbelt, I climbed on the seat so as to have my hands above the cockpit and I jumped head down from the left side of the airplane. I did that just in time, because flames had reached me and I was about to faint for the heat.
>
> After the launch I briefly lost consciousness, but I woke up as soon as I felt a big bang and a pull on my chest. I looked up and saw that the parachute had deployed; but it was turning and leaning very much… I also saw my airplane burning and falling down in spirals, then exploding with a big bang while still in the air. My sense of security didn't last long. At the height of 3,000 meters I saw close to me an enemy Sopwith that… shot at me a burst of phosphorus munitions."

Costantini, standing next to the rudder of his Spad, shows how small that fighter was. Careful study of the original shows that the rudder has serial number S4693, but a small inscription on the rudder says VII 1331.

Hefty managed to reach the ground without further damage near Ballon-Kompanie 16 and was rewarded with a furlough. Two British fighters were witnesses to Costantini's lonely attack to the Austrian formation, and probably it was them who strafed the defenseless enemy pilot.

On 25 September Costantini went on leave, but he returned in time to take part in the final great battle that was then named after his hometown. At the end of the war he had flown about 180 combat missions (120 according to other sources) with 18 combats and 6 victory claims, all of them confirmed and assigned to him. In the past the dates of his victories were reported erroneously, confusing them with the victories of *tenente* Andrea Costantini, another gallant pilot who flew SVAs.

Costantini was decorated with two Medaglie d'Argento. His first one reported in its citation his victories of 25 and 26 October 1917; the second one referred to the combats of 20 (actually 23) and 30 November 1917, 12 and 22 August 1918.

On 31 January 1919 he asked for a leave to attend the family interest, probably to check what was left of his family properties in Vittorio. Then on 3 September 1919 he was sent on permanent leave.

His period as an airman was over, he returned to Turin to his original love, racing cars. He waited in vain for a Fiat 1500 that his friend ingegner Cappa (who designed many car engines and a beautiful 400 hp aviation engine) was preparing, so he opted for a Bugatti 1500 and in the workshop of a friend of his in Milan tuned and hotrodded it. He won some competition in its class, then he realized

Right: In full flying combination, Costantini is ready for take-off in his Spad

This close shot of Costantini in his Spad shows details of the windscreen and gunsight

Right: The remains of a DFW C.5 shot down by Costantini, now exhibited in the Museum of the Great War at Vittorio Veneto. The commemorative inscription on the propeller blade describes this relic as an Aviatik, the name applied to all German two-seaters by the Italians.

Below: The propeller of Costantini' Spad is turning but there are still blocks at the wheels. One supposes that they were pulled away using strings, and not crawling under the propeller

that the Bugatti was obsolete and in 1922 he raced for the Bianchi automobile company, without success. Then he bought an Amilcar 1100 and won first place in its class in the Coppa delle Alpi.

Italian car companies ignored him, but the famous constructor Ettore Bugatti appreciated him and in 1923 he called him to France. He moved to Molsheim where he was test pilot and sport director, although for the latter job he never received any wage.

His capacity became apparent in the years 1925-1926: he participated in twelve races, winning six of them, getting second place three times and third place once. In 1925 there was the Grand Prix of France on the Montlhéry racing track, that was marred by the death of the champion Antonio Ascari. While the Alfa Romeo team withdrew from the race as a sign of mourning, Costantini ran a fantastic race and won the first place in the category of cars without compressor. Also in 1925 he won the Targa Florio, in Sicily, beating the strong Peugeot of Boillot and he won again in 1926, when Bugatti cars conquered 1st, 2nd and 3rd place at the Targa Florio with Costantini, Minoia and Goux. In those years the Targa Florio was the car race with the largest prize money in Europe. But during that race a pilot died, Giulio Musetti, killed with his Delage. Costantini was much impressed, and the next year he kindly refused the invitation of Cesare Florio. He quit car racing. When a journalist friend asked him to write about his racing triumphs, he answered:

Costantini in his new vocation, as manager of the Bugatti car racing team.

"Your kind letter shows the soul of an old true friend, the memories of youth now, alas, long gone. So I am sorry that I have to answer you no. I swore to myself never to write anything for newspapers, nor to give out photographs. This, obviously, not because I despise the press, but because I hate bluff and I don't like the current emphasis for sportsmen who, if they are pure like me, do it just for their pleasure, and therefore they gain no merit, just a personal satisfaction. Forgive me, and write nothing about me. My sports career anyway is over, I am not going to race any more. I am too old, and racing doesn't amuse me any longer. I give myself fully to technique, and I don't want to be distracted."

Costantini and his Spad 13 with Roman numeral III in the second half of 1918.

He remained with Bugatti as a technician until 1937, and much of the fame of this glorious car company is due to his work. Those who knew him, described his personality as "self-controlled, precise, aristocratic".

In 1929 he was ready for graduation at the Polytechnic of Turin and, as far as is known, he never married and had no children. His last known address was Milan, where he lived in 1930. He still belonged to the reserve of the Air Force, reaching the rank of lieutenant colonel in 1935. In 1936 he moved from the Ruolo Naviganti (roll of flying personnel) to Ruolo Servizi (support personnel) exempted from the rank of military pilot.

In 1937, having left Bugatti he joined Alfa Romeo, called by *Ing*. Gobbato with the task of technical commercial inspector, then in 1939 he was responsible for sport preparation of the racing department of Alfa Romeo. In 1939 he was also drafted for a partial mobilization, but he was released and placed in absolute discharge that same day, possibly for medical reasons.

Costantini fell seriously ill and died soon after, in the "Principessa Jolanda" hospital of Milan on 19 July 1941, at the age of 52. He is now buried in the cemetery of Ceneda, a suburb of Vittorio Veneto, under the sky that he dominated with his fighter.

## Bartolomeo Costantini
## 6 confirmed victories

| Claim | Vict. # | Date | Sqn. | Aircraft | Hour | Place | Victim |
|---|---|---|---|---|---|---|---|
| - | - | 10.9.1917 | 91 | Spad 7 | 10h | Isonzo | Brandenburg |
| *[with ten. Olivari, 91ª Sq.] Zugsf. Josef Treidler UNH/Oblt. Rudolf Hiller WIA, Flik 35/D, Br. C.I 329.12* | | | | | | | |
| 1 | 1 | 25.10.1917 | 91 | Spad 7 | 12h20 | Tolmino | Aviatik |
| *[with cap. Ruffo, 91ª Sq.] Lt. d. Res. Paul Dotzenrodt KIA/Lt. d. Res. Robert Lessing KIA, FA.17 or: Lt. Walter Conrad KIA/Lt. Albert Tawainer KIA, FA (A) 232 or: Vzfw. Friedrich Moritz KIA/Lt. Franz Neumarket KIA, FA 39.* | | | | | | | |
| 2 | 2 | 26.10.1917 | 91 | Spad 7 | - | Castelmonte | Aviatik |
| *[with serg. Fanti, 76ª Sq.] unknown pilot/Lt. d. Res. Erich Lucas KIA, FFA (A) 219 or: unknow pilot/Lt. d. Res. Gustav Krannig, KIA.* | | | | | | | |
| 3 | 3 | 23.11.1917 | 91 | Spad 7 | 11h | Cornuda | German two-seater |
| *[with serg. Magistrini 91ª Sq.] Vzfw. Wilhelm Krauser KIA/Lt. d. Res. Paul Strathmann KIA, FA (A) 232* | | | | | | | |
| 4 | 4 | 30.11.1917 | 91 | Spad 7 | 11h30 | Rivasecca | German two-seater |
| *[with ten. Bacula & asp. Oliviero, 91ª sq.] Lt. d. Res. Hans Schulze, FA 14* | | | | | | | |
| 5 | 5 | 12.8.1918 | 91 | Spad 13 | 11h | S. Lucia di Piave | D.III fighter |
| *[with cap. Riva, 78ª Sq.] Oblt. Ludwig Stillmungus KIA, Flik 40/P, Phönix D.I 128.11* | | | | | | | |
| 6 | 6 | 22.8.1918 | 91 | Spad 13 | - | Mareno di Piave | Small two-seater |
| *Offstv. Friedrich Hefty WIA, Flik 42/J, Alb. D III 253.71* | | | | | | | |

# Leopoldo Eleuteri

Leopoldo Eleuteri during training at the Gabardini flying school, in front of a Caudron. This photo is dated 15 August 1916 and is dedicated to Zanibelli, chief instructor with Gabardini.

In Castel Ritaldi, a small hamlet around a castle perched on the historical hills near Spoleto, Umbria, there is a house built in 1200, that is now fitted with a swimming pool and used as a holiday center for art-loving tourists. There Leopoldo Eleuteri was born in 17 December 1894, in a middle-class family. Art and spirituality are the dominating factors in Umbria, and Leopoldo's sister became a nun, later renowned for her culture and her writings, three brothers were farmers, but the young man had a strong attraction for mechanics, and he followed industrial studies. First he attended the boarding school of the Salesiani fathers in Spoleto, then he moved to the Technical Institute "Caio Cornelio Tacito" of Terni (a town nearby, which then was Italy's major steel center) where he graduated in 1915, with a specialization in mechanics and metallurgy.

After the entrance of Italy in World War I, in August 1915 Eleuteri was drafted and assigned to the Artillery Direction of the Terni weapons factory, and then in March 1916 he was posted to the technical offices of artillery in Rome. Although he could easily have spent the war period safely engaged in weapons production, Eleuteri volunteered for the air service, overcoming the efforts of his superiors to keep him in the offices. He was sent to the Gabardini private flying school in Cameri, where he trained in a class of 200 pupils that included a large number of future fighter pilots and aces: Giannino Ancillotto, Renato Donati, Cesare Magistrini, Arturo Ferrarin, Romolo Ticconi, and more. He obtained his first pilot brevet on Gabardini monoplanes on 8 September 1916 and the brevet on Caudron on 9 November. He then moved to the Cascina Costa military school where on 26 March 1917 he obtained his military pilot license, flying the SAML.

He was assigned to the 73ª Squadriglia Caccia, a unit equipped with the SAML S.1 and S.2 two-seater, based in Verona with the task of defending the city from possible Austrian attacks from the North and of performing reconnaissance flights over the Garda lake and the mountains toward Trento. Very soon Eleuteri, now a *sottotenente*, demonstrated his qualities of brilliant pilot and technical expert. He developed a maneuver for spin recovery on the SAML, which nobody had attempted before, and used it brilliantly when engaged by Austrian fighters. Eleuteri's first war flight was on 1 April 1917. On 24 April, flying with machine-gunner Gresino, he had a duel with an enemy that was seen gliding back toward the Austrian lines.

He flew a total of 20 war missions between April and September 1917, in which his airplane was twice hit by enemy artillery fire during photo missions, and once he made a forced landing for engine failure. His knowledge and keen interest in the technical aspects of aviation is proved by the fact that he entered in correspondence with the SAML company at Monza, pointing out minor defects in the construction of their airplane.

On 14 September 1917, this time flying with machine-gunner Bottero, Eleuteri had another duel with an Austrian airplane, which faced the intensive fire of the Italians, who fired four 50-rounds magazines of their Fiat gun, and then flew away toward Trent. There is a possibility that their enemy was *St.Fw.* Josef Kiss of Flik 24/F flying KD 28.37, who claimed a SAML in the sky above Arsiero, for his 7th victory, and then crash-landed on Pergine airfield with a pierced radiator and a seized engine, wrecking his fighter.

On 20 September 1917 the 73ª Squadriglia changed its designation, becoming the 121ª Squadriglia. On 11 November 1917 Eleuteri lost the SAML 2499, that suffered an engine failure and crashed against a hangar with no harm for its pilot. On 18 November Eleuteri was flying a reconnaissance mission over the Asiago plateau, escorted by *serg.* Giovanni Bartolomeo Arrigoni, of the 1ª Sezione SVA, a very valiant pilot (with four confirmed victories, shot down and killed on 4 August 1918) who was flying one of the first SVA reconnaissance fighters at the front. A patrol of Austrian fighters bounced Eleuteri and hit his SAML repeatedly, a bullet going through the shoe of the observer, *ten.* Velo, but by carefully spinning his airplane, Eleuteri managed to avoid the enemy attack and returned to base safely. The enemies were the "Kaiserstaffel," the successful trio of aces Arigi, Maier and Kiss of Flik 55, who then proceeded to shoot down in flames the escorting SVA of *serg.* Arrigoni, who survived unscathed. The Austrians also claimed the spinning and damaged SAML as shot down.

Eleuteri flew one more alarm mission on 30 November 1917, with gunner Arturo Marzari, and then he was posted to fighter school in Malpensa, near Milan. He converted on Nieuport and Hanriot fighters in January 1918, and then he staged through the Furbara aerial gunnery school near Rome.

In February he was assigned to the 70ª Squadriglia Caccia, a unit that had gone through a deep crisis after the Caporetto retreat and that had changed five commanders (Buzzi Gradenigo, Fougicr, Martinengo, Avet and Lombard) through that dramatic winter. Eleuteri arrived when the unit, then based in San Pietro in Gu, was finally regaining its combat readiness, and he flew his first combat mission on 22 February 1918 at the controls of a Hanriot Hd.1 fighter.

On the 24th he escorted some SAMLs, on the 25th he strafed the enemy airfield at San Fior and on the 26th he flew a patrol during which his wingman, *serg.* Ghislanzoni, got lost in the clouds and was shot down by an Austrian patrol led by Kurt Gruber of Flik 60/J.

In March 1918 Eleuteri flew 16 war missions, most of them patrols but also two escorts to Caproni bombers. On 3 April he attacked a kite balloon of Ballon-Kompanie 17 at Conegliano, hitting it with 48 bullets but not flaming it. Then he flew more offensive patrols until 17 April, when he scored his first official victory. On that day a formation of the 70ª Squadriglia, joined by *serg.* Cosimo Rennella of the 78ª clashed with a group of eight Austrian fighters escorting a two-seater. The Italians claimed three victories, two of them over fighters, assigned to Bocchese and Eleuteri, shared with Avet, Rennella and Resch and one over the two-seater.

One week later, 25 April 1918, while engaged on a leaflet dropping mission over Conegliano, Eleuteri together with other pilots

The Italian word for fighter pilot, *cacciatore*, also means hunter, and here Eleuteri plays both roles, holding a hare in front of a Nieuport fighter.

Below: The personal aircraft of Eleuteri with the 121ª squadriglia was this SAML S.1 serial 2887 with the insignia of an Indian chief.

of his 70ª Squadriglia clashed with enemy fighters, two of which were claimed shot down, but without confirmation, as the combat took place deep behind enemy lines.

On 1 May 1918 during an offensive patrol over Sernaglia Eleuteri had an inconclusive combat with an enemy two-seater. After a few more offensive patrols, he didn't fly from mid May to mid June, probably being absent on a furlough. He was back in action, however, on 15 June 1918, when the final Austrian offensive began all along the Piave River, and he flew a Caproni escort mission. Eleuteri flew intensively during the battle, three times crash-landing due to engine failure, on the 16th, 19th and 24th June. In a later flight on 19 June he had a combat with an enemy fighter, that managed to put some bullets through his Hanriot. He also began ground attack missions, dropping spherical bombs against enemy positions on many occasions.

Overall, in the second half of June Eleuteri flew 18 combat missions, and 14 in the first half of July, including three bombing attacks. Despite this high level of activity, he didn't meet the enemy in the air until July 15, when in a patrol with *serg.* Bocchese and *ten.* Avet he met a strong patrol of ten enemy fighter aircraft escorting two two-seaters. They had a long combat, Eleuteri engaging some fighters with his flight leader, while Bocchese dueled alone with five enemy fighters, one of which he claimed to have seen going down spiraling. Two enemy fighters were claimed as shot down, one over Sernaglia and one over Pieve di Soligo, on the Austrian-held side of the Piave River, both confirmed by the ground observers.

After two more missions, Eleuteri on July 19 took off from the airfield of San Luca, home of the 79ª Squadriglia, and he met an enemy two-seater that he shot down near Barbisano. This victory was assigned also to *serg.* Edmondo Lucentini of the 79ª. After other fruitless patrols, Eleuteri met the enemy again on the 29 July, when escorting a Caproni bombing mission. He defended a straggling bomber from repeated attacks, but he had met some gallant foes, and the Italian's Hanriot was riddled with bullets. Incendiary rounds put his fighter on fire, but Eleuteri managed to get out of the scrap and put out the fire. *Zugsführer* Max Kauer, of Flik 28 was perhaps his opponent, as this Austrian pilot claimed a Hanriot and a Caproni bomber, rather optimistically, as no Caproni was lost that day. Another possible opponent was *Oblt.* Vuk von Hreljanovic of Flik 56/J who, flying Alb. D.III 253.33 tangled with an escort fighter and claimed a probable.

The next day, during a Caproni escort, the Italian pilot had his revenge, fighting against an enemy fighter and forcing it to land on its own airport, Susegana. This victory, however, was not confirmed. For the rest of the summer Eleuteri was very active, but he didn't score, despite a combat with a two-seater on 1 August and an attack against a balloon on 9 August.

On 14 August Eleuteri tried a new fighter, the Ansaldo Balilla, that after a very long gestation was assigned in very small quantities for evaluation to all the frontline units, despite the very negative opinions of top Italian fighter pilots, like Piccio, who had test-flown it. By this time, Austrian aviation had almost ceased to exist

Eleuteri at the 73ª Squadriglia near the cockpit of a SAML.

as an aggressive force, and Eleuteri's missions were fruitless. In September he tested the Spad, which had been assigned also to the 70ª Squadriglia, that now flew two different types, Hanriot and Spad, plus one single Balilla.

On 4 October 1918 an unlucky enemy fighter, Albatros D.III 253.51 of Flik 56, was bounced by many Italian fighters and was shot down, victim of Eleuteri together with, again, sergeants Lucentini and Reali of 79ª Squadriglia. The enemy pilot, *Korp.* Otto Berggold, took to his parachute but it failed and he lost his life. The next day, 5 October, Eleuteri flew a combat offensive patrol with the Balilla, and on 8 October he scored the only victory ever gained on that type on the Italian front, when he claimed an enemy fighter out of a formation of six together with Avet, Lucchini and Beduschi, all pilots of the 70ª who flew Hanriots. Eleuteri with his faster airplane was the only one who could reach the escaping enemies, and he dueled with one of them until his victim fell by S. Lucia di Piave.

When the final offensive began, at the end of October, Eleuteri flew frantically in strafing and bombing attacks in support of the advancing army, and on 28 October he forced an enemy fighter to land on Arcade airfield. Eleuteri landed alongside his victim. This victory was assigned also to Avet and Bocchese, and to Ramberti of 76ª Squadriglia. All the Italians had their guns jammed, and they forced the enemy down without firing a shot. That was not the end of activities, his last mission was on 2 November 1918, when he attacked the Caldonazzo railroad station, braving an intensive anti-aircraft fire. His final war flight, an offensive patrol north of Trento, on 4 November, was uneventful.

Overall, during the war, Eleuteri flew 156 combat missions, 43 with the 73ª/121ª Squadriglia and 113 with the 70ª, and 164 frontline non-combat flights, which included a large quantity of test flights, showing how his technical abilities were put to good use. He was reported as having liked the Balilla (generally disliked by Italian pilots) and spent a long time trying to master its difficult aerobatics qualities. After the war, many of the surplus and unwanted Balillas were sold to Poland, where they were also built under license. In the war that opposed the new Republic of Poland to Bolshevik Russia, American volunteers of the Kosciuzko Squadron flew the Balilla with good results.

Leopoldo Eleuteri's wartime exploits were described in the wartime press by an article titled "The ace with no decorations" and actually just a bronze medal adorned his breast. After the war, *maggiore* Matteucci, commander of the X Gruppo, took action against this oversight and obtained the award of two Silver Medals for Bravery, as had been proposed in July and November 1918. A third silver medal was awarded later, for which the citation is unavailable.

On 14 September 1917 Eleuteri with gunner Bottero met the Hungarian ace Joszef Kiss, who in the combat suffered damages and crashed his Brandenburg KD serial 28.37 at Pergine arifield, with this result. *(Photo Höfling)*.

In December 1918 Eleuteri was assigned to the 90ª Squadriglia SVA that was getting ready to move to Libya, but it is doubtful that he actually went to that North African colony. In fact, after the war, Eleuteri left the army and became a university student, studying engineering at the Milan Polytechnic. He wanted to remain a pilot, and he re-enrolled in the Gabardini School at Cameri for a refresher course. There he flew the Gabardini 110-hp biplane, instructing other pilots in mock dogfights. He graduated as an engineer in May 1922, but he was still fond of aviation. When the Regia Aeronautica, the independent air force of Italy was formed, in 1923, it included a Corps of Aeronautical Engineers, and Eleuteri was admitted as a captain, on 31 October 1923. Under the strict and demanding leadership of General Guidoni, the G.A.R.I. (*Genio Aeronautico Ruolo Ingegneri*, Aeronautical engineers roll) was a small and exclusive group, that included many future designers like Stiavelli, Pallavicino and Tibaldi.

In 1923 then Eleuteri returned to flying service with the 1ª Squadriglia Sperimentale, at Furbara, near Rome, the experimental establishment of the air force, where he flew Hanriots, SVAs and Nieuport 10s. In 1924 he test flew also the Fiat BR.1 bomber and the Ansaldo A.300, and in August 1924 he was commended for his ability in saving an airplane when a pyrotechnic device had exploded. He gained the reputation of a capable, strict and inflexible

Leopoldo Eleuteri seated over a Spad without any marking, but with ribbons, probably indicating a flight leader, attached to the inner wing struts.

A line-up of the fighters of the 70ª Squadriglia. The first Hanriot shows the flag of Sardinia, with a red cross and four heads of Moors, the third Hanriot is the striped personal airplane of Avet, the fifth one, in dark camouflage and with number 4, is the airplane of Eleuteri. Almost invisible at the end of the line, behind some Spads, there is an Ansaldo Balilla.

The pilots of the 70ª Squadriglia posing in front of a Hanriot and a Spad. Standing, seventh from left, Avet, then Resch. Sitting down, third from left Eleuteri, fifth Bocchese.

Leopoldo Eleuteri in front of Hanriot Hd.1 11411, the personal airplane of Flaminio Avet.

Eleuteri, third from left, after the war at Montecelio experimental establishment with a war-booty Fokker D-VIII, still carrying German black crosses on its wings.

test pilot. In 1925 he test flew the new fighters under evaluation for the Regia Aeronautica competition: thus he first flew the Fiat CR on 2 January, and the Nieuport 29 on 16 March. On 19 January 1926 Eleuteri first flew a radically new and different type, the Ansaldo AC.2, Italian version of the French Dewoitine D.1, a metal high wing monoplane. On that day, with the fighter serialled 1177, Eleuteri flew two test flights of 30 minutes each. He took off for a third flight and engaged in a mock dogfight with another fighter, an Hanriot piloted by *serg.* Corrado Augias: suddenly at 4.10 p.m. the two fighters collided, crashing to earth at the airfield boundaries, with the death of both pilots.

His former squadron commander, Flaminio Avet, was deeply moved by his death, and proposed that the new 70ª Squadriglia be named after him, like the 91ª was named after Baracca, but his suggestion was not picked up. After a solemn funeral, Leopoldo Eleuteri was buried in his native village, where a monument to him now stands in front of the castle. The memory of this reserved and daring fighter and engineer has not been lost.

Eleuteri as a test pilot, in the cockpit of the beautiful and fast SIAI S.52 metal biplane fighter.

## Leopoldo Eleuteri
## 9 victory claims - 7 confirmed victories

| Claim | Vict. # | Date | Sqn. | Aircraft | Hour | Place | Victim |
|---|---|---|---|---|---|---|---|
| - | - | 24.4.1917 | 73 | SAML | - | Trentino | EA |
| *Possibly: Zugsf. Schuster UNH/Hidegkéety UNH, Flik 21, Br.C.1 129.27* | | | | | | | |
| - | - | 14.9.1917 | 73 | SAML | - | Trentino | EA |
| *Stfw. Josef Kiss UNH, Flik 24, KD 28.37* | | | | | | | |
| 1 | 1 | 17.4.1918 | 70 | Hd 523 | - | Valdobbiadene | Fighter |
| *[with ten. Avet, Resch, serg. Bocchese, 70ª Sq. & serg. Rennella, 78ª Sq.] Maybe: Oblt. Michael Gassner- Nordon von Laudon KIA, Flik 42/J, Alb. D.III 153.152* | | | | | | | |
| 2 | - | 17.4.1918 | 70 | Hd 523 | - | Valdobbiadene | Fighter |
| *[with ten. Avet, Resch, serg. Bocchese, 70ª Sq. & serg. Rennella, 78ª Sq.] Maybe: Oblt. Richard Fitz,UNH, Fik 42/J, Alb. D.III, returns to own airfield with damages.* | | | | | | | |
| 3 | - | 25.4.1918 | 70 | Hd | - | Conegliano | Fighter |
| *[with ten. Avet, Guglielmotti & Croci, 70ª Sq.] Maybe: Korp. Rudolf Kusyn UNH/Lt. i. d. R Julius Knüppel UNH, Flik 4/D, Br. C.1 169.82* | | | | | | | |
| 4 | 2 | 15.7.1918 | 70 | Hd | 10h30 | Sernaglia/Conegliano | Fighter |
| *[with ten. Avet & serg. Bocchese, 70ª Sq.]* | | | | | | | |
| 5 | 3 | 15.7.1918 | 70 | Hd | - | Moriago | Fighter |
| *[with ten. Avet & serg. Bocchese, 70ª Sq.]* | | | | | | | |
| 6 | 4 | 19.7.1918 | 70 | Hd | - | Barbisano | EA |
| *[with serg. Lucentini, 79ª Sq.] Offstv. Hans Huka WIA/Oblt. Josef Karner UNH, Flik 67/S, Br .C.1 169.138* | | | | | | | |
| - | - | 30.7.1918 | 70 | Hd | - | Susegana | Fighter FTL? |
| 7 | 5 | 4.10.1918 | 70 | Hd | - | Fontigo | Fighter |
| *[with serg. Reali & Lucentini, 79ª Sq. ] Korp. Otto Berggold KIA, Flik 56, Alb. D.III 253.51* | | | | | | | |
| 8 | 6 | 8.10.1918 | 70 | A.1 16558 | - | S.Lucia di Piave | Fighter |
| *[with ten. Avet, serg. Lucchini, ten. Beduschi, 70ª Sq.]* | | | | | | | |
| 9 | 7 | 28.10.1918 | 70 | ? | - | Arcade | Fighter |
| *[with ten. Avet & serg. Bocchese, 70ª Sq., serg. Ramberti, 76ª Sq.]* | | | | | | | |

# Guglielmo Fornagiari

High in the Apennine Mountains behind Bologna, in Central Italy, the little village of Lizzano in Belvedere is a place of rural beauty, proud of its preserved heritage and environment, and enjoying its sophisticated VIP tourism. Its suburb of Pianaccio, among pure streams and chestnut woods, may look picturesque today, but one hundred years ago it was a place of absolute poverty, of hard toil and long, cold winters. Guglielmo Fornagiari was born there on 11 March 1892, in a stones house that now looks beautiful, thanks to modern plumbing, heating and electricity, but that must have been a simple peasant house in his times. His relatives are called alternatively Fornagiari or Fornaciari, depending which kind of spelling was applied at the moment of registration of newborns by the local registry officer, and even in the wartime Squadron logbooks and documents his name is spelled alternatively in the two ways (sometimes even in the same single sheet!).

Fornagiari was a little, impish country boy, running barefoot in the woods and tending sheep. He had a passion for tree climbing that almost cost him his life. Once a branch broke, and he fell heavily on a paved road, loosing blood from his mouth and ears, and the local doctor doubted for his life. He recommended cold compresses, so his friends went to the mountains, to gather the remaining snow that was applied in a pig's bladder. A couple of days later, the boy was already on his feet and running, but his face remained scarred, and possibly his later health was affected. Fornagiari attended only elementary school, and then immigrated to France, looking for work, as so many Italian men were doing, to avoid the poverty of the countryside. In France he must have dabbled with mechanics, because he managed to pass for an experienced engineer when he was drafted, in 1912, and he was sent to the Battaglione Aviatori, the Italian air arm, at Mirafiori, Turin.

He was very short, just 1.60 meters, but of irrepressible energy. He rose in the ranks, becoming a corporal in October 1913 and a sergeant in November 1914.

On 24 May 1915, when Italy entered the war, he was assigned to the war zone, as an airfield mechanic, and he worked with the first Caproni bombers at Aviano. Obviously he had the dream of becoming a pilot, and in November 1915 he was posted to the flying school at San Giusto, near Pisa. He was licensed as a pilot in April 1916, and then he was sent to the Cascina Costa for a quick training on the Nieuport.

On 6 May 1916 Guglielmo Fornagiari, since March a *sergente maggiore*, was assigned to the new 77ª Squadriglia Caccia, that was forming at Istrana, and soon moved to Cascina Farello, near Aquileia. No data is known of his brief service with that unit, for at the beginning of September Fornagiari was assigned to the 78ª Squadriglia, a new unit that had just been formed, commanded by *cap.* Bolognesi, and based at Istrana, operating on the Asiago front. His first combat flight with the 78ª was a patrol to Feltre on 14 September, and his log for the rest of 1916 was 12 takeoffs on alarm, for a total of just 10 and half flight hours.

On 8 January 1917, while engaged in an escort to Caproni bombers, he had his first combat with the enemy, over the Asiago plateau. Then for the first six months of 1917, Fornagiari flew his

Guglielmo Fornagiari, the poor mountain boy who became a deadly fighter ace.

Nieuport in uneventful offensive patrols and occasional escorts to Capronis or Caudron and Farman reconnaissance airplanes. He flew a total of 53 combat missions in the first semester.

On 19 June 1917 he met the enemy in the air for the second time, and he scored. During an escort flight to Caproni bombers he attacked a Brandenburg C.1 of Flik 17, serial 29.73, as it was shooting at the bombers. The bullets of the Italian pilot killed the pilot, *Oblt.* Alfons Rakovits and the airplane fell at Braida, behind the Austrian lines, the observer *Oblt.* Rudolf Lassmann was wounded. Other Austrian aircraft suffered attacks in this occasion, Br. C.1 129.25 of Flik 15 and Br. C.1 29.60 of Flik 24, both force landed at Pergine, so the real identity of Fornagiari's victim is still conjectural.

In June 1917, trying to break free from the stalemate of the Isonzo front, the Italian command launched a bloody attack on the Ortigara mountain, in which the Alpini, the Italian élite mountain troops, were slaughtered without gaining any ground in one of the most impassable terrain in which man had ever fought. This battle brought intense air activity, however, in which the Austro-Hungarian Luftfahrtruppe suffered heavy losses, and the Italians gained the ascendancy in the air. One reason for this success was the introduction of the reliable and well-armed Nieuport 17 and Spad 7, but the 78° still flew the obsolete Nieuport 11.

Fornagair with a Nieuport 11 that seems to show the markings of the 79ª Squadriglia. As Fornagiari didn't fly with that unit, this is probably one of the aircraft detached to reinforce the Isonzo front during the spring of 1917 *(Photo Franz Selinger)*.

On June 27 a peculiar tragedy happened. *Serg.m.* Fornagiari had a motor failure and had to land his Nieuport serial 1664 at Fossalunga. Having fixed the engine, he took off again but suddenly he saw a woman and two children who had strayed onto the airfield. In order to avoid them the pilot took off too soon, stalled and fell over a vineyard, killing two boys who were hiding there, looking at the airplanes.

Fornagiari flew many long missions in that period, and on 11 July he met two enemy airplanes, attacked one and drove it away, shooting 110 rounds. In August, as fighting had stopped on the Asiago area, the 78ª Squadriglia, located in a secondary part of the front, was shifted to Borgnano, near Gorizia, to support the new major offensive, the 11th Battle of the Isonzo. The unit, however, was still equipped mostly with old Nieuport 11s, quite war-weary, but it scored a certain amount of victories. Fornagiari's first duel on the new front took place on 10 August, and an enemy airplane was driven away over the Sabotino. On 22 August he flew an offensive patrol mission over San Marco, and he met an enemy that he claimed shot down over Ternova. No corresponding Austrian loss is known. The combat was described in the press as an attack against a two-seater escorted by three fighters, which put some bullets into Fornagiari's airplane. For this combat, he received his first Medaglia d'Argento: "On 22 August 1917, he faced three enemy aircraft, attacking decidedly one of them, and while being counterattacked closely by the other ones, he shot it down after a long and lively combat."

Action was very intensive in those days, with Fornagiari often flying two missions each day, and on 31 August he again shot down an enemy airplane over Lokve. There is no known corresponding data from Austrian sources, but the authors found photos of a wrecked Brandenburg captioned as "the airplane shot down on 31 August on the Bainsizza". On 2 September he force landed for an engine failure, but the next day he was back in the air. The small and indefatigable pilot was gaining a reputation, and on 20 September 1917 his squadron commander proposed him for the award of the Medaglia d'Argento.

On the first day of October 1917, Fornagiari met an enemy that again escaped him, but also managed to put 12 bullets into the *sergente*'s airplane, and almost to bring his career to an end.

The next day, however, Fornagiari got his revenge. He attacked and shot down an enemy aircraft over Podmelec. Possibly his victim could be a German two-seater of Fliegerabteilung (A) 219. The duel is described as a combat that started at 4,800 meters and going down until 400, with the enemy gunner put out of action. Flying at treetop level, the enemy pilot stalled trying to avoid a rock and crashed in the Iudrio River.

On 3 October he attacked and shot down an enemy over mount Globogiag, together with a Spad. His victim most likely was Brandenburg Br. C.1 329.20 of Flik 53, shot down by *maggiore* Piccio for his 15th victory with the participation of Fornagiari. Piccio claimed his victory near Auzza (today Avce) which is a village at the foot of the Globogiag hill.

He had further duels in October, twice having to retreat because of a jammed gun. On 16 October he reported a combat over Mount Matajur, that he had to abandon for the failure of a camshaft and of the gun synchronizer: this event is surely the combat that ended with a German two-seater shot down by his squadron mate *serg.* Magistrini over Mount Nero. The village lying between the two mounts, Caporetto, was soon to become sadly famous, and the curiosity of the German reconnaissance airplane should have sounded a warning.

On 26 October he flew an escort mission for the Savoia Pomilios that tried to attack the advancing Germans, then his logbook stops, and we have no data for his action during the retreat, until 13 November 1917. Having set fire to the hangars of the 78ª Squadriglia, Fornagiari escaped under enemy fire. During his retreat he found the wreck of an airplane and the corpse of *ten.* Veronesi, his flight commander who, trying to save a fighter, had taken off despite the fog and had crashed against a hill, or, possibly, was shot down by a German fighter pilot, *VzFw*. Hipper of Jasta 39, who claimed a fighter on that day and in that place. Fornagiari brought the news of his tragic death when he regained his unit. He had gone through days of suffering and anguish, surviving on a diet of burned wheat, and he had been posted as missing.

On 13 November he was back at the controls of a fighter escorting Savoia Pomilios, but this time the area of his action was Ponte della Priula, in the Piave area, as the best part of two Italian regions had been lost to the enemy. His squadron, the 78ª, in fact, had abandoned twelve brand new Hanriots and retreated to Aviano, Arcade and finally back to Istrana, its original airfield, where at this time it was equipped with just a handful of Hanriots.

After several ordinary missions over various sectors of the front, Fornagiari took part in the "battle of Istrana", when a large formation of German two-seaters, attacking in two waves and strangely not covered by its fighter escort, ran the gauntlet of all the fighters of the four squadrons at Istrana, 76ª, 78ª, 81ª and 82ª, losing a total of eleven aircraft. It is difficult, in that veritable "turkey shoot", to determine who shot down whom. Fornagiari shot down one aircraft over Falzè, in cooperation with ace Guido Masiero and soldier pilot Panero, and another one north of Musano, together with *ten.* Comandone of the 82ª Squadriglia. For this action, he was again recommended for another Silver Medal, that was awarded with a citation that quoted his victory of 2 October and the two victories over Istrana.

At the end of 1917, he had logged 132 combat flying hours. Fornagiari began the New Year with a cruise above Cismon, during which he dueled with an enemy aircraft, shooting 60 rounds without success.

On 27 January 1918 Fornagiari flew protecting a SIA 7b two-seater of the 32ª Squadriglia when he claimed an enemy aircraft shot down north of S. Gaetano. *Cap.* Riva described the events in a letter protesting the confirmation of victories:

> "On 27 January only one victory was confirmed to me, located at Val Brenta, instead of two in collaboration with serg. magg. Fornagiari of my squadron. On that day I took off at 2.40 p.m. with serg. maggiore Fornagiari and Germino to escort a SIA in the area Gallio-Cismon. Nine enemy aircraft attacked us above Sasso Rosso; serg. magg. Fornagiari and I disengaged the SIA (serg. magg. Germino was isolated and was shot down and wounded) then we attacked the enemy, first shooting down a fighter near S. Martino (which is the one confirmed to me as at "Val Brenta") and a two-seater (which is the one confirmed to serg. maggiore Fornagiari as at "S. Gaetano")."

And then Riva asked that both victories be confirmed to both pilots. That day the pilots of the 78° Squadriglia most likely had a combat with the Albatros fighters of Flik 55/J in which the aircraft of Josef Kiss and Gottlieb Munzar were forced to land, with the

The flight from the 78ª Squadriglia detached for the protection of Padua: from the left, *serg.* Fornagiari, *ten.*Giacomo Brenta and *serg.* Cesare Magistrini. Nieuport 11 serial 1675 with personal markings, unfortunately obscured, shows an engine cowling in a solid light color.

great Austrian ace badly wounded in the abdomen. Other authors have credited the successful action against Kiss to Capt. M. B. Frew, RAF No. 45 Squadron, but the British pilot claimed a victory in a different part of the front, near Conegliano, very far away, while the action of the Italians took place close to Pergine, where the badly wounded Kiss landed.

Month after month, Fornagiari remained at the front, fighting and scoring. His next victory came on 18 February 1918, when during a S.P. escort mission he attacked and enemy and shot it down near Mount Lisser. Together with him fought another Hanriot of the 78ª, with pilot *serg.* Leonida Schiona, who after the war became a great airline commander. Again, there is a plurality of Italian claims on that day and in that area, against known Austrian losses of one Phönix of Flik 39/D and a Brandenburg of Flik 16/D, so that it is impossible to determine who Fornagiari's victim may be.

Just three days, but seven combat missions later, on 21 February, Fornagiari scored again, for what was his 10th claim. Once more, he was escorting a Savoia Pomilio when he shot down an enemy over Gallio, in cooperation with his squadron mate *serg.* Antonio Chiri and with the crew of the Pomilio itself, Fusco and Calì of 33ª Squadriglia. Their victim could possibly be *Korporal* Franz Lahner, who was forced to land after a duel with Italian airplanes at Rodighieri, a hamlet one mile from Gallio. Lahner, of Flik 55/J, was flying Albatros 153.158. The Italian press, with a bit of an overstatement, wrote that 200 bullets riddled the airplane at Gallio and that the dead pilot was a 22 victories German ace. These events brought him his third medal, a bronze:

> "On 27 January 1918 escorting one of our reconnaissance planes, attacked by eight enemy fighters, he held in the uneven fight shooting down one of the attackers and saving the escorted airplane. On 21 February again escorting a reconnaissance airplane attacked by five enemy fighters, with his quick and dashing intervention he freed from the menace the airplane assigned to his protection shooting down one of the attackers and dispersing the enemy group."

The next day Fornagiari flew his 200th combat mission. After the end of February, enemy air activity decreased, and Fornagiari had no further combats, despite his intense activity. On 10 May he attacked a kite balloon, but the soldiers on the ground pulled it down, without excessive damage. Meanwhile, his unit moved from Istrana to Nove di Bassano and then, in March, to San Luca. In June Fornagiari was promoted "*aiutante di battaglia*" which is the highest honor rank for a non-commissioned officer.

The 78ª Squadriglia was heavily engaged during the Austrian offensive that began on 15 June, and on that day Fornagiari scored his 11th victory. Escorting a SAML of the 114ª Squadriglia, he shot down an enemy two-seater in the early morning near San Giovanni. His possible victim was Brandenburg C.1 329.48 of Flik 26 with the crew of *Fw.* Rudolf Berger and *Oblt.* Ernst Petric, both killed in action. Serg. Leonida Schiona and serg. Francesco Gritti took part in this mission, but they had no credit for this victory.

Fornagiari claimed a victory on 31 August 1917 for which no corresponding Austrian record has been found. However, there are photos like this one captioned: "The airplane shot down on 31 August on the Bainsizza". To make matters worse, souvenir hunters removed the serial from the fuselage! There are similar instances of incontrovertible Austrian photos of shot down Italian airplanes, for which these authors have failed to find correspondence in Italian records.

Proud and elegant, Guglielmo Fornagiari sits on the wheel of an Hanriot, with his pet dog next to him.

This is his battle report:

> "During an escort to the 114ª Saml at about 6h45 I saw an enemy aircraft at a lower level ( about 1500M). I attacked it from the tail shooting about 120 rounds. The enemy observer shot a burst against me but missed. I saw the plane first raise its nose, then fall down breaking apart in the air. I reached the Saml maintaining the escort with the other components."

Above: Fornagiari in a low quality photo, probably after the Armistice. The Hanriot next to him shows one of the typical pennants of his 78ª Squadriglia, and a solid dark color cowling. The last Hanriot shows the red star on a white field insignia of the 80ª Squadriglia.

Left: This close portrait of Fornagiari shows many details of the cockpit and front fuselage of the Hanriot *(Photo Franz Selinger)*.

Below: The sort of photograph that enthusiasts appreciate: the pilot Fornagiari next to his Hanriot, not obscuring the insignia, the pennant, nor the serial numbers, N° 60 French, 6092 Italian, of his Hanriot.

## Gugliemo Fornagiari
## 13/14 victory claims - 7 confirmed victories

| Claim | Vict. # | Date | Sqn. | Aircraft | Hour | Place | Victim |
|---|---|---|---|---|---|---|---|
| 1 | - | 19.6.1917 | 78 | Nieuport | - | Fassa | EA |
| *[with ser. Macchi, 71ª Sq.] Maybe: Oblt. Alfons Rakovits KIA/Oblt. Rudolf Lassmann WIA, Flik 17, Br. C.I 29.73* | | | | | | | |
| 2 | 1? | 22.8.1917 | 78 | Nieuport | - | Ternova | Two-seater |
| 3 | - | 31.8.1917 | 78 | Nieuport | - | Lokve | EA |
| 4 | 2? | 2.10.1917 | 78 | Nieuport | - | Podmelec | EA |
| *Maybe: Uffz. Sauerwin WIA, FA (A) 219* | | | | | | | |
| 5 | - | 3.10.1917 | 78 | Nieuport? | 16h+ | Globogiag | EA |
| *[with magg. Piccio, 91ª Sq.] Korp. Josef Czerip WIA, POW/Lt. Bruno Bélohlàvek WIA, POW, Flik 53, Br. C.I 329.20* | | | | | | | |
| 6 | 3? | 26.12.1917 | 78 | Hd | - | Musano | EA |
| *[with ten. Masiero, 78ª Sq., s.ten. Scaroni & Fucini, 76ª Sq., ten. Comandone, 82° Sq.]* | | | | | | | |
| 7 | 4? | 26.12.1917 | 78 | Hd | - | Falzé | EA |
| *[with ten. Masiero, 78ª Sq. & sold. Panero, 82ª Sq.] Maybe:<br>Lt. Kessler WIA/Lt. d. Res. Edebohls KIA, FA 2, DFW C.5 or:<br>Uffz. Petzold KIA/Lt. d. Res. Voigt KIA, FA 2, DFW C.5 or:<br>Vzfw. Uecker KIA/Lt. d. Res. Pfeiffer KIA, FA 2, DFW C.5 or:<br>Vzfw. Pohlmann KIA/Lt. Schlamm WIA, FA (A) 219, DFW C.5* | | | | | | | |
| 8 | 5? | 27.1.1918 | 78 | Hd | 15h+ | S.Gaetano | Two-seater? |
| *[with cap. Riva, 78ª Sq.] Maybe: Korp. Gottlieb Munzar, Flik 55/J, Alb. D.III FTL, or: Offstv. Josef Kiss WIA, Flik 55/J, Alb. D.III* | | | | | | | |
| 9 | - | 18.2.1918 | 78 | Hd | - | M.Lisser | EA |
| *[with serg. Schiona, 78ª Sq.] Maybe: Zugsf. Alois Weinert MIA, Flik 39/D, Ph. D.I 128.08, or:<br>Zugsf. Anton Hollriegel UNH/Lt. Bruno Fichtinger UNH, Flik 16/D, Br. C.I 369.20* | | | | | | | |
| 10 | 6? | 21.2.1918 | 78 | Hd | 10h55 | Buso/Frenzela | Fighter |
| *[with serg. Chiri, 78ª Sq.] Korp. Franz Lahner WIA, Flik 55/J, Alb. D.III 153.158.* | | | | | | | |
| 11 | - | 15.6.1918 | 78 | Hd | 6h45 | S.Giovanni | EA |
| *[with serg. Schiona & Gritti, 78ª Sq.] Maybe: Fw. Rudolf Berger KIA/Oblt. Ernst Petric KIA, Flik 26, Br. C.I 329.48* | | | | | | | |
| 12 | 7? | 18.6.1918 | 78 | Hd | evening | S.Salvatore | EA |
| *Maybe: Zugsf. Josef Grimm WIA, Flik 44, Br. C.I 369.42* | | | | | | | |
| 13 | - | 23.8.1918 | 78 | Hd | - | Falzè di Piave | EA |
| *Zugsf. Gottlieb Konecny KIA/Lt. i. d. R. Rudolf Hruschka WIA, Flik 67/DS, Br. C.I 429.27* | | | | | | | |

# Mario Fucini

In the preface of his book he described himself as "a soldier who just did his duty, nothing more and nothing less." Mario Fucini was not an ace who lives forever in legend, yet he deserved to, if for no other reason for the fact that he fought at the front through most of the war. From the first air actions supporting the Carso offensives in the legendary 25ª Squadriglia Voisin, to the combats in the final offensive of Vittorio Veneto in 1918, he took part in all the toughest battles of the war, without respite and without faltering. He was a witness and an active participant in the development of Italian aviation from the start of the war to the day of victory.

Mario Fucini was born on 1 February 1891 in Empoli, a town in the province of Florence. His family was wealthy, belonging to the middle class, and his grandfather Renato Fucini was a well-known and successful writer, author of books of poems and short stories describing contemporary life in Tuscany. Love of literature was a family trait, because grandson Mario became a prolific writer on aviation subjects.

Mario's love of flying began at an early age. When he first saw an airplane flying he was excited. Pioneer pilot Mario Cobianchi flew his Farman at San Rossore Park, near Pisa. Fucini went there with his brother, and saw the daring exhibition ending with a sideslip and a non-fatal crash, which did not dampen his passion. Fucini became an avid reader of all news concerning the new conquest of man, flight. With his brother he built a "concoction" that was supposed to be a glider, or, using the French language as was customary with aviation at the time, a *"Planeur"*. He jumped from a cliff with his device made of canes, string and silk, with the only result of getting himself a hernia for the effort. Adding insult to injury, he flunked school because instead of studying he had wasted time on aeronautical design.

The draft board rejected him on account of that hernia, so when Italy entered World War I he stayed home, admiring his brother who went to the front. He was then a university student majoring in Chemistry. He had a surgical operation to get rid of his hernia, and then volunteered, obviously opting for the Battaglione Aviatori, much to the worry of his parents. They, and everybody else, considered airmen a sort of daredevils who were madly risking their lives.

A simple soldier, he was posted to the Mirafiori flying school in Turin, near the Fiat works, where he became airplane pilot on 1 December 1915. His training period was wonderful, with good friends among whom stood out for his eccentricity Guido Keller, the mad artist. Fucini recalled when the populace presented to the command its outraged protest because Keller played "satyr and nymphs" with the washing women, meaning that he got stark naked and ran after the girls all the way from the river to the village, much to the scandal of the citizens.

Winter set in, and Fucini obtained his military pilot brevet only on 1 February 1916. He was commissioned as *sottotenente di complemento* on 2 March 1916; he had a short leave then he was assigned to an operative unit, the 5ª Squadriglia Voisin that became 25ª Squadriglia in April 1916. Commander *capitano* Stanzani and

The first steps of an ace: young Mario Fucini proudly showing the glider he built and unsuccessfully tried to fly.

prince *ten.* Ignazio Lanza di Trabia welcomed him. The latter, totally forgotten today, became a legend in Italian aviation during the war.

The squadron operated the Voisin, a slow and clumsy airplane, but incredibly strong and reliable. Fucini had flown it just a few times, and his commander wisely had him fly a few familiarization flights before sending him out in a combat mission.

He had his baptism of fire on 1 April 1916, a photo-reconnaissance mission with observer *ten.* Armando Fabretti. They had to fly 20 km in enemy territory to take photographs of important targets. Anti-aircraft fire rocked their plane a few times, but they returned safely, after having seen the takeoff from Aisovizza of a yellow dot, a Fokker monoplane out to intercept them. After the landing they discovered that enemy guns had been accurate, like usual: their wing was pierced and their engine was dented.

Fucini became friend to all the components of the unit, but his closest friend was a recently arrived observer, *sottotenente* Giacomo De Carlo, who two years later gained fame for his daring undercover missions behind the enemy lines, for which he received the Medaglia d'Oro al Valor Militare. Eight days later he had a first-hand experience of Fokker fighters. He was piloting for *capitano*

The Voisin was not totally defenseless, and it was an incredibly sturdy airplane. Here Fucini is in the nacelle of a Voisin of the gallant 25ª Squadriglia.

*osservatore* Djalma Juretigh, an "irredento", that is to say an Italian born in land under Austria, who deeply hated the Dual Monarchy and, not satisfied with reconnaissance, always added a souvenir for his enemies. In this case it was a bomb dropped on the village of Loque. The bomb went off in the middle of town but Fucini didn't share Juretigh's satisfaction, thinking of whom may have been hurt. He had better reasons to worry soon after:

> " ...now I heard a weird sound, that seemed to follow behind me. It was a series of dull sounds that seemed to repeat the rhythm of the engine. I tapped on my companion's knee: say, can't you hear this sound? Look what happened to the engine or the propeller.
>
> No' I don't see anything. But the sound went on, clearer now. It's a machine gun!
>
> I turned and saw a few dozen meters behind my Voisin the sharp profile of a fighter monoplane approaching fast."

The enemy fighter kept shooting at the poor Voisin; Fucini tried his best to maneuver the two-seater and Juretigh standing in the cockpit returned fire with his Mauser pistol, because their machine gun could not fire backwards. The first Fokker got away, probably out of munitions, but a second one replaced it, until it also expended all its munitions and got away. The "sandbags wall", as they called the Voisin, was damaged but brought back to base its occupants.

Each mission was very dangerous, but the crews, as if to lighten the tension, often pulled practical jokes at the enemy, like throwing down old shoes, pipes for Emperor Franz Josef, or standing up in the cockpit to urinate on the soldiers below. The command forbid them to drop bombs on their personal initiative, so Juretigh dropped stones, which he had personally signed.

Fucini met the Fokkers other times especially when flying with Juretigh, and on 28 July 1916 he had a very close call. As they had struggled to drop some bombs that didn't get off their hooks, they met two Fokkers that mauled their Voisin, so much that it was written off. A Nieuport fighter intervened and drove the Fokkers away, probably saving their lives. Their successful attacker was *Stbfw.* Stefan Huzjian, flying Fokker A.III 03.51. For this event Fucini got his first medal, a Medaglia d'Argento.

Fucini getting out of an Isotta Fraschini-powered Voisin, serial 1292, of the 25ª Squadriglia, which used this model to the end, October 1917.

Instead of running away, on another occasion Fucini engaged the fighter that had attacked him, and drove it away, describing it as "a little biplane", probably one of the new Brandenburg KD.

Fucini gained a reputation as an excellent pilot, and his friends advised him to ask for conversion to fighters. His request was accepted, but first he had two close calls in a few days. On 1 February 1917 as he was flying with *tenente* Giacomo Macchi (another remarkable observer, with the highest number of photographs to his credit) anti-aircraft fire ripped his radiator. Boiling water poured out, and they couldn't stop it. With an overheated engine, they force landed in a field near Medeuzza, just across the lines. A few days later, he sampled all the problems of a reconnaissance pilot. On 16 February he was flying with De Carlo for a difficult mission to the Gulf of Sistiana, the Hermada hill and Castagnevizza, that is to say the most contested positions of the front, barring the road to Trieste.

Left: Fucini, left, here with *serg.* Emilio Zampieri, decorated his Nieuport 11 at the Sezione Difesa di Grottaglie with a white lightning, but there was not much to do in that remote part of Southern Italy. That air base however became very important in World War II and during the Cold War.

Now also the Austrians had enough fighters, and the old Voisins always had a Nieuport for escort. An enemy fighter attacked them frontally over the Gulf, approaching them so quickly that the observer didn't have time to man his gun. The wing of the Voisin was hit; the Nieuport stepped in and after a brief dogfight the enemy fighter retreated. The Nieuport pilot indicated that his gun was jammed and flew away, Fucini recklessly decided to continue the mission alone. Then also ground fire hit the fuselage; the two airmen completed their photos, and were about to head for home when a new threat appeared:

> "...And then a volley of machine gun sounds quite close to my back, hitting something below the nacelle. He's aiming right! A small biplane, the same as before? Its threatening nose is aiming at us, a few feet back. Also this time there is nothing to do. The enemy is in such a position that it would be impossible to shake him off with any maneuver, but I think I'll try something crazy, a sideslip. I would like to tell De Carlo, that we are not falling because I'm hit, but there is no time for such niceties. I drop my plane that was already leaning left... Nothing! The fighter regains immediately its position, shooting again. This time he is aiming true. I smell gasoline and liquid on my feet: the fuel tank! The nacelle is flooded with gasoline; the wind is seeping through and sprays it in my eyes... I'm almost blinded. The level of fuel rises, now all my legs are wet. I turn to De Carlo: shoot, shoot, I cry. He tells me that he can't, he would fire through our wings. The enemy is still there, ready for another volley...."

Fucini tried all possible maneuvers, with no result. With his heel he smashed though the floor of the cockpit letting gasoline out, but it ignited and the Voisin caught fire. He cut the contact of the engine, but then he realized that they would land in enemy territory, so he restarted the engine, which worked briefly, before dying out. Gliding their smoking airplane, the two Italians managed to cross the lines and to land in a small field near Monfalcone. Fucini was bruised but unhurt, De Carlo had broken his teeth. They tried to savage some equipment, especially their precious photographs. Some soldiers approached, and hearing Fucini say "fire extinguisher", an unknown word for them, they ordered "hands up". Fucini answered them with a volley of Tuscan swear words, which proved their nationality, then he loaded the soldiers with the cameras and got away just in time before the Voisin exploded. This air victory is not listed among Austrian fighter claims.

This episode marked the end of his career in reconnaissance; he was tired of being the "hare", now he wanted to be the hunter. On 12 March 1917 he left his friends of the 25ª Squadriglia.

In Turin he was told that he was tasked to convert not on the Nieuport but on the new Pomilio reconnaissance two-seater. His old commander *cap.* Stanzani, who happened to be in Turin for an assignment, stepped in and managed to get him posted back to Nieuports, at the Malpensa flying school.

"Two little wings stuck to your shoulder"; this is how Fucini enthusiastically defined the Nieuport, noticing at the same time the difficulties of such a small, light and maneuverable airplane for pilots coming from sluggish two-seaters. Having got his brevet, he then went to the Pisa S. Giusto gunnery school, near his hometown and then to Arcade for aerobatics practice. On 2 June 1917 he was assigned to the 84ª Squadriglia da Caccia, but we don't know if he actually reached that unit, because events took a bad turn for him. On 16 June 1917 he was ordered to go to Grottaglie, in Apulia, the heel of Italy's boot, to be commander of the local Sezione di Difesa (defense flight). There was nothing to do there, really, but Grottaglie was an airship base and it had a small flight equipped with Nieuport

The Nieport 11s of the Sezione Difesa Grottaglie at rest while an M Class airship hovers above. Three of them show the white lightning insignia.

When Fucini was assigned to the real front, in the 76ª squadriglia, he again painted a lightning on his aircraft. His French Hanriot with Italian serial 6647 is here pictured at Istrana in the winter of 1918.

11s for its protection. Fucini remained there three months, in which he spent most of his time writing letters asking to be recalled to the front.

When a replacement finally came (*ten.* Luigi Giovanardi) he returned to Turin for a new surprise: an assignment to the 87ª Squadriglia, a new unit that was in formation with the Ansaldo SVA, beautiful and fast but unfit for fighter service. After some initial misgivings Fucini got to like this new airplane, and his period in training at Ponte S. Pietro was useful, making him dream of great deeds. The rout of Caporetto changed all that, as on 27 October all pilots were suddenly ordered to reach the front and to report at Udine.

As he went there, he crossed retreating and scattered soldiers. Before Udine, an officer of the Carabinieri stopped them and forced them back at gunpoint: the Germans had already taken the city, and the pilots had to join the columns of retreating troops. In Conegliano he found his friends Lanza and Macchi, who told him the fiery end of his old squadron, sent to slaughter in a futile attempt to stop the enemy. German fighters had intercepted them and tore them apart, four Voisins took off on 26 October and only one came back.

Fucini got to Padua, where he was ordered to bring an envelope to commander Moizo in Pordenone, and he had to ride in an old unarmed SP with a rookie pilot. They reached their destination with many difficulties, and Fucini finally got an assignation to a fighter squadron. He met for the last time his friend Lanza who, no longer having a unit, was volunteering for all sorts of missions. A few days later he took off on a Caproni, that was also intercepted and shot down by the Germans, with the death of the gallant prince.

His destination was the 76ª Squadriglia that had retreated to Arcade. It had its pilots, but with just a few efficient Nieuport 17, a type that Fucini had never flown, so initially he had to wait for his turn. His first mission was together with another newcomer, Silvio Scaroni, who was also flying his first fighter sortie, and the result was unsettling. Three enemy fighters attacked them, they faced them but after a brief dogfight Fucini found himself alone in the sky, with no trace either of Scaroni or of the enemy.

The situation rapidly improved with the arrival of food, clothing, and especially new fighters, the excellent Hanriots that replaced the worn-out Nieuport 17s. " Wonderful, exhilarating, terrible in combat thanks to its exuberance of agility even at high level", this is how Fucini described the French fighter, license built by Macchi at Varese, a real thoroughbred that gained glory not in its native land but in Italy.

The 76ª Squadriglia with commander *cap.* Alberto De Bernardi moved to Istrana. It had no aces yet, but many of its pilots would soon become aces, like *tenente* Mecozzi and *sottotenenti* Scaroni and Michetti. Activity was intense in November, and Fucini escorted reconnaissances to the new Austrian lines along the Piave. He attacked a two-seater together with Donati, saw it diving into a cloud but couldn't have any confirmation.

His first victory arrived on 13 November. As he was cruising with *sergente* Gastone Razzi he saw an enemy two-seater. There were two other Hanriots in the area, and they all attacked the enemy that had no chance, it fought well but soon it crashed in flames near Arcade. It was one of the combative German crews, the pilot *Leutnant* Eduard Schutze of Fliegerabteilung A 232 was found burned in the wreck, it was reported that he was "much decorated". The observer fell out of the burning cockpit and fell at a distance. This victory was assigned to Fucini, Razzi and two pilots of the 78ª Squadriglia, *sergente* Antonio Chiri and *tenente* Guido Masiero.

Fucini's second victory came a few days later, on 17 November, during an escort, together with Scaroni and Masiero, to a two-seater and a Nieuport 11 equipped with rockets of pilot Gino Allegri that had to attack a kite balloon. They had a hectic dogfight with a patrol of German fighters that came to protect the balloon, and Fucini shot at one of them, seeing it apparently out of control. Ground observers gave notice of two enemy fighters falling down, and these victories were credited, one each, to Scaroni and to Masiero and Fucini. Actually it seems that only one German fighter was shot down, corresponding to Scaroni's victim.

A few days later, the actual date however is unknown, Fucini attacked a two-seater and ground observers saw it crossing the Piave trailing a long plume of smoke, but this victory was not confirmed.

The Austrian offensive subsided, and bad weather set in. On Christmas Day 1917 the fighters flew just to cheer the infantrymen on the Grappa. The next day, however, there was the famous Battle of Istrana, which began with the German retaliation for the strafing attack of Captain William Barker at San Fior airfield on Christmas Day. In the confused free-for-all Fucini claimed two victories, an airplane flamed and crashed and another one that turned over on the ground. Italian official documents however credited him with only one victory, together with others. He skillfully described that hectic action:

> "Shadows of fleeting wings are everywhere: black crosses, cockades. A wing passes by me, it's a miracle that we don't collide, but its wing rocks my plane and forces me to correct my aim, while under enemy fire. Somebody helps me, a comrade also engages the enemy aircraft that now rocks and banks, looking for safety with irregular maneuvers: you are not going to cross the Piave! Here it is at last, it drops a wing, goes down out of control, falls in a fields, flames engulf it immediately. That's one. Where is another one? Another one! Today it's difficult to choose. As I turn around looking for a prey, I'm targeted by fire. It's a German observer who is shooting at me from the side. Not for long, because one of our fighters attacks it. I leave them alone and I attack another one that gets itself in front of me, in my field of fire. And now, what's that? I see long gray shadows passing in front of me. Bombs. Will they hit me? I don't have time to think about that: I look at the bomber above, the explosions below: missed. They just dig holes in a field. I resume my attack, I aim carefully, and I shoot. ...I'm back against the enemy. As I fire, the enemy gunner opens his arms and slumps across the fuselage, shaken by the wind. I can already feel the excitement of a new victory .. yes, this one also is falling. But I lose it from sight below my wings. Turning to look for it, I get another one of ours in front of me, I avoid crashing with a sudden maneuver. The enemy airplane has landed badly in a field, turning over. "

Fucini then, together with Comandone, who had helped him in his first victory, went to visit the two German airmen in a hospital. This way, they missed the interception of the second attack of the day, and he had to look at the combats from the ground.

After this luckless enterprise, the Germans bombed the Italian airfields by night, with good results, as they destroyed several hangars and aircraft and forced the Italians to disperse their squadrons to many airfields. Air activity by day was now scarce, and Fucini scored again only on 28 January 1918 when he intercepted a German two-seater south west of the Montello. As he attacked, also Scaroni joined, and the enemy fell down over Biadene, within the Italian lines. To his horror, Fucini saw the German observer, Dietrich Stapefeld, desperately hanging from the overturned airplane and then falling down. Fucini received another Medaglia d'Argento al Valor Militare for this victory.

After a brief period of leave, on 11 February 1918 Fucini was shifted to the 78ª Squadriglia, as flight commander. The 78ª was based at Nove di Bassano and its commander was the future ace *capitano* Antonio Riva while two good friends of Fucini, *tenenti* Mecozzi and Brenta commanded the other flights. Many pilots of the 78ª had a pennant painted all along the fuselage as a personal insignia and also Fucini painted one on his Hanriot. Commander Riva had a red pennant, Mecozzi a blue one and Brenta a green

One of the many German DFW C.5s shot down at Istrana on 26 December 1917.

Right: A close portrait of Fucini next to his Hanriot, showing details of the lightning and of the shadow shading of the serial number.

Left: When Fucini moved on to the 78ª Squadriglia he also flew a Hanriot with a beautiful red wing growing out of the central red dot of the national roundel. Right: While visiting General Pirzio Biroli is interested in the cockpit of the Hanriot of Fucini, who is standing on the right, we can see that the red wing motif was applied also on top of the horizontal tailplanes of Hd 7549.

one, so Fucini adopted white, or silver, as he said using the official term of heraldry, adding as a personal touch a black skull for each of his victories.

Initially his service with the 78ª was not productive: he flew very much, but he didn't meet any hostile aircraft until the arrival of spring. The tasks of a flight leader included leading pilots of lower rank, but whenever possible he flew in offensive patrols with his friends Mecozzi and Brenta, but on account of the famous bad luck of Mecozzi this trio of fighters never managed to engage enemy aircraft.

So far Fucini was credited with four confirmed victories, so he was yearning to score one more and to become an ace, but he had to wait until 17 May 1918 for a new air combat, of which his handwritten combat report survives:

> "Time of departure 10h01
> Time of landing 10h21
> Attained height about 3800
> Type of aircraft HD
> Machine gun Wichers
> Bullets fired about 100
> Report
>
> I took off at 10h01 and I climbed toward the Piave, reaching about 3800 meters. As I followed three enemy fighters very high I spotted below me to the left a reconnaissance airplane that I believe was under the escort of the fighters, in the direction Cimadolmo –Treviso. I attack it with a first burst shooting vertically while it …. I attack it with a second burst higher and from its tail. The aircraft after a clumsy maneuver went down spiraling. I follow it occasionally firing short bursts, stopped three times by gun jams, and then I saw it still spiraling hit the ground with its wings and breaking apart in a field south of Maserada. (During the descent of the enemy plane I saw another Hanriot arriving, sporting two sloping bands from the wing leading edge to the trailing edge).
>
> War zone, 17 May 1918.
> Tenente Mario Fucini."

The enemy two-seater belonged to Flik 12 and the commander of its escort, the Austrian ace Franz Gräser was also shot down in flames by the Italians. When he returned to base Fucini flew barrel rolls for victory and his companions cheered him. He described this action as his fifth confirmed victory, and actually it was originally credited to *sergente* Cosimo Rennella with the participation of Fucini and *tenente* Flaminio Avet, and the enemy crew declared that three fighters had shot them down. Later on, however, and for unknown reasons, this victory was credited only to *ten.* Avet of the 70ª, that is to say the Hanriot with sloping bands described in Fucini's report.

Fucini later went to the hospital to meet the captured crew and he had a long talk with the Czech pilot Valenta, who suffered from an amputated leg. He openly complained about the behavior of escort fighters who, he claimed "run away all the times" and about the fact that NCO pilots like himself always got the worst aircraft, while officer pilots got the newer ones. The transcription of his interrogation, preserved in Italian archives, describes well the growing fatigue and disillusionment seeping among the ranks of Austro-Hungarian troops. They discovered they had a common friend, pilot Max Brociner, who also had been wounded in an air combat and had lost a leg. After World War II the two pilots remained in contact and wrote to each other often.

Then came the Austrian offensive that on 15 June 1918 flamed the entire front, from the mountains to the sea. On the first day Fucini took off with his patrol, saw many aircraft and combats in the distance, but couldn't join any of them. The next day as he was flying with *cap.* Riva and *soldato* Venier he attacked at 10h30 over the Montello a Brandenburg two-seater and shot it down across the Piave, seeing it wrecked against a hedge between Malborghetto and Fontigo. The enemy crew described one of its attackers as having "blue-white-red wings and a black skull on its fuselage" a description that fits the tricolor below the wings of the Hanriots and Fucini's little skulls. A few moments later he saw another two-seater, that after a tough combat he forced down near Pilonetto, close to the banks of the river. He returned to his airfield at low level, targeted by the rifles of enemy troops, where he was welcomed by *ten.col.* Piccio, who congratulated him.

*Soldato*, then *caporale* Venier was the wingman of Mario Fucini. His Hanriot with obscured serial numner, ??26, shows one of the many variations of the pennants of the 78ª Squadriglia.

After a whole day of flying, at 6 p.m. he decided to go on offensive patrol alone, but *serg.* Ermenegildo Bocca of his flight asked to come along. When they reached the lines Fucini saw two enemy fighters, one of which dove down on the Hanriot of his wingman, who was attacking a two-seater. He attacked, saw the two-seater falling in flames but also noticed that Bocca, clearly hit by his adversary, was no longer reacting. He shot from a distance at the fighter and its pilot, concentrated in shooting at Bocca, didn't notice the tracer bullets passing around him. Only when Fucini was behind him he realized that he was under attack and turned. It was a very tough dogfight at low level, and the Austrian was an excellent pilot. Fucini tried to gain the advantage of height, his adversary followed, both fighters getting higher to gain the ascendancy:

> "...but now I feel near the end, my Hanriot is no longer responsive. Soon I'll slide down against the Austrian. Will he be able to avoid me? He has understood, overcome by obstinacy. Seeing that I am about to fall upon him, he maneuvered to avoid the collision. So? That's what I wanted. You avoid the crash, but not my volley. I regain speed and get behind him at full throttle. A few instants of tension to aim well, and off the tempest of bullets. Hit, at last.
>
> His small black-crossed wings turn full circle. Did I miss? Is he coming back to dogfight? No, it's over. The circle disfigures into a violent bank. The vanquished passes next to me like a meteor, head down, flutters once or twice, falls down among the trees breaking up on the hills above Nervesa."

When he returned to base he was told that Bocca was missing. He was admitted in a hospital hit by five bullets, the last of which passed through his nose. He had fainted in the cockpit but luck and the qualities of the Hanriot saved him: the little biplane had landed itself. His Austrian enemy was almost certainly *Leutnant* Wolfschutz, missing in action in that area. The rudder of his Albatros is now preserved in the Museo della Scienza e della Tecnica in Milan. Two of these three victories were credited to Fucini; the first one was assigned to Riva alone. He also got a prize of 500 Lire for the second one and 1,000 Lire for the third one.

In the following days Austrian aviation, that had worn itself out, was hardly active above the battlefield, so, despite the bad weather, Italian fighters tried to help the ground troops strafing and bombing. On 23 June, flying with *serg.* Chiri and *sold.* Venier, he attacked and hit an enemy aircraft, that was seen going down out of control across the lines, but that was lost from sight. Despite the testimonials of ground observers, this victory also was not confirmed. A few days later, after the battle was over, he went as representative of his Squadriglia to the funerals of *maggiore* Baracca.

In the next months encounters with the enemy became very rare, as the Italians had gained mastery of the sky. On 25 July Fucini with *caporale* Venier had to escort a reconnaissance SP across the Piave but due to a misunderstanding the two-seater flew over their airfield without waiting for its escort. They took off after it and met a two-seater, but not their own, an enemy one, and they attacked it with good coordination, so that despite many gun jams they saw it falling down trailing black smoke and crashing against a line of trees near Moriago. Back to base, Fucini painted the eleventh skull on the white pennant on his fuselage.

The twelfth one was late in coming. It happened on 31 August, when he flew with two remarkable wingmen, *sergenti* Rennella and Chiri, aces both. It was a relatively easy combat. The enemy crew spotted them too late. The enemy airplane described as a "small two-seater" was attacked and shot down among the Austrian trenches near Mandre. Data about this combat are confused: Fucini said it happened at 13h30, while for the logbook of Chiri it happened in the evening. The latter time is more likely, as it would correspond with data from Austrian sources, which report the loss of an aircraft in that sector of the front near sunset.

The final mount of Fucini was this Hanriot in dark camouflage, emblazoned with a white pennant showing a little black skull for each of his victory claims. This fighter is armed with two machine guns.

Later on, Fucini improved visibility of the black 78 squadron number, adding a thin white outline to it. On the left, one can see a couple of American soldiers.

*Sergente* Ermenegildo Bocca, wingman of Mario Fucini on 15 June 1918, was hit by five bullets. The serial number of his Hanriot is 13229. The first aircraft of that production batch reached the front in June 1918.

*Ten.* Giacomo Brenta and *ten.* Mario Fucini, two flight leaders of the 78ª Squadriglia, in a moment of rest.

One day Fucini was invited to Padua to represent his squadron in a large parade of Allied troops, reviewed by all the Italian top brass, and he had a lot of fun watching Scottish troops with their kilts:

> "At the end there was a dance of Scotsmen, who shook their little skirts and showed their forms for more than half an hour. Somebody had to approach the piper and tell him to stop, because there was the final ceremony. The dancers were so excited that they kept on dancing without the pipe music, so the end was delayed."

In July 1918 the Austrian ace Linke-Crawford had compelled to land on his own airfield a Sopwith Camel. The Italians learned through intelligence that this fighter was intact, and fearing that the enemy could use it, issued an order that all Camels were to carry two canvas ribbons at the struts. If a Camel didn't carry the ribbons, it was the one with an Austrian pilot and it had to be shot down. So one day Fucini stumbled into a Sopwith Camel flying over the Montello without the proper markings, doubly suspicious because its pilot was taking notes. Fucini was puzzled because the intruder seemed too much unconcerned, so instead of shooting he followed it. When, after a lot of wandering, the Camel landed at an allied airfield Fucini was relieved. *Cap.* Riva soon clarified the matter: a British Squadron had not received that order, and had kept flying without the new identification marks. The Allies, for thanks, invited Fucini and his commander to their base. Riva, who spoke perfect English, enjoyed a never-ending party, but Fucini, who didn't speak one word of English, had to stand a succession of horrible food and most peculiar entertainment, that included singers in drag, and was bored to death. When they finally got away, his comment was: "I should have shot the Englishman down!"

The final offensive of the war brought the last adventures for Fucini. On 26 October as he was cruising above the Piave an artillery shell ripped away the landing gear of his fighter, and stuck a wheel into the lower wing. He managed to return to base and to perform a perfect belly landing. The next day he scored his final victory. As he was flying toward the Livenza River admiring the battle raging below him, he saw five airplanes approaching that he mis-identified as SVA, but then he noticed the black crosses. The enemies attacked him with élan and remarkable ability: they were flying fast and maneuverable fighters. Fucini defended himself as best as possible, then while four of them were all around him he attacked from below the fifth one, who was rocking his plane, probably trying to see him. He hit it from below and saw it spinning and crashing. Meanwhile *sergente* Luigi Capparucci and other Italian pilots had arrived and engaged the other Austrian fighters, and the combat moved away.

Fucini in his postwar memories dated this combat on 28 October, but the squadron papers dated it on the 27th. The enemy fighters, which he identified as "Pfaltz", probably were Aviatik Berg D.1. The body of the enemy pilot was found after the retreat still resting in the wreck of his fighter, but it was not possible to determine who had shot him down.

Within the joy of victory, Fucini had the sorrow of losing his wingman in the final days of the war. On 28 October he went out for a fighter mission with *tenente* Carlo Pasquinelli, but lost him in the clouds. On his return he learnt that his wingman had returned seriously wounded: he had engaged an enemy two-seater, but a group of fighters had bounced him, hitting him repeatedly. Pasquinelli died of his wounds on 9 November in a field hospital.

At the end of the war Fucini had claimed a total of 13 victories, but in 1919 the 5ª Sezione admitted only 7 of them, cutting down drastically the claims of the pilots of the two Squadriglie of the XXIII Gruppo, 78ª and 79ª. The day before armistice Fucini, possibly worn out, was briefly admitted in a field hospital. He was decorated with a silver medal for his combat of 28 July 1916, a bronze medal for his adventure of 16 February 1917 and for a second Medaglia d'Argento that reported in its citation the four victories from November 1917 to January 1918.

A few days later he was assigned to the Direzione Tecnica Aviazione Militare in Turin. He later had different commands and after the war he lead fighter squadrons, and was promoted to *Capitano* in 1926. On 15 January 1928 he was assigned to the General Staff of the Regia Aeronautica. In 1932, now a Lieutenant Colonel, he was commander of the Fighter School and in December 1933 he became editor of the Rivista Aeronautica, the official magazine of the air force, and he kept this post until 16 August 1935, when he became commander of the 2° Stormo Caccia.

In the early Thirties he published a very good book of war memories, called *"Voli sul nemico"* (Flights above the enemy) that sold very well. He was promoted to the rank of colonel in February 1936 and went on special leave one year later, in September 1937. In April 1943 he became *Generale di Brigata Aerea* and after the war, in 1954, *Generale di Divisione Aerea.*

He became a widower on 21 May 1953; he married again on 23 October 1954 with Emilia Capobianco in Rome where he lived, in a small house full of relics and memories of the war, until his death, on 1 September 1977. One of the authors of this book had the honor of meeting him personally.

Right: Fucini had a long and successful career in the Italian Air Force. Here he stands in front of a Macchi Nieuport 29 of the VII Gruppo at Ciampino airfield, Rome, about 1927.

## Mario Fucini
## 13 victory claims - 7 confirmed victories

<table>
<tr><th>Claim</th><th>Vict. #</th><th>Date</th><th>Sqn.</th><th>Aircraft</th><th>Hour</th><th>Place</th><th>Victim</th></tr>
<tr><td>1</td><td>1</td><td>13.11.1917</td><td>76</td><td>Hd</td><td>15h45</td><td>Arcade/Bavaria</td><td>Two-seater</td></tr>
<tr><td colspan="8">[with serg. Razzi, 76ª, serg. Chiri & ten. Masiero, 78ª Sq.] Lt. Eduard Schulze Watzum KIA/Lt. d. Res. Kamillo Ehrler KIA, FA 232</td></tr>
<tr><td>2</td><td>2</td><td>19.11.1917</td><td>76</td><td>Hd</td><td>-</td><td>Falzé di Piave/Vidor</td><td>Fighter</td></tr>
<tr><td colspan="8">[with ten. Masiero, 78ª Sq. and s.ten Scaroni. 76ª Sq.] maybe: Vsfw. Kaspar Rahier KIA, Jasta 31, Albatros D.V, or: Oblt. Deindl WIA, Jasta 1</td></tr>
<tr><td>-</td><td>-</td><td>??-??-1917</td><td>76</td><td>Hd</td><td>-</td><td>Ciano di Piave</td><td>Two-seater</td></tr>
<tr><td>3</td><td>3</td><td>26.12.1917</td><td>76</td><td>Hd</td><td>9h</td><td>Musano</td><td>Two-seater</td></tr>
<tr><td colspan="8">[with s.ten. Scaroni 76ª, ten. Masiero & serg. Fornagiari, 78ª, ten. Comandone 82ª Sq.]</td></tr>
<tr><td>4</td><td>-</td><td>26.12.17</td><td>76</td><td>Hd</td><td>9h</td><td>Camalò</td><td>Two-seater</td></tr>
<tr><td colspan="8">[with s.ten. Scaroni & Michetti, 76ª, cap. Riva 78ª, ten. Avet, serg. D'Urso, Benvenuti, Teobaldi & ten. Comandone 82ª Sq., 2nd Lieuts. Jarvis, Frayne & Mulholland, 28ª Sqn RFC]<br>Possible victims:<br>Lt. Kessler WIA/Lt. d. Res. Edebohls KIA,F A 2 , DFW C.V<br>Uffz. Willi Petzold KIA/Lt. d. Res. Wilhelm Voigt KIA, F A 2, DFW C.V<br>Vzfw. Karl Uecker KIA/Lt. d. Res. Heinrich Pfeiffer DOW, FA 2, DFW C.V<br>Vzfw. Pohlmann WIA/Lt. Schlamm WIA, FA (A) 219, DFW C.V</td></tr>
<tr><td>5</td><td>4</td><td>28.1.1918</td><td>76</td><td>Hd</td><td>-</td><td>Pederiva di Biadene</td><td>Two-seater</td></tr>
<tr><td colspan="8">[with ten. Scaroni, 76ª Sq.] Vzfw. Max Schreiber KIA/Dietrich Stapelfeld KIA, FA (A)219, DFW C.V</td></tr>
<tr><td>6</td><td>-</td><td>17.5.1918</td><td>78</td><td>Hd</td><td>10h15</td><td>Maserada</td><td>Two-seater</td></tr>
<tr><td colspan="8">[with ten. Avet, 70ª Sq. & Rennella 78ª Sq] Fw. Johann Valenta WIA, POW/Lt i. d. R. Viktor Herzog, POW, Flik 12/Rb, Br. C.1 229.30.</td></tr>
<tr><td>7</td><td>-</td><td>16.6.1918</td><td>78</td><td>Hd</td><td>10h30</td><td>Malborghetto/Fontigo</td><td>Brandenburg</td></tr>
<tr><td colspan="8">[with cap. Riva & sold. Venier, 78ª Sq.] Korp. Bruno Koziarski UNH/Oblt. i. d. R.. Julius Tiefbrunner UNH, Flik 38/D, Br. C.1 369.61</td></tr>
</table>

| Claim | Vict. # | Date | Sqn. | Aircraft | Hour | Place | Victim |
|---|---|---|---|---|---|---|---|
| 8 | 5 | 16.6.1918 | 78 | Hd | 10h30 | Pilonetto | Two-seater |
| 9 | 6 | 16.6.1918 | 78 | Hd | 18h | Nervesa/Bavaria | Fighter |
| *Lt i. d. R. Hans Wolfschütz MIA, Flik 41/J, Alb. D.III 153.222* | | | | | | | |
| 10 | - | 23.6.1918 | 78 | Hd | - | S. Michele | EA |
| *[with serg. Chiri & cpr. Venier, 78ª Sq.]* | | | | | | | |
| 11 | 7 | 25.7.1918 | 78 | Hd | - | Moriago | Two-seater |
| *[with cpr. Venier, 78ª Sq.]* | | | | | | | |
| 12 | 8 | 31.8.1918 | 78 | Hd | 13h30 | Mandre | Small two-seater |
| *[with serg. Rennella & Chiri, 78ª Sq.] Maybe: Stfw. Franz Prochàzka UNH/Lt. i. d. R. Emil Bauer DOW, Flik 22/D, Ufag C.I 161.107* | | | | | | | |
| 13 | - | 27.10.1918 | 78 | Hd | 16h50 | Vazzola | "Pfaltz" |
| *Possibly: Korp. Josef Podpetschan MIA, Flik56/J or: Oblt. Stefan Horvath MIA, Flik 56/J or: Korp. Adolf Wissinger MIA, Flik 74/J* | | | | | | | |

Fucini was credited with 7 confirmed victories, but this research shows he had eight confirmations. It is impossible to ascertain which one was deleted from the list of Aces of 1919.

## Attilio Imolesi

The people from the region of Romagna in Italy are stereotyped as being hard working, fun loving, with a healthy passion for wine, women and song, and an immoderate passion for dancing and for fast engines of any kind. Attilio Imolesi was born on 11 October 1890 in Cesena, a quiet, wealthy and artistic town in the heart of Romagna, and we don't know how many traits of the "romagnolo" he inherited: he was a good hearted, considerate young man, hard working and familiar with music, and surely he liked motors, because he was an automobile mechanic, a very rare specialization in pre-war Italy. This skill facilitated his acceptance into aviation, where motorists were eagerly looked for. He volunteered in 1915 and was sent to Infantry, but after only one month he was sent to the aviation motorists school in Mirafiori, Turin.

In November 1915 he was assigned to the air base of Aviano, where he worked on the first Caproni three engine bombers. On 31 December 1915, however, he was accepted to pilot training, and was sent to the Malpensa School. He got his first level brevet on 1 April 1916, flying Farman N. 913. In July 1916, still with the rank of soldier, he was assigned to the newly formed Sezione Difesa di Rimini, (Rimini Defense Flight) that flew Farmans for the protection of the coast of Romagna, under the threat posed by the Austrian seaplanes from across the Adriatic. Despite the assignment to his native land, Imolesi disliked the relatively quiet assignment to home defense, and he repeatedly requested to be sent to the front. On 25 November 1916 he was licensed to fly the Nieuport fighter and five days later Imolesi, now a corporal, was assigned to the new 79ª Squadriglia, that was formed at the end of the year in Arcade and that was mobilized in January 1917 on Istrana airport, for the service of the 2nd Army, covering the northern and rear part of the Isonzo front, equipped with Nieuport 11. That section of the front was quiet, and the first encounters with the enemy took place only in June 1917, when the Italians launched a terrible offensive on the Ortigara mountain, that turned into a slaughter of Alpine troops. Meanwhile, however, in the first half of 1917, several pilots of the 78ª and 79ª Squadriglie were repeatedly detached to reinforce the crucial Isonzo front. With these pilots, large escort formations were organized to protect missions of the Caproni bombers. At the same time, the 70ª Squadriglia at Santa Caterina was training chosen pilots of the other units on the Spad fighter. Imolesi was sent to S. Maria la Longa airport on 24 April, to fly the Nieuport 110 hp.

On 26 April 1917, Imolesi participated in an attack against an Austrian two-seater conducted by *cap.* Francesco Baracca of the 70ª Squadriglia from an altitude of 4,000 meters. The long pursuit ended with the destruction of the enemy, Brandenburg 129.17 of Flik 35, with the death of its crew, *Zugsf.* Majsai and *Leut.* Treer, and the diary of the 70ª reports: "Serg. Imolesi took part in this fight". Baracca gave to Imolesi a prize of 1,650 Lire for his part in this victory.

This is the combat report:

"Aircraft Nieuport 110 hp 2142. Machine gun Wickers and Lewis. At about 11h10, as I was in a patrol mission over the Carso at a height of 3500 m. I saw an enemy aircraft higher than me that was over Gradisca engaged by a Nieup. I gained height and I approached it just when it was trying to run away from the Nieup. It was coming toward me. When it got close, at a height of about 4300 m. it noticed me and turned left and I, turning right, got on its tail and at the same time I fired two bursts with the Lewis and the Wickers. So I forced it to turn back toward the Nieup. It immediately began to go down and following it I realized that besides me there were two Nieuports against it. In the pursuit always descending, I shot two bursts at the same time, and then as the Wickers had jammed I finished the magazine of the Lewis. The enemy aircraft going lower and lower entered a cloud at about 2000 meters and I had to give up the pursuit in order not to crash into a Nieuport in the cloud. Because I was very close to it."

Attilio Imolesi, one of the sergeant pilot aces of the 79ª Squadriglia.

In the final post-war evaluation of victory claims, this combat is placed above Asiago, with the remark that no confirmation exists in the papers of the 79ª Squadriglia. In fact, of course, the confirmation can be found in the combat diary of the 70ª. It is about this time that dates Imolesi's promotion from corporal to sergeant. In May he remained on the Isonzo front, being temporarily attached to the 44ª Squadriglia, a reconnaissance unit equipped with Caudrons.

Imolesi described his life as a fighter pilot in vivid letters to his relatives:

"At the moment of the attack we meet, fighter versus fighter, life versus life, and it is a moment that nobody could describe, he who is ready for battle turns into a different being, freezes on the controls of the airplane, the eye on the gunsight of the machine gun, and at the right moment you squeeze the lever that controls the gun and it sings ta.. ta... ta... Also yesterday I had three combats, driving down the enemy and getting back to base safe, I and the two officers on the reconnaissance airplane that I was escorting".

On 7 May 1917 Imolesi was posted from S. Caterina to nearby Medeuzza, where a fighter section commanded by *ten.* Bartolomeo Costantini was formed to protect the slow and defenseless Caudrons of the 43ª Squadriglia heavily engaged in the 10th Offensive of the Isonzo. On 10 and 18 May Imolesi was praised for his escorts to a reconnaissance airplane above Reifenberg, a hotly defended railway center, and for an escort to a photographic mission in bad weather. On 23 May he took part in the massive Italian air attacks in support of the 10th Battle of the Isonzo, escorting large formations of Capronis that bombed the enemy trenches, and he emptied all rounds of his Vickers gun on the Austrian troops.

On 24 May Imolesi took part in a large escort mission to two British monitors, H.M.S. *Earl of Petersborough* and *Sir Thomas Picton* that in the high Adriatic Sea fired at the Austrian coastal positions and at the Prosecco airport. The Austrian seaplanes took up the challenge and flew against the allied formation. Imolesi dove against the attackers but he was surprised by a flying boat fighter that maneuvered skillfully, but that dogfight had no winners. His adversary was surely Gottfried Banfield, flying seaplane A11, who reported a dogfight with a "French" airplane. Other Italian fighters shot down the seaplanes L136 and L137, whose crews were captured.

Imolesi in Nieuport 11 serial 1648, a first batch airplane that here appears without armament.

The next day, 25 May 1917, Imolesi took off to escort another two-seater off the coast of Trieste, but he missed the Italian airplane, that had been driven away by the arrival of the Austrians. Imolesi attacked an enemy, but got too low over the enemy batteries, and had to quit the field, surrounded by ack-ack fire. A similar mission the next day was not so lucky, and the enemy gunners hit his fighter, putting the engine out of action. Imolesi glided down over the sea, and managed to crash-land his fighter on the Italian

Imolesi sitting astride Nieuport 11 serial 2125. The diagonal white bands identified the flights within the 79ª Squadriglia.

cause the fighter had run away. My companion started firing right away, I pressed the enemy closely from the side without shooting for the precaution of not hitting my colleague, but I saw him after a burst banking on one side and abandon his prey. His gun had jammed, bus as I was following him closely with my eye to the gunsight and the fire lever in my hand, all I had to do was to press and my girlfriend sang tan tan tan and I gave it the coup de grace. Then I heard a burst behind my back and when I turned I saw the fighter high above me, but giving a wide berth: he had understood that it is dangerous to get in harm's way..."

Attilio Imolesi in front of his Nieuport 27, the airplane on which he lost his life on 10 March 1918.

Despite his victories may portray Imolesi as a dashing loner, the quality that was most appreciated in him was his reliability as a wingman, and he usually flew together with *cap.* Bertoletti, commander of 79ª Squadriglia.

Time was running out for Imolesi: on 10 March 1918 he died in an accident with the Nieuport 27, a fighter that equipped the 79ª Squadriglia and that was considered dangerous by the Italians, who much preferred the all-forgiving Hanriot. Probably the rudder bar detached itself, a common failure of the Ni.27, and Imolesi crashed while flying near the ground, his skull being broken by the collision against the machine gun. He died in Marostica hospital. In 1921 his body was brought back to his native Cesena, that has named a street after him.

## Attilio Imolesi
## 6 confirmed victories

| Claim | Vict. # | Date | Sqn. | Aircraft | Hour | Place | Victim |
|---|---|---|---|---|---|---|---|
| 1 | 1 | 26.4.1917 | 79 | Ni.17 2124 | 11h10 | Gradisca | Two-seater |
| *[with cap. Baracca & serg. Gorini, 70ª Sq.] Zugsf. Josef Majsai KIA/Lt. Emmerich Treer KIA, Flik 35, Br. C.I 129.17* | | | | | | | |
| 2 | 2 | 26.9.1917 | 79 | Ni.17 | 10h | Asiago | Two-seater |
| *Zugsf. Walter Horny KIA/Oblt. Artur Klose-Kuschel KIA, Flik 21, Br. C.I 129.29* | | | | | | | |
| 3 | 3 | 26.9.1917 | 79 | Ni.17 | 10h+ | Asiago | Fighter |
| *[with ten. Mazzini, 79ª Sq.] Fw. Lajos Telessy UNH, Flik 21, Alb. 53.29* | | | | | | | |
| 4 | 4 | 13.12.1917 | 79 | Nieuport | - | Ghertele | Two-seater |
| *[with serg. Ciotti, 79ª Sq.]* | | | | | | | |
| 5 | 5 | 11.1.1918 | 79 | Nieuport | 11h55 | Crosara | Fighter |
| *[with serg. Ciotti, 79ª Sq.] unknown pilot KIA/Lt. Hans KIA, unknown unit* | | | | | | | |
| 6 | 6 | 14.1.1918 | 79 | Nieuport | 15h05 | Valstagna | Two-seater |
| *[with serg. Reali, 79ª Sq.]* | | | | | | | |

# Giulio Lega

"Stretched infantry" was the nickname assigned to the soldiers of the Reggimenti Granatieri (grenadiers) and judging from the photos of Giulio Lega, 182 centimeters tall, it was fully justified, considering that the Italian population in the early 20th century was on the average much shorter than today. In the few known photos in which he appears, he towers above his colleagues, easily recognizable also for the particular collar tabs on his uniform.

The Granatieri were an élite unit, heir to one of the most ancient regiments of the army of Piedmont, and Lega must have been proud of his original service, considering that he adopted its symbols, a grenade and the collar tab as his personal insignia on his Hanriot.

Among his colleagues in the squadron he was nicknamed "the doctor", because when war broke out he was a student in the Medical School of the University of Bologna. Giulio Lega was born in Florence on 12 November 1892, but when he was drafted he was living in the region of Emilia, reporting to the military district of Ravenna. He was recruited as a soldier on 20 July 1913 but his service was deferred, probably because he was a university student. When was broke out he was recalled and was admitted to officer training in the prestigious military school of Modena on 23 June 1915.

He was sent to the front with the rank of sottotentente on 26 September 1915 with the 2° Reggimento Granatieri. Soon after the Supreme Command launched the Third and Forth battles of the Isonzo. During the latter offensive the Brigata Granatieri was severely engaged trying to overcome very strong enemy positions at Oslavia, north of Gorizia. Its attack was launched under a pouring rain and in very cold weather. Many of the soldiers had been suffering from a stomach ailment with symptoms similar to the ones of cholera that had struck most of them drastically reducing the number of able men. In this situation Lega distinguished himself particularly on 20 November 1915 when, despite being sick, he returned to the line of battle to take up command, setting his example to his soldiers and leading them to the attack of the well defended enemy trenches. Again on 30 May 1916 he fought gallantly in the combat at Malga del Costo, when he managed to protect and bring back to the lines a group of soldiers that he was leading. For this action he was decorated with his first Medaglia di Bronzo al Valor Militare with this citation: "In a violent and brief hand-to-hand combat he showed courage and valor, managing to avoid capture of himself and of some subordinates."

Soon after he was admitted to the Battaglione Aviatori, to which he was posted on 29 June. Lega got his military pilot's brevet on 31 January 1917 and he was assigned to the 21ª Squadriglia da ricognizione. This was a new unit, just formed and equipped with Savoia Pomilio SP.2, a new model of pusher two-seater technically obsolete even before entering service. Together with the commander *capitano* Tasselli, there were five pilots including Lega and originally only two observers.

The 21ª Squadriglia operated supporting all ground actions in the spring and summer 1917, performing good reconnaissance work and very often sending its planes low to strafe and harass the enemy. In June one of its crews was lost, shot down by ground fire, and many airmen were wounded in action. Lega was awarded a Medaglia d'Argento for his service in this period, but all data pertaining his flying activities are lost. The citation of his medal reports:

> "Daring airplane pilot, in many reconnaissance and bombing missions, ignoring danger and full of enthusiasm, he flew very low over the enemy to strafe; he shot excellent photographs; although enemy fire several times hit his plane, he didn't leave the area of danger until his job was done, being a great example of firmness and valor. Sky of S. Marco and Bainsizza, May, June, August, September 1917."

Then came the 12th battle of the Isonzo and the retreat. The worn out 21ª Squadriglia was withdrawn from the front and sent to Centro Formazione Squadriglie at Ponte San Pietro, its commander was (unfairly) relieved of command. Lega was promoted to *capitano*

A group portrait of the pilots of the 76ª and 81ª Squadriglia. Lega is the tallest, seventh from the left, and next to him is Mazzucco, commander of the 81ª. Second from the left is *cpr.* Astolfi, two victories, who shot down the Austrian ace Frank Linke Crawford.

on 31 October, and, as he no longer had a unit, he was sent to fighter training.

On 16 November he was at the Malpensa training airfield for conversion on the Nieuport. Then he went to Furbara on 23 December for aerial gunnery school, where his results were, according to an official report, "*mediocre*".

On 2 February 1918 Lega was assigned to the 76ª Squadriglia equipped with Hanriot Hd.1 fighters, which on that very day moved from Istrana to Isola di Carturo, and then moved two weeks later to Casoni, its final base. These movements of units were due to the effective night bombing raids of German and Austrian aircraft against Italian airfields, which caused considerable damage. The 76ª had been very successful recently, mostly for the ability of *sottotenente* Scaroni, who at that time had a log of 13 air victories. With his rank of captain Lega theoretically could aspire to the command of a Squadriglia, but being an officer of the reserve and with no experience at all as a fighter he didn't get it.

Lega didn't fly very much when in the 76ª Squadriglia, at least if we compare his activity with the other pilots, and one could guess that he was also employed in administrative tasks. Only one month had to pass, anyway, before he claimed his first victory. On 17 March 1918 he was tasked to escort a SIA 7b two-seater of the 22ª Squadriglia that had to fly an important reconnaissance mission in the area of the plateaus. He was flying with *tenente* Ludovico Censi and *sottotenente* Renato Donati (future world height record holder), two pilots who often flew together. When near Mount Spitz they met a big gaggle of enemy fighters, probably ten, that attacked them. Lega first of all took care of the SIA two-seater and protected its return to the lines, and then when he saw that it was safe, he counterattacked and claimed one fighter shot down over Col d'Astiago. Donati also claimed one victory, but it is not possible to determine if this is a duplication of Lega's claim or it is another fighter.

A few days later, on 25 March, at 11.30 a.m. Lega scored his second victory as he shot down together with *sottotenente* Giuseppe Retinò and *sergente* Luigi Fanti an enemy airplane out of a patrol of four, over the Montello. Retinò also claimed another fighter that he pursued across the Piave and saw it falling down near S. Pietro di Feletto after having fired 150 rounds at it.

After these two events Lega didn't meet more enemy aircraft until 24 June, that is to say when the final Austrian offensive had worn itself out. That day he was flying in formation with two exceptional pilots, *tenente* Scaroni and the just as able *tenente* Michetti, two expert fighters. Scaroni described this combat in his memories:

> " ...Unwillingly Michetti brought that unfortunate Austrian reconnaissance airplane just in front of the muzzles of my guns. He, with Lega, flies a bit behind and above me in our usual triangle formation when I, leader of the flight, go and attack. It is noon of 24 June. First Michetti shoots one or two bursts and the enemy pilot dives at full throttle to escape, but in doing so he passes right in front of my fighter. I aim and I pull the trigger, but my first burst misses. The enemy aircraft vainly tries to defend itself, some minor bank to the right and to the left are all of his defensive maneuvers. I immediately realize that the enemy pilot is not an expert of air combat, and he must not be much of a pilot either. In such conditions it is natural that his career come to an end immediately as soon as he meets an enemy fighter. As I failed my first burst because I shot too soon, I get below the tail of the enemy and at the same time I fire a second burst of some twenty rounds. From that position the impression is more of shooting at a machine than at men. The usual sinister flame explodes from the belly of the enemy plane, in a second due to the action of the wind also its tail flames and the whole plane turns into a burning pyre.
>
> In its fall the plane turns over, its observer is thrown out and falls in the bed of a little stream that flows from the Grappa while the ball of fire falls near Possagno, close to the tomb of Canova, where it soon burns itself out and disappears.
>
> We got here and all we found was the engine, turned into a block of rubbish. All the documents that we found on those poor victims, a captain and a 1st lieutenant, we brought to our airfield. The next day we collected them in a parcel, tied to a long colored ribbon, and I dropped them on the enemy airfield at Feltre."

The airfield of Casoni in the early spring of 1918, with Nieuport 27 and Hanriot fighters of the 81ª Squadriglia. The frontline was across the mountains looming over the airfield.

"Bigio" and "Toni", armorer and fitter of the 76ª Squadriglia in front of the Hanriot of Lega, armed with two Vickers guns.

An unknown person in front in front of the French-built Hanriot of Lega. The cowling is very dirty and the propeller blades show reinforcements to protect in case of a failure of the guns synchronizers.

Lega in front of his Hanriot during maintenance: the tail is missing. His personal insignia showed the collar tabs and the grenade of his Grenadiers regiment. *(Photo Franz Selinger).*

A gathering of officers, many of them fighter pilots. Identified by numbers are: 1 Piccio, 2 *cap.* Tacchini, 3 *ten.* Ranza, 4 *cap.* Ruffo di Calabria, 5 *cap.* Lega, 6 *cap.* Mazzucco, 7 *s.ten.* Ferrarin, 8 *ten.* Resch. Eighth in the back row there is also *ten.* Mecozzi. *(Photo Franz Selinger).*

Lega scored his second victory together with *s.ten.* Giuseppe Retinò, who flew this Hanriot of French production, serial number 529.

Scaroni didn't say what was Lega's part in this combat. It is interesting to remark that all Italians, not just Scaroni, regularly confused the ranks of their victims, thinking that the three stars shown on the lapels of Austrian sergeants (feldwebel) meant the rank of captain, as in the Italian army. In fact their two victims were *Feldwebel* Pachta, that is a sergeant, and *Oberleutnant* Persoglia, a lieutenant, and they were flying a Brandenburg C.1, the real workhorse of the Austrian air force.

The next day, 25 June, Lega could claim two more victories over enemy fighters, both were confirmed and he entered the roll of Aces. He took off that day with his squadron in an escort mission to two formations of Caproni bombers going out to attack Orsago, but due to a mistake only the first group got the escort, while the second one had the bad luck of crossing the lines with the escort of just three fighters, the ones of Scaroni, who was the leader despite his lower rank, *sergente* Romolo Ticconi and *cap.* Lega.

Several flights of Austrian fighters rising up from the airfields of S. Fior and Godega attacked the Italian bombers, and the three pilots of the 76ª Squadriglia were sorely taxed to repulse them. At the end of that combat two enemy fighters were shot down, and also Lega was credited with them. Scaroni in his memories quotes Ticconi as his precious partner during the dogfight but he doesn't talk about Lega sharing in the victories, actually from his description it appears that he was separated from his two companions:

> "I see Lega much below me as he furiously pursues an enemy fighter a few dozen meters from its tail. In his dash he is getting dangerously low. Stop it, Lega, climb back to us, don't stray too much, don't get too low, you might get bounced any minute... don't stick too long with the same enemy, when there are so many of them buzzing around, I shout to him in my mind.
>
> But Lega obviously cannot hear me and he keeps on the pursuit. By instinct I dive toward him, as much as my adversaries allow me. But this cannot last; now there are two fighters above me... I am forced to counter-attack and I can't think about Lega any longer."

Italian ground observers across the Piave saw two enemy fighters falling vertically near Mareno di Piave and both these victories were credited to all three fighter pilots who took part in the combat; actually the second fighter was credited also to *sergente* Pietro Mondini also of the 76ª Squadriglia.

After this event Lega remained in the squadron until the end of the war. During the summer period also he didn't fly very much, from 13 July to 8 October he flew just 24 combat missions, without ever having combats with enemy aircraft.

Many years later the free-talking Fucini writing an article for a magazine described him, without specifically quoting his name: "... as a pilot, he was a good doctor...".

Overall during the war Lega had had only five air combats with enemy aircraft, that brought to him five victory claims, all of them confirmed after the war, and that are all specifically quoted and dated in the citation to his second Medaglia d'Argento.

On 1 December 1918 he was transferred to the 81ª Squadriglia but on 17 March 1919 he was briefly hospitalized in the Brescia military hospital. Then he was assigned to the 91ª Squadriglia Baracca, but for a very short time, then he was sent back to the Regimental Depot on 4 May 1919 and then he was posted to the military district of Bologna, in order to complete his university courses that he had to stop on account of the war. On 3 July 1919 he was allowed on temporary leave pending for discharge, which arrived on 19 September 1919.

Lega graduated from medical school in 1922, and in the following years he maintained his pilot rating flying tourism airplanes. He was reported as rated in the years from 1928 to 1934. On 1 November 1936 he was called back to arms for a period of three months, assigned to the Italian Institute of Malaria, and attached to the Comando Aeronautico A.O.I. (Africa Orientale Italiana, Italian East Africa), to be discharged on 10 February 1937.

When World War II broke out he was recalled again, but only with administrative tasks, and he was posted to Ciampino Sud airport, near Rome, with the SAS, Servizi Aerei Speciali, the transport element of the Italian air force. When armistice came he was sent in unlimited leave, and he presented himself to the command again on 4 June 1944, when the Allies had liberated Rome from Nazi troops.

His final discharge from the Aeronautica came on 5 June 1956, and he was considered unfit for flight on account of his service. He became the medical consultant of the Italian Chamber of Deputies, and he died in Rome on 11 July 1973.

Lega with his Hanriot now showing dark painted cowling and wheel covers.

## Giulio Lega
## 5 confirmed victories

| Claim | Vict. # | Date | Sqn. | Aircraft | Hour | Place | Victim |
|---|---|---|---|---|---|---|---|
| 1 | 1 | 17.3.1918 | 76 | Hd | - | Col d'Astiago | Fighter |
| *[with s.ten. Donati & ten. Censi, 76ª Sq.]* | | | | | | | |
| 2 | 2 | 25.3.1918 | 76 | Hd | 11h30 | Montello | EA |
| *[with s.ten. Retinò & serg. Fanti, 76ª Sq.]* | | | | | | | |
| 3 | 3 | 24.6.1918 | 76 | Hd | 12 | Possagno | Two-seater |
| *[with ten. Scaroni & Michetti, 76ª Sq.] Fw. Franz Pachta KIA/Oblt. Wilhelm Persoglia KIA, Flik 2/D, Br. C.I 369.112* | | | | | | | |
| 4 | 4 | 25.6.1918 | 76 | Hd | 11h+ | Mareno di Piave | Fighter |
| *[with ten. Scaroni, serg.Ticconi & Mondini, 76ª Sq]* | | | | | | | |
| 5 | 5 | 25.6.1918 | 76 | Hd | 11h+ | Borgo Villa | Fighter |
| *[with ten. Scaroni, serg.Ticconi & Mondini, 76ª Sq] Maybe: Oblt. August Selinger UNH, Flik 42/J, Alb. D.III 153.202 FTL* | | | | | | | |

# Alvaro Leonardi

Every air force had its own élite squadrons, where the most famous and successful pilots flew. In Italy there was the 91ª Squadriglia, with the top aces like Baracca, Ruffo, Piccio, Ranza. But there were also many squadrons that didn't gain notoriety but did their part, fulfilled their duty with diligence and brought their contribution to attain mastery of the sky. One such unit was the 80ª Squadriglia da caccia, which had only one official ace, Alvaro Leonardi.

Alvaro Leonardi was born in Terni, an industrial city in the central region of Umbria, on 16 November 1895, son of Ubaldo and Virginia, he a cloth merchant, she coming from the Sopranzi, a wealthy family of landowners from Stroncone.

We can guess what attracted this young man to aviation: the first rudimentary airfields were appearing everywhere, one of them in Spoleto, close to Terni, in 1911, while the newspapers reported the deeds of the first military aviators in the war of Libya. Leonardi was drafted on 18 January 1915, when Italy was still neutral in the European war. He was assigned to the 6° Reggimento Genio Ferrovieri (railroad engineers regiment). He asked for attachment to the Battaglione Aviatori as soon as possible and he reached the Cameri flying school of Gabardini on 11 May 1915. This small private school in the heath of northern Italy was to become very large and important for aviation and for the life of Leonardi himself.

The school was equipped with three types of single-seat monoplanes, starting with the one powered by a 35-hp Anzani engine, that was used for ground runs, of no more than two minutes. The pace of training was slow, so the students compensated by wandering around with the motorcycle of Natale Palli. As they couldn't run after Austrian airplanes, they ran after the local girls. Years later, a fellow student described Leonardi as short-built and thin, with a halo of curly and unruly light brown hair, an aquiline nose, two mischievous eyes with the typical lively and penetrating look of intelligent and strong-willed people.

Leonardi completed his flying training on 1 November 1915, obtaining license n. 61 piloting an 80-hp Caudron. During his period at Cameri, on 26 October 1915 he had the honor of piloting a Caudron G.3 for an exceptional passenger, Guglielmo Marconi who was experimenting with radio transmission from and to the air.

On 11 December he reached the front, assigned to the 2ª Squadriglia Aeroplani per l'Artiglieria, a unit equipped with the Macchi Parasol and based at Medeuzza. He remained there briefly, as on 26 January 1916 he was shifted to the 1ª Squadriglia Artiglieria based at Risano and equipped with Caudron G.3s. There are no data about his activity with this unit until 5 May, when he was assigned as a pilot to the Scuola Osservatori (observers school) of Centocelle, Rome. He was promoted *caporale* on 7 July and then *sergente* on 31 October, but he kept presenting requests to be moved to fighters, and finally he got there.

On 28 February 1917 he was posted to the new 80ª Squadriglia caccia, then in formation at Arcade. Its commander was *capitano* Mario Gordesco, a leader of great capacity and popularity. His squadron was equipped with Nieuport 11s built by Macchi in Italy, one of which, serial Ni 2160, was ferried by Leonardi from Arcade to S. Maria la Longa on 9 March 1917, flying together with four other pilots, one of them another future ace, *sergente* Michele Allasia. Three days later the squadron was finally operative, with a strength of ten airplanes.

The beginnings of his career as a fighter were not very auspicious, like for many other future aces. One reason was bad weather. What he did was many missions escorting reconnaissance planes, some attempted interception on alarm that failed to bring combats, and two airplanes written off in landings, on 16 and 19 March. The latter was due to the bad quality of the landing ground that caused Nieuport 2160 to turn over.

On 19 April Leonardi finally met an enemy fighter, he attacked it but it refused combat and flew away toward Prosecco. The spell was broken, and other combats followed, all of them inconclusive, sometimes on account of a jammed gun, sometimes because the Nieuport was simply outrun by the new faster Austrian two-seaters and fighters.

The 80ª Squadriglia moved to Aiello, near the sea, and took part in the large Italian offensive of May 1917, the 10th battle of the Isonzo. Many pilots painted on their airplanes different versions of Happy Hooligan, a cartoon character then very popular in Italy with the name *Fortunello*, Lucky, and at least for Leonardi, this insignia really brought good luck.

Alvaro Leonardi, standing in the cockpit of a Gabardini trainer at the flying school of Cameri.

Leonardi in the capacious cockpit of a Gabardini monoplane.

On 24 May 1917 he finally scored his first victory. The 10th battle was near its close, and the Italian troops were exerting one final terrible effort to conquer the Hermada hill, the last enemy defense position on the road to Trieste. He took off from Cascina Farello advanced airfield at 6.25 a.m., and flew over the Gulf of Panzano providing air cover to two British monitors, H.M.S. *Earl of Petersborough* and H.M.S. *Sir Thomas Picton*, which were shelling Austrian positions at Prosecco and Opicina. The situation was desperate for the Austrians; a formation of seaplanes of the Naval Aviation lead by commander Banfield attacked the ships to drive them away. Leonardi saw an Italian flying boat attacked by an enemy seaplane, he rushed to its help, hit repeatedly the enemy and forced it to alight. The enemy crew was captured. The ace Luigi Olivari of the 91ª Squadriglia shot down another seaplane, and also an Italian FBA flying boat took part in this combat. Banfield, forced into dogfights with other Italian fighters, could not protect the bombers. He was probably engaged by *serg*. Imolesi of the 79° Squadriglia who had a combat with a fighter flying boat that *"maneuvered in a wonderful way"*. The two shot down flying boats were Lohner L 136 and L 137, both crews were taken prisoner.

As the Italian pressure did not subside, the 80ª Squadriglia was more actively engaged. Offensive patrols and strafes joined the usual escort missions. Despite the reduced number of efficient airplanes, Leonardi had a combat on 25 June together with Guido Keller and another one on 13 July when alone he faced five enemy fighters over Dosso Faiti saving a Voisin two-seater from likely destruction.

Leonardi with some Nieuport fighters at S. Maria la Longa airfield.

Right: 24 May 1917: the first victory! A happy Leonardi points at his good-luck charm, Happy Hooligan.

At the Observers school of Centocelle, Rome, Leonardi with the observer Bravi, who defines himself "a grateful post package". Bravi during the war earned two Medaglie d'Argento.

A smiling soldier of the *Corpo di Sanità* (Sanitary corps) in an image that shows clearly the "Fortunello" on Nieuport 11 serial 2123.

A Nieuport 17 of the 80ª Squadriglia at Aiello. Leonardi is the second from right, with his hands in his pockets.

In July the 80ª Squadriglia began re-equipping with the Nieuport Ni.17, and its pilots at last could face the enemy with an equal mount. Other combats followed, and in one of them two enemy fighters placed a few bullets into Leonardi's airplane.

Only in the sad days of the Caporetto retreat did Leonardi find victory again. As his squadron was preparing for retreat, on 26 October he shot down another enemy seaplane, K212 on Doberdò Lake together with *mar.* Cabruna and *ten.* Ancillotto, also of the 80ª, and *ten.* Francis Lombardi of the "sister" 77ª Squadriglia. When his unit settled at Arcade, across the Piave, he was back in the air trying to stem the enemy advance. On 6 November together with *serg.* Cosimo Rizzotto of the 77ª Squadriglia he shot down a two-seater at Conegliano. His feat had an exceptional witness, *magg.* Baracca, who had just shot down an Albatros fighter and then a German two-seater nearby. For this victory, his third, and the two other ones, Leonardi received his first Medaglia d'Argento.

In that period Leonardi began a lifelong friendship with another future ace, *brigadiere dei Carabinieri* Ernesto Cabruna. Even when Cabruna was shifted to the 77ª Squadriglia, that shared with Leonardi's 80ª the same airfield of Marcon, the two pilots were together whenever possible, on the ground and in the air.

On 27 November 1917 Leonardi scored his fourth victory when, after an escort flight to an SP.3 he engaged a fighter above Zenson and shot it down on the left bank of the Piave at his third firing pass. A possible identification of this victim is *Leutenant* Härtl of Jasta 1, who on that day was wounded in combat near Treviso.

On 26 December Leonardi had a close call that almost put an end to his career. Detached at Padua airfield for an escort mission to Capronis, he crashed against another airplane on landing, but he didn't suffer any injury, while both planes were wrecked.

With the New Year, the 80ª Squadriglia received a new fighter that soon became its only equipment, the excellent Hanriot Hd.1.

On 5 February 1918 Leonardi claimed a victory over an enemy reconnaissance two-seater shot down over Fagarè di Piave, behind enemy lines, with *sergente* Rodolfo Piermattei of his squadron.

Leonardi was highly appreciated, not so much on account of his victories but for the fact that he was a complete pilot, not only a good dogfighter but also an excellent leader, reliable toward his wingmen and his mission. For example, we know that reconnaissance crews in that area appreciated having him as leader of their escort.

A second friendship was born in those days, although its beginnings were turbulent. *Tenente* Carlo Alberto Conelli, Conte de' Prosperi was a nobleman from Piedmont, very wealthy, which mattered, even in those periods of general hardship. During a combat Leonardi had got him out of troubles, so Conelli de' Prosperi tried to show his gratitude hiding an envelope stuffed with money under the pillow of his mate. Leonardi didn't appreciate the gesture, and the argument grew into a fistfight. After that, they became excellent friends. Conelli de' Prosperi, who among other qualities was an excellent cards player, later married an American girl and moved to the States.

Leonardi scored again on 24 March 1918 when he shot down an Austrian fighter over Follina. One of his squadron mates, *serg.* Corti was less lucky, as the enemy forced him to land, but it is not clear if the two events are related.

For one month Leonardi did not fly, probably he was enjoying a well-deserved leave. He rose in rank to *Aspirante Ufficiale*, and on 23 May 1918 he claimed his seventh victory, shooting down a two-seater over Fossalta together with other pilots of his 80ª Squadriglia, *ten.* Conelli de Prosperi and *s.ten.* Antonio Pascoli. This feat earned him his second Medaglia d'Argento, its citation crediting him with four victories.

During the offensive of June, Leonardi claimed another victory over an enemy two-seater, on 18 June at 14.30 p.m. without being able to determine the lot of his opponent that took in a burst of 15 rounds and then dove into some clouds and disappeared. Actually the enemy plane, a Brandenburg of Flik 19/D managed with some luck to get back to its base badly damaged. Austrian groundcrews counted many bullet holes, showing that Leonardi had aimed well. The Austrian airmen claimed to have shot down their enemy, identified as a Spad, but Leonardi returned to base with no damage.

On 15 July he bounced and rapidly shot down a lone enemy airplane (which he reported as an "Albatros B-3", which could mean either an Alb. D.III or some sort of Brandenburg C.1) on the left bank of the Piave, near the bridge of San Donà. He generously shared the victory prize with *caporale* Testa, his wingman who had protected him from above.

Commissioned as a *sottotenente* on 28 July, Leonardi claimed his last victory in the war, his tenth, on 20 August 1918 when he attacked an enemy fighter near Levada. After his first attack his gun jammed, but then his wingman *soldato* Ottavio Sottani stepped in and finished off the enemy airplane that crashed near Campobernardo. Finally the next day, 21 August, he together with *cpr.* Vincenzo Bonanni had a duel with an enemy fighter that was seen diving vertically over Negrisia di Piave.

Leonardi served in the 80ª Squadriglia until 14 October 1918. By that time all those combat missions had worn out the nerves of this pilot. Fearing battle fatigue, the commands assigned him to the 122ª Squadriglia, a second-line fighter unit defending Milan, far from any likely threat. Probably he never took service there, as he was immediately assigned to the aviation park at Poggio Renatico. To all effects, he was sent to recuperate away from the frontline, where he remained until 28 October. Perhaps at the start of the final offensive he returned to the front, but this is not known. He had claimed 10 or 11 victories, but the official list published in 1919 cut down this number to eight.

After the armistice Leonardi remained with the 80ª Squadriglia until it was disbanded, in February 1919. As for many other pilots, return to civil life was probably not easy for Leonardi, who tried somehow to remain connected with aviation. In the early postwar period he was first owner of *"Trasporti Aerei Turistici"*, a company that flew joyrides over Florence with an old Caudron G.3, then a pilot with *"Trasporti Aerei Gastone Dalla Noce & C"*, a company owned by Baron Dalla Noce, an aviation pioneer from Florence.

In late 1920 he returned to Cameri as instructor with the Gabardini flying school, where he tutored to their pilot's license about 300 students. In 1921 he flew in the Circuito Aereo di Brescia,

Leonardi in the cockpit of his Nieuport 17 serial 3664, September-October 1917. In that period the Star of Italy became the insignia of the squadron.

The remains of an enemy airplane shot down at Follina, 24 November 1917.

Right: Leonardi and five other pilots of his squadron in front of an Hanriot, with the customary pet mutt.

Below: Leonardi examines one of the first Hanriots of Macchi production, serial 6112. There are no roundels over the upper wings.

Bottom: A line-up of Hanriots of Italian production of the 80ª Squadriglia.

A thoughtful Leonardi. The magazine in his hands depicts a kite balloon.

Leonardi ready for flight. His daughters still preserve his leather helmet.

Leonardi in His Hanriot, a French-built machine. On the fuselage there is the red star, unit marking, and his personal number, a lucky 13. The number appears also over the top wing. The Hanriot is a late French production model, with the Vickers machine gun moved to a central position.

Leonardi on Lake Maggiore, in front of the villa of his friend Conelli de' Prosperi.

Five officers during World War II in front of a Caproni Ca. 164 trainer. Leonardi is the second from left.

an aviation meet of some importance, piloting a Gabardini that he took to first place in its category.

In 1923, when the Regia Aeronautica was created as an independent air force, good pilots were needed and Leonardi didn't miss that opportunity. On 17 July 1923 he joined the new service with the rank of *sottotenente* and was assigned to the 71ª Squadriglia Caccia, based at Cinisello, Milan.

On 28 June 1924 in Cameri he married Maria Maddalena Galli, a local girl, and soon after he was assigned back to the flying school of Cameri. His family life was happy, and two daughters were born. To summarize the kind of life of a pilot in the interwar years, there are the memories of Leonardi's daughter: *"...those were good times..."*. When old friends came to visit, it was always a happy meeting, particularly if the visitor was his friend Ernesto Cabruna.

When his native city, Terni, decided to celebrate him with the award of a gold metal, it was Cabruna who was the master of ceremonies, during a great party held in the local Teatro Verdi. After two years in the reserve he returned to active service in 1928. He was first at Centocelle Nord airport, Rome, then in Pisa as manager of flying training for the courses of the Regia Accademia Aeronautica, the Academy of the air, then located in Livorno (Leghorn), near Pisa. So those were years of moving from one city to the other, with tasks always concerning flying training.

In 1934, by that time with the rank of *Capitano*, Leonardi returned to Cameri as commander of the airport and manager of courses at the local *Scuola Militare di pilotaggio di 1° grado* (primary training military school) that had taken the place of the dissolved Gabardini private school. He also was with the *Squadriglia Propaganda del Trasporto Personalità* (VIP squadron) reporting directly to the Ministry of Aviation.

In 1938 he had the rank of *tenente colonnello*. At the start of World War II he was posted as vice Commander of the *Scuola di Bombardamento Terrestre* (bomber school) at Malpensa, then back to the Regia Accademia Aeronautica now at Caserta. On 8 September 1943, when Italy surrendered, he was commander of the airport of Venaria Reale, near Turin. Without orders and without information, when German troops surrounded the airport he decided to give in, stopping young officers who wanted to organize armed resistance. Probably it was a wise decision, but after the war he was put on trial for that.

He remained in hiding to avoid deportation to Germany. He worked as a farmer in his father in law's farm near Cameri, and he secretly collaborated with the Resistance. With the permission of the local Committee of National Liberation and of the CANSA company, he prepared the runway of Cameri for heavy aircraft landing, so that on 2 May 1945 ten American airplanes landed there.

After the war he was placed in permanent leave, but later he was raised to the rank of colonel, dating from 1953.

Alvaro Leonardi spent the last years of his life in Cameri, the place that marked all the main events of his life. Here he had first flown, he had married and he had raised his family, and here he died on 1 January 1955, in the house where his daughters still preserve with affection the war memories and the photographs that belonged to their father.

## Alvaro Leonardi
## 10/11 victory claims - 8 confirmed victories

| Claim | Vict. # | Date | Sqn. | Aircraft | Hour | Place | Victim |
|---|---|---|---|---|---|---|---|
| 1 | 1 | 24.5.1917 | 80 | Ni.11 2123 | - | Gulf of Panzano | Seaplane |
| *Fähnr. i. d. R. Alfred Riesner POW/Fl.Gast. Karl Plasil POW, SFS Triest, L136 or: Stb.M. Hirnickl POW/Stb.M. WirknerPOW, SFS Triest, L137* | | | | | | | |
| 2 | 2 | 26.10.1917 | 80 | Ni.17 | - | Doberdò Lake | Seaplane |
| *[with ten. Ancillotto & Cabruna, 80ª, Lombardi, 77ª Sq.] Frg. Lt. Maximilian Kramer Edler von Drauberg WIA, POW/Matr. Marcello Anasipoli POW, SFS Trieste, K212.* | | | | | | | |
| 3 | 3 | 6.11.1917 | 80 | Ni.17 | 12h | S. Michele di Conegliano | Two-seater |
| *[with serg. Rizzotto, 77ª Sq.] Zugsf. Josef Feiler KIA/Lt. i. d. R. Othmar Schwarzenbach POW, DOW, Flik 12D, Br.C.1 229.24* | | | | | | | |
| 4 | 4 | 27.11.1917 | 80 | Ni.17 | - | Follina-Zenson | Fighter |
| *Maybe: Lt. Erwin Härtl WIA, Jasta 1, Alb. D* | | | | | | | |
| 5 | 5? | 5.2.1918 | 80 | Hd | afternoon | Fagaré di Piave | Two-seater |
| *[with serg. Piermattei, 80ª Sq.] maybe: Zugsf. Rudolf Horatschek UNH/Mg. Luftsch. Andreas Kerepetzky UNH, Flik 22/D, Br. C.1 269.38 or: unknown pilot/Lt. Wilhelm von Pichs UNH, Flik 22/D. Br. C.1 169.22 FTL* | | | | | | | |
| - | - | 5.3.1918 | 80 | - | - | Lower Piave | Kite balloon |
| 6 | 5?/6? | 24.3.1918 | 80 | Hd | - | Follina | Fighter |
| *Maybe: Oblt. Guido Hauger, Flik 42/J, Alb. D.III 153.138 FTL* | | | | | | | |
| 7 | 6?/7? | 23.5.1918 | 80 | Hd | 19h | Fossalta | Two-seater |
| *[with ten. Conelli & s.ten. Pascoli, 80ª Sq.] Fw. Heinrich Brejla UNH/Oblt. Lázár Illics KIA, Flik 22/D, Av. C.1 37.56 or: Korp. Leopold Kubalek KIA/Lt. i. d. R. Karl Reiss KIA, Flik 49/D, Br. C..1 369.58* | | | | | | | |
| 8 | - | 18.6.1918 | 80 | Hd | 14h35 | Meolo | Two-seater |
| *[with serg. Pascoli & Corti, 80ª Sq.] Korp. Franz Frantal UNH/Lt. i. d. R. Karl Roessler UNH, Flik 19/D, Br. C.1 369.41* | | | | | | | |
| 9 | 7?/8? | 15.7.1918 | 80 | Hd | 7h45 | Ca' Denego | "Albatros B-3" |
| *[with cpr. Testa, 80ª Sq.] Maybe (?): Zugsf. Franz Pospichal UNH/Beob. Oblt. i.d.R. Hans Ritter von BeckerUNH, Flik 49/D, Br. C.1 429.01 FTL* | | | | | | | |
| 10 | 8? | 20.8.1918 | 80 | Hd | 6h20 | Campo Bernardo | Albatros fighter |
| *[with sold. Sottani, 80ª Sq.]* | | | | | | | |

Alvaro Leonardi claimed 10 or 11 victories, 8 of which were credited to him in 1919. Only 7 confirmed claims have been found, and it is not possible to know for sure which one was the eighth one.

# Carlo (Francis) Lombardi

Francis Lombardi was one of the most lively and delightful personalities in the history of Italian aviation: pilot and ace of World War I, recordman, designer and industrialist, he was always present to all air meetings and conventions of old pilots, where he invariably grasped all the attention thanks to his memories and his qualities of story-teller. He was really a man who marked the progress of Italian aviation, thanks not only to his accomplishments, but above all to his spirit.

Carlo Lombardi was born in Genoa on 21 January 1897. When Carlo was a year old, his father Felice, a rice producer, moved the family to Vercelli, a foggy and sedate town in the northern region of Piedmont, which was then center of the rice growing industry. His father was a stern man, who was the manager of Genoa's harbor, and later the largest rice merchant of Italy, and mayor of Vercelli.

The third of five brothers, all of whom became pilots, Carlo Lombardi was a rather unruly teenager. Felice Lombardi had many passions: his work and his family first of all, but also politics, (he was a Liberal), dogs, and speed. His De Dion car of 1903 was the first ever registered in the province of Novara, followed by a De Dion 1907 that, at the age of 13, Carlo drove into the pond of the family house, getting sent to boarding school as a punishment. This stopped his construction of an amateur glider, as flight, by this time, had supplanted automobiles in the dreams of young men.

Carlo didn't finish high school but studied motors and mechanics on his own, and was a complete sportsman, skier and swimmer, and he was proud of having played in the Pro Vercelli, the football team from his hometown that was champion of Italy at the beginning of the century. He joined the armed forces as a volunteer in the air service when he was 18, just as Italy entered the war. He was sent as a ground serviceman to the 11ª Squadriglia Farman, based at Chiasiellis. There he had his first flight as a passenger. He was commended for the daring rescue of an airplane that had force landed in no man's land and then, as he had a good level of schooling, he was used as a clerk. Meanwhile, his strenuous efforts succeeded, and he was admitted on 18 May 1916 at the Venaria Reale training field, near Turin, where he started to fly the 35 hp Bleriot monoplane, with a Portuguese instructor named Suarez, according to his memories. Chief instructor was *ten.* Francesco Starrabba di Rudinì, a Sicilian prince son of a former Prime Minister, who was considered an excellent leader, (the only case we know of a WW1 Italian pilot with eyeglasses) and who died, shot down on 3 October 1917 probably by *Lt.* Schroeder of Jasta 1.

Carlo Lombardi was licensed as a pilot on 14 October 1916. From Venaria he went to Cascina Costa, where he was licensed as a military pilot on 14 February 1917. His excellent qualities as a pilot turned against him, because he was employed as an instructor at Cascina Costa since 1 April, flying SAML and SIA 7 two-seaters. Only in late July 1917 he managed to get away from training establishments, thanks to the efforts of his mother, who used all her acquaintances to support her unusual request of moving her son from a safe place to the frontline. He went to Malpensa on 27 July to train on the Nieuport fighter, (he flew Ni.11 1683, formerly of 76ª Squadriglia) then on 15 August he passed to San Giusto, near Pisa, for a quick training in air gunnery. Finally on 23 August 1917 he was assigned, now with the commission of *sottotenente*, to an operational fighter squadron, the 77ª Squadriglia at Aiello.

Formal portrait of Carlo "Francis" Lombardi with the rank of *sottotenente* and the military pilot's crowned eagle on his sleeve.

Equipped with Spads and sporting a red heart as a unit emblem (derived from its first commander captain Piccio) the 77ª was a highly effective unit, with an excellent combat spirit. Its commander, *cap.* Mario Gordesco, also lead the XIII Gruppo, formed by the 77ª and the 80ª Squadriglia, flying for the 3rd Army, the one holding the lower part of the front, near the Adriatic sea. Gordesco was an excellent leader, always praised for his psychological qualities, who in the spring of 1918 was assigned the task of creating a real Fighter School in Furbara, near Rome.

In his first days in the squadron, Lombardi flew only familiarization flights with thc Nieuport 17. His first combat flight was a cruise on 7 September 1917 on a Nieuport, but about mid month he started to train on the Spad 7. In the last decade of September, Lombardi flew 22 missions, half of them with the Nieuport and half with the Spad, including photographic reconnaissance flights.

Official documents still reported his name as Carlo (another Carlo Lombardi was a valiant observer in the 32ª Squadriglia) but he liked to call himself Francis, until his chosen name replaced the real one. In the squadron, he was nicknamed "piccinin", tiny, or kid. He was not at all short, however, being 1.72 meter tall.

In late September 1917 Lombardi, now with the rank of *tenente*, started flying the Spad 7, a fighter that he liked very much. Besides many test and escort flights, he also flew photographic reconnaissance missions. No flights of his are recorded between 2 and 21 October, when he was on leave. He was back in action as the enemy offensive developed, and on 26 October 1917 he scored his first victory, shooting down together with *ten.* Ancillotto and *serg.* Leonardi of 80ª Squadriglia an Austrian flying boat that fell in Lake Doberdò.

The next day, 27 October, Lombardi scored again shooting down an enemy by Ranziano (now Rence, Slovenia) and shooting at another, which he did not claim in his logbook, but referred to in his later recollections. The Austrians suffered several losses that day in that area near the sea: his victims could be either two Brandenburgs of Flik 101/G, or maybe Lombardi shared with Ancillotto the destruction of a seaplane. Another possibility could be a fighter pilot, *Zugsf.* Udvardy, an enemy ace who was seriously wounded while protecting the attacking seaplanes. The fighters of Flik 42/J escorting the seaplanes claimed two Nieuports forced to land, but such losses cannot be determined through the skimpy surviving Italian records.

During the Caporetto retreat, the 77ª Squadriglia moved back in good order from Aiello to La Comina, then Arcade and finally Marcon, near Venice, where it remained until the end of the war.

On 3 November Francis Lombardi scored his third victory, shooting down by Oderzo a German DFW two-seater, with the death of *U.Off.* Wilhelm Noss and Richard Monnighoff. Again, his partner in this victory was Giannino Ancillotto.

The next day, 4 November, Lombardi is assigned a victory on a German fighter over Cividale (a city that, at this time of the war, after the Austro-German advance, was way behind the enemy lines). His logbook reports a long Caproni escort mission, but no victory claim.

After this hectic decade in which he claimed five victories, Lombardi flew a lot, but without further combats. The 77ª Squadriglia was a "workhorse" unit, constantly engaged in unrewarding escorts, alarm flights and photoreconnaissance missions,

Francis Lombardi, right, when flying as an instructor at Cascina Costa with a SAML. This two-seater of original German design was used at the same time both as a trainer and as a combat airplane, and it long survived its intended successor, the Pomilio P.

so it did not have the freedom to roam the sky in offensive patrols looking for combats, like the privileged 91ª could. Lombardi flew a lot (nine combat missions just in the last week of the year 1917) but he had only one combat, over Conegliano on 27 December.

In January 1918 also Lombardi again flew several different missions, mostly two-seater escorts and photographic reconnaissance missions. In February 1918 he started using 5-kg bombs, that were held in containers within the cockpit of the Spad, and he strafed enemy troops several times, after fruitless cruises. From 26 February to 16 March he was on leave. He made up for the lost time flying many missions in late March and April, but with a single combat, on 24 April above Ponte di Piave. He attacked two enemy aircraft and fired 200 rounds, without advancing any claim. Another fruitless dogfight took place on 1 May, again over the Piave River.

In this period, Lombardi flew intensively in all kind of missions. When the last Austrian offensive began, on 15 June, he was ready. In his first mission, over Susegana, he had two dogfights, in the second one, over Grave di Papadopoli and San Donà, he had two more. The third mission in the same area brought one more dogfight. His fourth mission of the day was an alarm take-off over Villa Cucca, and this time an enemy fell under his gun (that fired no less than 400 rounds) and crashed near San Biagio. His victim was

Left: Francis Lombardi in the cockpit of his Spad showing the red heart of the 77ª Squadriglia.

a fighter, identified by a black hand painted on the fuselage by the many observers on the ground.

The next day of the offensive, 16 June, Lombardi attacked a kite balloon and shot down an enemy that fell near Pezzon di Mela. The rather unreliable postwar reconstruction of Aces' victories assigns to him one victory against a fighter defending a ballon on the 15th, a victory against a two-seater on the 16th and finally a victory against one fighter out of a swarm of 23 fighters a few days later, therefore bringing his total to eight. His rather sketchy logbook, however, shows only the victory of the 16th, that could be over a Brandenburg.

His activity was ceaseless: in the last ten days of June he flew 11 combat missions and two ferry flights. On 1 July he dropped leaflets and had a dogfight against an enemy fighter that escaped him with a steep dive, then he flew many strafing missions. In July he flew a total of 31 combat missions, three of them being standing patrols to protect troops engaged in parades and ceremonies. The new policy of the Italian Army, after the harsh discipline of the command of General Cadorna, was to take care of the psychology of soldiers, and celebrations, usually with the participation of Allied delegations, were a good way to foster the self-esteem of the troops.

On 9 August Lombardi was sent on a one month leave for rest and medical examination, and then he had one month of furlough. On 25 October he was considered fit for flying again, but he seems not to have returned to his squadron before the end of the war. Overall, Lombardi flew 295 combat missions for a total of 480 hours, had 21 combats and scored eight victories.

Brandenburg C.I 369.02 of Flik 101/G was possibly the victim of Francis Lombardi on 27 October 1917. *(Photo Höfling).*

He was awarded three Medaglie d'Argento (Silver medals) with these citations:

"Expert and bold fighter pilot, he distinguished himself in many flights over the enemy. He flew reconnaissances over 100 km from our lines, facing the threats of anti-aircraft fire and of the many fighter airports in the region, bringing back

Pilots of the 77ª Squadriglia at ease at Marcon in 1918. From the left, sitting, Lombardi, Bozzetto, Sambonet, Allasia (also sitting). The man leaning on the wall with a cigarette is squadron commander Marazzani, and the officer with a bonnet leaning is Mario Gordesco, commander of the XIII Gruppo.

Francis Lombardi with standard pilot leather jacket and helmet, in front of a Spad that shows on its fuselage both the national roundel and the red heart.

information of high military value. Sky of Gorizia, 5-13 December 1917." The second, assigned in June 1918 was: "Expert fighter pilot, he had many combats, all with winning result. On 27 October 1917 he shot down an enemy aircraft near Ranziano. On 3 November 1917 he shot down after a long and daring combat a German airplane near Ravarè (Treviso)." And the third one was: "Fighter pilot of special ability and daring, in the first days of the Austro-Hungarian offensive, with magnificent dash and unbeatable contempt of danger, he had seven air combats forcing two enemy fighters to land in national territory. Sky of the Piave, 15-17 June 1918."

After the war, the dynamic Lombardi took some time before settling to civilian life. In 1919, he was assigned as commander to Novi Ligure, an airfield where now useless SVAs were stored. In the cloak-and-dagger romantic atmosphere of that period, he performed a special mission: on 7 October 1919 he flew a SVA 10 two-seater from Novi Ligure to Fiume with a special passenger, a young and rising politician named Benito Mussolini, who wanted to fly to Fiume for a secret meeting with the hero poet d'Annunzio, who had established his revolutionary "free state" in the contested city of Fiume.

As he was officially AWOL, Lombardi followed Gabriele d'Annunzio in the exciting adventure of the patriotic city that challenged the government of Italy and the decisions of the Allies, and survived for one year and a half as a haven of patriotic buccaneers.

Lombardi, by now universally called Francis, finally seemed to settle in Vercelli, and joined his father in the rice trade. In December 1921, however, he attended a refresher flying course at the flying school of Gabardini in Cameri, where he shared the course with two other aces: Bedendo and Eleuteri. In 1923 he got married, and he occasionally flew with an Hanriot that he kept at nearby Cameri airport. In 1927 Vercelli became a provincial capital, and built its own airfield. Lombardi was one of the founders of the Aero Club of his town, that became one of the most active of Italy, second only to Milan in the number of trained student pilots.

After competing in the Air Tour of Europe, 1929, he had the idea to fly to Somalia using one of the Fiat AS.1 lightplanes of his Club, I-AANH, of which the club, that bought three, was the launch customer. As he didn't have time nor money to get visas, he planned his route flying only to Italian colonies, which meant a 1,500 miles leg between Tobruk, Libya, and Massawa, Eritrea, not bad for an 80-hp airplane with an endurance of four hours at 100 mph. He got Fiat to install a 500 liters tank in the passenger seat, and a special arrangement for filling up the oil valve. He flew with no subsidy from the government nor from any company, just with the promise of free Esso fuel and Castrol lubricating oil if the flight succeeded.

He took off from Rome on 12 February 1930. Stopping at Tripoli, Tobruk, Massawa and Bender Kassim, he reached Mogadishu in six days, welcomed as a hero by the local community, that saw for the first time an airplane joining Somalia with the mother country.

A few weeks later, back in Italy, he was contacted by Umberto Savoia, the director of the aviation branch of Fiat, who requested some special record flight around Europe, to boost the reputation of the AS.1 lightplane that had suffered from some accidents. Francis Lombardi accepted, proposing instead a flight to Australia or Japan. As the season prohibited Australia, Lombardi organized a flight to Japan through the Soviet Union, to be completed in seven days. He chose to carry a motorist, Gino Cappannini, who had already flown to Japan with Arturo Ferrarin in 1920, with a SVA biplane, during the trail-blazing Rome-to-Tokyo epic. In just one month, the flight was prepared, the permissions from the Russians were easily obtained and Fiat AS.1 I-AAVE was prepared.

The raid began on 13 July 1930, Lombardi flying from Vercelli to Vienna, Austria. The next day the two Italians flew to Warsaw and then to Smolensk, USSR. In the following days they flew to Moscow, Kourgan, Novosibirsk, Omsk, Verkuei Udinsk, Nijisne Udinsk and Tchita, always warmly welcomed by the Soviets, who were remarkably open and cooperative. Their next stop was Harbin, in then independent Manchuria, and then to Japanese-held Mukden. Here, the warm hospitality that they had received along the way disappeared, and they were simply put in a cell by the Japanese military, whom Lombardi described as "*the most hateful people that ever existed.*" Without maps, without a flight plan, because the Japanese had forbid them to fly over restricted zones, but at the same time had declined to explain which were such restricted zones (because just pointing them out would have revealed a military secret) they somehow managed to find the airfield of Ain Yu, in Korea, and then Seoul. Flying over to Hiroshima, Japan, Lombardi skirted the forbidden island of Tsu Shima. He landed at the city that later became dramatically famous, and then he was diverted to Osaka, to meet the Italian ambassador. There he was photographed by Japanese journalists, who gave him films to bring to Tokyo for their newspapers. When he landed at Tokyo, in darkness and rain, at a military airfield, he was promptly arrested as a spy with Cappannini, and the film was held as evidence against him!

The two Italians were finally set free, and got to Tachiwawa airport, Tokyo, for their reception. Their AS.1 however was seized by the Japanese, so they had to return to Italy by rail. Back in Rome, he was received by Mussolini, who hadn't met him since their flight in 1919, who discussed about Russia with him. Cappannini, born in 1900, was later a motorist in the 1930 mass flight of S.55 seaplanes across the Ocean, from Italy to Brazil, and he was Italo Balbo's motorist in his Siai S.55X during the 1933 mass flight to America. He died, together with Italo Balbo, shot down by friendly fire on 28 June 1940. In 1973, when Alitalia opened its service from Rome to Tokyo, Lombardi was invited for the inaugural flight, on a DC-8.

Later that year, Lombardi was involved in another crazy idea, a flight around Africa organized by Franco Mazzotti and Mario Rasini, two scions of rich business families of Milan. The three daredevils got from Fiat three AS.2, I-AAVR, AAVQ and AAVS, that had just flown in the Air Tour of Europe, and were followed by Marino Battaglia, the motorist of Vercelli Aero Club. Privately organized, this raid started on 28 October 1930 from Catania, Sicily, and followed the airports of Libya, Egypt, Sudan, Kenya, Rhodesia, South Africa, Namibia, Angola, Congo, Gabon, Cameroon, Nigeria, Ghana, Ivory Coast, Senegal, Mauritania, Spanish Sahara, Morocco, Algeria and Tunisia, and finally Naples, that was reached on 9 January 1931. In over two months of travel, the four Italians had the most incredible adventures, including a flip-over of Lombardi's Fiat, that was damaged but repaired by Battaglia.

Lombardi was proud of his three outstanding raids in a single year, that he recollected in a delightful book, "*Tre raid del 1930*" privately published in 1980:

> "I wrote down my memories in July 1980. I have reached the age of 83, and now I am on the waiting list. I admit that I was a very lucky man, belonging to a very lucky generation. The generation that saw the birth and development of all that forms the environment of modern civilization. Everything, from bicycles to cars, to airplanes, electricity, television, computer, rockets and satellites, man on the moon and space exploration.

Spads of the 77ª Squadriglia at Marcon, late in the war. Top wing roundels have different sizes, and the frist Spad shows individual number XVI, serial number 1673. The pilot is sergente Carlo Ferrario, who was shot down by ground fire and captured on 2 November 1918. The second Spad carries individual number II on its fin.

And in my generation, I belong to a small group of men who could and, let's admit it, were able to exploit favorable opportunities to contribute to the progress of aviation, getting also the opportunity to know the world."

Francis Lombardi in his Fiat AS.1 I-AANH showing the shield of the Aero Club of Vercelli and the racing number of the European Challenge of 1929, having just touched down at Mogadishu, Somalia, 18 February 1930.

In January 1932, the irrepressible Lombardi organized a new flight to Africa. Using a Caproni 105 single engined monoplane and flying with sportsman pilot Leonida Robbiano, he planned to beat the speed record from Europe to Capetown. The two Italians took off from Rome on 21 January, and flew to Tobruk, Libya, and then Wadi Halfa, Egypt and Malakal. Their engine started giving troubles, however, and the Caproni was more and more behind schedule, until the two pilots decided to give up the attempt, and having reached Tabora, Tanganika, they stopped the flight and returned to Rome. Lombardi was daring, but not reckless, and he planned every flight carefully. He used to say that pilots must have the courage to be afraid.

He was back in Africa in 1933, however, flying to Addis Abeba, Ethiopia, with a Caproni 101 I-ABBW, together with pilot count Alberto Marazzani, his former commander at the 77ª Squadriglia, who had gone to Ethiopia himself three years before, bringing a Breda 15 monoplane as a gift for the coronation of Emperor Haile Selassie. Their official goal of the mission was to study a mail route to the African nation, their secret one was to support explorer baron Franchetti, who was trying to bribe Ras Hailu, a local chieftain. The cloak-and-dagger operation failed, and the Italians, having enjoyed the hospitality of the Emperor, returned home safely.

In 1934 Francis Lombardi attacked the Atlantic Ocean. A SIAI S.71 airliner, I-ABIV, was prepared for a record fast flight to Brazil. Its crew included co-pilot Franco Mazzotti, engineer Marino Battaglia and radio operator Davide Giulini (who, had flown in the 1933 Atlantic cruise with Balbo). Franco Mazzotti, one of the heroes of the 1930 round-Africa flight, was an accomplished sport flyer, and he was the creator of the famous "Mille Miglia" car race. Their S.71 took off from Montecelio experimental airfield, near Rome, on 27 January, stopped at Casablanca, Morocco, and then reached Thiès, Senegal, the next day. After a short stop for refueling, it flew toward Brazil but lacking proper radio connections with the French ground station in South America, it flew off course, until it force landed on a beach in Ceara, Brazil, getting written off. The flight was a partial failure, as it failed to reach its final destination, Buenos Aires, Argentina, but it was still a remarkable achievement.

In 1934, Lombardi prepared for the McRobertson Trophy, the London to Melbourne race that was eventually won by a de Havilland Comet. He was to fly a special racing monoplane, the single-engined Caproni PL.3 (standing for Pallavicino, the designer, and Lombardi) powered by a Fiat A 59 engine, a licence-built Pratt & Whitney Hornet. His co-pilot was that airline pilot Vittorio Suster, but the Italian team didn't participate, as their airplane, that carried race number 61, was not quite ready, and the race prizes were not very tempting anyway.

On 9 November 1934 Francis Lombardi was chosen, together with airline pilot Vittorio Suster, to return to Somalia, for a special mail flight on the occasion of the visit of king Victor Emmanuel III to the colony. This time, he flew another SIAI S.71 three engined airliner, I-ALPI, and the flight took only three days. The airplane however was severely damaged landing at Mogadishu airport.

One year later, Lombardi was hired by Ala Littoria, the Italian airline, to fly on the new so-called Imperial Line, from Italy to Eritrea where the invasion of Ethiopia had begun. He flew the route from Benghazi to Sollum, Cairo, Wadi Halfa and Kassala with the Caproni Borea light twin engined monoplane.

By this time, Lombardi had done feats that could fill many lifetimes, but it was just the beginning. He became an aircraft designer, and together with engineer Bottini he designed the LB, a pusher monoplane with twin booms that was built by SAIMAN of Rome, and he became the distributor in Italy of Tipsy of Belgium, Zlin of Czechoslovakia and Stinson, USA. In 1937 he set up his own aircraft company, AVIA (Anonima Vercellese Industrie Aeronautiche) in Vercelli, famous above all for the FL.3, a light trainer monoplane with low wing and side-by-side seating, that became one of the most popular Italian aircraft ever.

Right: Lombardi in the cockpit of Siai S.71 I-ABIV ready for take-off from Montecelio airfield on 27 January 1934, headed for Brazil.

The unforgettable AVIA FL-3, one of the prettiest and most loved Italian airplanes. This is a batch produced in wartime for the Croatian Air Force.

A fascinating talker, here Francis Lombardi discusses his aviation experiences with one of the authors of this book in September 1980.

The original L.3 was built in a school: it was a side-by-side small low wing monoplane, powered by a 60 hp engine, that was first flown in late 1938 by Francis Lombardi himself. It was series produced for the Italian Air Force, for a total of 335 aircraft. Initially however the FL.3s were not liked, Italian basic flying schools being accustomed to biplanes, so that most of them were stored, and were seized in large quantities by the Luftwaffe, after Italy's surrender.

Besides AVIA, Lombardi influenced Italian aviation industry in another way. He had his own Hanriot, which he had fitted with a Fiat A 80 radial engine. This little biplane was a joy to fly, so when the CANSA company in nearby Cameri, a factory of the Fiat group, had to produce a new primary trainer, the FC.5 was designed, that had exactly the dimensions and wing layout of the World War I Hanriot. Lombardi's Hd survived until 1944, when it was seized during the German occupation and disappeared.

After the war, the FL.3 remained in production, fifty more being built, and it was one of the most common Italian lightplanes, remaining in service until recent times, and a few still fly. In 1947, five of them were bought by Gino Rovere, who organized at Rome Urbe airport a flying school for the newborn Israeli Air Force. Other products of AVIA were the LM.02, a low wing glider dive bomber, designed to be towed over Gibraltar for no-return missions, the pilot getting interned in Spain after the attack; the L.4 twin-engine tourist airplane, and the beautiful LM.5, a gull-winged lightplane that first flew in 1945, one of which is still in the Italian register, and the LM.7 of 1949.

After the war, AVIA branched out in car manufacture, producing special versions of Fiat compact cars. Lombardi was a peculiar businessman, in love with his work and with the world, who always strived to produce good products, such as his planes and cars, making them as cheap as possible, so that also young people could afford them. He introduced many innovations in coach building, such as curved glasses, that he didn't patent. He made his own the saying of St. Paul: "Money is a good servant, but a bad master".

The company had a period of success, with its Grand Prix and Lucciola, special versions built on Fiat chassis gaining several orders, but in 1976 it went bankrupt. Francis Lombardi paid all his debts, and took care of all his employees, but financially he was ruined.

The old Francis Lombardi was a sort of Italian Jimmy Doolittle: not just an old important pilot, but a man who represented in his person all the aviation history of his nation. He was a fantastic story teller, and he attended all aviation meetings in Italy. In June 1978 he organized a reunion of the surviving pilots of the 77ª Squadriglia, at their old Marcon airport, sixty years after the battle of the Piave, and put down his memories in a delightful booklet, "Gli Amici di Marcon", that was also published by "Cross & Cockade Great Britain" magazine, issue 3/1981. In 1981 he organized the Reunion of Air Aces of World War I in Paris, together with John M. Kent and Charles C. Dent, United Airlines captains, who later was the leading force in the construction of Leonardo da Vinci's bronze horse that was given in 1999 to the city of Milan.

In Paris Lombardi partied in the exceptional company of American, British and German aces. The parchment with the signatures of Douglas Campbell, George Vaughn, Lord Balfour of Inchrye, Gottfried von Banfield and the many other aces was one of his fondest possessions.

He died on 5 March 1983 in his elegant home in Vercelli, full of the memories of an exceptional life.

## Carlo (Francis) Lombardi
## 7 victory claims – 8 confirmed victories

| Claim | Vict. # | Date | Sqn. | Aircraft | Hour | Place | Victim |
|---|---|---|---|---|---|---|---|
| 1 | 1 | 26.10.1917 | 80 | Ni 17? | - | Doberdò lake | Seaplane |
| *[with serg. Leonardi, m.llo Cabruna and ten. Ancillotto, 80ª Sq.] Frg.Lt. Maximilian Kramer Edler von* Drauberg WIA, POW/Matr. Anasipoli POW, *Seeflugstation Trieste, K212* | | | | | | | |
| 2 | 2 | 27.10.1917 | 77 | Spad | - | Ranziano | EA |
| *Zugsf. Josef Bangerl WIA/Lt. i. d. R. Alois Wasmer WIA, Flik 101G, Br. C.1 369.02 or:*<br>*Fähnr. Richard Gröger POW/E. Frwg. Alois Mericka POW, Seeflugstation Trieste, K366* | | | | | | | |
| 3 | 3 | 27.10.1917 | 77 | Spad | - | Doberdò | Fighter |
| *Zugsf. Ferdinand Udvardy WIA, Flik 42, Alb. D.III or:*<br>*Zugsf. Johann Trnka UNH/Lt. i. d. R. Johann Halir UNH, Flik 101G, Br. C.1 329.23* | | | | | | | |
| 4 | 4 | 3.11.1917 | 77 | Spad | - | Ravarè | German a/c |
| *[with ten. Ancillotto, 77ª Sq.] Uffz. Wilhelm Noss KIA/Lt. d. Res. Richard Monninghoff KIA, FA 14, DFW C.V* | | | | | | | |
| 5 | 5 | 4.11.1917 | 77 | Spad | - | Cividale | Fighter |
| 6 | 6 | 15.6.1918 | 77 | Spad | - | San Biagio | Fighter |
| *Maybe from Flik 41* | | | | | | | |
| 7 | 7 | 16.6.1918 | 77 | Spad | - | Pezzon | Two-seater |
| *Maybe: Oblt. Walter Gwiggner POW/Oblt. i. d. R. Georg Garabb KIA, Br. C.1* | | | | | | | |

Carlo Francis Lombardi appears with eight credited victories in the List of Aces; however his papers show only seven victory claims.

# Cesare Magistrini

Cesare Magistrini was the embodiment of a fundamental category of Italian pilots: the sergeant pilot from Northern Italy, the expert, assured pilot who gets the reputation of the "*manico*", the expert. But Magistrini took these virtues to the highest accomplishments, becoming one of the greatest Italian pilots, almost unknown to the general public, but highly regarded among his peers.

He was born on 26 January 1895 in Maggiora, a village in the northern province of Novara, and he had the typical characters of an Italian born in the regions just below the Alps: strong, modest, of few words, religious and with deep family values, patriotic but devoid of any form of jingoism. Although in love with flight since, as a boy, he had heard of Geo Chavez's crossing of the Alps, at the beginning of the war he was drafted and opted for the Railway Engineers, as his parents were opposed to aviation. But at the draft board, he learnt that there were no opportunities on the railroad, while an option could be aviation, which he readily took up. In January 1915 he joined the Battaglione Aviatori, and spent time at the Gnome engine factory in Turin. In June he volunteered as a pilot, but for a disciplinary violation, instead of being sent to flight training in France, he was posted to the 2ª Squadriglia Artiglieria at the front, an unit that was flying the dangerous Macchi Parasol. There Magistrini flew for the first time, as a passenger with Mario Stoppani, the future test pilot. When the 2ª Squadriglia was reformed on Caudrons, Magistrini placed a second request for pilot training, and was recalled to Turin, and then assigned to the first course organized by the Gabardini flying school in Cameri, a few miles from his own town. In March 1916 he qualified as a pilot, with ranking of third out of 100 in a class that produced may other aces and outstanding pilots. Rushing his training, in 20 days he got his second license, as military pilot, then he went to Cascina Costa for combat training on the Farman, but of course he was fascinated by the local Nieuports, and he got a transfer to the fighters. From Cascina Costa, he once flew over nearby Maggiora, his hometown. How he described that flight, best shows his character as an airman:

> "During that flight for sentimental reasons, I realized that I had a quality that now, as I am an expert, believe is indispensable for an airman: calm, that is control over one's nerves, both in the moments of danger and in the times of highest enthusiasm. In every situation one must be present to oneself, never forgetting that "you are flying". After a turn above the city at high level, I flew back, satisfied that I was not carried over by the wish to do some stunts, dangerous and useless."

On 1 August 1916 Magistrini, now a corporal, was declared a fighter pilot, and was assigned to the 78ª Squadriglia of *cap.* Bolognesi, that was forming in those days on La Comina airport on Nieuport 11s. On 3 September 1916 the new squadron moved to its battle station, Istrana, near Treviso, an area rather remote from the

Cesare Magistrini in training at the Gabardini school in Cameri, sitting in Caudron G.3 serial C1067.

main front, the Isonzo, and closer to the Asiago plateau. In September Magistrini began combat missions, with patrols over enemy territory in the Trento area. The unit partially re-equipped with the Nieuport 17, and was engaged mostly in escorts to Caproni bombers. During the 10th offensive, May 1917, Magistrini was temporarily assigned to the Isonzo front. Possibly this assignment was due to a storm that wrecked the hangars and all the aircraft of the 70ª Squadriglia on 9 May.

On 10 May 1917 therefore he was flying from Santa Caterina, near Udine, with his Nieuport 11 serial 1658 in a scramble. He reported the dramatic event that followed:

> "I took off from S. Caterina airfield on account of an alarm at Cividale. When I reached that town I saw an enemy airplane flying towards Gorizia; I reached it near the Podgora Mount at

Right: Pilots of the 78ª Squadriglia deployed on the Isonzo front in the spring of 1917: from the left, *serg.* Guglielmo Fornagiari, *ten.* Giacomo Brenta and *serg.* Cesare Magistrini.

a height of 4,000 meters. I approached it and got onto its tail firing a magazine of 50 rounds from a distance of a bit more than 50 meters. After my burst I saw the enemy diving but as I was in pursuit of it I met two enemy fighters that jumped me firing repeatedly. I had to stop my combat because a bullet hit me in the back. As I tried to escape the fire of the enemy fighters I noticed that two other Nieuports went after the airplane that I had attacked before, and I lost them from sight at a height of 2000 meters. As I felt blood running down my spine I had to land at Bolzano airfield, where a medical captain took care of me."

We don't have records for an Austrian claim of this victory.

The Trentino front heated up in June 1917, when the Alpine troops were thrown in an operation on the Ortigara mountain, that ended with no gains for the Italians, but with terrible losses inflicted by the skillful Austrian defense. In the air, however, there were many combats and the Luftfahrtruppen units were mauled by the Italian fighters. On 14 June 1917 Magistrini fought against an enemy aircraft, that got away toward Marcesine flying irregularly. On 17 June, flying Nieuport 3126, he claimed a probable above fort Luserna. This is his Combat Report:

"Tasked with a standing patrol I took off at 6.30 a.m. heading to the Plateau. At 7.20 I saw an enemy Albatros type air-

Sergente Cesare Magistrini in a studio portrait.

Below: Magistrini flew, among others, also this French Nieuport 17, serial N 3122.

plane that was flying from Cima 12 to Asiago. I flew immediately in that direction and despite my lower altitude I engaged it the first time at 7.25 over Asiago (at 600 meters) and then as meanwhile I had gained altitude I attacked it another time at 7.30 over Schio, and a third time at 7.35 in Valdassa. In this last combat I eventually saw the enemy aircraft,which had always vigorously reacted with its weapon to my attacks, hitting the spar of my lower right wing, that stopped fighting and flew irregularly down to 600 meters, and then it suddenly banked left and I lost it from sight near Luserna."

On that day Magistrini was awarded his first Silver Medal in the field:

"... As he attacked an enemy aircraft, although attacked by two more and wounded on his right hip, he kept on fighting strenuously, until forced to land due to the considerable loss of blood. – Sky of Podgora, 10 May 1917. – He daringly attacked a powerfully armed enemy aircraft, forcing it to land after a long and lively combat – Sky of Val d'Assa, 17 June 1917."

One month later, on 18 July, he scored his first confirmed victory: together with *sergente* Guido Nardini, also from 78ª Squadriglia, he shot down over Asiago Brandenburg 229.112 of Flik 24, with crew of *Korp.* Volisin, POW, and *Lt.* Viktor Maly killed in action. The two noncommissioned officers later got to the frontline to pick up instruments and guns from the wrecked airplane, and were reprimanded for this risky breach of rules. On the other hand, Magistrini paid homage to his fallen adversary and attended his funeral with full military honors.

On that occasion the Austrian Brandenburg had as an escort two brand new Albatros D.III fighters that had just arrived at Pergine airfield, but their baptism of fire was not auspicious. Two Italian fighters, identified as Spads, attacked them, and they were both hit, so that both had to land in emergency at their airfield, while the two-seater was shot down. One of the two Austrian pilots was one of the best officers, *Lt.i.d.R.* Josef Friedrich, on Albatros D.III 53.33 also of Flik 24, credited by the Austrians with seven victories, the other one was *Fw.* Schwarz on 53.38.

In August the 78ª Squadriglia was moved to Borgnano, close to the Isonzo front, for the big summer offensive, but its fighting capacities were low, because all its Nieuport fighters were worn out. Magistrini was constantly in action, but he didn't score again until 16 October when, together with a Hanriot, almost surely the one of Guglielmo Fornagiari, he shot down an enemy recce biplane (maybe a German from FA 232, with crew of *Unteroffizier* Bolte and *Lt.* von Glass, both KIA).

Meanwhile, since May, Magistrini had been assigned to the 91ª Squadriglia, the élite fighter unit of the Corpo Aeronautico Militare, (a meaningful honor for a sergeant) but Cdr. Bolognesi managed to have him remain with the 78ª, this showing how much he was appreciated. In the hectic days after the rout of Caporetto Magistrini at last moved to the 91ª, where he took service on 1 November 1917, together with seven other pilots. He was trained to fly wingman by Italy's ace of aces, Francesco Baracca, and on one of those flights, in which Baracca scored a victory, Magistrini noticed a strange phenomenon: his eyesight was so sharp, 12 tenth, that he misjudged distances, and saw enemy aircraft closer than what they actually were. After many escorts, he had his first combat on 18 November, but his gun jammed.

On 23 November 1917 Magistrini shot down his third enemy, identified as a German DFW C.5, that he attacked at low level together with *cap.* Costantini and that was forced down inside Italian lines. The enemy however, hit Magistrini himself in the leg and he force-landed his Spad close to the enemy. He could see the dead pilot, and felt very depressed at the tragedy that war had caused. He was given a machine gun from the shot down airplane, but quickly gave it away for an exhibition. Only the return to combat flying

The military funeral of *Leutenant* Viktor Maly, shot down and killed on 18 July 1917. His winner, *serg.* Cesare Magistrini, is the first on the right.

could dampen the turmoil that he was feeling. Meanwhile, one of his brothers had been killed in combat on the Isonzo, and the other one also was killed in the final days of the war. On 29 November, flying together with *ten.* Novelli and the Argentinean *aspirante* Olivero, at the end of a long patrol lasting two hours he attacked a German two-seater above Oderzo, at a height of 4,000 meters firing 150 rounds into the escaping enemy airplane but he had to break off the attack when jumped by three enemy fighters. The enemy possibly force landed beyond Roncadelle, but Magistrini did not claim this event as a victory.

Two more enemy reconnaissance aircraft fell to his guns before the end of the year, a German shot down in cooperation with Ranza and *ten.* Novelli on 30 November and an Austrian Brandenburg together with Ranza by San Pietro di Bassano on 7 December, making him an ace and winning him a second silver medal, also in the field:

> "...He brilliantly fought in twenty air combats against well armed enemy aircraft, effectively sharing in the destruction of four of them, He was wounded when during an escort many enemy aircraft attacked him, Sky of Monte Nero, Cornuda, Moriago and Nove di Bassano, 7 December 1917."

On 18 February 1918 four fighters of the 91a Squadriglia escorted a SIA 7b recce biplane over Marcesine. Enemy fighters attacked, and during the dogfight Magistrini was wounded in a leg, and made a forced landing in Nove. It was his third combat wound. He was back in action after a few days, however.

In the spring of 1918, the 91ª Squadriglia re-equipped with the Spad 13 and moved from Padua to Quinto di Treviso. On 17 April 1918 Magistrini was flying in a patrol with *tenenti* Novelli and Keller when five fighters jumped them hitting Novelli's wing. They were not Austrians, however, but British, in a not uncommon example of friendly fire.

On 11 May 1918 together with Novelli, Bacula, D'Urso and Olivero, Magistrini had a confused combat with the Austrians, two of which were claimed as probably forced to land in Moriago and Quero, but were not confirmed to them, accordingly to the strict rules of the 91ª Squadriglia. Magistrini, who as usual received bullets in his airplane, did not specifically claim a victory in his logbook, but Austrian data perfectly report an airplane forced to land after combat with Spads on that date and place.

On 17 May there was another confused combat, in which a patrol of five Spads joined the Hanriots of the 78ª Squadriglia and tackled a flight of nine Austrian fighters, one of which was shot down near Pero by the combined action of Magistrini, Nardini and *ten.* Novelli. Their victim was *Lt.* Franz Gräser, of Flik 61, a gallant Austrian ace who lost his life flying Albatros D.III 153.221.

During the hectic days of the battle of the Piave, mid June 1918, Magistrini strafed enemy troops and escorted bombers, but he didn't score any victory. He was the kind of pilot who flew a lot, but didn't

Nieuport 17 serial 3615 of the 78ª Squadriglia at Borgnano.

Right: Magistrini near the end of the war, sporting on his breast the ribbons of his decorations. The winged propeller in his sleeve was the insignia of non-officer pilots.

Spads in the snow. This unidentified picture could posibly show fighters of the 91ª Squadriglia at Padua airfield in the winter of 1917/18.

show off. One of his usual duties was to ferry new or war-weary fighters to and from the Aviation Park of Poggio Renatico. On 12 July 1918 he was involved in the hectic battle between Austrians and a mixed formation of Italian and British fighters, in which Silvio Scaroni was seriously wounded, and he shared with a Camel in the damage that put an Albatros out of control.

On 23 August 1918 he claimed as possibly shot down an enemy "Albatros" that he saw going down out of control by Oderzo, behind the enemy lines, but this victory was not confirmed. After that combat, he was probably war-weary, and didn't fly very much: only three escort flight on 14, 15 and 25 September. He was singularly absent from the large mass missions of the final days of the war, when his activity was limited to a few ferry flights, one abortive attack to kite balloons on 1 October and one Caproni escort on 26 October. Ranza however proposed him for a further Bronze Medal, that was assigned in 1919. He was diagnosed an exhaustion, due also to the grief for the loss of his last remaining brother, and spent some time in medical rehabilitation. His accomplishments had been remarkable: 1,700 hours of combat flying, six confirmed victories, two Medaglie d'Argento and one Medaglia di Bronzo al Valor Militare, a Croce di Guerra.

The Comando Superiore d'Aeronautica in July 1919 wrote reports about all the pilots of the 91ª Squadriglia Baracca, and this was their description of Magistrini:

> "Sergente maggiore Cesare Magistrini was assigned to the 91° squadriglia on 1 November 1917, coming from the 78ª squadriglia, when he already had a record of nine air combats and two victories to his credit. He is a very able pilot, a daring and enthusiastic soldier, never the second in the most risky

In the famous version of this photo showing the aces of the 91ª Squadriglia, Magistrini is invisible, hidden by Costantini. In this alternate shot he appears, fourth from the left. From the left: D'Urso, Aliperta, Novelli, Magistrini, Costantini, Ruffo, Piccio, Keller, Baracca, Ranza, De Bernardi, Bacula, Nardini and Olivero.

Airline commander Magistrini in front of a Junkers F 13 of the airline Transadriatica in the late 1920s.

Magistrini, in full Ala Littoria commander uniform, in front of a Junkers Ju 52 of the Italian national airline.

enterprises. He increased his score of victories shooting down four more enemy airplanes within our lines. He has flown over one hundred and fifty patrols, fifty escorts, twenty strafes and a few reconnaissance missions, and this numbers alone show the enthusiastic activity of this valiant solider. Escorting a Pomilio in a photo reconnaissance over the Tagliamento River, he earned praise by the Comando Superiore d'Aeronautica. His decorations include two medaglie d'argento al valor militare, one war cross and the Serbian silver cross, and there is the proposal for a medaglia di bronzo."

And yet this humble, solid and reliable pilot blossomed into one of Italy's most admired pilots. Magistrini was demobilized in 1919, then in 1923 he returned to aviation, now the independent Regia Aeronautica, and remained in service until 1927, when he became a commercial pilot. He flew the reliable Junkers monoplanes of the Transadriatica airline connecting Germany with Italy, then when this airline folded, he joined SAM, that later became Ala Littoria, the national carrier. Commander Magistrini was the natural pilot, quiet, projecting trust to his crews and passengers, a natural talent who never had an accident. He got a notoriety when he was the first Italian pilot to reach 1,000,000 km of flight in June 1935, then he easily reached the limits of 2,000,000 km in April 1940, and then 3,000,000 km.

When the Second World War started, he joined the SAS, Special Air Services, the Italian air force transports component, as a Captain and flew unarmed transport planes over the Mediterranean for a total of 1,350 hours, 450 of them combat missions. In 1959, he was awarded a second Bronze Medal for his service in World War II. After the war, he kept flying with the new LATI, an airline that flew SIAI Sm.95 four-engined airliners over the South Atlantic, then with SAIDE, an Egyptian airline formed with Italian capital, and finally he flew Dakotas in far-away Yemen, until 1953. In 1954 he was awarded the Gold Medal for Long Air Navigation: his log shows a fantastic score of 17,000 flight hours. He suddenly died on 26 October 1958, aged only 63, as a result of a surgical operation.

"A man of an imposing moral stature, incredibly modest, as great and valiant. He never talked about himself, his adventures, and his actions: instead, he told about Baracca, Piccio, Ranza, and so many others, heroes just like him. In his private life he was always an example of honesty, modesty, sobriety, generosity. He taught me how to cross the Alps and how to pass through the narrow valleys of Pontebbana and Fellah with our generous but weak single Junkers engine that didn't make it above 3,000 meters in bad weather. Magistrini was a master of flight; he had in his blood a prodigious feeling for navigation in the air. All his life was dedicated to his country, his family and aviation."

This is how another great airline pilot, Cdr. Bertocco, remembered Magistrini in his obituary.

In the summer of 1995 all the surviving SAS pilots met at Rome Fiumicino airport for a celebration. Every one of them was a remarkable pilot, with pioneering and epic flights (including the only flight from Europe to Japan during the war) but among the living and the missing, only one was called "legendary": Cesare Magistrini.

Magistrini and his crew with a Siai Marchetti Sn.95 four engined airliner of the Egyptian airline SAIDE.

## Cesare Magistrini
## 8 victory claims – 6 confirmed victories

| Claim | Vict. # | Date | Sqn. | Aircraft | Hour | Place | Victim |
|---|---|---|---|---|---|---|---|
| - | - | 17.6.1917 | 78 | Ni 17 3126 | 7h25 | Val d'Assa | Two-seater |
| 1 | 1 | 18.7.1917 | 78 | Ni 17 | - | Asiago | Two-seater |
| *[with serg. Nardini, 78ª Sq.] Korp. Stefan Volosin POW/Lti.d.R. Viktor Maly KIA, Flik 24, Br. C.1 229.12.* | | | | | | | |
| - | - | 18.7.1917 | 78 | Ni | - | - | Fighter |
| *[with serg. Nardini, 78ª Sq.] Fw. Wenzel Schwarz UNH, Flik 24, Alb. D.III 53.38 FTL* | | | | | | | |
| - | - | 18.7.1917 | 78 | Ni | - | - | Fighter |
| *[with serg. Nardini, 78ª Sq.] Lt. i. d. R. Josef Friedrich UNH, Flik 24, Alb.D.III 53.33 FTL* | | | | | | | |
| 2 | 2 | 16.10.1917 | 78 | Hd? | - | Monte Nero | Two-seater |
| *Vzfw. Bolte KIA/Oblt. von Glass KIA, FA 232.* | | | | | | | |
| 3 | 3 | 23.11.1917 | 91 | Spad | 11h | Cornuda German | two-seater |
| *[with cap. Costantini, 91ª Sq.] Vzfw. Wilhelm Krauser KIA/Lt. d. Res. Paul Strathmann KIA, FA(A) 232, DFW C.V* | | | | | | | |
| 4 | 4 | 30.11.1917 | 91 | Spad | 13h40 | Moriago | German two-seater |
| *[with ten. Ranza & Novelli, 91ª Sq.] Uffz. Kurt Kieling KIA/Lt. d. Res. Hugo Huck KIA, FA (A) 232, DFW C.V* | | | | | | | |
| 5 | 5 | 7.12.1917 | 91 | Spad | 12h15 | S.Pietro | EA |
| *[with ten. Ranza, 91ª Sq.] Korp. Johann Frick KIA/Lt. Johann Knezevic KIA, Flik 39, Br. C.1 369.21* | | | | | | | |
| - | - | 11.5.1918 | 91 | Spad | 11h10 | Quero | EA |
| *[with ten. Novelli & Olivero, 91° Sq.] Korp. Anton Bickel UNH/Oblt. Georg Vucetic UNH, Flik 2/D, Br. C.1 169.07* | | | | | | | |
| 6 | 6 | 17.5.1918 | 91 | Spad | 9h | Pero | Fighter |
| *[with ten. Novelli & serg. Nardini, 91ª Sq.] Lt. Franz Gräser KIA, Flik 61/J, Alb. D.III 153.221.* | | | | | | | |
| 7 | 7 | 12.7.1918 | 91 | Spad | 9h | Monte Santo | Fighter |
| *[with ten. Keller, 91ª Sq. & British pilots] Unknown pilot, Flik 30/J, Alb. D.III 153.259 damaged* | | | | | | | |
| 8 | - | 23.8.1918 | 91 | Spad | - | Oderzo | Two-seater |

Cesare Magistrini is listed in the official list of Aces as having six confirmed victories, but during the war he was specifically credited with seven, including the one of 12 July 1918.

## Federico Carlo Martinengo

The Martinengo are an ancient family of counts, who during the centuries gave to Milan and Venice many military commanders and military engineers. Outstanding among them was Nestore Martinengo, who serving the Republic of Venice distinguished himself during the siege of Famagusta, Cyprus, being captured by the Turks after the surrender, in 1571. This nobleman managed to escape from captivity and return to Italy after adventures worthy of a book. It seems that in the genes of this family there was a propensity for military and naval destinies.

Federico was born in Rome on 18 July 1897, son of a Naval officer. He attended the Torquato Tasso High School, then in September 1911 he joined the Regia Accademia Navale at Leghorn. The young man was an excellent student, and for his qualities he obtained the privileges of paying half of the tuition fees and to wear on his cadet's uniform the Royal Initials. During the traditional summer cruise of the Naval Academy he took part, albeit briefly, in the war between Italy and Turkey, on board RN *Flavio Gioia*, an old 2,000-tons cruiser used by the Instruction Division of Vice Admiral Del Bono. After two other periods at sea, with the RN *Amerigo Vespucci* and RN *Etna*, two other old ships of the same Division, Martinengo when the Great War broke out was on the battleship RN *Dante Alighieri*, where he received his commission as *Guardiamarina*, ensign, on 25 May 1915, the day after Italy joined the conflict.

The Italian Navy was lagging behind the other nations in the sector of aviation, and it had a desperate need for personnel for its air service, particularly officers. Martinengo, described by those who knew him as an active and strong-willed young man, was quick to answer to the call, and he was admitted to the Aviation School at Taranto on 27 December 1915, now freshly promoted to the rank of *Sottotenente di Vascello*.

There he trained as a pilot flying Curtiss flying boats, and on 16 September 1916 he was posted to the Stazione Idrovolanti (seaplanes station) of Venice, where the next day he flew his first training flight on Macchi L.1 serial 118. Venice was then the main base of naval aviation; its seaplanes were engaged to contrast the aggressive Austrian *Seeflieger* and to attack the enemy naval bases and forces across the Adriatic Sea.

On 16 October the order came to send three aircraft to bomb Pola. Three FBA flying boats took off but the first one, carrying *Tenente di Vascello* Giuseppe Garassini Garbarino, who will be the first Italian naval aviator awarded the Medaglia d'Oro al Valor Militare, the gold metal, developed troubles and had to abort the mission. The two other aircraft went on, one was FBA 408, with as pilot *2º capo* Pietro Valdimiro and as observer Martinengo, the other one was FBA 410. Bad weather closed in and the two seaplanes lost contact. Martinengo, realizing that it was impossible to reach Pola, decided to attack three torpedo boats that were visible in the haze. The FBA dropped its four bombs but immediately it had to face the attack of an Austrian seaplane. Valdimiro maneuvered placing his FBA behind and above the enemy, offering Martinengo a perfect target. He opened fire and the Austrian plane was hit, and

*Tenente di Vascello* Federico Martinengo with two other fighter pilots of the Miraglia Naval Air Station, Venice: *2° capo* Guido Jannello and *marò* Giuseppe Pagliacci.

then went down in a glide, with its engine stopped. Martinengo ordered his pilot to get down for a second attack against the ships, but two more Austrian seaplanes engaged the FBA. The Italian airmen ran out of ammunition, but they managed to get out of that bad situation using the cover of the clouds and returned to Venice. Their ordeal was not over yet, because 25 miles off the coast their engine quit, and Valdimiro had to alight over the stormy sea. Luckily the torpedo boat "15 OS" was quick to recover the two airmen and their seaplane and to bring them back to Venice.

Martinengo was credited with five victories, although it is not possible to identify all of them. One of them most likely was his share in the destruction of Austrian seaplane L 138, shot down during an attack against Caorle by the French pilot Paul Xavier Garros, of *Escadrille* N.392 of Venice, the crew of *SeeKadet* Franz Viranyi and *E. F.* Hoch being captured, on 23 October 1916. The Italian officer was flying Macchi L.2 serial LC 252 with *2° capo* Daniele Minciotti.

On 11 November 1916 *STV* Martinengo was assigned as commander to the 1ª Squadriglia, soon to be re-named 253ª Squadriglia (not to be confused with the 1ª Squadriglia Idrovolanti of the Army). Based at Grado, it was the most advanced unit of the Navy, practically in view of Trieste, the main Austrian naval air station.

Martinengo, a dashing soldier, turned out to be also an excellent commander, well prepared, with an excellent gift for organiza-

L101 was a Macchi Lohner L.1 on which Federico Martinengo flew at Venice.

Martinengo flew this Macchi L.2 during the air combat on 23 October 1916. The code L.C. 252 stood for Lohner Corsa, or "racing" Lohner-type flying boat.

tion but also friendly and open. The monthly reports that he sent to the higher commands show his deep knowledge of men and of aircraft, and are full with good advice for the improvement of the service. For instance, he wrote asking that a certain spot in the canal of the island of Gorgo be broadened, because it posed serious limitations to operations, and take-offs and alighting had become "*an obsession*" for his pilots.

In the short time from his arrival at Grado to the end of the year, Martinengo flew 50 times, 9 of them combat missions. His first war flight in 1917 took place on 12 January, over Nabresina. On 12 April he flew an escort mission for some monitors piloting Macchi L.3 3407. In this period Martinengo flew exploration, visual and photographic reconnaissance and bombing missions. He flew bombing raids on 25 May, on 6 June over the enemy army's rear, and on 29 June at Salvore.

On 7 July 1917 Martinengo flew a search and rescue mission for FBA 4281 that went into a spin and crashed near Grado, but his efforts were vain, because the two airmen, *2º capo torpediniere* Guido Niccolai and motorist Mario Cattaneo were dead.

In the summer the unit, now called 253ª Squadriglia, was heavily engaged in reconnaissance and bombing raids. On 5 August Martinengo, while remaining with the unit, passed his command to *Tenente di Vascello* Aldo Pellegrini, his superior in rank. On 20 August he marked in his logbook his first mission with the new Macchi M.5 fighter, an escort flight over Trieste, and in the following days he flew alternatively the M.5 and the L.3.

In the autumn of 1917 production of the new flying boat fighter was sufficient for the Regia Marina to form a squadron equipped exclusively with this type. Martinengo reached the new 260ª Squadriglia that was in formation at Venice on 28 October, in the days of the Caporetto retreat.

Preparation took place at a frantic pace, and in the first days of November the 260ª Squadriglia was operative, but its baptism of fire was not auspicious. On 5 November the squadron escorted a

Macchi L.3 serial 3412 of the 1ª Squadriglia at Grado, personally flown by Federico Martinengo.

Macchi L.3 that was going to bomb a pontoon bridge near Latisana, but a team of three Austrian aces, *Hptm.* Brumowski, *Hptm.* Linke-Crawford and *Oblt.* Szepessy-Sokoll of Flik 41/J bounced it. Despite the efforts of Martinengo and of his wingman *Guardiamarina* Paolo Morterra to disengage the bomber, thc L.3 was shot down in flames, with the death of *ten.* Arnaldo De Filippis, commander of the 259ª Squadriglia, and of *TV* Francesco Cappa. Martinengo and Morterra in their dismal combat reports remarked that the Albatros of the flight leader was painted bright red.

The situation allowed no respite, and escort, fighter and patrol missions followed one another, Martinengo took part flying first Macchi M.5 number 2, and then number 10. In December *Tenente di Vascello* Luigi Bologna was appointed commander of the new Gruppo Idrocaccia, (seaplane fighters group) so Martinengo replaced him as commander of the 260ª Squadriglia. His qualities were apparent also in the new post, and on 21 February 1918 he advanced in rank to *Tenente di Vascello*. On 1 March he flight-tested the floatplane version of the Ansaldo SVA, but this aircraft proved disappointing as much as the landplane version was successful.

With the return of spring activity became more intense. On 4 May 1918 Martinengo guided the escort to Macchi M.5 number 8, tasked with a photographic reconnaissance of the harbor of Trieste. Austrian observers at Punta Salvore spotted the Italian formation, and three Austrian seaplane fighters took off from the station near the Arsenal, closely followed by the commander himself, *LschLt.* Gottfried Banfield flying A82. The two formations met above the Tagliamento River and broke up into single, confused dogfights. The young Austrian pilot *Egast.* Franz Boros, at his first combat mission, noticed A 91 of pilot *Fhr. I.d.R.* Josef Niedermayer alighting at sea, so he took down his A 78 to check the situation of his comrade. Doing this, he noticed an Italian seaplane low and circling the stricken enemy, probably to show his positions to the torpedo boats, and he went to the attack. His target seemed a tempting easy shot, but Boros was lacking in experience, and soon the hunter turned into the quarry. Four more Macchis bounced him from above and Boros paid for the mistake of not watching his back. He alighted in a storm of tracing bullets, and he didn't even have the time to scuttle his fighters, that was captured by a torpedo boat, which triumphantly brought it to Venice. Also the other fighter, A 91, was captured together with its pilot, who a few days before had had a dramatic and unlucky encounter with the fighters of the 261ª Squadriglia. Also Banfield was forced to alight, allowing the Italians to claim three victories without losses. The other pilots of the 260ª Squadriglia who shared with Martinengo the success of this mission were *STV* Calvello, *2° capo* Rivieri and Pagliacci and *marò* Jannello, the future pilot of the 1919 Schneider Trophy race at Bournemouth.

On 8 June 1918 Martinengo, after almost two years at the front and 172 combat missions, was assigned to the Flying School at Bolsena, on the lake of Bolsena, in Central Italy, as Director of Special Courses, that is to say chief instructor on fighters. Bolsena

Federico Martinengo sitting on the prow of a Macchi L.3 flying boat.

Right: The Austrian flying boat L138 was shot down on 23 October 1916 by French fighters of Escadrille N.392 with the collaboration of Federico Martinengo.

naval air station became the largest seaplane flying school, attended by Italian and American students. Martinengo remained there three months, then on 14 September he was shifted to Otranto, in the heel of Italy, as commander of the local Stazione Idrovolanti (seaplanes station).

Martinengo remained in that little town in Apulia until the end of the war, then on 7 December, as the Station of Otranto was closed down, he was assigned as commander to the new air base at Curzola, in Dalmatia, formerly *Flugstützpunkt* of the Austrian naval aviation. In February 1919 Martinengo was shifted to command another formerly Austrian naval air station, Sebenico, northeast of Spalato (Split) and he stayed there until October. There he wrote down in his logbook his last flight, on 14 September, at the controls of Macchi M.8 number 1.

He had received two Medaglie d'Argento for his aviation service: one was issued in November 1916, and in its citation it reported his action against an Austrian seaplane that was forced down; the second one was issued in October 1917 and reported his action as squadron commander and his bombing missions during the summer of 1917.

While many Italian naval airmen remained in the flying services and joined the new independent air force at its formation, Martinengo remained in the Regia Marina and was frequently at sea, with successive commands of ships. From 1931 to 1933 he was in command of the naval detachment at Tientsin, a small Italian concession in China. On 17 May 1938 he married in Rome Angela Maria Lante della Rovere, from one of the most important Italian noble families, but their honeymoon was short, because on 29 May Martinengo, now *Capitano di Vascello*, reached as commander the RN *Muzio Attendolo*, a light cruiser of the *Condottieri* Class.

The personal insignia of Martinengo, a winged rat. The wooden fuselage of the Macchi M.5 and probably the artistic influence of Venice generated remarkable personal insignia among the fighters of the Miraglia Air Station.

When Italy entered World War II Martinengo was still in command of that beautiful warship, and he lead it into the Battle of

Federico Martinengo proudly standing in front of the Austrian fighter A.91, captured on 4 May 1918.

Punta Stilo (the Battle of Calabria for the British) and in many other naval actions.

On 12 December 1940 Martinengo had completed his period of command of the cruiser and was assigned to the Commando of the *Dipartimento Marittimo dello Jonio e Basso Adriatico*, (naval department of the Joinian and Lower Adriatic seas) first as Joint Chief of Staff then, since 23 January 1941, as Chief of Staff. It was a very demanding task, because Taranto was the main Italian naval base, and in that period that area was the theater of the war against Greece. Advancing to the rank of Rear Admiral on 18 May 1942, Martinengo moved from Taranto to La Spezia on 12 March 1943, now tasked with the Higher Command of Anti-submarine forces.

In the summer, Mussolini was ousted, and the new Italian leader, General Badoglio, secretly negotiated surrender with the Allies. When, at 7.45 p.m. of 8 September 1943, the radio suddenly broadcast the voice of Badoglio announcing the surrender and the cryptic remark that Italian forces should lay their arms to the Americans but resist attacks "from any other direction", Martinengo was driving from Rome to la Spezia.

At dawn of the 9th, when Martinengo arrived, he learnt what had happened and he found an empty harbor, because the *Squadra Navale da Battaglia*, the battle fleet, had sailed at 3 at night with its escort, heading for Malta and surrender. Martinengo took the situation under control, gave orders to scuttle the ships that were unfit to sail and for the remaining ships under his command to get ready for a quick departure. It was a difficult task, because the depots were empty after supplying the main battle fleet. As large German forces were reported approaching the city, many ships sailed quickly without fuel and supplies, so the Rear Admiral gave the general order to stop at Portoferraio, the harbor of Elba Island, and then later to continue the trip South. By mid morning the Germans were besieging the harbor and Martinengo decided to go out at sea with his two ships, "Vas 234" and "235", two small anti-submarine vedettes of just 70 tons, that still had not loaded their munitions. When they reached the island of Gorgona, the leading ship "234", carrying Martinengo, spotted two ships that turned out to be German. They were two minehunters of the 11th *Räumboot-Flottille* that had sailed at dawn from Leghorn to intercept Italian warships sailing South. "R 212" and "R 215" approached the Italian formation and opened fire. Martinengo, aware of the inferiority of his ships, ordered full speed ahead. It was a furious combat, at ever-closer range, that lasted one hour. At 13h45 a shell hit "R 212" causing a fuel loss that could not be stopped. Then German fire killed the helmsman of "234" and Martinengo replaced him, but a few moments later another hail of bullets hit the bridge and Martinengo fell down, killed by a bullet in the brain. Still the two small ships kept fighting, until they ran out of munitions. "234" was burning and it was grounded near Cala Scirocco, its crew abandoned it just before fire reached its tanks and the ship blew up. The German ships retreated, and the defenseless "235" escaped.

On 14 September the sea returned the body of Martinengo, who was buried with military honors in the small military cemetery on Gorgona. For his gallant final combat he received the highest Italian decoration, the Medaglia d'Oro "in the field".

In the late Eighties the remains of Federico Martinengo were moved from Gorgona to the naval cemetery at La Spezia, as his wife requested.

## Federico Carlo Martinengo
## 5 confirmed victories

| Claim | Vict. # | Date | Sqn. | Aircraft | Hour | Place | Victim |
|---|---|---|---|---|---|---|---|
| 1 | 1 | 16.10.1916 | FBA 408 | - | - | Rovigno | Seaplane |
| *[pilot 2° capo Pietro Valdimiro]* | | | | | | | |
| 2 | 2 | 23.10.1916 | LC 252 | - | - | Caorle | Seaplane |
| *[with 2° capo Daniele Minciotti and Paul Xavier Garros, Escadrille 392] Skdt. Viranyi POW/E.F. Hoch POW, L 138* | | | | | | | |
| 3 | 3 | 4.5.1918 | 260 | M.5 # 1 | - | Gulf of Trieste | Seaplane |
| *[with STV Calvello, 2° capo Jannello, Rivieri & marò Pagliacci]Fähnr. i. d. R. Josef Niedermayer POW, A91* | | | | | | | |
| 4 | 4 | 4.5.1918 | 260 | M.5 # 1 | - | Gulf of Trieste | Seaplane |
| *[with STV Calvello, 2° capo Jannello, Rivieri & marò Pagliacci]Flgst. Franz Boros POW, A78* | | | | | | | |
| 5 | 5 | 4.5.1918 | 260 | M.5 # 1 | - | Gulf of Trieste | Seaplane |
| *[with STV Calvello, 2° capo Jannello, Rivieri & marò Pagliacci] Lschlt. Gottfried von Banfield UNH, A82 FTL* | | | | | | | |

## Guido Masiero

A fighter ace, a volunteer in three wars, trail-blazer in the history of aviation, aircraft designer and test pilot, holder of four Medaglie d'Argento and one Bronzo al Valor Militare. Such are the outstanding qualifications of Guido Masiero.

He was born in Padua on 24 August 1895 from a middle class family. Young Guido got his Technical Industrial High School diploma at Belluno in 1912, and at the end of 1913 he volunteered in the reputed cavalry regiment of the Lancieri di Novara. Following his technical experience, he asked to be admitted to pilot training, and he was accepted on 1 February 1915. He took his first flying lesson at Mirafiori airfield with instructor Petazzi on 15 March, on a Farman with Renault 70-hp engine. He got his pilot license on 1 August 1915 and his military pilot brevet on 15 November 1915, on the Voisin biplane.

At the end of 1915 Masiero, now with the rank of *sergente* and sporting the winged propeller insignia of a military pilot, was sent to the War Zone, assigned to the 7ª Squadriglia Voisin at S. Maria la Longa airfield. Masiero was active participating in the operations of that unit, which flew reconnaissance and bombing missions in the Carso sector, and which became the 26ª Squadriglia since 15 April 1916.

At the 26ª Squadriglia he met other outstanding pilots who later became fighter aces: Alessandro Resch and Flavio Torello Baracchini. On 20 April 1916 Guido Masiero fired his first shots in anger during a cruise flight between Monfalcone and Cormons with gunner Portaluri. On 25 April he attacked an enemy kite balloon. On 25 June it was an Austrian seaplane that attacked him over the Gulf of Trieste. On 28 July, flying with observer *ten.* Massimo Vitale he faced the attack of a Fokker Eindecker, serial 03.51 that placed five bullets in his Voisin. This event was claimed as a victory by the Austrian pilot, *Fw.* Stefan Huzjan. The same crew had another duel on 3 September, in which an enemy two-seater hit their Voisin.

As his log of missions grew, together with his experience, he rose in the ranks as *aspirante ufficiale*, in the *Arma del Genio* (Corps of engineers), on 30 September, and then *sottotenente di complemento* (reserve 1st Lieutenant) on 11 December 1916.

On 14 January 1917 Masiero flew a standing patrol with gunner Procchietto between Mount Cone and Nad Logen that was terminated by Austrian cannon fire that riddled his aircraft. Despite the serious damage, the pilot managed to regain his airfield, and was commended, but the Voisin was written off.

On 4 March 1917 the 26ª Squadriglia was disbanded, and its components were scattered. With that unit Masiero had flown 117 flights, 58 of them combat sorties and he had earned a Bronze Medal, recognizing his service over the Carso in January-February 1917. He was very briefly posted to the 103ª Squadriglia Difesa at Brindisi, a defense unit in Southern Italy, also equipped with the Voisin, but he remained there only a very short time, and then in May 1917 he went to the Malpensa flying school for conversion on the Nieuport. After the aerial gunnery training at S. Giusto, where he was graded as "good", Masiero returned to Malpensa for instruction on the Ansaldo SVA, after which he went to Ponte San Pietro, an airfield used for the formation of new units, assigned to the 181° Squadriglia, soon re-named 86ª Squadriglia SVA.

Tenente Guido Masiero in the summer 1918, in a signed portrait.

Men and airplanes were rushed to the front in the emergency after the enemy attack had broken the Italian lines at Caporetto, and thus Masiero went for a few days to the 91ª Squadriglia of Baracca, and then on 2 November 1917 he was posted to the 78ª Squadriglia, based at Istrana airfield. This unit had lost most of its equipment and its new Hanriot fighters during the retreat, but it was quickly provided with new aircraft. Masiero shared his room with Renato Fucini, another great pilot, with a frank and frisky character, who quickly became a good friend.

On 7 November Masiero engaged some German aircraft that took off from Aviano, and claimed one shot down over Cividale, but this victory did not obtain confirmation. On 13 November, however, he did score a victory that was confirmed: together with his friend Mario Fucini and another ace of the 78ª, *sergente* Antonio Chiri, and *serg.* Gastone Razzi of the 76ª, he shared in the destruction of a German two-seater that crashed in flames near Arcade, behind Italian lines, killing its crew.

Just six days later and then Masiero scored another victory and was recommended for a Silver Medal. On 19 November *s.ten.* Gino Allegri, later one of the dearest companions of d'Annunzio, set out to attack the kite balloon at Valdobbiadene going with the escort of

One of the DFW C.5s shot down in the countryside around Istrana on 26 December 1917. As this photograph, and the relics preserved in the Vittorio Veneto museum show, the workmanship of this aircraft was outstanding.

Masiero goofing for the camera in front of a single-seat SVA, serial 6839. This aircraft belonged to the 2ª Sezione and was struck off charge on 24 May 1918.

Masiero, Fucini and Scaroni. They went with a SP of the 22ª Squadriglia to fool Austrian ground observers pretending to be its escort, but close to the target the two-seater turned for home while the Hanriots and the Nieuport of Allegri dove to the attack. The Le Prieur rockets missed, ground fire suddenly ceased and the Italians realized that enemy fighters had arrived and were joining the fray. Masiero's flight logbook curtly reports the following events: "Attacked by three enemy fighters engaged one of them bringing it down between S. Gregorio and Colderone (time 11.20) Himself attained in the top wing and longeron, 6 m.g. bullets. Proposal for the award to Military Valor (silver)." *Leutnant* Bucholz is a possible choice for the man who placed his bullets into Masiero's fighter, as this German commander of K-Flak 101 anti-aircraft battery claimed a bit optimistically a "Sopwith" north of Fontigo, while the Italian's victim could be Albatros D.V of *Oberleutnant* Otto Deindl, Jasta 1, who was wounded, or possibly the same victim of Scaroni, *Vizefeldwebel* Rahier. Back to base they found Allegri despondent for his failure to hit the balloon, but the crew of the SP confirmed that two enemy aircraft had gone down.

On 20 November Masiero claimed an enemy two-seater shot down over Moriago which, according to some documents, may be the same combat in which *serg.* Chiri claimed a victory, at Mosnigo, which is another village close by. His third confirmed victory was on 10 December: during an offensive patrol between Priula and Vidor he shot down together with Silvio Scaroni and *serg.* Chiri, a two-seater, probably Aviatik C.1 37.24 of Flik 58, with the death of its airmen, *Korporal* Josef Bantz and *Leutnant* Felix Kanitz. On 14 December Masiero claimed another victory over a two-seater between Valle and San Siro, which got no official confirmation.

He did get his first Medaglia d'Argento in the field, which in its citation quoted his three victories of 13, 19 and 20 November, specifying that the first one was against a German airplane, the second one against a fighter from a flight of three, and the third one against a two-seater that crashed in flames near Moriago.

When the German bombers arrived to attack Istrana airfield in the morning of 26 December, Masiero was in the air and he shared in the destruction of two aircraft, a DFW C.V, together with *s.ten.* Scaroni and *ten.* Fucini, and another one with *serg.* Fornagiari and *sold.* Clemente Panero of the 82ª Squadriglia, the first one over Musano and the second one at Falzè.

On 10 January 1918 Masiero, according to an article in the contemporary press, claimed a victory in a combat over Mount Lisser, but this event is not reported in his flight logbook. In this period there were the first contacts between Italian pilots and the airmen of the Royal Flying Corps, which had sent three Camel squadrons, a RE.8 squadron and a Bristol Fighter flight as reinforcement to the hard-pressed ally. Relations were limited on account of language, and also of different habits. Masiero admired their equipment, but had strong reservations about their food. On 17 January 1918 he got his commission as *tenente*, backdated to July 1917.

On 23 January 1918 Masiero was on offensive patrol near Vidor and as he didn't find anything in the air, he went down to strafe enemy troops marching. Five days later, 28 January, flying over the Asiago plateau as battle of Col del Rosso was raging on the ground, he discovered a formation of a two-seater with three escorting fighters, and engaged them, claiming one of the fighters shot down near Pedscala, some ten km west of Asiago. Later on in the same day he flew as escort to a Savoia Pomilio, and he claimed another aircraft shot down, in collaboration with another fighter, probably one of the many combats in which Italian fighter pilots claimed several victories, in that hectic and confused day. Both victory claims, however, are reported only in his logbook, and not in the daily War Bulletin nor in the postwar "official" list of victories.

On 18 March 1918 Masiero claimed his final victory with the Hanriots of the 78ª Squadriglia, a kite balloon shot down near Conegliano.

On 29 March 1918 Masiero, now with the rank of *tenente*, was transferred to the 3ª Sezione SVA, at Isola di Carturo. With the 78° Squadriglia he had flown 78 combat sorties, for a total of 102 hours 23 minutes of flight. The 3ª Sezione (nominally of the 86° Squadriglia, but this unit never operated as a whole) was one of the new strategic reconnaissance flights assigned to each Army command, that used the SVA not just for photographic reconnaissance, exploiting its high speed and range, but also for ground attack. Masiero, who flew his first flight with this unit on 4 April, approached his new role with the character of a former fighter pilot, so he used every opportunity to engage in air combat. On 20 May he returned to base from a long reconnaissance mission all the way to Bolzano with the glass plates of 36 photographs and the claim of a red-painted fighter driven down near Trento, after a combat with four which had tried to intercept him.

An Ansaldo SVA of the 3ª Sezione, with its beautiful black cat unit insignia. The pilot in the cockpit may seem to be Guido Masiero.

During the Austrian offensive of mid-June he flew many ground attack missions, first dropping bombs and then going down low to strafe enemy troops near the Piave. After the offensive was stopped, photo reconnaissance flights were resumed and on 1 July Masiero flew to Feltre to drop leaflets, when two Austrian fighters challenged him. In the ensuing dogfight he claimed one of them shot down, although Austrian lists of known losses don't confirm this claim. His final missions took place in the preparation of the Italian final offensive, and included bombing missions in zones behind the front.

Guido Masiero received two more Medaglie d'Argento, one for his reconnaissance flights over the Alps with the 3ª Sezione SVA and another one, also in the field, for his total activity as a fighter pilot, with a citation crediting him for six aircraft and one kite balloon destroyed, a total that is probably closer to the truth than the five victories credited him by the so-called Bongiovanni Commission in early 1919. He later received another Silver Medal, his fourth one, for his combat activities after World War I.

When the war was over, Masiero was chosen as a member of the Italian aviation mission to France that was welcomed at the Villacoublay experimental establishment. He then took part in the ELTA aeronautical exhibition at Amsterdam and then, on 23 November 1919, he was discharged from the active service. His period of leave was rather brief, however, because on 14 February 1920 he was back at the controls of a SVA, taking off from Rome together with another young but already highly regarded pilot, Arturo Ferrarin, with a most special destination, Tokyo, Japan.

This flight had had a long gestation. In the spring of 1919, the poet Gabriele d'Annunzio, weary of the post-war inactivity, planned a flight from Italy to Japan by a squadron of five SVA, that were to fly over 10,000 miles at the very optimistic rate of 1,000 miles per day. The Direzione Generale di Aeronautica, the commanding agency of the air force, endorsed this project, and organized the trip during the summer of 1919, in the hope that d'Annunzio would get out of Italy and out of harm's way. Instead, in September 1919 the restless hero organized the take-over of Fiume, a contested town at the border between Italy and the Kingdom of the Serbs and Croats, where he set up his own republic, surrounded by officers who were technically rebels against Italy, and who included aces Cabruna, Lombardi, Stoppani and Ancillotto.

The raid went on anyway, as its organization had been exceedingly expensive, with support bases set up all along the route, from the Middle East to the Far East, equipped with spare engines, while spare SVA airplanes were shipped to Basra, Calcutta and Canton. However, with d'Annunzio dropping out, also some of the most reputed pilots, like Locatelli and Ancillotto, also quit. Instead of the original single flight of SVA, a total of eleven aircraft took part

Right: Arturo Ferrarin and Guido Masiero in China, during their long flight toward Japan.

in the raid: a Caproni Ca.4 triplane, two Caproni Ca.3 and one Ca.5 biplanes, and seven SVAs, two as a vanguard, with Masiero and Ferrarin, and five as the main formation.

First to leave was the Ca.3 23180 of *ten.* Edorado Scavini and *s.ten.* Carlo Bonalumi, on 8 January 1920, followed on 18 January by the Ca.4 5380 of *tenenti* Luigi Garrone and Enrico Abba, with motorists Alfredo Mono and Alfredo Rossi, then Ca.5 11928 of *tenenti* Virginio Sala and Innocente Borello, with motorist Antonio Sanità, and finally on 2 February the Ca.3 23181 of *ten.* Leandro Negrini and *s.ten.* Giovanni Origgi, with motorist Dario Cotti, but all of the big bombers had to abandon the race after different accidents in the Middle East.

On 14 February 1920 two lonely SVAs took off from Rome's Centocelle airport. Guido Masiero piloted SVA 9 13157, with motorist Roberto Maretto, while Arturo Ferrarin, formerly instructor and fighter pilot with the 82ª Squadriglia, piloted SVA 11849 with motorist Gino Cappannini. They flew to Southern Italy, (after a brief diversion north which made the top brass fear that they were defecting to d'Annunzio) then Albania, Greece and Turkey. On 18 February Masiero had to land in the middle of Anatolia, where war was raging between the Greeks and the Turks, to fix a leak in his radiator. In Adalia, Turkey, the two pilots separated, Ferarrin left and in the next days flew to Aleppo, Baghdad, Basra and Bushir. He was at Bandar Abbas by the end of the month, while Masiero had to stay to fix his engine.

When Masiero reached Basra, he crashed his SVA on landing, and used the spare SVA that had providentially been sent there. From Bandar Abbas, he flew directly to Karachi, skipping the stop at Chiaubar, Pakistan, and there he rejoined Ferrarin. On 3 March Masiero flew from Karachi to Delhi, but when he took off from that city's racing grounds his engine failed him, and the SVA was wrecked. So Masiero reached Calcutta by train, and there he picked up yet another spare SVA. There, together with Ferrarin, he had to stay and wait for the SVA squadron that had left Rome on 11 March, so he gave joyrides to local authorities and the press. On 31 March they left Calcutta, flying to Akyab, Burma, and then Rangoon. Masiero took off alone to fly to Bangkok, but his engine suffered an oil leak, so he was forced down in a rice paddy, where he fixed the trouble. His flight to Ubon, Thailand, was also cut short by a magneto failure. By mid April the two Italian aircraft were in Hanoi, then on 21 April they took off to go to Canton, braving the monsoon season. They found the field totally flooded: when they complained with the Italian officer who had organized the stop, he apologized saying that he was a seaplane pilot. The Chinese were enthusiastic of the raid, and the Italian airmen were feted and decorated. Take off from flooded Canton was not easy, however. Ferrarin managed it, but Masiero, trying it the next day, 28 April, hit a tree and totally wrecked another SVA. He and Maretto then had to reach Shanghai by train, there picking up one more spare SVA. When Ferrarin reached that Chinese city by air, celebration parties lasted seven days. The two SVAs left separately for Tsingtao, then a Japanese possession, on 8 May, then to Beijing, for other long celebrations, Mukden, Manchuria, by the end of May, Shuingishu, Korea, and then Seoul, where they flew in formation again for the final legs to Osaka, 30 May, and finally to Tokyo, 31 May 1920. The Japanese welcome was beyond compare, 42 days of celebrations. Only in September were the Italian airmen allowed to take a boat to return home.

Italian officers in Somalia, 1936. From the left, Guido Masiero, an army officer, Felice Porro (wartime commander of the 48ª Squadriglia), Ferruccio Ranza and Antonio Locatelli.

The "official" SVA squadron, meanwhile, had departed on 11 March. The pilots were *cap.* Mario Gordesco, commander, and Giuseppe Grassa on SVA 13179, *cap.* Umberto Re and cameraman Bixio Alberini on 13195, *cap.* Ferruccio Ranza and motorist Brigidi on 13197, the ace *ten.* Amedeo Mecozzi and *ten.* Bruno Bilisco on 13200 and *ten.* Ferruccio Marzari (a colleague of Masiero in the élite 3ª Sezione SVA) and motorist Giuseppe Damonte on 13196. *Cap.* Re quit in Valona, Marzari wrecked his SVA in Adalia on 26 March, Mecozzi was the victim of sabotage in Aleppo, on 29 March, and so two airplanes only remained: Ranza, who took along Marzari, and Gordesco with Grassa. The latter two were killed in a landing accident at Dushir, Persia, on 12 April, while Ranza, after an epic adventure as a prisoner of Arab rebels in Syria, reached Baghdad, Dushir, Bandar, Ciaubar, Delhi, Allahabad and Calcutta, where he crashed and finally had to call it quits.

Masiero and Ferrarin reached Tokyo after 105 days, 25 of them of flight, flying 109 hours at an average speed of 100 mph, braving the most extreme weather, primitive local conditions, the presence of all kinds of war, rebellion and piracy all along the Asian continent, and the general climate of post-war relaxation that marred the performance of Italian authorities and of other military detached in the various support missions. All this was compensated by the eagerness and courage of the Italian airmen and by the enthusiasm and cooperation or British, French and Chinese authorities and peoples.

The Rome-to-Tokyo flight was harshly criticized at home right after its success, as it had cost the extravagant sum of 20 million lire and it had been more a political than a technical enterprise. Yet it was a romantic, trail-blazing adventure that thrilled the Chinese and Japanese and that later was rightly praised. For Arturo Ferrarin, it was the stepping stone for a career that made him Italy's most loved pilot. Masiero instead did not capitalize on it. Moreover, since he rode a train from Delhi to Calcutta he was not technically a hero like Ferrarin, who had flown all the way.

His name may have been forgotten by most, and yet when in 1970 the fiftieth anniversary of this epic flight was celebrated, his son was invited to Japan, flying there in a much faster and more efficient way, on a Douglas DC-6 of the Aeronautica Militare. By the way, it took five years for another airplane to fly from Europe to Japan, with the French pilot Pelletier-Doisy who reached Tokyo with mechanic Bésin at the controls first of a Breguet XIX and then of a Breguet 14, that is to say an all-metal airplane with twice the power of the SVA.

Now a reserve officer, first in the ranks of the Army and then, after its creation, of the Italian Air Force, Masiero attended other professions in the civil society, and was first a farmer and then a building businessman. In July 1935 as Italy did not conceal its armament and the coming invasion of Ethiopia and tension in Europe was rising, Masiero, now *capitano*, returned to the active service flying with the 20º Stormo B.T. (bombing wing). Freshly married and expecting a child, he volunteered and sailed to Massawa, Eritrea, on 3 November 1935. He was assigned to the 108ª Squadriglia, a unit operating the Romeo Ro.37 bis biplane on the front of Somalia supporting the troops of Gen. Graziani moving northward for the conquest of Ethiopia. Masiero flew 65 sorties in the first six months of 1936 from a variety of advanced landing grounds, most of them ground attack and strafing missions. His logbook records a weird entry: a flight on 30 May 1936 from Djidjiga to Dire Dawa carrying *Monsieur* Henri de Monfried, who was a then well-known French explorer and writer, connected, in the opinion of some, in shady dealings in Ethiopia.

Masiero remained in Africa exactly one year and one day, sailing back from Mogadishu on 4 November 1936. He flew many combat sorties with Romeo Ro.1, Ro.37 and Caproni 101 aircraft, in many bombing and ground attack operations.

Back in Italy, now a *maggiore*, he was assigned first to the 5º Stormo, then to the 50º, two wings of the new *Assalto* specialization, the tactical ground attack units formed according to the doctrine developed and strongly championed by Amedeo Mecozzi, another ace and mate of Masiero at the 78ª Squadriglia.

He was now over forty, with a young daughter, so Masiero could legitimately think that he had already given enough to his country. On the contrary, he volunteered again as a pilot in the Aviazione Legionaria, the Italian air corps in the Spanish Civil War. He departed on 1 October 1937, with the cover name of "Guido Magoni". He converted on the Breda 65 attack bomber at Tudela airfield. The Breda was a mediocre and dangerous fighter-bomber, equipping the 65ª Squadriglia, and Masiero flew ground attack missions together with pilots twenty years younger than him. In the period from 29 November 1937 to 28 March 1938 he flew 22 combat sorties, most of them strafing and ground attack missions over the roads in the region of Teruel. He then returned from Spain to Italy and on 31 May 1938 he was discharged. At the end of the year his second child was born, a son, who was also called Guido.

On 14 December 1939 Masiero became chief test pilot of the Breda aviation works at Sesto San Giovanni. In this period he also designed his own airplanes: the M.1 was the project for a twin-engined light aircraft, probably with tailbooms, in two versions, with Alfa 110 and 115 engines, four or two seats; the M.2 was a version of the same project with a single pusher Alfa 110 engine. The models were tested in the wind gallery at Guidonia in 1940, but there was no further development. It seems that Masiero also designed a dive bomber. When Italy entered World War II he was 45 years old, and yet he volunteered for conversion on the Junkers Ju 87 Stuka, but they left him with Breda. On 31 December 1940 he reached the rank of *tenente colonnello*.

At Breda, Masiero test flew hundreds of Breda 65 and 88 and Macchi C.200 and C.202. The job of a test pilot, for a man with his experience, who had braved the guns of Austrian Albatros and Spanish Republican Ratas, may seem uneventful, but danger was never far away. In the morning of 24 November 1942 Masiero took off with a Breda-built Macchi MC.202. It was a foggy day over Bresso airfield and his fighter, probably serial 9589, collided with MC.202 serial 9597 which had another exceptional pilot, Francesco Agello, who in 1934 had established the absolute speed record with the Macchi 72 racing floatplane. The two fighters crashed near Cinisello, killing both occupants.

# Guido Masiero
# 13 victory claims – 5 confirmed victories

<table>
<tr><th>Claim</th><th>Vict. #</th><th>Date</th><th>Sqn.</th><th>Aircraft</th><th>Hour</th><th>Place</th><th>Victim</th></tr>
<tr><td>1</td><td>-</td><td>7.11.1917</td><td>78</td><td>Hd?</td><td>10h40</td><td>Cividale-Aviano</td><td>German a/c</td></tr>
<tr><td>2</td><td>1</td><td>13.11.1917</td><td>78</td><td>Hd?</td><td>15h45</td><td>Arcade</td><td>German two-seater</td></tr>
<tr><td colspan="8">[with serg. Chiri, 78ª Sq., ten. Fucini & serg. Razzi,76ª Sq.] Lt. Eduard Schulze Watzum KIA/Lt. Kamillo Ehrler KIA, FA (A) 232</td></tr>
<tr><td>3</td><td>2</td><td>19.11.1917</td><td>78</td><td>Hd</td><td>11h20</td><td>S.Greg./Colderone</td><td>Fighter</td></tr>
<tr><td colspan="8">[with s.ten. Scaroni & ten. Fucini, 76ª Sq.] Vzfw. Kaspar RahierKIA, Jasta 31, or Oblt. Deindl WIA, Jasta 1.</td></tr>
<tr><td>4</td><td>-</td><td>20.11.1917</td><td>78</td><td>Hd</td><td>-</td><td>Mosnigo</td><td>Two-seater</td></tr>
<tr><td colspan="8">[with serg. Chiri, 78ª Sq.] maybe: Offstv. Sokolowski WIA/Lt.d.Res. Kaiser WIA, FA (A) 219 or: unknown pilot/Lt. d. Res. Binder KIA, FA (A) 219</td></tr>
<tr><td>5</td><td>3</td><td>10.12.1917</td><td>78</td><td>Hd</td><td>-</td><td>Villa Ancillotto</td><td>Two-seater</td></tr>
<tr><td colspan="8">[with s.ten. Scaroni, 76ª and serg. Chiri, 78ª Sq.] maybe: Korp. Josef Bantz KIA/Oblt. Oswald Kanitz KIA, Flik 58, Av. C.1 37.24</td></tr>
<tr><td>6</td><td>-</td><td>14.12.1917</td><td>78</td><td>Hd</td><td>-</td><td>Val Brenta</td><td>Two-seater</td></tr>
<tr><td>7</td><td>4</td><td>26.12.1917</td><td>78</td><td>Hd</td><td>-</td><td>Musano</td><td>German two-seater</td></tr>
<tr><td colspan="8">[with s.ten. Scaroni, ten. Fucini, 76ª Sq., serg. Fornagiari, 78ª Sq., ten. Comandone, 82ª Sq.]</td></tr>
<tr><td>8</td><td>5</td><td>26.12.1917</td><td>78</td><td>Hd</td><td>-</td><td>Falzé</td><td>German two-seater</td></tr>
<tr><td colspan="8">[with serg. Fornagiari, 78ªSq. & sold. Panero, 82ª Sq.]<br>Possibly: Lt. Kessler WIA/Lt. d. Res. Edebohls KIA, FA 2, DFW C.V<br>Uffz. Willi Petzold KIA/Lt. d. Res. Wilhelm Voigt KIA, FA 2, DFW C.V<br>Vzfw. Karl Uecker KIA/Lt. d. Res. Heinrich Pfeiffer DOW, FA 2, DFW C.V<br>Vzfw. Pohlmann WIA/Lt. Schlamm WIA, FA (A) 219, DFW C.V.</td></tr>
<tr><td>9</td><td>-</td><td>28.1.1918</td><td>78</td><td>Hd</td><td>-</td><td>Pedscala</td><td>Fighter</td></tr>
<tr><td>10</td><td>-</td><td>28.1.1918</td><td>8</td><td>Hd</td><td>-</td><td>West of Asiago</td><td>Fighter</td></tr>
<tr><td colspan="8">[with another pilot]</td></tr>
<tr><td>11</td><td>-</td><td>18. 3.1918</td><td>78</td><td>Hd</td><td>18h</td><td>Conegliano</td><td>Kite balloon</td></tr>
<tr><td>12</td><td>-</td><td>20.5.1918</td><td>3ª Sez.</td><td>SVA</td><td>-</td><td>NE of Trento</td><td>Fighter</td></tr>
<tr><td>13</td><td>-</td><td>1.7.1918</td><td>3ª Sez.</td><td>SVA</td><td>-</td><td>Val Frenzela</td><td>Fighter</td></tr>
</table>

# Amedeo Mecozzi

There is a stereotyped idea of the fighter pilot from films and books: a young man who, when not in the air shooting down enemies, spends his time playing cards, running after girls or reading the sports pages sitting in an easy chair, waiting for his mission. Nothing like this in the life of Amedeo Mecozzi, who all through his life had a penchant for study and analysis, becoming an important theorist of the role of aviation when the new arm had reached its maturity. He was a peculiar personality, best described by Mario Fucini, one of his squadron comrades, first in the 76ª and then in the 78ª Squadriglia:

> "Then there is Mecozzi, tireless, always everywhere. You go up flying, and there you find him, at four five thousand meters, alone. You approach him, having seen his personal markings, and he hardly answers your greeting. Of course he is burning with rage because he doesn't find anything, always unlucky. The Austrian comes when he, after two hours of patient and tiring hunt, has landed with a single furious dive. You go for a nap in a warehouse and there you find him, giving orders to the men, organizing, and writing notes to get the burden off his commander. On the airfield, as we wait for our time to fly chatting or, standing on alarm, we spend the endless wait for the enemy with various futilities, he walks up and down in front of his airplane, already with his fur on, reading, studying something. I saw him many times, when the alarm was sounded, sticking the book in his pocket and jumping in the cockpit with eyes flaming for a fever of victory that always eluded him, staring at a suspect dot in the sky, taking off like a meteor, first to go, last one to return, disappointed once more: a gun jam, the escape of the enemy, and sometimes a terrible lesson from five or six black crossed fighters that riddled his airplane... He always got close to success and missed it, and yet he was obstinate, he was able, an excellent pilot, at the same time enthusiastic and reflexive, and above all he applied most the quality that he most lacked: patience."

Amedeo Mecozzi was born in the old heart of Rome on 17 January 1892, near the popular market of Campo de' Fiori. His parents died of tuberculosis when he was a little boy, and this event may have influenced his character, turning him to introversion and to steadiness. His grandparents raised the little boy and, worried about his education, considered sending him to boarding school. Amedeo retaliated running away from home, and finally the police picked him up in Naples, sleeping under the colonnade of Piazza Plebiscito.

Mecozzi volunteered at the end of 1913 in the 6º Reggimento del Genio Ferrovieri (railroad engineers) from which he was assigned to the Regio Corpo di Truppe Coloniali (colonial troops) sailing to Massawa, Eritrea, on 20 March 1914. He remained in Africa over one year, returning to Italy soon after its entry in the war. Now with the rank of corporal, he rejoined his regiment. On 2 September 1915 he was accepted as a student pilot, training at Malpensa flying school and getting his pilot license on 1 February 1916, flying Farman model 1912.

Pilots of the 78ª Squadriglia: from the left *cap.* Antonio Riva, *ten.* Giacomo Brenta, *ten.* Amedeo Mecozzi, *ten.* Mario Fucini and another officer, not a pilot.

On 27 March 1916 Mecozzi reached the front, assigned to the 46ª Squadriglia, a reconnaissance unit equipped with Farmans, with as commander *capitano* Riccardo Adorni. The young pilot immediately faced the harshness of those missions, and on 22 May he returned from a sortie in the zone Biaena-Martinello with his aircraft hit by ground fire. Mecozzi then was assigned to other reconnaissance units, the 48ª and 45ª Squadriglia, but on New Year's Day 1917 he was back at the 46ª Squadriglia at Castenedolo airfield. On 8 January 1917 during a mission anti-aircraft artillery hit his engine and a splinter struck his flying helmet. For this event he received his first award, a Medaglia di Bronzo, which in its citation commended his courage in flying over a battery at very low level and his ability in nursing the damaged airplane back to base.

On 19 February as he was flying with observer *ten.* Attilio Zoppi in an artillery-spotting mission Mecozzi met an enemy fighter aircraft that engaged him over the Pasubio and damaged his Farman, forcing him to abandon his sortie. A second Bronze Medal in the field acknowledged his accomplishment. Known Austrian data don't report the name of Mecozzi's opponent, who might fairly have put forth a victory claim. Again on 18 March in the same area Mecozzi, now flying with the 49ª Squadriglia and having the rank of *aspirante*, met an Austrian airplane that easily damaged his Farman and drove him away. Another pilot of the 49ª, *serg.* Simonetta, also had a similar combat that day. *Zugsf.* Julius Kowalczik of Flik 24, flying Brandenburg 29.71, claimed a Farman forced down near Asiago at 10 a.m. and then a Voisin forced to land near Grigno 15 minutes later.

Mecozzi remained at the 49ª two months, and at the end of April he was assigned to the 50ª Squadriglia, another Farman unit operating on the Asiago plateau/Trentino front. On 19 June he was flying with *s.ten.* Giulio Mazzocchini, on a bombing mission to Albergo Ghertele. Two Austrian fighters attacked them hitting the

Mecozzi shared with serg. Capparucci two victories on 26 May 1918. Capparucci flew an Hanriot, serial 7479, with a diamond (possibly light blue) as a personal marking.

radiator, Mecozzi yet came back to drop the last bomb and then force landed in a field near Schio. It was once again Julius Kowalczik of Flik 24 flying Albatros D.II serial 53.02: the Austrian ace placed 29 bullets in the aircraft that he claimed as a "Caudron" shot down, this being the first victory for the Albatros D fighter on the Italian front.

On 20 July 1917, for reasons unknown, Mecozzi was admitted to a field hospital, soon after being commissioned as a *sottotenente*. Surely he had had enough of playing the target for Austrian gunners and pilots, thus on 13 September he was able to move to the Cascina Costa training airfield for conversion on the Nieuport fighter. After instruction he was posted back to the front with the 76ª Squadriglia on 12 October, and thence in late November he passed to the 78ª Squadriglia, both units then being based at Istrana.

It was a well-sorted team, and the witty Giorgio Michetti nicknamed Mecozzi "Catone" after the Roman stern moralist Cato, because he was always reflexive, gave good advice and studied a lot. But soon he also got the reputation of having bad luck: he flew many sorties, but always failed to find the enemy, or at least to engage it. He had his first success only on 26 December, during the Battle of Istrana, shooting down an enemy airplane near Volpago, which crashed in the barn of some farmers. When Mecozzi drove there, he found them much annoyed, so Michetti could joke that he wasn't able to shoot down an airplane correctly.

His bad luck was remarkable also on 3 February 1918, during an offensive patrol with *sergente* Oreste Codeghini and *caporale*

*Serg.* Capparucci later in the war flew another Hanriot, which was emblazoned with a diamond, apparently white.

Fortunato Celio. The Italians engaged a formation of six enemy aircraft over the Brenta River. Codeghini, flying above, realized that his companions were fighting with jammed guns and were in bad trouble, so he joined the dogfight, he made possible their escape and claimed an enemy shot down in the Val Cismon, returning with three holes in his Hanriot. Their enemies possibly were fighters from Flik 60/J, who claimed two Nieuports shot down, one by *Oblt.* Linke Crawford and one by *Zugsf.* Karl Teichmann for no losses. Both sides, it seems, in the heat of combat had overclaimed.

Poor Mecozzi had to wait almost two months to claim a victory, although in this period he logged 38 offensive patrols and 11 escort flights. He often flew with Fucini and with *ten.* Brenta, gaining the reputation of spreading his bad luck to his wingmen too.

Things changed on 26 May 1918. This was his combat report:

> "At about 8.40 a.m. I was flying a standing patrol with sergente Capparucci at an altitude of 5200 meters over Arcade and I spotted nine enemy aircraft, one of them a two-seater and the other ones fighters. Some of them were above me and all were cruising more or less above Lancenigo. I flew straight at them followed by sergente Capparucci and I saw them turning back and disbanding. I attacked the two-seater that was closer firing a single burst of 100 rounds and I saw it going down out of control. I couldn't follow its fall because right away I attacked a fighter that dove toward its lines. I fired in more bursts all the remaining bullets of my gun and I followed it down to 800 meters over enemy territory and finally I saw it first spinning then going down to crash against trees between San Miche and Ormelle. As I was very low and machine guns on the ground were shooting at me I quickly returned to our lines flying over the Grave di Lovadina."

Mecozzi was flying Hanriot serial 11435. There is no way to identify his victims through Austrian data; on the other hand Italian contemporary papers reported three crashes on that day and the recovery of three corpses: the one of *Zugsf.* Hoftsaedtler, shot down

Hanriot fighters of the 78ª squadriglia, the second airplane is the one of Mecozzi, showing its insignia of two blue bands and a circle with a question mark.

by Piccio, the one of a fighter pilot near Postioma and the one of a two-seater airman near Maserada, the latter two victims of Mecozzi. On that day there were multiple claims in a combat that saw the participation of Baracchini, 76ª Squadriglia, Piccio, Novelli and Keller, 91ª, Mitchell, Hudson and Hallonquist, No. 28 Sqn., RAF. The Austrians lost two fighters of Flik 42/J, one behind Italian lines, victim of Piccio, and one forced to land. In Austrian documents there is no trace of the two-seater. Later on Mecozzi and Capparucci received a prize of 500 lire for each aircraft shot down, and this would prove that they fell in Austrian land, as the prize for airplanes brought down behind own lines was 1000 lire.

Mecozzi got another confirmed victory on the first day of the mid-June offensive. In the morning of 15 June Mecozzi flew over the Piave with *cap.* Riva, his squadron commander and personal friend, and found a Brandenburg over Maserada. Both pilots had to attack intermittently, as they both had problems with their guns, until finally the aircraft crashed in the river, in front of Cimadolmo. Mecozzi added to his combat report: "The locality of this combat and of the crash is the one where the land battle is most intense. Machine guns from the ground fired at me. "

19 June was another hectic day for the Roman pilot, who together with *cap.* Alberto Moresco claimed two airplanes shot down between Villa Jacur and Casa Mercadelli and a kite balloon at Barco. Witness to the first combat was Lieutenant A.G. Cooper, RAF No. 28 Squadron, because the two Italians drove off an Austrian airplane that had got behind his tail:

> "While on offensive patrol in the locality of the Montello I saw 8 enemy machines, led by a Scarlet coloured D.5. attacking some Italian 2 seaters and Scouts. During the fight I observed two Italian machines drive down two D.3's one of which was attacking me while I was engaged in combat with another enemy scout. I observed the two E.A. which were driven down by the two Italian scouts, crash. One about 1 Kilometer West Nervesa, and the other one, on the other side of the River. Signed: A.C. Cooper Lieut. No.28 Squadron R.A.F."

"Catone" felt he had dispelled his supposed curse. In the morning of 27 July he scored another victory: "I saw the enemy airplane coming in from Moriago so I kept just below the rim of the clouds and I let it cross the lines, following it at a lower altitude. Then it noticed me and turned back, but I cut its path and engaged it, following it down shooting. " The enemy was described as a "large and fast reconnaissance airplane " which fell "close to a farm, Casa Musil, I believe, not far from Sernaglia."

There were more combats in October. On 6 October Mecozzi was flying withg *cap*. Riva and *serg*. Lino Gandini, and the three Italians engaged a formation of seven "D.5" fighters escorting a two-seater. In the tough combat Mecozzi's fighter was hit, but two enemy fighters were claimed as shot down, one near Susegana and the other one at Col Fosco. We have the combat report of the wingman, *serg*. Gandini:

> "In a patrol with my Captain and with ten. Mecozzi we spotted among anti-aircarf bursts and we attacked a group of eight enemy aircraft. The combat was very violent for the reaction of the enemy fighters and it lasted all along the way from Mount Salvor to Susegana and it ended with the fall of two fighters. .. Two of the enemy fighters, the more aggressive ones, were marked with the insignia K and Y. The hardest part of the dogfight was above the north-eastern slope of the Montello."

Insignia with capital letters could indicate Flik 56/J, but no losses for this Austrian unit are recorded on that day. On the other hand, two pilots of that unit, *Korp*. Podpetschan and *Korp*. Podzena, each claimed a fighter shot down in flames over the Montello, which could be the Hanriot of Mecozzi that was hit. In any case it is likely that this is yet another case of mutual overclaiming.

In a separate combat also on 6 October Mecozzi, either at 7 a.m. or at 7 p.m., shot down a two-seater behind the castle of Susegana. If the right time was 7.p.m. his victim could be a Phönix of Flik 57/Rb whose pilot jumped with a parachute but who was killed when it failed to deploy properly.

For his service as a fighter pilot, Mecozzi received just one award, a Medaglia d'Argento which in its citation credited him with more victories than the five ones admitted by the post-war "Bongiovanni Commission":

> "Fighter pilot of high valor, he gave constant and admirable proof of a high sense of duty, daring and strong will. On 26 May 1918, in collaboration with another pilot of his squadron he attacked nine enemy aircraft and after a long and tough combat he shot down two of them, bringing his total of victories to six. – Sky of the Piave, 1 November 1917-26 October 1918."

After the war Mecozzi together with other chosen pilots was a member of the Italian aeronautical mission to Villacoublay, France, a particularly fit destination considering his scholarly character. He returned to Italy in October 1919. He was chosen as a pilot in the patrol of Ansaldo SVAs that were to fly from Rome to Tokyo, and that took off on 11 March 1920. While only two pilots, Guido Masiero and Arturo Ferrarin, reached their destination, Mecozzi lacked good luck, not capacity. His adventure ended on 4 April 1920 when his SVA serial 13200 crashed during take-off from Aleppo, Syria, after hitting a metal tube filled with concrete hidden in the tall grass, probably an act of sabotage. He sailed back home and was assigned to the Direzione Sperimentale d'Aviazione, the experimental establishment at Montecelio, (later re-named Guidonia), where he tested many German and Austrian airplanes delivered to Italy as war booty.

Serg. Codeghini and Capparucci were the protagonists of a daring rescue from behind the enemy lines in the final days of the war. Here they recreate it for the camera with the Hanriot of *serg*. Codeghini.

On 14 July 1921 he married in Rome miss Lucia Taroppi. He remained in the air force, since 1923 the independent Regia Aeronautica, and he became the protagonist of a heated debate on the doctrine, and almost the philosophy, of the air force, opposing the strategic theory of "Dominion of the air" of Gen. Douhet as a paladin of a tactical air force, with his theory of *"Tendenza dell'Assalto"* the trend of ground attack. Countless pages were written in this debate, trying to determine what an air force should, or could, do. According to Mecozzi, an opponent of attacks against

Mecozzi getting out of a SVA during a post-war air race, the Baracca Cup of 1921.

In illustrious company: Mecozzi during an air exercise. Second from the left is Air Minister Italo Balbo, then General Giuseppe Valle, Chief of Staff of the Air Force, General Lombard and Major Mecozzi.

At Rome's Ciampino airport Mecozzi discusses his countless experiments in ground attack with General Valle and German General Milch. The decrepit Ansaldo AC.3 shows camouflage and a device attached to its landing gear.

civil populations, aviation should be used in the tactical field, destroying enemy capacities close to the frontline, and not in the "masses" that Douhet thought should be unleashed in a strategic concept against the vital points of the enemy nation.

Mecozzi accompanied his writings with a daily experience as commander of units specialized in ground attack. In October 1929, with the rank of *maggiore*, he lead first the VII Gruppo and then the 5° Stormo at Rome Ciampino. He shaped these units, equipped first with the Fiat CR.20 biplane and then with the Ansaldo AC.3 (license-built Dewoitine) high-wing metal monoplane fighter, that remained in service for a decade, up to 1936-37. Mecozzi organized countless experiments and tests of equipment, while still enjoying the beauty of flight. During the huge "Giornata dell'Ala" air show at Rome, on 8 June 1930, he flew as a soloist through smoke rings drawn in the sky by his pilots.

In April 1937 he rose to the rank of General, and in that same year he went very briefly to Italian East Africa, but somehow his career had entered a blind alley, and he was removed from the command of combat units. When Italy entered World War II he was appointed president of the R.U.N.A., Reale Unione Nazionale Aeronautica, the Aero Club of Italy. He was then chief of the Publishing Department of the Air Ministry. When Italy asked for an armistice he kept his nerves, hiding money and documents from the German occupants, and leaving his home. He kept in hiding, changing his lodgings frequently, and did not join the Repubblica Sociale of Mussolini.

As soon as the Allies liberated Rome, Mecozzi reported to the command of Aviation to resume service. He was now 53 years old, and on 5 December 1945 he was discharged on account of a physical inability due to his service. After the war he was in charge of the Associazione Culturale Aeronautica which published the magazine Rivista Aeronautica, which is now the official magazine of the air force. He had the unusual habit of going to work dressed in flying overalls.

In 1970, on the occasion of the 50th anniversary of the Rome-Tokyo flight, he finally reached the capital of Japan, not as a pilot of a SVA but as a passenger in a DC-6 of the Aeronautica Militare. He died of a sudden illness on 2 November 1971.

## Amedeo Mecozzi
## 11 victory claims, 5 confirmed victories

| Claim | Vict. # | Date | Sqn. | Aircraft | Hour | lace | Victim |
|---|---|---|---|---|---|---|---|
| 1 | 1 | 26.12.1917 | 78 | Hd | 8h40 | Volpago | German two-seater |
| *[with serg. Benvenuti, 82ª Sq.] Possibly: Lt. Kessler WIA/Lt. d. Res. Edebohls KIA, FA 2, DFW C.V or: Uffz. Willi Petzold KIA/Lt. d. Res. Wilhelm Voigt KIA, FA 2, DFW C.V or: Vzfw. Karl Uecker KIA/Lt. d. Res. Heinrich Pfeiffer WIA, DOW, FA 2, DFW C.V or: Vzfw. Pohlmann WIA/Lt. Schlamm WIA, FA (A) 219, DFW C.V* | | | | | | | |
| 2 | 2 | 26.5.1918 | 78 | Hd 11435 | 8h30 | Virago | Two-seater |
| *[with serg. Capparucci, 78ª Sq.] Unknow a/c fallen behind Italian lines* | | | | | | | |
| 3 | 3 | 26.5.1918 | 78 | Hd 11435 | - | S. Michele/Ormelle | Fighter |
| *[with serg. Capparucci, 78ª Sq.] Maybe: Zugsf. Ferdinand Udvardy, Flik 42/J, Alb. D.III FTL* | | | | | | | |
| 4 | 4 | 15.6.1918 | 78 | Hd | 9h55 | Cimadolmo/Maserada | Two-seater |
| *[with cap. Riva, 78ª Sq] Possibly: Korp. Franz Hyks KIA/Oblt. Alexander Gherardini WIA, Flik 62/D, Br. C.1 69.61 or: Fw. Alfred Brandt/Oblt. Karl Hiemesch WIA, Flik 19/D, Br. C.1 369.69 or: Zugsf. Wilhelm Watzl MIA/Lt. Sylvester Walitza MIA, Flik 5/F, Br.C.1 369.126..* | | | | | | | |
| 5 | - | 19.6.1918 | 78 | Hd | 11h20 | Villa Jacur | Fighter |
| 6 | - | 19.6.1918 | 78 | Hd | 11h20 | Villa Jacur | Fighter |
| *[with cap. Moresco, 78ª Sq.]* | | | | | | | |
| 7 | - | 19.6.1918 | 78 | Hd | - | Barco | Balloon |
| *[with cap. Moresco, 78ª Sq.] Maybe: kite balloon, BK 20.* | | | | | | | |
| 8 | 5 | 27.7.1918 | 78 | Hd | 10h18 | Casa Musile/Sernaglia | Two-seater |
| *Fw. Ferdinand Czerwenka UNH/Lt. i. d. R. Karl Zinner WIA, DOW, Flik 52/D, Br. C.1 169.141* | | | | | | | |
| 9 | - | 6.10.1918 | 78 | Hd | 15h50 | Auggio valley | Fighter |
| 10 | - | 6.10.1918 | 78 | Hd | 15h50 | Col Fosco | Fighter |
| *[with cap. Riva & serg. Gandini, 78ª Sq.]* | | | | | | | |
| 11 | - | 6.10.1918 | 78 | Hd 11411 | 7h? 19h? | Susegana | Two-seater |
| *Fw. Karl Kallasch MIA/Oblt. Johann Dopita MIA, Flik 57/Rb, Phönix C.1 121.61.* | | | | | | | |

# Giorgio Michetti

Giorgio Michetti was a rather peculiar ace, because he is best remembered not so much for his undeniable qualities of fighter, but for his merry and playful character. World War I pilots who wrote their memories didn't fail to describe his personality and the pranks and tricks that he organized. Giorgio Michetti was born on 29 May 1888 at Francavilla a Mare, a beautiful town along the southern Adriatic coast, near Pescara. His father was the painter Francesco Paolo Michetti, much appreciated at that time and at the peak of his fame when Giorgio was born. In 1871 and 1875 he had held exhibitions in the Salon of Paris and he was much admired in France. He experimented in different techniques, and finally became an admirer of Japanese visual arts. Since 1877 he was honorary professor at the Institute of fine arts of Naples. In 1880 he met Gabriele d'Annunzio and with him and musician Tosti he formed the Trio of Abruzzo, a group of co-operating artists. The great poet lived a long time in the convent turned into a home and workshop of Michetti, and he even wrote some of his works there. The place is now a wonderful museum. In 1900 Michetti partly abandoned painting and concentrated on a new form of art, photography. In 1909 King Victor Emmanuel appointed him Senator of the Kingdom.

The influence of such an important father was great on young Giorgio, who grew up and was educated quite differently from other young men at that time. His father wanted him to grow up independently, getting his experiences from the school of life, so his three children, Giorgio, Alessandro and Aurelia, never went to school but were privately educated learning, according to legend, nothing but foreign languages and the works of Dante. Growing up in an environment dominated by art, Giorgio was not at all illiterate. He is remembered as a man of high culture, with a biting spirit, a rebel against self-righteousness, bureaucracy, and dull normalcy.

His military career, which was opposed to the radical anti-militarism of his father, began on 2 November 1905, when he was enrolled in the 13° Reggimento Artiglieria. In the following years as a sergeant he passed to other regiments, and on 31 December 1914 he was in the Milizia Mobile of the 34° Reggimento. Just before Italy entered the war, he was called back to the 27° Reggimento Artiglieria, but he remained there briefly because he was fascinated with aviation and on 1 August 1915 he managed to get accepted in the Battaglione Aviatori, the air service.

His training process went smoothly, he received his pilot license at the Gabardini flying school in Cameri and on 16 March 1916 he was posted to the 41ª Squadriglia da ricognizione, a reconnaissance unit equipped with twin tailboom Caudron G.3 tractor biplanes.

Not much is known about his war service with this squadron, as his personal papers are missing. To fly up and down the frontline with a slow Caudron, always flying straight so that the observer may do well his job and at the same time accepting patiently the shells of the accurate Austrian anti-aircraft guns surely was not the dream of Michetti. His explosive character wanted something more than piloting an unarmed two-seater, and yet it was a time of meaningful learning. A that time the function of a reconnaissance crew was not fully appreciated, and there were many skeptics who still had doubts about aviation, like an artillery colonel who told him:

> "When I ride on a train and look out of the window at the rocks along the tracks or the hedges that border the railway, I cannot tell apart one rock or one bush from another one, and yet I am a few feet away and the train is running at a speed of 40 km per hour. And then how can you claim that you can distinguish batteries, roads, even moving troops from an alti-

Giorgio Michetti in the cockpit of his Hanriot Hd.1 serial 6254, Casoni airfield, spring 1918. His personal seahorse insignia appears also on the horizontal stabilizers, as was common practice in the 76ª Squadriglia.

tude of 2,000 meters, and flying at a speed of over 100 km per hour?"

What Michetti really couldn't stand were the missions of stereoscopic photo reconnaissance. Stereoscopic photos were very accurate and provided all details of the terrain but in order to be successful they required straight and level flight, and this of course was what the Austrian gunners needed to target effectively an airplane above. Those missions were flown at low altitude, and one can imagine the feelings of a crew in the flimsy nacelle of a slow Caudron. It was the rule that an airplane returned from such missions with some damage, and then Michetti would comment his usual way, cursing whoever had invented stereo photography.

*"A bizarre spirit"*, thus with a few effective words Silvio Scaroni described him, who later became Michetti's best friend and Italy's second ace.

In early 1917 Michetti was finally commissioned as a *sottotenente*, and then on 11 June he was assigned to the flying school at Cascina Malpensa for conversion on fighters. He completed that course with ease, and on 29 July 1917 he was posted to the 76ª Squadriglia da caccia, then based at Borgnano, very close to the Isonzo front. That squadron was one of the top Italian fighter units; its new commander was *magg.* Calori and it included the famous ace Flavio Torello Baracchini, who was badly wounded in combat a few days after the arrival of Michetti, when his score stood at 13 victories. Two months before another six victories ace, *ten.* Olivi, had been killed in an accident. In that period the 76ª Squadriglia was assigned to a new command, the Sottogruppo Aeroplani Borgnano and it received a new commander, *ten.* Gastone Novelli, also a future ace. It had mixed equipment of Nieuport 11s and a few Spad 7s, and since 18 August 1917 it started receiving the first French-built Hanriot Hd.1 fighters. The ones built in Italy by Macchi started reaching the units only a few months later. As was customary, the Hanriot had its teething troubles, particularly the landing gear, that was too weak, and initially there were not enough spare parts, so they were not used very much at first.

Michetti flew his first fighter sortie on 20 August escorting a formation of Capronis. It was the period of the offensive called the 11th Battle of the Isonzo, that was fought from 18 August to 12 September and that brought the conquest of the Bainsizza plateau. He had his first combat on 25 August, which a gun jam cut short, and the Austrian plane got away. Michetti flew very much, once or twice each day, but he had only three other air combats, all of them fruitless, until 1 October, when he went on leave. On 4 October he was promoted to the rank of *tenente*.

On 24 October 1917 the joint attack of Austrians and Germans broke the Italian front. It seems that during the tragic days of the retreat Michetti didn't fly, or at least no mission was recorded, but those were confused days, the Italian Squadriglie were forced to burn their hangars and airplanes and to join a stream of soldiers and refugees who were moving without an aim, without orders. During the retreat Michetti found his good friend Silvio Scaroni at la Comina air base, as he was looking for somebody to tell him which squadron to report to. Michetti quickly solved his problem and got him accepted immediately in the 76ª Squadriglia, which meanwhile had moved back to Arcade.

German and Austrian airplanes were operating all over the battlefield, there was plenty of action, but Michetti was not lucky, he had four combats, but three times his gun jammed again and he was forced to turn back.

Then came 26 December and the large attack of German bombers against Istrana airfield, then the base of the 76° Squadriglia and of many other units. According to reports, Michetti was the first one to climb into his Hanriot, sporting his insignia of a sea horse, and to take off while bombs were falling down and the German crews were strafing anything that moved on the field. In the confused combat that followed, he managed to get behind a German DFW C V and, together with other fighters, he shot it down within the Italian lines, near Camalò. Scaroni described the final phase of this combat:

Below: A line-up of fighters of the 76ª Squadriglia at Casoni, summer 1918. The Hanriot of Michetti is the fourth from left. It shows its personal insignia with inverted colors, to show up against the dark camouflage finish.

*Capitano* Alberto De Bernardi, commander of the 76ª Squadriglia, flew this Hanriot emblazoned with an eagle and a hunting horn, referring to the symbols of the Alpini mountain troops to which he belonged.

"A lightning zooms over my head and aims straight at my adversary: I recognize Michetti. Good, Giorgio, come here, there is plenty to do for another one too. The sooner we dispatch it the sooner we can get at the other ones, which are now turning back, to go home. Michetti fires and misses, I go after him and I miss too; the enemy flies straight to home safe and sound. I go down once again and then Michetti a second and a third time. The enemy waits until the fighter is real close and then opens fire at close range. This business is getting serious, it seems that our bullets and the German ones too are made of clay, nobody manages to place a final strike and nobody gets hit... As I bank to get again in firing position the enemy observer, standing up in the turret ring, shoots a whole burst at me from a few meters. By instinct I shut my eyes: it's impossible to miss at this close range, but he must be very excited, because he aims badly once again... I see him quickly unfastening a belt of bullets wound around his waist and inserting it in the machine gun, and now he 's ready to shoot again. He must be an expert in this movement, as he did it in the quick moment when I couldn't fire. The guy knows his business. ... Michetti and I are convinced that we are dealing this time with a couple of experts: this crew has shown us a hundred times how to avoid the fire of machine guns, even our well aimed fire."

Finally with more attacks they brought it down and it crashed and turned over in a field near Camalò. The observer got out of the wreck unscathed then burned the airplane with the pilot still inside, and getting burned himself, as flames spread to his clothes. He was, they reported, a German Major, and when they visited him in hospital he firmly stated to the two Italians that he was a victim of British fighters. Scaroni and Michetti thought that this was just German arrogance, but probably he had identified in good faith their Hanriots as Sopwith Camels.

Relations with the British pilots also based at Istrana with the 76ª Squadriglia were excellent, and Michetti talking excellent French extolled the "qualities" of local girls to the appreciating Englishmen. That winter's snow briefly put a halt to flight activities but only by day, because in the night enemy bombers regularly raided Italian airfields, keeping everybody awake and causing considerable damage.

On 12 January 1918 Michetti together with Scaroni and *serg.* Gastone Razzi had another duel over Mount Cismon, that was not at all easy, Michetti alone firing 350 bullets. The enemy was forced to land behind its lines and was wrecked.

On 12 March he had a peculiar accident. His favorite pastime was making jokes of his friends, he could never keep his mouth shut especially if he could make fun of *"Cato"* Mecozzi, famous for his chronic bad luck and then, after Mecozzi was shifted to the 78ª, of his other companions such as Retinò who had long and curly hair (Michetti was almost bald) or Donati who was so small that when he took off it looked as if the Hanriot was flying with nobody in the cockpit. On 12 March the tables turned. Landing back together with Scaroni after another fruitless patrol, he banked sharply, gestured to his mate in a very impolite way and then he landed so suddenly that he didn't realize that in the other direction there was a SIA 7b of the 22ª Squadriglia coming in to land. The two air-

planes collided in the middle of the runway, spreading cloth and wood all over, totally destroyed. Scaroni landed and found his friend Michetti absolutely unhurt, who was cursing in his native dialect the pilot of the two-seater. Luckily none of the three airmen (the observer of the SIA was *cap.* Lodi, the squadron commander) was hurt, according to Scaroni. According to archive papers Michetti did have some wound, which didn't stop him from flying.

Five days later Michetti was back in the air, still escorting the SIA of the 22ª, he was not officially disciplined but he paid a heavy price in terms of jokes: "See George, you finally shot down an airplane, in our territory...and all alone too...Baracca couldn't do better... this one is going to be confirmed for sure..." His friends kept rubbing salt in his wounds for a long time concerning this misadventure.

Michetti used to go out on patrol with his friend Scaroni, who wanted to shoot down an enemy reconnaissance airplane which each morning, at the same time, appeared over the lines in the area where his brother Annibale was based, commanding a gun battery. So each day, before sunrise, Scaroni dragged a skeptical Michetti out of bed, forced him to dress up in his warmest clothing and to take off heading for the mountains. But for some unknown reason when they were ready for their ambush the Austrian plane didn't show up, and Michetti solemnly promised that it was the last time that he got up so early to look for the "ghost airplane" produced by the sick fantasy of the Scaroni brothers. But the next day he reneged on his promise and took off again with his friend for one more patrol over the Grappa. On 21 March, during one such hunt for the ghost plane, they found instead six Austrian fighters heading for an Italian kite balloon.

They followed them, and when the lowest fighter was about to attack the balloon the two Italians pounced the enemy formation. Scaroni shot down in flames the leading fighter just as it was about

Fighters of the VI Gruppo, 76ª and 81ª Squadriglia, at Casoni airfield.

The Hanriot of *serg.* Raimondo Di Loreto.

to attack, while Michetti maneuvered crazily and so confused the five escorting fighters that they quickly turned back. They were fighters from the famous Flik 55/J, one of the best Austrian units. They reported a combat with *"Sopwiths"* and *Feldwebel* Behounek even claimed one of them as probably shot down, although both Hanriots of the 76ª safely returned home. As was then customary, both pilots were credited with this victory. Their victim was captured totally unharmed, and spent the rest of the war in prison camp.

On 3 May 1918 Michetti claimed another victory, as together with Baracchini he shot down an enemy airplane near Sernaglia, having fired 150 bullets. From official documents however it seems that he was not credited with this victory. On 22 May again an enemy reconnaissance airplane fell under the guns of Michetti and Scaroni despite the presence of four escort fighters. Michetti had another combat the next day, over the valley of Alano, but the enemy outran him and returned to its own lines.

In June again Michetti flew as many sorties as possible, but he failed to meet enemy aircraft until the great Austrian offensive began. In the night of 14 June, from a makeshift observation post he watched with Scaroni thousands of enemy artillery pieces throwing up their barrage, all the way from the Asiago plateau to the sea. Something had to be done, and they decided to take off way before sunrise, and Scaroni in the darkness barely missed an Italian balloon. As the sun came up, flying over the Grappa they spotted an enemy fighter crossing the Piave and going to attack the balloon that they had just missed. They went after the fighter in a long dive, the Austrian saw them and tried to turn back and get away, but they were too fast, and they quickly sent it down in flames to crash on the banks of the Piave River, south of Moriago.

Michetti then had some fruitless combats, in the afternoon of 15 June, on 18 June and then the next day, when over Nervesa an Austrian observer placed an incendiary bullet in his seat, which was still burning when he landed. Looking for revenge, in the afternoon he went after the Austrian kite balloon at Susegana, but once more he was on the losing side, and returned with his wings riddled with bullets. That was the hardest day for the Italian army, and all the air force was in the air bombing and strafing the Austrians with all available aircraft. Michetti himself flew four combat missions that day.

On 24 June, flying with Scaroni and *cap.* Lega he attacked an Austrian two-seater that crashed in flames near the tomb of the sculptor Canova at Possagno, within the Italian lines. Both occupants died, the body of the pilot was found beheaded in a tree, close to the wreck of the airplane. The next day the Italians dropped a message with the documents of the fallen enemies on Feltre airfield.

When the huge battle was over, Michetti kept on flying in routine sorties. On 14 August, flying with the trusted *sergente* Romolo Ticconi as a wingman, he shot down an enemy fighter near Rasai, in the Val Stizzone, but that was a rare occasion, usually he failed to meet enemy aircraft, and he usually flew escorting airplanes that were dropping leaflets to undermine the morale of the multi-ethnic Austrian army. Michetti, just for fun and to break the routine, once also dropped on their trenches a sack full of mice "captured" at

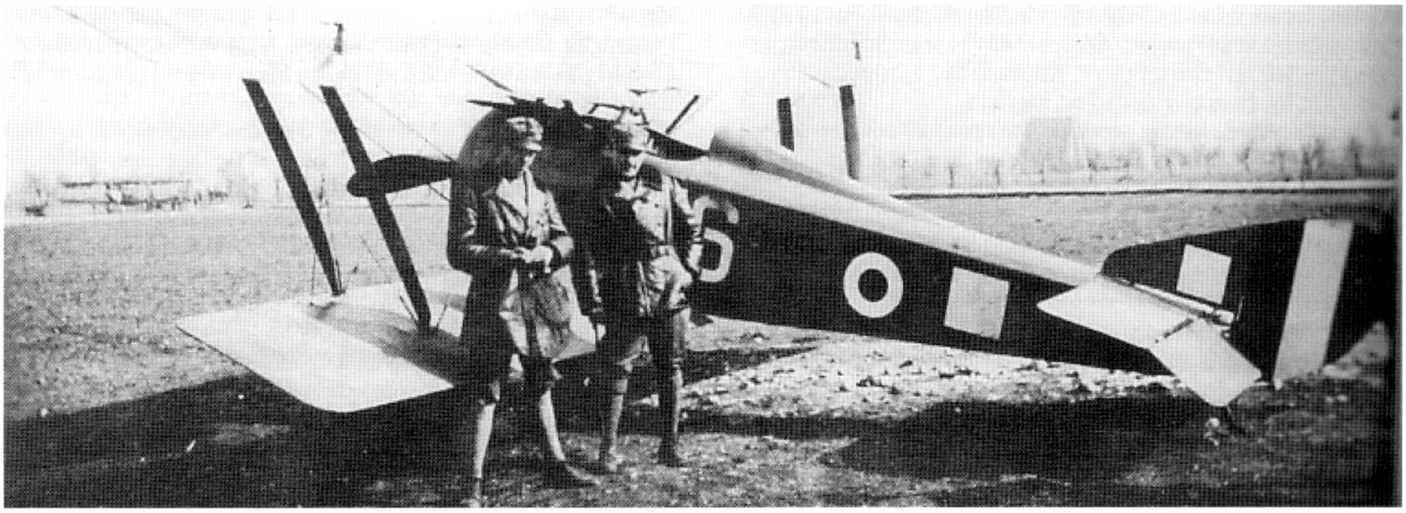

Michetti and Scaroni in front of Hanriot 6239 – Scaroni's personal mount.

Casoni airfield, although there are no details of how he fitted those creatures in the narrow cockpit of his Hanriot.

On 3 August he had been posted to the Commissariato Generale d'Aeronautica, the political agency responsible for aviation, but he managed to delay his departure, and his final war mission was logged on 1 September, a patrol over Alano and Valstagna. On 24 September Michetti was posted to the *Scuola di Tiro Aereo* (school of aerial gunnery) at Furbara, near Rome. It was a large training airfield, which many American student pilots also attended, and expert instructors were needed to prepare new fighter pilots, as the war was expected to last until 1919. The surrender of the Central Empires came sooner than expected, and Michetti didn't return to the front. Globally during his service as a fighter pilot he had flown 209 combat sorties and had 21 air combats, being credited with five confirmed victories. He was awarded two Medaglie d'Argento, one for his participation in the air battle of 26 December 1917, and the other one for the total of his activities, which in its citation reports his six victories of 12 January, 21 March, 3 May (this one actually not confirmed), 22 May, 15 and 24 June 1918.

In February 1919 Michetti was assigned together with Scaroni to the 350ª Squadriglia, a special unit forming the Italian Aeronautical Mission to Argentina, to foster the sale of Italian aircraft in Latin America. The elite pilots of the Mission flew a variety of trail-blazing flights with their Ansaldo SVA, Balilla and flying boats, and Michetti himself flew a raid from Buenos Aires, Argentina, to a Montevideo, Uruguay. He then sailed back to Italy and on 3 October 1919 he was sent on permanent leave.

He moved to Rome, and for a period he was engaged in political activities. As a war hero and the son of a famous artist, the high society of the capital warmly welcomed him. He did not pursue a military career, as others, like Scaroni, did, because he understood that it obviously was not his vocation. In 1925, following his father's footsteps, he was one of the founders of the Istituto Luce, an Italian state agency for photography and movies, which still exists and has carefully preserved a large and precious collection of aviation images and movies.

On 19 July 1931 Michetti was dropped from the rolls of army officers and he was placed as a captain of the reserve in the rolls of the Regia Aeronautica. He still did fly, because papers show that he had refresher flights on tourism airplanes in 1932, 1933 and 1938. He rose in rank to *tenente colonnello*, but we don't know how he served in World War II, if at all. In June 1942 he was moved in the reserve, and he was placed on permanent leave on 16 November 1954.

He married twice, once with a lady of the aristocracy of Venice, but very little is known of his later private life. Giorgio Michetti died in Rome on 4 February 1966.

Right: Pilots of the Italian aeronautical mission to Argentina, 1919. Top row, Gaio, an airship pilot, *STV* Luigi De Riseis, Royal Italian Navy, Giorgio Michetti, Antonio Locatelli. Sitting: Enrico Burzio, the commander *cap.* Enzo Giovannardi, unknown, De Luca and Silvio Scaroni.

Another Hanriot showing the checkers markings of Di Loreto, with silver dope finish – in an embarassing position.

## Giorgio Michetti
## 8 victory claims – 5 confirmed victories

| Claim | Vict. # | Date | Sqn. | Aircraft | Hour | Place | Victim |
|---|---|---|---|---|---|---|---|
| 1 | 1 | 26.12.1917 | 76 | Hd | 9h | Camalò | Two-seater |

*[with s.ten. Scaroni & Fucini, 76ª Sq., cap. Riva, 78ª Sq., serg. D'Urso, Teobaldi, Benvenuti, ten. Avet & Comandone?, 82ª Sq., 2nd Lieuts. Jarvis, Mulholland & Frayne, No.28 Sqn. RFC] Possible victims:*
*Lt. Kessler WIA, POW/Lt. d. Res. Edebohls KIA, FA 2, DFW CV or:*
*Uffz. Willi Petzold KIA/Lt. d. Res. Wilhelm Voigt KIA,, FA 2, DFW CV or:*
*Vzfw. Karl Uecker/Lt. d. Res. Heinrich Pfeiffer DOW, FA 2, DFW CV or:*
*Vzfw. Pohlmann WIA, POW/Lt. Schlamm WIA, POW, FA(A) 219, DFW CV*

| Claim | Vict. # | Date | Sqn. | Aircraft | Hour | Place | Victim |
|---|---|---|---|---|---|---|---|
| 2 | - | 12.1.1918 | 76 | Hd | - | Cismon | Two-seater |

*[with ten. Scaroni & serg. Razzi, 76ª Sq.] Zugsf. Josef Passer UNH/Lt. i. d. R. Heribert Fraess Ritter von Ehrfeld UNH, Flik 16/D, Av. C.I 114.04 FTL*

| Claim | Vict. # | Date | Sqn. | Aircraft | Hour | Place | Victim |
|---|---|---|---|---|---|---|---|
| 3 | 2 | 21.3.1918 | 76 | Hd | - | Cascina Zocchi | Fighter |

*[with ten. Scaroni, 76ª Sq.]Korp. Gottlieb Munzar POW, Flik 55/J, Alb. D.III 153.100*

| Claim | Vict. # | Date | Sqn. | Aircraft | Hour | Place | Victim |
|---|---|---|---|---|---|---|---|
| 4 | - | 3.5.1918 | 76 | Hd | - | Sernaglia | Fighter |

*[with ten. Baracchini, 76ª]*

| Claim | Vict. # | Date | Sqn. | Aircraft | Hour | Place | Victim |
|---|---|---|---|---|---|---|---|
| 5 | 3 | 22.5.1918 | 76 | Hd | - | Quero | Two-seater |

*[with ten. Scaroni, 76ª Sq.]*

| Claim | Vict. # | Date | Sqn. | Aircraft | Hour | Place | Victim |
|---|---|---|---|---|---|---|---|
| 6 | 4 | 15.6.1918 | 76 | Hd | 6h15 | south of Moriago | Fighter |
| 7 | 5 | 24.6.1918 | 76 | Hd | - | Canizza | Two-seater |

*[with cap. Lega & ten. Scaroni,76ª Sq.] Fw. Franz Pachta KIA/Oblt. Wilhelm Persoglia KIA, Flik 2/D, Br. C.I 369.112*

| Claim | Vict. # | Date | Sqn. | Aircraft | Hour | Place | Victim |
|---|---|---|---|---|---|---|---|
| 8 | - | 14.8.1918 | 76 | Hd | - | Rasai | Fighter |

*[with serg. Ticconi, 76ª Sq.]*

# Guido Nardini

Among all aces Guido Nardini is one of the least well known, despite the fact that he spent all his life in aviation. Even his personal papers were lost in the archives of the air force, and part of his career has to be reconstructed through guesswork. He was born in Florence on 30 July 1881, son of Giovanni and Natalizia Guidi. He was christened with three names, Guido Giulio Cesare, and some books in the past quoted a pioneer airman Giulio Nardini, creating the confusing impression of two different pilots.

Nardini went to France and became a pilot there, receiving pilot's license number 590 at Bétheny aerodrome, sometimes in the summer of 1911. He then obtained the Italian license number 128. He was to all effects a member of the wonderful and exalted circle of the pioneers of aviation. He remained in France a rather long time, and in 1912 he was chief pilot of a company called *"Navigation Aerienne"* of Paris, and he distinguished himself as the first Italian to cross the Channel by air, on 12 May 1912. When the European war broke out he returned to Italy, and on 24 May 1915 he volunteered in the Battaglione Aviatori. In those times those few citizens who could pilot an airplane, or even drive a car, could volunteer as privates in the Army, that welcomed them and placed them in a special roll. Luigi Olivari was one of them. Likewise the Navy had its "volontario motonauta", owners of motorboats who patriotically offered them to the state. This system was discontinued in 1916, the volunteers becoming regular service personnel, most of them quickly converted to sergeants or officers. Nardini at that time was 34, much older than other pilots, and certainly more expert than most of them. He still had to get his regular military pilot license, but this process was obviously very quick, and he completed it by 1 October 1915, flying a Bleriot monoplane, probably at Mirafiori.

More time had to pass before he could be assigned to an operational unit, maybe in this period he was an instructor or he tested airplanes. In any case he converted to the new Nieuport fighters arriving from France. On 26 May 1916 he reached his unit, the 75ª Squadriglia da caccia. This was a new unit formed a few days before and operating from Tombetta airfield, a suburb of Verona, with commander *capitano* Matteo Scarpis. Its main task was the protection of that city, where many commands were located, and that was also a center of distribution of material to the front. It was equipped with the Nieuport 11, an excellent fighter license produced by Nieuport-Macchi of Varese, which had no comparable Austrian fighter opposing it.

Air activity in that sector was very limited: Verona was 45 km from the frontline, and the closest enemy air base, Pergine in Valsugana, was 70 km away. Still, every so often some Austrian plane would fly a reconnaissance over the city to spot railroad movements and to use the opportunity to drop a few bombs, with minimal military results, but of considerable psychological effect for the civilians. Thus when the Nieuports of the 75ª began to intercept Austrian airplanes and to drive them away back to their lines, the whole city rejoiced.

Guido Nardini, a pioneer of aviation, in a signed portrait.

*Sottotenenti* Mario De Bernardi, the future great test pilot, and Alessandro Buzio were the first ones, on 11 May. On 16 May, as the cannon fired its alarm shots, Buzio, *cpr.* Consonni and *soldato* Nardini took off to intercept a raider. As the three Italians attacked it, the Austrian maneuvered violently and dropped its camera, that fell in a field in the city. The Military Division of Verona commended the three pilots for this action. In the following days Austrian raids continued, and they even attacked the airfield at Tombetta, a bomb falling near its hangars but producing no damage.

On 27 June 1916 Guido Nardini scored the first victory, together with Buzio, De Bernardi and Consonni. They attacked a Hansa Brandenburg C.1 of Flik 21 coming from Gardolo, near Trento and forced it down near Arsignano. Its pilot and observer, both wounded in this combat, were captured. Again the Commander of the Fortress of Verona commended them: "...I therefore pray the command of the 3° gruppo aeroplani to present to the four airmen, and particularly to pilot Nardini, my warmest praise." The victory was equally credited to all four pilots, but this message seems to indicate that Nardini was the main participant.

The destruction of the first enemy airplane raised the enthusiasm of the whole city. The four pilots received a prize of 1,750 Lire each, offered by the population, but they all turned over this sum to a charity for disabled veterans. Later on, 19 November, there was a big ceremony in which General Bompiani, commander of Verona, decorated them personally with a Medaglia di Bronzo al Valor Militare, issued in the field.

At the end of 1916 Nardini, still with the lowly rank of corporal, was reassigned to the 78ª Squadriglia Caccia at Istrana. Although operating in a more active role than the second-line 75ª, also the new unit did not offer many opportunities for combat on that northern sector of the front. On 8 January 1917 Nardini flew with his whole squadron as escort to the Caproni bombers.

On 1 March 1917 Nardini reached the rank of *sergente*. Probably in this period he was moved away from the front and assigned to conversion on the Pomilio, because on 4 May 1917 he was qualified as pilot for the Pomilio 200 hp, the new fast two-seater for which the best pilots were hand-picked, and on 18 May he was assigned to the first unit with this new type, the 131ª Squadriglia, but this assignment did not take place, and he returned to the 78ª Squadriglia. Some sketchy biography in the Thirties report that before being assigned to the 131ª Squadriglia he was an instructor at Coltano and then test pilot with the Bauchiero firm in Turin.

Almost one year passed before Nardini could score again. On 14 June 1917 as he was flying a patrol over the Asiago plateau and the Valsugana he engaged at 8.20 a.m. an enemy airplane over the upper Val d'Assa, and brought it to a forced landing. Soon after, at 8.40 a.m., he had another combat and another victory, bringing down an airplane that crashed in flames in the Valsugana. It is remarkable that he scored this double victory flying a Nieuport 11 serial 3213, with a 80 hp engine, a fighter that was competitive one year before, but which in 1917 was to all effects obsolete, slow and under-gunned. For this feat, Nardini was awarded a Medaglia d'Argento in the field, which remained his highest decoration, as his career was rather under-appreciated.

Likely his first victim was *Feldwebel* Schuster of Flik 21, who was badly wounded and went down in a glide to a crash landing near Cima Vezzena, destroying his Brandenburg KD 28.35 fighter. The crew of the second airplane, a Brandenburg C.1 of Flik 24, perished as the flaming plane crashed on the slopes of Mount Armentera. According to Austrian sources they had been flying a standing patrol in Valsugana and they had shot down a Nieuport (which Italian data don't report) before another one attacked and shot them down.

As the Italians were still using the obsolete Nieuport 11, the Austrians didn't fare any better: their only real fighter was the Brandenburg KD, available in small numbers, an unstable and demanding airplane, and more often they used, as in this case, the jack-of-all-trades Brandenburg C.1, an excellent aircraft but not a fighter.

The outstanding qualities of Nardini stood out in another hectic day of air combat, 18 July 1917. With the "indirect" cooperation of *serg.* Magistrini he shot down another Brandenburg of Flik 24 near Asiago, in Italian territory. The pilot was wounded and became a prisoner, the observer was killed and the Italians gave him military honors and a solemn burial, attended by the two pilots whom war had forced to cause his demise. More things had happened, which the Italians didn't realize: the Austrian two-seater had as escort two brand-new Albatros D.III fighters, which had just arrived at Pergine airfield, but their operational debut was unauspicious.

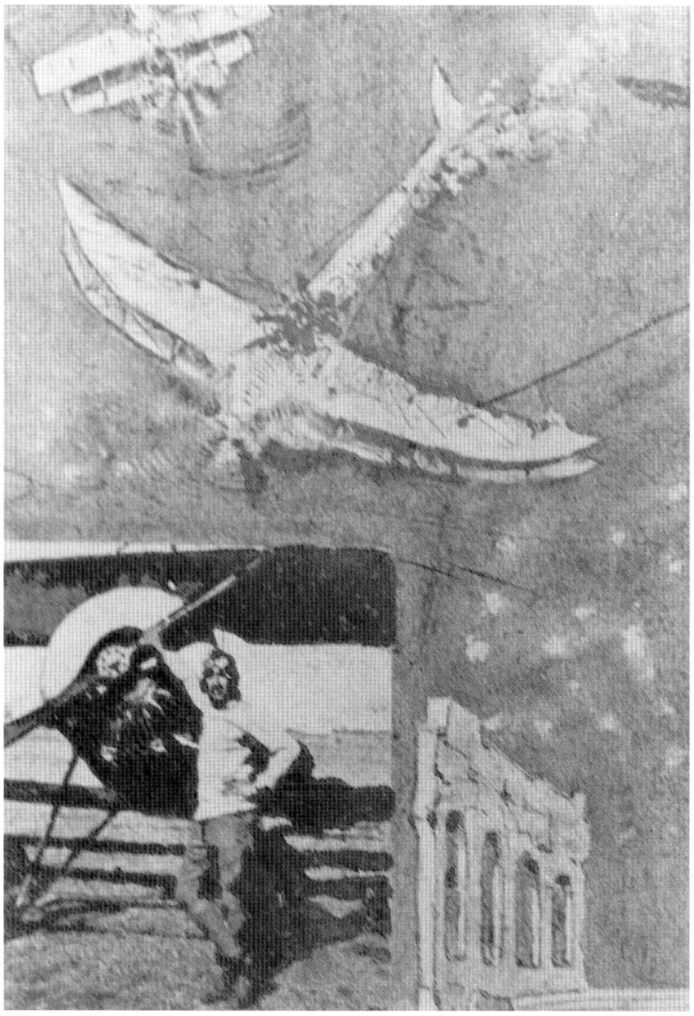

The destruction of the first enemy airplane over Verona on 27 June 1916 was a big affair, and this rather crude postcard was printed for the occasion.

The pilots reported an attack by two Italians, wrongly identified as Spads, and both were hit and force landed at their airfield, while their protégé was shot down. One of them was one of the best Austrian pilots, *Leutnant I.d.R.* Josef Friedrich, of Flik 24, a seven-victory ace according to the Austrians. Reconnaissance squadrons of the Double Monarchy unlike the ones of all other nations during the war usually had a mixed complement of two-seaters and a few single-seat fighters. As the war progressed the scarcity of aircraft forced the reduction of fighters assigned to recce units, in order to ensure a full complement of aircraft at least to pure fighter squadrons.

In the summer of 1917 the 78ª Squadriglia was still equipped with the Nieuport 11 and it kept them when it moved to the front of the Isonzo on 10 August, despite the fact that commander Bolognesi complained about the inadequacy of its aircraft. But when the unit settled at Borgnano airfield it received its first French-built Hanriot Hd.1, and then this excellent fighter completely re-equipped the 78ª.

Nardini painted on his own airplane a little devil making gestures, and maybe with this Hd he scored new victories on 6 September. Sources are scarce and contradictory on this event. In 1919 the magazine "Il Cielo" reported that Nardini shot down two out of

Nardini the aviation pioneer in a Deperdussin monoplane with Anzani engine, France 1912.

Guido Nardini with a Nieuport 17 of the 78ª Squadriglia.

a group of 15 airplanes bombing Gorizia, one in flames over S. Gabriele and another one over Ternova, and soon after he shot down another one, German. That press article may be, as often happened, over-inflated, but he did claim at least one victory. Nardini himself however was wounded, an incendiary bullet striking his thigh, and for many months he was away from the front. It's not clear when he returned to the front, possibly late December, anyway he didn't get back to the 78ª, he went to the "squadron of aces", the famous 91ª Squadriglia.

January 1918 brought little air activity for the squadron, but then things started to move. 10 February was a good day for Ranza and Bacula, who shot down two enemy airplanes, it was a bad one for Nardini. He took off for an engine test flight but, against his orders, he strayed away from the airfield, and when he finally came back his engine seized, and as a result Spad 7 serial 5807 was wrecked. Nardini, besides the scare, got a negative report and a fine of 100 Lire, a hefty sum for a sergeant in those years.

Maybe he didn't develop a feeling for the Spad, because one week later he went to S. Pelagio airfield to pick up Nieuport 27 serial 5888. Only two Nieuport 27s were assigned to the 91ª Squadriglia. Probably he got his own on a special request and it remained as his personal airplane. In the night of 21/22 February enemy bombers struck the hangar of the squadron, and no less than six Spad 13s and seven Spad 7s were written off, but not Nardini's Nieuport.

No airplanes were met during the month of March, as weather was bad and enemy activity was limited. In April bad weather again covered the airfield of Padua, reducing flight activities, until 11 April when the 91ª moved to a new location, Quinto di Treviso, closer to the frontline.

On 17 April good weather finally returned and Nardini at last had an eventful day. He took off at 12.30 with *serg.* D'Urso, a newcomer in the squadron, for an offensive patrol over the Piave, and an enemy fighter bounced him, fired a burst and then flew away to its lines. Soon after, at 1.10 p.m. he saw six fighters over Ponte di Piave and engaged them. He shot at one, but another one got in firing position and riddled his top wing, luckily with no further consequences. This was the only air combat of the 91ª during the whole month.

This combat was much embellished later on by the press which, with much fantasy, identified the enemy as the "Tango squadron" of Manfred von Richthofen and dubbed Nardini "the ace who dueled with Richthofen". Even in rubbish there may be a grain of truth. Probably what Nardini did was to clash with fighters of the famous Flik 41/J with its leader, the great ace Godwin Brumowski, who flew an Albatros all painted red, like the Fokker of the great German ace. Actually that day Brumowski had a combat exactly in that area and together with *Oberleutnant* Navratil he claimed a "Sopwith" shot down, his 31st victory that was confirmed by Austrian authorities, probably less strict than the cantankerous lt. Col. Piccio, the commander of Italian fighters (unless their victim was Camel B 6342 of 2nd Lt. W. G. Hargrave, Sqn. 28, shot down and killed that day).

On 3 May Nardini scored his fifth confirmed victory. He took off as wingman to *magg.* Baracca who described the combat in a letter to his mother:

> "At 11 I had been in a patrol for one hour with sergente Nardini; an enemy airplane had flown above us at 5,000 meters coming from Treviso but we were unable to reach it. There was an intense movement of airplanes across the lines. We bumped into an Austrian big fighter formation of six "Albatros"; they had the advantage of height and we attacked driving them off. Five minutes later another flight of five enemy fighters, we attacked one of them with brief sharp attacks not to let the other ones bounce us, it didn't seem to go down, maybe it was forced to land. They also went away and we found an isolated two-seater. Sergente Nardini attacked, as he was closer, but his

A very striking perrsonal insignia, a funny little devil for the Hanriot Hd.1 of Guido Nardini, French serial 18, Italian serial 6614.

A line-up of the fighters of Flik 61/J. Besides the third airplane, that is an Aviatik Berg D.I, all the other ones are Albatros (Oeffag) D.III. Nardini shot down the top ace of this squadron, Lt. Franz Gräser, on 17 May 1918.

gun jammed. Then I also attacked and after a burst the "Albatros" that was putting up a good fight, went down to 1,000 meters and waited again, accepting combat over the Grave di Papadopoli. I arrived at high speed attacking it again from the side, and the Albatros turned very quickly but the pilots probably were hit and after a last attack at 1,500 meters the enemy caught fire and crashed in the bed of the river, in front of our lines."

The gallant pilot and observer of the Brandenburg of Flik 19/D perished. The wrecked airplane remained there, on the Italian bank of the river, in full view of the enemy artillery. According to Austrian documents the two-seater had the escort of the fighters of Flik 51/J, possibly the ones that previously tangled with Nardini and Baracca. Their commander *Oberleutnant* Fiala von Fernbrugg claimed one Spad shot down during this mission, a victory that finds no confirmation in Italian data.

The next day in the morning Nardini with six other pilots flew to Verona to fly an escort mission for the Capronis going to bomb the electric power plant at Dro, but his engine refused to start and the pilot had a day of forced rest.

Air activity was now picking up and on 17 May the ace scored another victory. The 91ª sent off from Quinto di Treviso a patrol formed by four Spads, one Hd and one Nieuport 27 with *serg.* Nardini as pilot. The flight soon lost *serg.* D' Urso, who suffered an engine failure and returned to base. They found a group of enemy aircraft and got ready to attack. *S.ten.* Olivero and *serg.* Aliperta flew toward the Piave to engage some fighters while *ten.* Novelli, *serg.* Magistrini and Nardini turned to the other enemies that were flying over Italian territory. Nardini described the events in his combat report:

"I took off at 9 a.m. with tenente Novelli (leader) and other four pilots for a patrol between Quero and Ponte della Priula. I saw at about 10 anti-aircraft shots in the direction of Treviso. I went in that direction with the whole flight at an altitude of about 3,800 m. and at 10.10 a.m. I saw a patrol of several enemy aircraft about 200 meters higher than me. Tenente Novelli and sergente Magistrini who were at the same altitude of the enemy aircraft engaged a fighter that maneuvered to get free of their attack and went down to my altitude. I then attacked it and with a few maneuvers I managed to get behind its tail at a distance of 20 meters and I fired about 200 rounds of machine gun, after which I saw the enemy aircraft catch fire and crash vertically near Pero. Almost in that same moment a camshaft of my engine broke and the cowling flew off, and after a few seconds the engine seized completely, so I had to force land in a small field some 300 meters from the point where the enemy airplane had crashed. In landing my plane turned over."

His airplane was probably Nieuport 27 serial 5888, the only one then in force with the 91ª. Victim of Nardini was the enemy flight leader, *Leutnant* Franz Gräser, who was killed. At the time he was credited with 18 victories. Although not all of them can be confirmed from Italian data, he certainly was a very good pilot, and his loss was badly felt by the enemy airmen. When this victory had to be confirmed, all reports stated that Nardini was the real winner, as it had been his gun that had destroyed the enemy fighter, but also the two other pilots of the 91ª, Novelli and Magistrini, got credit for it. Even more pilots claimed credit for this victory: *sergenti* Chiri of the 78ª and Reali, 79ª, *cpr.m.* Lucentini also of the 79ª. These three had taken part in the attack to the enemy patrol, but in 1919 they were denied their claim in the demise of the enemy ace, probably rightly so. In fact the *208ª batteria antiaerea*, (AA battery) phoned that a single fighter pilot had shot down the enemy aircraft that fell between Breda di Piave and Pero. Also a combat report of one of Gräser's wingmen stated that the ace was attacked and shot down by three fighters, obviously the three pilots of the 91ª, while the other pilots dueled with other Italian airplanes. The two-seater that the fighters of Flik 61/J escorted was also shot down by other Italian fighters within their own lines. This event was another proof that mastery of the air was now firmly in Italian hands.

An enemy offensive was expected, and all the Italian air force intensified its actions. The 91ª Squadriglia flew escort missions for

On 4 July 1918 the Spad 13 of Nardini marked with an individual number XII on its fuselage had an engine failure followed by a bad crash landing.

Nardini the barnstormer: this postcard advertises passenger flights, drops of leaflets, photography from the air in Florence, with a "fast and trustworthy SVA" beginning 26 February 1922.

the Capronis that almost daily bombed the enemy airfields. Belluno, Godega, Motta di Livenza were effectively bombed but the Austrian fighters based there did not rise up to meet the challenge. The airfield at Pramaggiore was also effectively hit on 9 June, but in landing Nardini turned over, wrecking Spad 1580.

When the offensive was launched, he was immediately in action. In the morning of 15 June he spotted over the Montello at 9.15 a.m. an Albatros D.III flying at 1,600 meters and he engaged it together with four other Hanriots. He fired 81 bullets and saw his quarry going down toward Moriago, trailing smoke and flames. He did not get confirmation for this claim, although the Squadron Diary of the 91[a] considered it as such. However after the war he received his third and last decoration, another Bronze Medal, which in its citation, praising him as a bold and untiring fighter pilot, credited him with three victories on 3 and 17 May and 15 June. Citations were usually written copying word by word the proposals of the squadron or group commanders, while the compilers of the "Bongiovanni Commission" were particularly strict.

Those were days of constant hard missions of patrols, strafings, always in a horrible weather. The enemy on the ground was throwing all its forces trying to break through the Italian defenses, and the intervention of airplanes was paramount. The airplanes returned from strafing missions riddled with bullets from ground fire. On 17 June, returning from one such mission in the evening, Nardini got lost in the rain and low clouds, and force landed in the countryside. The squadron lost its commander, Baracca, but actions went on.

Another problem was the kite balloons directing the fire of enemy artillery from across the river. Nardini tried to shoot one down using incendiary bullets on 21 June, but also some Le Prieur rockets were available. He tested them in a couple of flights, then on 22 June he set out against one of those "flying sausages" floating over Nende, near S. Polo. At 11 a.m. he swooped down on the "draken" but his action was only partially successful. The Austrians, as they saw the incoming fighter, started to hoist down the balloon and when Nardini fired his rockets it was already close to the ground. His rockets missed but one of them hit another balloon that was on the ground deflated and set it on fire. There were no witnesses unfortunately so this sort of unusual victory, his last claim, didn't get confirmation.

Three days later, as he was on patrol across the front with *ten.col.* Piccio he had a hectic dogfight over Oderzo against a strong formation of enemy fighters, but with no results. As they were coming back, ground fire slightly hit the airplane of commander Piccio. In the following days Nardini was tasked with the usual escort flights to the Capronis and with long offensive patrols. The Italians now challenged the Austrians over their airfields, but they usually refused to come up. The very hard losses of June had left their mark, and air combats were few. Danger still lurked in the form of failures of the Itala-built Hispano Suiza engines of the Spad, a real thorn in the side for the Squadron of Aces. Nardini's turn came on 14 July, when the engine of his Spad 13 failed and he made a forced landing, completely smashing the fighter but coming out of it unharmed.

On 25 July he left the squadron on leave. He returned on 16 August coming from Turin and ferrying with *tenente* De Bernardi two Ansaldo A.1 Balilla fighters, serials 16560 and 16551. The pilots of the squadron soon tested them, but they were not particularly satisfied and they kept on flying with their Spads.

23 August was a fine day, with haze over the plains of Veneto. During the morning there were offensive patrols, and *serg.* Magistrini shot down an enemy aircraft across the lines. At noon the order arrived to move a flight to Verona, probably to escort the Caproni bombers. This mission was for *cap.* Ruffo, *tenenti* Ranza, Keller, Bacula and *sergenti* Magistrini and Nardini. The last one could not take part in the mission: in the morning he had gone to S. Luca airfield on a motorcycle, and coming back he had an accident, so bad that he was sent to a hospital.

We don't know what his wounds were, but they were serious enough, because when Nardini returned to his unit he didn't fly very much any longer. When the Italian troops were launched in the final offensive, since 24 October, the 91ª Squadriglia was much engaged in fighter and ground attack missions against the faltering Austrian army, but Nardini flew just about ten standing patrol and test flights.

After the war, as we have seen, he got a second Medaglia di Bronzo, which, together with the Argento, probably didn't honor enough the gallant Florentine pilot. In 1919 the notes of the Comando Superiore d'Aeronautica described him this way:

> "Very gallant, resolute and enthusiastic pilot, in the squadriglia Baracca he was remarkable for his good will. In a short period of time he flew very many escort and patrol missions. He had many air combats and shot down two more enemies, bringing his total of victories to six. In the offensives of June and October 1918 he went down to strafe enemy troops twenty times."

The official list of aces issued in February 1919 credited him with six victories out of some eleven claims. In August 1919 he was dismissed, and like many other pilots he tried somehow to make a living out of aviation. He took part in many air shows all over Italy, flying a SVA with civil registration I-EOLO and he had a special air show in his hometown Florence. In 1923, when the Regia Aeronautica was born, he applied to be admitted to service and on 1 July he got back his rank of *sergente* and was sent to Ghedi for conversion on the Hanriot. In a short time he was qualified again on the Hanriot and the Spad, he was first assigned to the 78ª Squadriglia in August and then in November back to the 91ª.

A late photograph of Guido Nardini with an Ansaldo AC.3 fighter at Ciampino airport.

In early 1925 he was assigned to the 36° Squadriglia, and then in May he was back with the 91ª. He remained there until 10 February 1927, being assigned to the 84ª Squadriglia of the 7° Gruppo Caccia at Rome Ciampino, flying the Fiat CR.20 biplane fighter. He had attained the rank of *maresciallo di seconda classe* (warrant officer 2nd class) when fate caught up with the old pioneer pilot: on 26 January 1928 his fighter crashed to the ground at Ciampino, and Guido Nardini was killed. His family was reported to have immigrated to the United States, and no known relative remains in Florence.

## Guido Nardini
## 11 victory claims – 6 confirmed victories

| Claim | Vict. # | Date | Sqn. | Aircraft | Hour | Place | Victim |
|---|---|---|---|---|---|---|---|
| 1 | 1 | 27.6.1916 | 75 | Ni 11 | - | M.t Verena | Two-seater |
| *[with s.ten. De Bernardi, s.ten. Buzio & cpr. Consonni, 75ª Sq.] Zugsf. Josef Holub WIA, POW/Fähnr. i. d. R. Friedrich Edler von Langer WIA, POW, Flik 21, Br. C.1 26.11* | | | | | | | |
| 2 | - | 14.6.1917 | 78 | Ni 17 | 8h30 | Val d'Assa | EA |
| *[Maybe with serg. Cerutti, 79ª Sq.] Maybe: Fw. Richard Schuster WIA, Flik 21, KD 28.35* | | | | | | | |
| 3 | 2 | 14.6.1917 | 78 | Ni 17 | 8h30 | Valsugana | EA |
| *Korp. Franz Dostal KIA/Lt. i. d. R. Paul Rotter KIA, Flik 24, Br. C.1 26.29* | | | | | | | |
| 4 | 3 | 18.7.1917 | 78 | Ni | - | Asiago | Two-seater |
| *[with serg. Magistrini, 78ª Sq.] Korp. Stefan Volisin POW/Lt. Viktor Maly KIA, Flik 24, Br. C.1 229.12.* | | | | | | | |
| - | - | 18.7.1917 | 78 | Ni | - | - | Fighter |
| *[with serg. Magistrini, 78ª Sq.] Fw. Wenzel Schwarz UNH, Flik 24, Alb. D.III 53.38 FTL* | | | | | | | |
| - | - | 18.7.1917 | 78 | Ni | - | - | Fighter |
| *[with serg. Magistrini, 78ª Sq.] Lt. i. d. R. Josef Friedrich UNH, Flik 24, Alb.D.III 53.33 FTL* | | | | | | | |
| 5 | 4 | 6.9.1917 | 78 | Hd? | - | S. Gabriele | Two-seater |
| *Maybe: Zugsf. Josef Skubal UNH/unknown observer UNH, Flik 35/D, Br C .1 129.43 damaged* | | | | | | | |
| 6 | - | 6.9.1917 | 78 | Hd? | - | Ternova | EA |
| 7 | - | 6.9.1917 | 78 | Hd? | - | - | German a/c? |
| 8 | 5 | 3.5.1918 | 91 | Ni 27 | 11h | Salettuol | Two/seater |
| *[with magg. Baracca, 91ª Sq.] Zugsf. Josef Friedrich KIA/Oblt. Karl Rosenbaum KIA, Flik 19/D, Br. C..1 369.28* | | | | | | | |
| 9 | 6 | 17.5.1918 | 91 | Ni 27 | 9h | Pero | Fighter |
| *[with serg. Magistrini & ten. Novelli, 91ª Sq.] Lt. i. d. R. Franz Gräser KIA, Flik 61/J, Alb DIII 153.221* | | | | | | | |
| 10 | - | 15.6.1918 | 91 | Spad? | 9h05 | Moriago | Albatros D.III |
| *[with four pilots of Hd]* | | | | | | | |
| 11 | - | 22.6.1918 | 91 | Spad? | 11h | Nende (S.Polo) | Kite balloon |

# Giovanni Nicelli

Giovanni Nicelli was born on 27 October 1893 at Lugugnano Val d'Arda, a serene and wealthy agricultural in the province of Piacenza, next to Fiorenzuola d'Arda, home of another ace, Ranza. Coming from a farmer family that, if not wealthy, at least had a good standard of living, Giovanni, called Giannetto, could attend school, not a common event for many peasant boys in Italy one century ago. He didn't like school too much, and after four years he quit. His family placed him as an apprentice in mechanical works at the Caccialanza e Bronzini works of Piacenza.

When in March 1911 the pioneers Umberto Cagno and Giuseppe Rossi first flew over Piacenza, Nicelli watched them and was fascinated by aviation. Using his mechanical skills, he built airplane models. When his father died, Nicelli decided to turn a passion into a career and at the end of 1912 he volunteered, and was accepted in the Battaglione Aviatori as a mechanic. At that time in Italy before the age of 21 one was still a minor, so Nicelli needed his mother's permit to volunteer.

As soon as possible he volunteered for pilot training, and he was eventually posted to the Malpensa flying school for instruction. In January 1917, a pilot and with the rank of *caporale*, he was assigned to the 79ª Squadriglia, which in those days was completing its formation at Arcade airfield and then moved to Istrana, to operate over the Asiago northern sector of the front.

Nicelli flew his first combat sortie on 20 January, a standing patrol between Asiago and Cima 12. He flew eight more combat sorties, on 4 March he dropped over Treviso leaflets extolling the National Loan, but only on 4 April he had the first encounter with the enemy, which immediately flew away.

During the spring of 1917 he didn't fly very much, just 13 more combat sorties until 14 June, but on that day he had his first air combat and he immediately proved his qualities. He took off flying Nieuport 17 serial 3134 on an offensive patrol that brought him to Enego and then to Cima 11, where at 9 a.m. he found an enemy two-seater crossing the lines heading to Valsugana. In his Combat Report he wrote:

> "...I was at an altitude of 3,600 meters. I went in a dive toward the enemy aircraft and I reached it at 2,400 meters near m. Verena, exactly a bit south of the mountain, almost over Val d'Assa. At a distance of 200 m. I fired a first burst of some 40 rounds. The enemy aircraft did a tight turn, I managed to get back onto its tail and as it was descending I followed it down to 2,200 m. firing again some 50 bullets more. I couldn't get any lower because I was just above the peaks of the mountains. The enemy aircraft went on diving in the direction of the upper Val d'Assa."

This was credited to Nicelli as his first victory, and two days later he received the rank of *sergente*.

On 29 June he was assigned to the Sezione Difesa (defense flight) of Padua together with *serg.* Attilio Imolesi, another future ace. It was not the post where it was likely to meet enemy aircraft;

An intense portrait of Giovanni Nicelli standing in front of a Nieuport 17.

Nicelli flew 48 standing patrols over that beautiful city, three of them on 21 August alone, but they were all uneventful. He returned to his 79ª Squadriglia presumably in early September.

In the morning of 11 September flying over the Asiago plateau he met a two-seater with an escorting fighter returning from a mission over Italian territory and from an encounter with Imolesi. He attacked first, but the fighter went straight at him frontally from above and opened fire. Nicelli returned fire but his gun jammed immediately, so he started maneuvering and placed himself behind the tail of the Austrian who got out of the supposed danger diving away from him. He had other fruitless duels in the following days. On 17 September Nicelli found a two-seater near Enego, he had hardly the time to fire a few rounds and then another airplane bounced him. The Italian pilot did an Immelmann turn and got into the tail of his new enemy, firing 130 rounds at him. The Austrian escaped him and flew away to Marcesine; Nicelli was running out of fuel and had to let him go. Two days later, on 19 September, Nicelli had another combat with two Austrian fighters.One of them got away immediately, but the other one engaged Nicelli in a long dogfight, which ended when the Austrian quit the field. Nicelli remained over that spot in case the enemy returned, then he ran out of fuel and force landed near Casella d'Asolo with no damage for his fighter. Similar duels without results took place on 22, 23 and 30 September and on 1 October. In this period Nicelli, like the other sergeant pilots of the 79ª Squadriglia, Imolesi, Reali and Cerutti, was alternating between detachments at Padua and patrol or escort missions over the mountains, frequently encountering the enemy.

On 25 October, as the Austrians and Germans were advancing, Nicelli attacked over Mount Lisser a two-seater that was crossing the lines toward the Valsugana. He fired 270 rounds, and saw his enemy going down out of control, but this victory was not confirmed, this fact making him annoyed and unhappy.

On 2 November 1917 the 79ª Squadriglia moved from Istrana to Nove di Bassano. Nicelli took off from that airfield in the early afternoon of 7 November, as aircraft were spotted above Primolano. He engaged Brandenburg C.1 29.71 of Flik 24/F that was on a bombing sortie and forced it down near Fonzaso. The Austrian airplane turned over in the crash landing but its occupants, *Zugsführer* Friederich Schieg and *Leutnant* Friederich Wirth, were unharmed and became prisoner. The engine of Nicelli's Nieuport faltered and also the Italian force landed, without further damage. At the airfield came a call from the Comando Truppe Altipiani announcing that a "Super Nieuport" had shot down an enemy and then had gone down, so there was fear concerning the destiny of the missing Nicelli. The commander of the 79ª Squadriglia, *tenente* Cesare Bertoletti, drove to the place of the fall, much worried. When he got there he found the enemy airplane picketed by the Carabinieri and Nicelli close by, happily sitting next to his fighter. The mixture of joy and anger overwhelmed Bertoletti, who punched Nicelli accusing him of not having called. Giannetto, while trying to fend off the blows of his enraged officer, serenely answered: "The other time they cheated me and didn't credit me with the plane I shot down, but this time they cannot steal it away from me!" For this victory, Nicelli received his first Medaglia d'Argento in the field, which in its citation at least reported also that he had forced an enemy airplane to land on 25 October.

Exactly one month later, in the morning of 7 December, Nicelli was tasked with the escort of a SP two-seater over Cismon and Asiago. He saw two enemy airplanes flying from the Grappa to Bassano, but he didn't want to leave the Savoia Pomilio alone. When the latter completed his mission and returned, he went back looking for them and found one near Mount Lisser. He fired 70 rounds, his gun jammed, Nicelli cleared it and attacked again near Roana, until finally the Austrian two-seater fell in the Val d'Assa, near the Verena.

On 13 December Nicelli claimed another victory near Asiago. While escorting a SAML in the Val Galmarara, he spotted an enemy patrol near Cima 11. As he had the advantage of altitude, he decided to attack so that the two-seater could get out of harm's way. His gun immediately jammed, however, and he was forced to dogfight against four enemies, but he managed to escape them. He then cleared his gun and, no longer encumbered with the escort mission, he crossed the lines again. Near Bassano he found an Austrian airplane and he attacked it frontally, then he turned and engaged again from behind, shooting it down near Asiago. The adversary of Nicelli possibly was either an aircraft of Flik 17/D, in which the observer *Leutnant* Robert Kabelac was wounded in combat against a "Sopwith" or an aircraft of FlG.1. Back to base, the Italian pilot found two bullets lodged in the tail of his fighter.

Nicelli with a Nieuport 11 of the 79ª Squadriglia.

On 16 December Nicelli was the powerless witness to a tragedy: he was escorting some SAML two-seaters of the 115ª Squadriglia when he saw one of them stray away heading for Fonzaso, and a group of enemy fighters pouncing on it. It was a group of crack pilots of Flik 55, the aces *Offstv.* Arigi, *Offstv.* Kiss and *Korp.* Lahner. Immediately Nicelli rushed at the enemies but his gun jammed and one of the expert Austrians got behind his tail. Nicelli maneuvered and got free, tried to attack the other two Albatros but still his gun refused to fire, and he saw the Italian SAML go down in flames near Feltre. The two Italian airmen, *sergente* Matteo Fabbian and *sottotenente* Orazio Giannini were dead, and Nicelli sadly returned to base.

In 1918 air activity remained intense in the difficult mountain area of the Asiago plateau, and Nicelli often flew with other aces of the 79ª Squadriglia, *serg.* Reali and Cerutti, the latter his companion in instruction at Malpensa. On 30 January 1918 "Giannetto" with Reali found a reconnaissance airplane with an escort fighter and he attacked the two-seater, which tried to maneuver but was hit, shed its wings and crashed near Costalunga. Possibly his victim was Brandenburg C.1 29.07 of Flik 45/D, its occupants *Zugsführer* Ernst Winkler and *Oberleutnant* Oskar Knoll were missing in action.

For this feat Nicelli received his second Medaglia d'Argento in the field, the citation commenting on his gallant efforts during January and on his victory of 30 January, reported as his seventh one.

On 4 February he was escorting a SP over the Plateau. Having completed his mission, as he was bringing back the two-seater across the lines, Nicelli spotted near Valstagna an enemy patrol climbing to altitude. As soon as the SP was safe, he turned back and attacked a fighter that went into a sideslip and a spin. He couldn't press on his attack because another fighter was on his tail, but maneuvering he turned the tables and forced the Austrian to fly away.

In the morning of 5 February 1918 Nicelli took off with Cerutti for a standing patrol between Valstagna and Gallio. The two aces met near Col del Rosso an Austrian Brandenburg crossing the lines with two escort fighters. They decided to let them go through and to attack them over friendly soil. Nicelli engaged the two-seater while Cerutti drove away the fighters and then returned with him. The fifth confirmed victory of Nicelli, Br. C.1 29.16 of Flik 45/D, was forced to crash land near Casa Gilardi and its two occupants were captured. An Italian document described this airplane as "painted dark green on the upper surface and gray below." As was customary, the Italians a few days later dropped a message from the now prisoner *Leutnant* Hugo Welser to his squadron commander.

In the morning of 13 February Nicelli claimed a victory over a fighter at Marcesina, which was not confirmed. Things were quite different on 24 February, when he escorted SP.3 4577 of the 26ª Squadriglia, with the crew *sergente* Gaggero and *tenente* Monti. The large pusher flew four times from Col Caprile to the Grappa and back spotting for the guns of a battery, with Nicelli and other pilots of the 79ª trailing it. During the fifth run an enemy patrol climbed to altitude near Asiago and headed for the Italians. *Sold.* Bozzetti, one of the Italian fighters, flew close to the SP and gesturing tried to warn its crew of the incoming Austrians, but they kept on flying to Enego. One moment later *Oblt.* Linke Crawford, flying Phönix 228.14, bounced the two seater and set its engine on fire with a long burst. Nicelli and the other pilots tried to get into action but the Austrian ace escaped them spinning his fighter, while the SP burned and one of its airmen jumped down to his death to avoid

This Hansa Brandenburg C.I with a spotted tail, 229.09 of FlugGeschwader I, was damaged in combat on 13 December 1917 and it could be a victim of Nicelli. (Photo Masajdek).

the flames. The Italians then engaged the other fighters, anxious for revenge, and after a confused and hectic dogfight one of the enemies went down, Phönix D.I 328.12 with pilot *Oberleutnant* Richard Hess, killed. Nicelli as usual had to struggle with a jammed gun, and also *ten.* Federico Comirato claimed credit for this victory, although he doesn't appear in the list of victories of 1919.

By now the young pilot was a national hero, the press talked about him and he received prizes and praise: the gold medal of the "Pirelli Prize" for the 2nd Competition *"Cacciatori del Cielo"* (hunters of the sky), another gold medal from the Lega Aerea Nazionale and the Belgian *Croix de Guerre* pinned on his breast by the king of Belgium, Albert, on Padua airfield.

On 4 May 1918 Nicelli scored his two final victories. He took off for an engine test but he realized that the artillery was firing, meaning the presence of enemy airplanes. Over the Montello he found a group of enemy fighters and he attacked them together with some Camels of No. 66 Squadron of the Royal Air Force. Two Albatros D.III s were shot down within the Italian lines, and both pilots were killed: 153.182 of *Oberleutnant* Karl Patzelt, the commander of Flik 68/J and a five victories ace, and 153.210 of *Korporal* Karl Fritsch, who was at his second combat sortie. The pilots saw Fritsch climbing over the wing to escape the flames and then jump down from an altitude of 100 meters. The unlucky pilot fell close to a large haystack and he survived two hours after his fall. One of the British pilots, Jerry Birks, 63 years later, during the Reunion of World War I Fighter Aces in Paris still recalled with horror that event.

Nicelli stands in fron of the wrecks of two Albatros fighters that he shot down on 4 May 1918.The next day also the Italian ace was dead.

Giovanni Nicelli in the cockpit of his Nieuport 17.

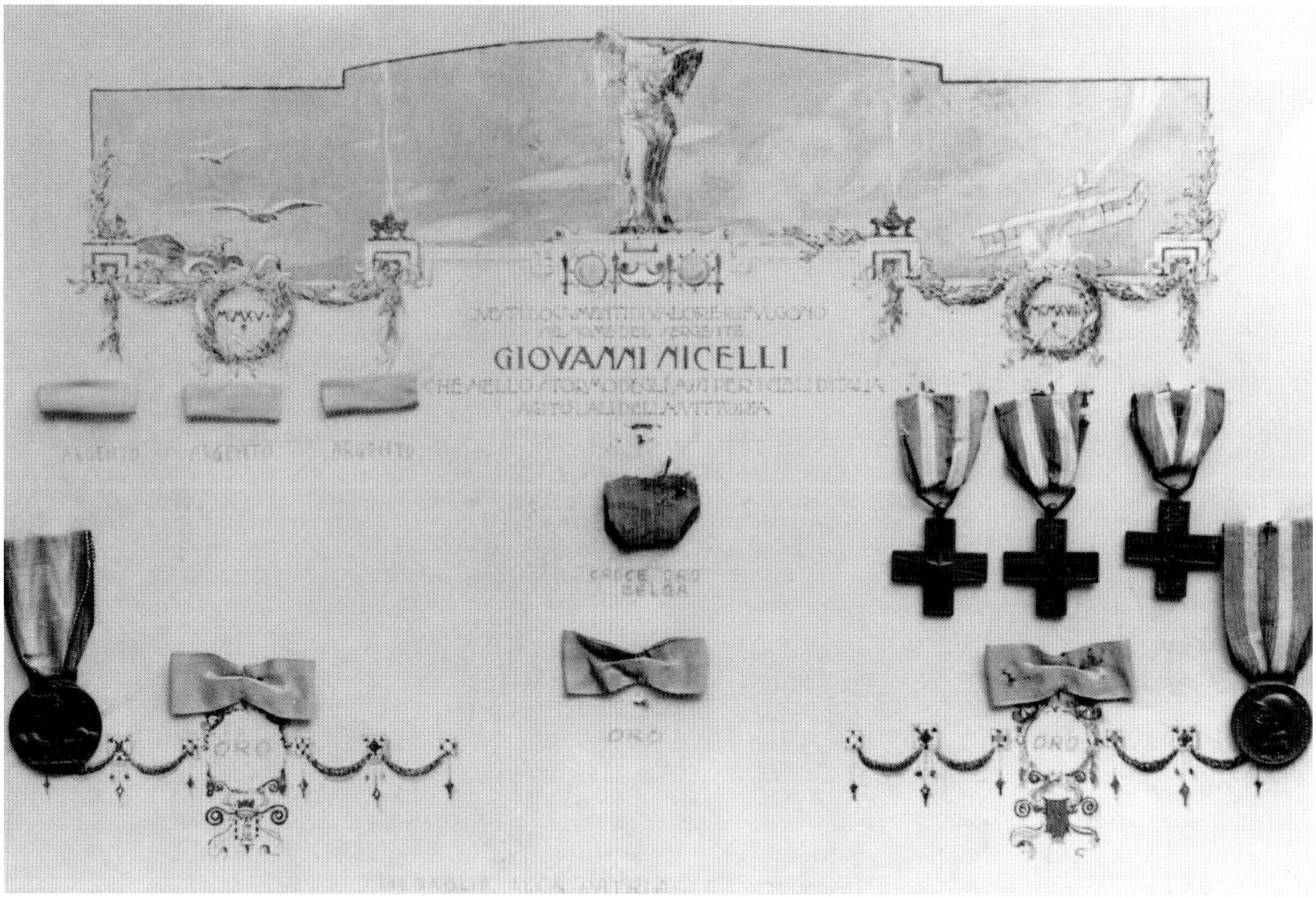


The medals of Giovanni Nicelli.The writing says:"These documents of valor shine in the name of sergeant Giovanni Nicelli, who in the squadron of aces moved wings of victory in the sky of Italy""All the gold and silver medals are missing, offered to the nation in 1935as a patriotic gesture, when Italy was subjected to sanctions by the Society of Nations.

Nicelli was much admired for his aerobatics talent: he used to come to a landing with his airplane banked vertically and then to straighten it thirty feet from the ground with a single smooth maneuver, and to touch down without bumps. The day after his double victory *cap*. Umberto Mazzini, commander of the 79ª Squadriglia, asked him to perform some aerobatics for visiting allied officers. Nicelli took off in the early afternoon but at the top of a loop his Nieuport 27 serial 11353 shed one of the lower wings. The pilot tried to control the fighter acting on the throttle but he crashed in a courtyard at Porcellengo, near the airfield, mortally wounded. That same day the papers to ask for a third Medaglia d'Argento for his prowess in scoring a double victory against fighters had been mailed.

The body of Nicelli was buried two days later in the cemetery of Porcellengo. General Pennella, commander of the 2nd Army, delivered his final decoration to his mother on 22 May. In 1927 the airfield at S. Niccolò al Lido, Venice, was named after him. His body was moved to his native town in June 1939, and the ceremony was attended by a large crowd and by Mazzini, his commander.

## Giovanni Nicelli
## 12 victory claims – 8 confirmed victories

| Claim | Vict. # | Date | Sqn. | Aircraft | Hour | Place | Victim |
|---|---|---|---|---|---|---|---|
| 1 | - | 14.6.1917 | 79 | Ni | 9h | Mount Verena | "Albatros" |
| 2 | 1? | 25.10.1917 | 79 | Ni | 8h | Mount Lisser | Two-seater |
| 3 | 2? | 7.11.1917 | 79 | Ni.17 | 15h | Fonzaso | Two-seater |
| *Zugsf. Friedrich Schieg POW/Lt. Friedrich Wirth POW, Flik 24/F, Br. C.I 129.71* | | | | | | | |
| 4 | 3? | 7.12.1917 | 79 | Ni | 11h | Val d'Assa | Two-seater |
| 5 | 4? | 13.12.1917 | 79 | Ni | 10/12h | Asiago | EA |
| *[with serg. Ciotti, 79ª Sq.] Maybe: Unknown pilot/Lt. i. d. R. Robert Kabelac WIA. Flik 17/D or: Zugsf. Josef Bangerl WIA/Oblt. Oppacher UNH, FlG.1, Br. C.I 229.09* | | | | | | | |
| 6 | 5? | 30.1.1918 | 79 | NI | 10h | Costalunga | Two-seater |
| *[with serg. Reali, 79ª Sq.] Zugsf. Ernst Winkler MIA/Oblt. Oskar Knoll MIA, Flik 45/D, Br. C.I 29.07.* | | | | | | | |
| 7 | - | 4.2.1918 | 79 | Ni | 11h | Valstagna | Fighter |
| *[maybe with serg. Ciotti, Lucentini & Reali, 79° Sq.]* | | | | | | | |
| 8 | 6? | 5.2.1918 | 79 | Ni | 9h | Colrosso | Two-seater |
| *[with serg. Cerutti, 79ª Sq.] Korp. Karl Müller POW/Lt. i. d. R. Hugo Welser POW, Flik 45/D, Br. C.I 29.16* | | | | | | | |
| 9 | - | 13.2.1918 | 79 | Ni | morn. | Marcesina | Fighter |
| 10 | - | 24.2.1918 | 79 | Ni | - | Cismon/Enego | Fighter |
| *[with ten. Comirato, 79ª Sq.] Oblt. Richard Hess KIA, Flik 60/J, Phönix D.I 328.12* | | | | | | | |
| 11 | 7 | 4.5.1918 | 79 | Ni.27 | 10h | Montello | Fighter |
| *[with Lieuts. Birks, Apps, Parker, No.66 Sqn. RAF] Oblt. Karl Patzelt KIA, Flik 68/J, Alb. D.III 153.182* | | | | | | | |
| 12 | 8 | 4.5.1918 | 79 | Ni.27 | 10h | Montello | Fighter |
| *[with Lieuts. Birks, Apps, Parker, No.66 Sqn. RAF] Korp. Karl Fritsch KIA, Flik 68/J, Alb. D.III 153.210* | | | | | | | |

# Gastone Novelli

Italy did not really have a proper officers caste, like the Junkers of Germany, but it was quite common at the turn of the 19th Century that an officer' s son should follow his father's footsteps. Gastone Novelli was born in Ancona, a nice city on the Adriatic coast, on 13 June 1895, the son of General Corrado Novelli and of Marta Pesci Feltri, a girl from a noble family. As was customary for children of well-to-do families, he was a student in a boarding school in Spello, and then he attended high school in a technical institute in Ancona, as he had a strong interest in scientific matters. His career then progressed toward the military, and Novelli first joined the Collegio Militare of Rome and then the Scuola Militare of Modena, which he left with the commission of *sottotenente* in February 1915, assigned to the Lancieri di Montebello Regiment, a renowned cavalry unit formed in 1859, proud of its Latin motto *"Impetu hostem perterreo"*, I rout the enemy with dash.

When Italy joined the European War on 24 May 1915 Novelli, not yet twenty, sported the green collar tabs of his unit, part of the 3rd Cavalry Division of Lombardy, assigned to the Supreme Command in the region of S. Vito al Tagliamento and Spilimbergo. It was soon clear that barbed wire and machine guns made any fast and dashing action impossible, so cavalry was forced into a minor role of surveillance, and much of its now useless personnel were transferred to other arms. Out of its 3,000 officers 800 of them were shifted, Novelli among them, posted to the 43° Reggimento Artiglieria. The young and aggressive officer hated the idea of wasting his time with the calculations of cannon fire, and he grasped that aviation could offer the emotions of a cavalry charge. Novelli volunteered as an observer and on 25 August 1915 he was assigned to the 11ª Squadriglia at Chiasellis, equipped with Maurice Farman 1914.

Novelli flew 17 combat missions with this unit, that included reconnaissance, bombing and patrol sorties. Again, he volunteered for pilot training and on 31 March 1916 he was posted to the Comando Battaglione Scuole Aviatori, the training command. Novelli received his flight training first at Mirafiori, and then at Busto Arsizio, where on 1 August 1916 he obtained his pilot license flying a Maurice Farman 14 with Fiat 100 hp engine.

As soon as 11 August 1916 he was back at Chiasellis, assigned to the 30° Squadriglia Farman, one of the best reconnaissance units, which was the breeding ground of some of the best Italian pilots. The squadron was very active in the sector of the Carso and the Bainsizza plateau, flying many bombing and reconnaissance missions, in which Novelli participated, often returning with his airplane hit by Austrian anti-aircraft artillery. In this period, he earned his first decoration, a Silver Medal, with its citation reading: "Aviator pilot, in many offensive reconnaissance missions over the enemy he always distinguished himself for his admirable daring and high sense of duty, often returning with a damaged airplane – September 1915 – January 1917."

As could be expected, he was not satisfied with the life of a recce pilot, and he managed to get himself transferred to fighter training. After a brief course on the Nieuport, on 20 April 1917

Gastone Novelli, elegant in his cavalry uniform, and with a rose in his breast pocket.

Gastone Novelli was assigned to the 81ª Squadriglia, a newly created unit that on that very day moved from Arcade, its formation airfield, to the frontline at Borgnano. Novelli was soon engaged with his entire unit supporting the new offensive launched on 12 May 1917, the so-called 10th Battle of the Isonzo.

He claimed his fist victory on 3 June near Aisovizza, the area where air activity was most intense.

It is likely that his victim was Brandenburg C.1 129.02 of Flik 4, that was attacked by *s.ten.* Luigi Olivari, supported by *serg.* Luigi Poli, and that Novelli took part in this action. The enemy aircraft was forced down and hit by ground fire that killed both occupants. Possibly an alternative in the identification of Novelli's victim could be another Br. C.1 of Flik 4, 129.41, that also had a combat over Gorizia and had to force land at Prosecco airfield badly damaged.

Novelli scored his second victory again over Aisovizza on June 19, together with another crack pilot of the 81ª Squadriglia, Flavio Torello Baracchini. Possibly their opponent was a Brandenburg of Flik 19 crewed by *Zugsf.* Fejes and *Lt.* Purer, which however, according to Austrian sources, was only slightly damaged and forced to land, not before claiming one of its Nieuport opponents destroyed and one out of control. Baracchini and Novelli scored again just three days later, on 22 June, together with *serg.* Piaggio of the 77ª Squadriglia. The Austrians lost Br.C.1 229.05 of Flik 35. Flying a reconnaissance sortie to Villa Starkenfels, between Gorizia and Aisovizza, the airplane met four Nieuports and was forced down in

Gastone Novelli, at the 76ª Squadriglia, has an important visitor, his father the General. From the left, Curzi (not an airman), Novelli, Gen. Novelli and *ten.* Luigi Olivi.

Novelli scored his first victory on 3 June 1917 together with Luigi Olivari. A possible victim could be Brandenburg C.I 129.02 shot down on that day with the death of its crew. (Photo Masajdek).

Italian land at Hill 190 of Stara Gora, with the observer *Oberleutnant* Johann Miklosi killed and the pilot *Korporal* Stephan Hegedüs fatally wounded.

Novelli flew twenty more combat sorties and had four more combats without success, then on 13 July he went on leave. When he returned, on 1 August, he was assigned to the 76ª Squadriglia, also based at Borgnano Langoris airfield.

When the Eleventh Battle of the Isonzo began, the big offensive that was also called the battle of the Bainsizza, the squadron was heavily engaged in offensive patrols and escorts. On 22 August an Austrian airplane attacked Novelli above the Bainsizza and wounded him seriously in the left arm. He managed to bring back his fighter and land at an Italian airfield, and he was sent to Hospital N. 245, remaining there until 7 September, and then going first to Treviso and then to Turin for further treatment. Strangely, we know of no Austrian claim corresponding to this loss.

For this event the Duke of Aosta, commander of the 3rd Army, assigned to Novelli a Medaglia d'Argento, his second, in the field, that reported in its citation two air victories and then his wound.

When in Turin Novelli heard of the route of Caporetto, he volunteered to rush back to the front despite his convalescence. On 3 November he managed to be assigned to the 91ª Squadriglia, then based at Padua airfield, the best Italian fighter unit, with as commander Baracca, the top ace. At the same time he was also transferred to another cavalry regiment, the "Novara Cavalleria", and from then on he sported the elegant white collar of that unit on his uniform.

Together with the other top fighter pilots of the 91ª Squadriglia Novelli was fully committed to battle, and scored. This is his combat report for his victory of 23 November:

> "I took off on offensive patrol with Mr. Maggiore Baracca, at 3.05 p.m. an enemy fighter flying over the Montello was spotted. The patrol leader attacked it at a height of about 3,500 meters and the enemy airplane defended itself maneuvering and trying to reach its lines. I engaged combat firing about 60 bullets. The enemy airplane reached the bed of the Piave River south of Falze di Piave where it remained upside down."

The enemy fighter, an Albatros D.III of Jasta 39, became the target of artillery and burned in front of the positions of the 8º Reggimento Bersaglieri. One of those soldiers recklessly reached the wreck and took away one of its machine guns. The unlucky pilot, *Vizefeldwebel* Karl Überschär was found dead in his cockpit.

On November 30 Novelli had another combat above the Piave:

> "I took off on a cruise at 12.25 with Tenente Ranza and Sergente Magistrini, at 13.40 we spotted an Aviatik airplane that from Mount Grappa was flying toward the Montello at 4,000 meters. The patrol leader attacked the enemy that dove toward the Piave; I reached it and attacked it beyond its lines firing about 25 rounds. I saw the enemy plane falling vertically, out of control, near Vidor. I followed it down to a level of 600 meters and then as enemy artillery was firing at me I returned to our territory and landed at 14.35."

In the first months of 1918 Novelli flew routine missions with no particular event besides two small accidents, one on 17 March when, during a standing patrol over the Piave, some British fighters attacked him and placed a bullet in his Spad's wing, and then on 4 May, when on take off he damaged Spad number 2478, as he had to turn and crash against the embankment of the railroad, to avoid hitting a Hanriot that was landing in the wrong direction.

Right: Brandenburg C.I 229.05 forced down by Novelli and Baracchini on 22 June 1917. (Photo Masajdek).

In the morning of 11 May 1918 Novelli and other pilots of the 91ª Squadriglia were cruising in the area of Quero-Ponte alla Priula when they discovered two reconnaissance planes with the escort of four fighters in the area of Pederobba. Novelli attacked one of the two-seaters and saw it going down out of control, but he had to make room for Magistrini and defend himself from an attacker described as a *"small two seater biplane"* that he then harassed together with *ten.* Adriano Bacula until he had to give up the chase, on account of the failure of two wing struts.

Friday 17 May was a bad luck day, as it marked the death of one of the most popular Austrian fighter pilots, 17-victories ace Franz Gräser:

> "I took off in a patrol at 9 a.m. with S.Tenente Olivero, Sergente Magistrini, Sergente Nardini, Sergente Aliperta, Sergente D'Urso over the area between Quero and Ponte alla Priula, and at 10 a.m. I spotted anti-aircraft bursts in the direction of Treviso. I went in that direction at full throttle and at a height of 4,000 meters I saw a group of about nine enemy airplanes above S. Biagio. Sergente D'Urso meanwhile was forced to land at our airfield for an engine failure. S.Tenente Olivero and Sergente Aliperta engaged toward the Piave some enemy fighters, as I attacked an Albatros in the group and Sergente Magistrini immediately followed me. The enemy airplane started going down as soon as we had fired our guns. At a level of 3,700 meters there was Sergente Nardini who was able to attack again and after a long burst he flamed the Albatros that crashed and burned near Pero."

The enemy fighter group was formed by five pilots of Flik 61/J, Gräser, Schrimpls, Ludwig, Deubler and Reinl. Other Italian pilots had joined the fray, *sergenti* Chiri, Rennella, Reali and Lucentini, but the Italian command credited the victory only to Novelli, Magistrini and Nardini. Italian soldiers found in the charred remains of Albatros D.III 153.221 the body of the Austrian ace, but nobody knows where he was buried.

On 26 May Novelli together with *ten. col.* Piccio and *ten.* Guido Keller shot down in flames near Ronchi, behind Itlian lines, Albatros D.III 153.230 of Flik 42/J killing *Zugsf.* Franz Hofstädtler, a straggler in a patrol of nine fighters that the three pilots of the crack 91ª had attacked without hesitation. In this combat Novelli fired 130 bullets.

Again with *ten. col.* Piccio, Novelli scored his last victory on 11 August shooting down in flames near Maserada one of the new-

Pilots of the 91ª Squadriglia in early 1918: from the left: unknown (probably Costantini), Gastone Novelli, Adriano Bacula, Ferruccio Ranza and Edoardo Olivero. Behind the Spad of Ranza there is a French-built Hanriot.

Above: The Spad 6364 of the 91ª Squadriglia with the clover insignia was described in a document as assigned to Bacula, but the squadron papers show that Novelli flew it constantly. Behind the plane stand Ranza, Bacula and Novelli.

Right: English artillery officers visiting the 91ª Squadriglia at Santa Caterina. From the left *asp.* Michele Riello, Novelli, fourth Ranza, fifth Sabelli, who signed himself *"Moi"* in French, sixth Baracca. *(Photo Franz Selinger).*

Below: An elegant Novelli in a formal portrait. The Nieuport 27 behind him was unusual in the 91ª Squadriglia, where only *serg.* Nardini used it.

Personnel of the 91ª Squadriglia around cap. Ruffo's Spad 13. On top, from left, Gastone Novelli, Ruffo di Calabria, Mario De Bernardi. Standing, second from left, Conelli de' Prosperi.

Novelli controls as his white nosed Spad is pushed to cover.

Novelli in front of his white nosed Spad 7 in the summer of 1919. The propeller is running, this is probably one of the last photos of Novelli before his fatal crash, on 3 July 1919.

est types of Austrian two-seaters, Phönix C.1 121.26 of *Oblt.* Albert Graf von Herberstein with the observer *Lt.* Otto Krakowitzer, both killed. The Austrians dropped a message behind Italian lines asking for information about their missing crew, and the Italian pilots dropped their message with the required sad information on 21 August.

Novelli flew very much, but all his missions were uneventful, and he had just one combat on 22 September, with no results. During the final offensive of Vittorio Veneto Novelli attacked Austrian troops with strafing and bombing missions. On 30 October he was part of a patrol that attacked a train as it was entering the station of Pordenone, blowing up two locomotives. Then the Italian fighters spotted five airplanes at the local airport, lined up and ready to start, and attacked them also. They returned to base to load up incendiary bullets, returned to Pordenone and emptied their guns on those five airplanes on the ground, one of which was already burning. As he returned, Novelli made a bad landing and turned over, damaging his Spad.

When the war ended Novelli, who was an officer in the active service, remained with the 91ª Squadriglia and flew very much. His exuberance cost him a high price, ten days of prison, because on 17 March 1919 he had performed low level aerobatics over the Parioli racetrack of Rome.

On 3 July 1919 he landed at Padua coming from Milan to refuel and then to fly on to Zaule airfield, near Trieste, where the 91ª Squadriglia was then based. Novelli took off at 7.20 p.m., but right after take off in a tight climbing turn the engine of his Spad 1714 failed. The pilot, instead of force landing straight ahead, tried to turn back and land on the runway, but the Spad stalled and crashed on the bank of a ditch. Rescue was immediate and Novelli was rushed to the hospital of S. Giustino but his conditions were obviously desperate: as he was wearing no helmet, he had probably broken his skull. He never regained consciousness and died in the hospital. Many people cried for his death, including his pregnant girlfriend and his good friend ace Sebastiano Bedendo who, faithful to their common promise, took care of the girl and married her.

Posthumously he received a third Silver Medal that in its citation reported his activities in the final part of the war, including the final strafing actions.

In 1921 the body of the ace was moved into the family vault in Ancona. In that city the local branch of the Associazione Nazionale Arma di Cavalleria, Cavalry Veterans Association, is named after him.

## Gastone Novelli
## 10 victory claims- 8 confirmed victories

| Claim | Vict. # | Date | Sqn. | Aircraft | Hour | Place | Victim |
|---|---|---|---|---|---|---|---|
| 1 | 1 | 3.6.1917 | 81 | Ni.11 | - | Aisovizza | EA |
| *[with s.ten. Olivari, 91ª Sq. & serg. Poli, 70ª Sq.] Zugsf. Anton Hoffman KIA/Oblt. Friedrich Wowy KIA, Flik 4, Br. C.1 129.02 or: unknown crew, Flik 4, Br. C.1 129.41.* | | | | | | | |
| 2 | 2 | 19.6.1917 | 81 | Ni.11 | 8h45 | Aisovizza | EA |
| *[with s.ten. Baracchini, 81ª Sq.] Zugsf. Stefan Fejes UNH/Lt. i. d. R. Josef Purer UNH, Flik 19, Br. C.1 29.63 FTL* | | | | | | | |
| 3 | 3 | 22.6.1917 | 81 | Ni.11 | - | San Marco | Two-seater |
| *[with s.ten. Baracchini, 81ª Sq. & serg. Piaggio, 77ª Sq.] Korp. Stephan Hegedüs DOW/Oblt. Johann Miklosi KIA, Flik 35, Br. C.1 229.05* | | | | | | | |
| 4 | - | 2.8.1917 | 76 | Ni | - | San Marco | EA |
| 5 | 4 | 23.11.1917 | 91 | Spad | 15h05 | Falzè | Fighter |
| *[with mag. Baracca, 91ª Sq.] Fw. Karl Überschär KIA, Jasta 39, Alb. D.III* | | | | | | | |
| 6 | 5 | 30.11.1917 | 91 | Spad | 13h40 | Moriago | "Aviatik" |
| *[with ten. Ranza & serg. Magistrini, 91ª Sq.] Uffz. Kurt Kieling KIA/Lt. d. Res. Hugo Huck KIA, FA (A) 232* | | | | | | | |
| 7 | - | 11.5.1918 | 91 | Spad | - | Moriago | EA |
| *[with serg. Magistrini, 91ª Sq.] Korp. Anton Bickel UNH/Oblt. Georg Vucetic UNH, Flik 2D, Br. C.1 169.07 FTL* | | | | | | | |
| 8 | 6 | 17.5.1918 | 91 | Spad | 9h | Pero | Fighter |
| *[with serg. Magistrini & Nardini, 91ª Sq.] Lt.i.d.R. Franz Gräser KIA, Flik 61/J, Alb. D.III 153.221.* | | | | | | | |
| 9 | 7 | 26.5.1918 | 91 | Spad | 8h40 | Ronchi | Fighter |
| *[with t.col. Piccio & ten. Keller, 91ª Sq.] Zugsf. Franz Hofstädtler KIA, Flik 42/J, Alb. D.III 153.230* | | | | | | | |
| 10 | 8 | 11.8.1918 | 91 | Spad | 10h50 | Maserada | Two-seater |
| *[with t.col. Piccio, 91ª Sq.] Oblt. Albert Graf von Herberstein KIA/Lt. Otto Krakowitzer KIA, Flik 12/Rb, Phönix C.1 121.26* | | | | | | | |

# Luigi Olivari

Olivari was an unlikely hero; an intense, thin young man who became Italy's first ace, scoring his first victories while still a simple soldier, in a squadron that included the most famous officers. Luigi Olivari was born on 29 December 1891 in La Spezia, from a middle class family. In his youth, he moved to Turin with his family, spending the summer in the nearby village of San Maurizio Canavese, where he built kites. He was described as very intelligent, active and strongly passionate in sports. He was an ice skater and ice hockey player and a good motorcycle racer, and he performed well in several long distance and hill climbing races in 1914. He also drove cars, and had a good knowledge of engines and mechanics, and of course aviation held a great fascination on him. He joined the civil aviation school of Mirafiori, and on 27 November 1914 he brilliantly obtained his brevet of airplane pilot, number 305 training on Bleriot monoplanes.

He was rejected by the army for a heart deficiency, but he volunteered after the outbreak of the war, and, having the rare qualification of a pilot, he was admitted as a "volunteer aviator" simple soldier to the Battaglione Aviatori on 19 May 1915, five days before Italy's entry in the war. After training, which began on 15 June, he received his military brevet as a Bleriot pilot on 1 September 1915 and on 6 September was sent to Busto Arsizio to train on Farmans, and ten days later to Pisa to fly the Caudrons. His qualities must have been outstanding, because on 8 October he was admitted into a selected group of pilots who were sent to France, to train on the new Nieuport biplane fighters at Le Bourget, from October 1915 to January 1916. In France, he also flew Caudrons, Voisins and Morane Saulniers, and was highly commended for his particular aerobatics skills. After his return to Italy, on 28 January 1916 he was assigned to the 1ª Squadriglia Caccia of Santa Caterina, where he joined company with prestigious career officers, who had been pioneers of aviation, the commander *cap.* Tacchini, *cap.* Scarpis and Bolognesi and *ten.* Baracca. It was typical of the camaraderie and democracy of early Italian aviation that he was accepted as a peer by this band of officers and gentlemen. It was later reported, however, that some higher rank resented the fact that a simple soldier was more successful and popular than regular officers were, and that they tried to obstruct his career.

The 1ª Squadriglia was assigned the protection of Udine, where the Italian supreme command was located, and it had often tackled the enemy, but victory had escaped so far. The unit was equipped with the Nieuport 10, known in Italy as "18 square meters", in single seater version and armed with a machine gun on the top wing, but in early 1916 the first Nieuport 11, or Bébé, armed with an unsynchronized Lewis arrived from France.

Olivari flew his first test flight on 28 January 1916, and his first patrol mission the next day. His first combat, after many fruitless patrols, took place on 31 March 1916, when he attacked an enemy airplane that escaped. On 2 April 1916 Olivari scored his first victory, when, despite the emotion of an unexpected encounter, he fired at an enemy airplane that was attacking the Italian positions. The enemy crashed beyond the Isonzo River, over enemy

Luigi Olivari, still a volunteer soldier, next to a Nieuport 10 single seater armed with a Lewis gun.

territory, and Olivari, who only saw his enemy diving away, did not claim this victory and of course it was not confirmed. His likely victim was Albatros B.I 24.61 of Fliegerkompanie 12, which reported an attack over the Isonzo by an enemy fighter "with dark wings" that riddled it with holes. The crew of *Korp.* Franz Kolodzinski and *Fähnrich* Eugen Csutka survived. On that day, Flik 12 lost another airplane to Italian fighters, Brandenburg C.1 26.02 that was shot down further south, near Aquileia, by a Nieuport 10 of 2ª Squadriglia, with the crew of *Zugsf.* Traub and *Kadett* Marterer falling prisoners after having burned their airplane. The Italian winner was *serg.* Barattini, with gunner Moretto. On 4 April again Olivari had two duels, firing first 100 and then 75 rounds at two enemies that escaped him.

None of these victories, however, was confirmed by the Italian command, leaving to Baracca, five days later, the honor of the first confirmed Italian victory. On April 7, during the ill-fated attack on Udine repulsed by the 1ª Squadriglia, Olivari scored his first confirmed victory, shooting down Brandenburg 61.59 of Flik 2 together with *cap.* Tacchini and Bolognesi, although it was Olivari who drove the first attack. The Austrian two-seater was burned on the ground by its crew of *Lt.* Burian and *Oblt.* Osterreicher, before they were taken prisoners. That day marked the coming of age of Italian fighters, and the names of the pilots of the 1ª Squadriglia were reported all over Italy, while the citizens of Udine were literally ecstatic.

Olivari in civilian garb next to a military Bleriot. The location is probably Mirafiori airfield before the war.

The crew of Brandenburg C.I 62.59 shot down by Olivari on 6 April 1916, *Oblt.* Oesterreicher and *Lt.* Burian. The bus in the back belongs to the 4ª Squadriglia Caproni of the Gruppo di Aviazione per il Comando Supremo.

One week later, the 1ª Squadriglia was re-named, becoming 70ª Squadriglia Caccia. Meanwhile, Olivari had an unlucky duel on April 10 while escorting a Caudron: his Lewis jammed, and it was the enemy that hit his Nieuport repeatedly.

On 4 May 1916 his gun didn't jam, and he shot 50 rounds at an enemy, without scoring a kill. On 16 May 1916 the Austrians launched another massive attack against Udine. Their tractor biplanes, and especially the excellent Hansa Brandenburg C.1, gave them a marked superiority over the Italians. Once again the Nieuports of the 70ª Squadriglia challenged them. Olivari engaged and over Gorizia he fired at a Lloyd C.III that managed to limp back to Aisovizza airfield, behind its own lines, with the crew unharmed. A Caproni crew also claimed a victory on that day, most likely the same one, while Baracca also scored his second victory, shooting down Lohner B.VII 17.42 above Gorizia.

On 9 July Olivari got his third victory over Salcano, a suburb of the contested town of Gorizia. First he attacked an enemy that escaped, and then he had a close combat with another one, that he clearly saw getting hits. Olivari had other inconclusive duels on July 16th, 18th and 27th. On 25 July soldier Olivari finally became *Aspirante*, a rank meaning a trainee for the officer corps.

On 9 August he escorted Caproni bombers that attacked Dornberg. His fourth victory arrived on 25 August by Medea, possibly identified with a Brandenburg of Flik 4 that was hit and forced down on Aisovizza airfield with a wounded observer. His flying log, however, only reports an attack against an enemy above Palmanova and Doberdò.

On 2 September, flying together with *cap.* Bosio, Olivari spotted and attacked an enemy and drove it away. It was typical that, out of a formation, it was always just a few pilots who first saw the enemy and attacked: the few who have the quality of an ace. On 13 September he escorted the Capronis that made a large-scale attack on the military installations of Trieste. On 16 September 1916 he destroyed an airplane described as a big Lohner, near the Starisky mountain. It actually was an enemy Lloyd engaged on a reconnaissance mission; Baracca and Olivari attacked it, soon joined by Ruffo, and with good co-ordination they hit the enemy, killing instantly its pilot *Zugsf.* Morozko. The observer, *Lt.* Anton von Csaby, managed to crash-land the stricken airplane, but he was seriously wounded and died soon after. This was Olivari's fifth victory, making him Italy's first ace. His accomplishment also earned him a silver and a bronze medal, both for the same combats.

Olivari on 1 October was commissioned as a *Sottotenente di complemento del Genio* (reserve sub-lieutenant of the Engineers).

His sixth victory came on 31 October, with a Brandenburg crashed in the valley of the Sava River. This victory deep behind enemy lines corresponds with the Austrian loss of Brandenburg 64.14 of Flik 16 shot down above "Dobrova" (Dobrava) It was not confirmed, until Baracca shot down another enemy aircraft of the same unit, 68.03 of Flik 16 on 25 November: on the captured papers the events were reported, and it was learned that the pilot killed by Baracca, *Korp.* Fritz Fuchs, was the same previously shot down by Olivari. The latter however, was very modest in his claims: his log reports only an enemy "forced down near the lake of Dupla Plassiva" which, if identified with Plessiva, would correspond per-

Right: Classical pose of Olivari in the cockpit of a Nieuport 11.

fectly. On the same day, Olivari flew another mission of Caproni escort, above Dottogliano.

Olivari was particularly renowned for his aerobatics, and his looping was considered unmistakable. It was rumored that some old style high officer didn't appreciate that an upstart, and from such an unglamorous arm as the Engineers, bettered active service officers of the Cavalry. After his sixth victory, Olivari was removed from the front and assigned to special assignments, probably test flying, exploiting his peculiar flying skill. He was no longer reported flying operations with the 70ª until 6 January 1917, when he ferried the Nieuport 17 3134. On 8 January he had a duel with an enemy, but he had to break off when his gun jammed. After that, the squadron papers report no flight of his until 9 March 1917, when Olivari escorted a Caudron and drove away an enemy airplane that was straying too close.

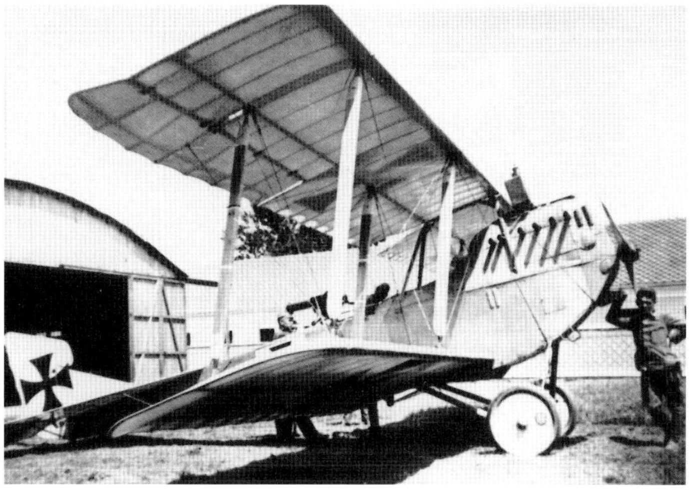

Brandenburg C.I 64.14 of Flik 16, forced down by Olivari on 31 October 1916.

On 15 March the two first Spad 7s were delivered to the 70ª Squadriglia, serials 4688 and 4689, and Olivari and Ruffo were the first ones to fly them, that very day.

Only on 18 March 1917 Olivari returned to victory. While on a cruise over Monfalcone at 9.35 a.m. he spotted two Austrian Brandenburgs that had bombed the area of Doberdò and Ronchi. He gained altitude and then pounced on them, one ran away but the other one stood its ground, accepting combat. The first attack went amiss for a jamming of the gun, but Olivari managed to repair it and attacked the Austrian frontally, hitting it with a full burst and sending it to crash near San Canziano. The duel had been witnessed by no less than the king and Gen. Cadorna, the Italian supreme commander, who drove to the crashed enemy and praised the ace, who had also arrived on the spot by car, shaking his hand. The unlucky crew of Brandenburg C.1 29.53 of Flik 23, *Korporal* Johann Alapy and *Oblt.* Egbert Heintschel Edler von Heinegg, had been hit and the observer was dead. This grieved Olivari, whose whish was to destroy airplanes without hurting men:

> "The enemy aviator lieut. Heintochl had his wife at the Prosecco airfield, who was anxiously waiting for the return of

Pilots and visitors at Santa Caterina: from the left, Olivari, unknown, Ruffo di Calabria, a Russian officer, unknown, *cap.* Tacchini, Francesco Baracca, unknown.

her husband from the dangerous flight. The whole day and night passed without any news of the missing Albatros reaching the Austrian lines. At dawn of the next day an enemy airplane, hidden by the morning mist, flew over our lines and dropped a letter, which a patrol of infantrymen picked up. The letter was written by Heintochl's wife and was addressed to her husband. It had, in the belief that he may be alive, encouragement for his misfortune and exhortations to be strong in captivity. A few hours later, one of our airmen dropped on the enemy lines the letter closed in an envelope, with a note informing the lady that her letter had not been delivered because Lieut. Heintochl was killed in action, in which he had shown a remarkable gallantry. Picked up wounded, he died two hours later in a hospital: military honors were attributed to his corpse."

Brandenburg C.I 29.53 of Flik 23 in service with the Austrian air force. (Photo Cejka).

The press widely reported his victory over this Brandenburg (as usual, called Albatros by the Italians) because of the royal presence. Olivari was awarded in the field a silver medal, his second one. It was also claimed that it was his 10th victory, making him the first Italian double ace: actually, it was only his seventh, and he had claimed far fewer.

The very next day Olivari had a new combat over Medea, but again his gun jammed. Again on 19 April and on 26 April Olivari attacked two enemy aircraft, but had to give up the fight with a jammed gun. On 24 April he performed a different kind of mission: a successful photoreconnaissance flight over Villach, during which three fighters attacked his Spad but he managed to escape thanks to its higher speed. The next day he flew an even longer mission to Hermagor and Arnoldstein, Austria.

Olivari was hugely popular with the local population of Udine that used to go watch the fighters take off from Santa Caterina. It was reported that Olivari was easily recognizable for his skillful flying and his loops.

On 1 May 1917 the 70ª Squadriglia was re-named 91ª, while a new, and different 70ª Squadriglia was formed. The best pilots went to the new unit: Tacchini, the commander, Baracca, Ruffo, Ranza, *serg.* Gorini and of course *s.ten.* Olivari. Thus the 91ª got to be known as *"la Squadriglia degli assi"*, the squadron of the aces. On 5 May 1917 Olivari pursued an enemy and saw it losing altitude by St.Veit. The next day he was posted to a flight of the 91ª attached to the 77ª Squadriglia, at Aiello, in the area of the 3rd Army, by the sea, soon to be engaged in the huge 10th offensive of the Isonzo River. He was very active, and had duels on 8 May and 10 May, both times being hit by the enemy. Then, the tables turned and on 18 May he shot down a large Albatros by Vojscizza, confirmed by ground observers. This victory is not confirmed by Austrian data, but the Italian official list of victories reports a victory on 14 May near the Hermada, that is to say, quite close to Vojscizza. Maybe Olivari's claim was on the 14th and it was recorded a few days later in the 91ª's war diary, as the pilot was away from base.

In the following days he had plenty of combats while engaged in escort missions for Voisins of the 25ª Squadriglia, and he strafed enemy troops. On 24 May 1917 Olivari took part in the large air battle that developed when the Austrians challenged two British monitors that were shelling their coastal positions. He was credited with the destruction of one of two enemy seaplanes, L136 and L137, while *serg.* Leonardi of the 80ª Squadriglia, also located in Aiello, brought down the other one. In the combat, an enemy seaplane fighter repeatedly hit his fighter. The enemy naval aviators were captured by an Italian destroyer, despite the shelling from Austrian coastal batteries that put at risk the rescue of their own men. Later, Olivari met his two victims, who told him: "Don't you know who was at your back, and was scoring hits in your airplane? It was Banfield, the famous Banfield. You are lucky to be alive". Olivari answered: "He is the luckier one of the two." For this action, he received his third silver medal in the field from Admiral Thaon de Revel, chief of the Italian Fleet. As the offensive subsided, Olivari returned to S. Caterina on 2 June 1917.

The next day, 3 June, flying his Spad 4700 he had one more victory, when he attacked in the mountains behind Gorizia an enemy two-seater, with the help of another pilot on a Nieuport 17, marked with a star, possibly the Swiss-born *sergente* Guido Poli. The enemy was forced down in a field behind the S. Marco hill,

Brandenburg 29.53 after being shot down by Olivari at San Canziano on 18 March 1917. Olivari, first left, accepts the congratulations of high officers, while camouflage covers the aircraft to hide it from enemy artillery observers.

The Austrian sepalane L137 in distress, forced to alight during the combat of 24 May 1917.

Olivari the test pilot, in the cockpit of an early SVA with two separate upper wings and external Vickers gun.

while four Austrian fighters tried to avenge it, but Olivari got out of the scrap with the help of the Nieuport. This is the combat report:

> "At 9.30 a.m. between M. Corada and M. Santo at 4200 mts I attacked an enemy aircraft that after my first burst went down steeply. My gun had jammed and I followed it in its dive and then I attacked it again. It landed in a field behind S. Marco in the plain between the castle and Aisovizza. In this combat I also saw a Nieuport 110 hp marked with a star that attacked. As I was flying low at 500 mt over the field where the aircraft laid, four enemy airplanes attacked me and in several combats, always with the help of the Nieuport I managed to drive them away."

Three days later, on 6 June, the whole 91ª Squadriglia moved to Istrana, to support the offensive on the Asiago front, the Ortigara battle. Just before leaving, though, Olivari joined a running combat between a Nieuport (flown by Baracchini of 81ª Squadriglia) and an enemy two-seater that broke up in the air and crashed near Gargano.

On June 17 Olivari was in Milan, to be awarded a commemorative medal by the Air National League together with Baracca and bomber pilots Ercole and Salomone. On 25 June Olivari had a long and heated duel with an enemy fighter "with crossing struts" (that is, a Brandenburg KD) that was inconclusive, the Italian's gun jamming after having fired 150 rounds.

On 6 July 1917 Olivari left the squadron, seconded to the Command of Aeronautics in Turin. He cooperated with the Ansaldo Company in the design of an indigenous fighter, derived from the large SVA, which was developed as the A.1 Balilla. On 5 August he was commissioned as a sub-lieutenant in the active service for wartime merits. He returned to his squadron on 21 August 1917 ferrying a new airplane, SVA serial 6755, together with *ten.* Guido Guidi (future designer and director of the Dornier-CMASA factory) with another one. Baracca in his diary commented: "Olivari has returned to the squadron, and I hope that he will score other victories. His health is very bad, and he cannot always fly as much as he would like."

On the next day he was back in action, attacking an enemy fighter over the Bainsizza and driving it off. He had more escort missions and inconclusive combats until 10 September 1917 when, together with *ten.* Bartolomeo Costantini, he attacked a two-seater escorted by two fighters. The combat was inconclusive according to the Italians, as Olivari's gun had jammed once more after just 20 rounds, but this combat perfectly corresponds to the action in which a Brandenburg of Flik 35/D was forced down.

In September Olivari was mostly engaged in introducing the SVA to the fighter pilots, who immediately disliked the big and

Olivari at Santa Caterina in front of his personal Spad with a red front cowling.

Olivari, now a much decorated sottotenente, with an unknown friend, probably at Genoa or Turin when test flying for Ansaldo.

Another formal portrait of Olivari, now with three ribbons of decorations.

apparently clumsy airplane, derided as a "camel". He was back in action on 28 September, with an escort flight to Capronis, and had a heated combat on 29th, saving a Nieuport from the attack of three fighters. The next day he had a duel with an enemy fighter that he hit and drove off. On 3 October he had another duel against a formation of four fighters, one of them was forced to break formation and escape. Olivari reported seeing an airplane falling down in flames; it was the fighter of *ten.* Di Rudinì, of the 76ª Squadriglia, who was killed. The winners were probably German fighters, freshly arrived on the Italian front.

On 13 October 1917 Olivari was sent together with Ranza on an escort mission to a Pomilio. The Squadron diaries report this event:

> "S.Tenente Olivari though with a nervous and excited disposition, taking off climbed dangerously his SPAD and stalled from 100 meters being killed instantly. Thus disappears one of the most famous personalities of the aviation world, daring pilot, very able, expert technician. He was on the eve of the introduction of the fighter that he had devised, called Balilla. He had shot down 12 airplanes."

He crashed in a way that was later described as mysterious. He had not had a leave in a long time being always engaged in technical missions, so it was reported that the call for the mission came as he was already packing his luggage, ready to see his family before returning to Genoa for the tests of the Ansaldo Balilla and this may have explained his nervousness. The press of the time confused things, reporting for instance that he had just returned from a leave, and much later, unsubstantiated accounts quote his family as having suspected sabotage. It was also written that his last words before his fatal take-off were about George Guynemer, the French ace that he so much resembled.

His funerals in Udine were very big, attended by all the citizenry that identified the young pilot as their defender. Baracca commented:

> "We have all been distressed for the loss of the good Olivari; we have been together for almost two years and ours was like a second family, so the sorrow that has struck us is really big. We cannot understand how an old pilot like him could have done such a wrong maneuver: this was due to a moment of distraction, or maybe of bad nerves. When the accident happened, I was flying. We have gone through very sad

Olivari at Santa Caterina, ready for action. Behind him stands *cap.* Costantini. (Photo Casirati).

days; Tuesday here was his funeral that was very impressive. Sadness will remain with us for a long time."

Then he wrote a very noble letter to Olivari's mother.

Olivari received three Medaglie d'Argento and three Bronzo. His first Bronzo was assigned in the field for the victory on 7 April (Baracca for a similar feat on the same day received an Argento, but Olivari was just a soldier then), his first Argento referred to his victories on 25 August and 16 September 1916 (the latter victory was also rewarded with a second Bronzo); his second Argento was also assigned in the field following his victory of 18 March 1917; the third Argento was also issued in the field, for the two victories of 18 and 24 May 1917; finally his last posthumous Bronzo was awarded after the war, with a citation that summed up his life as a pilot: "A fighter pilot, in every flight he showed his prowess and his valor. After having shot down 12 enemy airplanes and having had many air combats, he gave his young life to the fatherland on account of a flight accident. Santa Caterina, 13 October 1917." This lowly decoration was absurd, and one can only comment that young Olivari, who wore out his health in the cockpit and on the drawing board, deserved better.

In 1921, the airfield of Ghedi, then the operative center of fighters, today a base of Tornadoes, was named after him. He is buried in a chapel in the little cemetery of San Maurizio Canavese, next to Turin's international airport.

## Luigi Olivari
## 12 confirmed victories

| Claim | Vict. # | Date | Sqn. | Aircraft | Hour | Place | Victim |
|---|---|---|---|---|---|---|---|
| 1 | - | 2.4.1916 | 1 | Nieuport | - | Cormons | EA |
| *Korp. Franz Kolodzinski UNH/Fähnr. Eugen Csutka UNH, Flik 12, Alb. B.I 24.61* | | | | | | | |
| 2 | 1 | 7.4.1916 | 1 | Nieuport | 7 | Udine | EA |
| *[with cap. Tacchini and Bolognesi] Lt.i.d.R. Bogut Burian POW/Oblt.i.d.R. Johann Oesterreicher POW, Flik 2, Br. C.I 62.59* | | | | | | | |
| 3 | 2 | 16.5.1916 | 70 | Nieuport | - | Gorizia | EA |
| *[with cap. Carta, ten. Sala & sold. Saletta, 2ª Sq.] Korp. Karl Pichler UNH/Fähnr. i. d. R. Hermann Mehrfurt UNH, Flik 2, Lloyd C.III 43.65* | | | | | | | |
| 4 | 3 | 9.7.1916 | 70 | Nieuport | - | Salcano | EA |
| 5 | 4 | 25.8.1916 | 70 | Ni.11 | - | Medea | EA |
| *Oblt. Alfons Veljacic UNH/Oblt. Delimir Kahlen WIA, Flik 4, Br. C.I 64.01* | | | | | | | |
| 6 | 5 | 16.9.1916 | 70 | Ni.11 | - | Mount Stol | EA |
| *[with cap. Baracca & ten. Ruffo, 70ª Sq.] Zugsf. Franz Morozko KIA/Lt.Anton von Csaby KIA , Flik 16, Lloyd C.III 43.74* | | | | | | | |
| 7 | 6 | 31.10.1916 | 70 | Ni.11 | - | Sava Valley | EA |
| *Korp. Fritz Fuchs UNH/Lt. Ladislaus Pasiut UNH, Flik 16, Br. C.I 64.14* | | | | | | | |
| 8 | 7 | 18.3.1917 | 70 | Ni.17 3127 | - | S.Canziano | EA |
| *Korp. Johann Alapy WIA/Oblt. Egbert Heintschel Edler von Heinegg KIA, Flik 23, Br. C.I 29.53* | | | | | | | |
| 9 | 8 | 18.5.1917? | 91 | Spad | - | Vojscizza | Two-seater |
| 10 | 9 | 24.5.1917 | 91 | Spad 4700 | - | Adriatic | Seaplane |
| *Fähnr. i. d. R. Alfred Riesner POW/Fl.Gast. Karl Plasil POW, SFS Triest, L136 or:* *Stb.M. Hirnickl POW/Stb.M. W. Wirkner POW, SFS Triest, L137* | | | | | | | |
| 11 | 10 | 3.6.1917 | 91 | Spad 4700 | 9h30 | S.Marco | EA |
| *[with serg. Poli, 70ª Sq. & maybe ten. Novelli, 81ª Sq.] Zugsf. Anton Hoffmann KIA/Oblt. Friedrich Wowy KIA, Flik 4, Br.C.I 129.02* | | | | | | | |
| 12 | 11 | 6.6.1917 | 91 | Spad | 9h30 | Vodice | EA |
| *[with asp. Baracchini, 81ª Sq.] Korp. Veszprémy KIA/Lt. Ernst Pirnos KIA, FlG.1, Br. C.I 229.19* | | | | | | | |
| - | - | 10.9.1917 | 91 | Spad | 10h | Biglia | EA |
| *[with ten. Costantini, 91ª Sq.] Zugsf. Josef Treidler UNH/Oblt. Rudolf Hiller WIA, Flik 35/D, Br. C.I 329.12* | | | | | | | |

Although at the time only eleven of Olivari's victories were confirmed, he is reported in the official list of Aces as having twelve confirmed victories. A wartime magazine reported also a victory on 3 May 1917 over the Hermada, but no documents were found to confirm this event.

# Luigi Olivi

Luigi Olivi was born in Campobasso, in the central region of Molise on 18 November 1894 but his family moved to Ancona when he was still a boy. His father Torquato was a Colonel of the Carabinieri, the reputed military police, and Luigi was raised in an environment imbued with patriotism and sense of duty. He volunteered in the Battaglione Aviatori, the Italian air service, on 31 December 1913 and he was appointed corporal and then, on 31 July 1914, sergeant, and on that same day he was admitted to pilot training, getting posted to Aviano to fly Bleriots. It was the first class for reserve officers, but it was badly organized, and the students took a long time to graduate.

On 23 November 1914 Olivi began his flying instruction at Mirafiori air base, near Turin, and then on 20 March he went to S. Giusto airfield, near Pisa, where he got his pilot's brevet on 16 June 1915, when Italy had just entered the World War. In July he was posted to the training airfield at Somma Lombardo (Malpensa) where he had a landing accident with a Macchi Parasol and was slightly wounded. On 20 October he was qualified as a pilot of Caudron G.3 and on 7 November, now with the brevet of military pilot and the commission of *Sottotenente* in the Arm of Genio (engineers) he was assigned to the 2ª Squadriglia Aviazione per l' Artiglieria.

This artillery-spotting unit was in fact equipped with the Macchi Parasol, or P.M., a high wing monoplane that had caused a number of accidents and that was very much disliked. Not much is known about Olivi's activities with the 2ª Artiglieria, which must have been quite limited anyway, considering the winter season and the mediocre qualities of his mount. In April 1916 he got a promotion to the rank of *tenente*, while in those same days his unit changed its name, becoming the 42ª Squadriglia and also its equipment, as it then flew the trustworthy Caudron G.3, and used them to good effect for topographical survey of the frontline in the area of Gorizia.

In June 1916 Olivi left the 42ª Squadriglia and began fighter conversion on the Nieuport: his class included other pilots, Prince Fulco Ruffo di Calabria and Ferruccio Ranza, who were to be among the first aces of Italy. Olivi made his first flight on the new type on 13 June, and completed his conversion on 29 June.

On 1 August 1916 he was assigned to the 76ª Squadriglia da caccia, equipped with Nieuport 11s. This unit had reached the front only three months before, and was based near the village of S. Maria la Longa, a few miles from the beautiful walled city of Palmanova, one of the finest examples of Renaissance fortifications in Italy. Among its pilots one had gained distinction in air combats, *sergente* Mario Stoppani, already credited with two air victories.

Fighters used to take off on alarm when frontline observers gave notice of Austrian aircraft taking off from nearby Aisovizza, but not all pilots had the class of Stoppani. A few days after Olivi's arrival, on 24 August, there was a grotesque event: *caporale* Guido Iori, during an escort flight to a Voisin, didn't notice that he had ran out of fuel, and so he landed in enemy territory delivering to the Austrians at Prosecco airfield a brand new Nieuport 11, serial 1615, that they extensively tested giving it serial number 00.27.

Tenente Luigi Olivi

Olivi had plenty of class, and he distinguished himself very quickly. On 24 September he had four air duels, from which he returned with a wing riddled by bullets. On 3 October he attacked an enemy two-seater and saw it going down and force landing near Aisovizza air base. That airplane was a Brandenburg C.1 of Flik 2, and its observer was *Lt.i.d.R.* Leopold Hirth, who was badly wounded and died shortly after in a hospital. It was Olivi's first victory.

His second victory arrived a few days later, on 11 October. Austrian airplanes of Flik 19 were out in a bombing raid against the Italian positions at Lucinico, south of Gorizia. Olivi managed to intercept one of them together with Stoppani and he fired forty bullets at the enemy that was seen falling down near Biglia. Actually, the airplane managed to limp back to its airfield with a wounded gunner. At that time it was not clear when to consider shot down an aircraft, and from the ground any airplane that was seen diving out of control or flying irregularly in any way, was easily reported as shot down. Still, it was apparent that the gallant Austro-Hungarian aircrews were now paying a heavier and heavier toll.

On 28 October Olivi's logbook reports an experimental flight with a fighter equipped with Le Prieur rockets. Olivi flew to combat twice on 23 November and both times he intercepted the enemy. The first time he failed to score but the second time, in the afternoon, he shot down an enemy airplane that crashed near Biglia,

Olivi, left, with a Macchi Parasol. This airplane was unstable, so a stabilizing panel was added to the upper *cabane*, or flying wires truss.

The reason why the Caudron G.3 quickly replaced the Macchi Parasol: this is the accident of Luigi Olivi on 24 July 1915.

beyond the Italian lines. It was a Brandenburg of Flik 4 that soon before had forced down an Italian Caudron, but the two Austrian airmen had little time to rejoice of their success: the Nieuports attacked them and wounded them both: the observer died after a forced landing in their own territory, while Italian artillery opened fire to destroy the wreck. Olivi had shown outstanding marksmanship: he had fired only one magazine of 50 bullets, but the enemy airplane showed no less than 21 bullet holes.

His fourth victory arrived only two days later, when another Brandenburg, of Flik 23, was shot down near Schönpass. On the ground the battle raged, as the Italian soldiers finally managed to conquer Gorizia ("damned Gorizia" as they called it in a forbidden war song) and air activity was correspondingly very intense.

Up to the end of the year Olivi continued his combat sorties, and he didn't rest even in these days reserved for religious holidays. On Christmas Day 1916 he had a combat, but he had to break it off and force land at Borgnano, the Italian airfield closest to the frontline, as his engine had failed. His adversary maybe had been hit, because observers reported it gliding "abnormally" and landing at Aisovizza.

By the end of 1916 Olivi, with four credited victories in just five months, had become well known thanks to the coverage of the press, and he competed in terms of notoriety and victories with pilots as famous as Olivari, Baracca, Ruffo and his squadron mate Stoppani. At that time of the war, four victories were a big accomplishment, which earned Olivi the Medaglia d'Argento al valor Militare, with a citation that reported the combats of 3 and 11 October and 23 November.

As winter set in, air activities were reduced, Olivi flew almost every day but he had only one air combat in January and two in March, on the 17th and 20th, and both times he drove away an enemy plane. Meanwhile in February 1917 the 76ª Squadriglia moved from S. Maria la Longa to Borgnano. He had two similar engagements on 12 and 13 April. On 22 April the commander of the 76ª Squadriglia, De Carolis, was promoted to the rank of *maggiore* and rotated, so Olivi became commander of the 76ª. Such a position was usually assigned to officers of Captain rank, and of the permanent active service, not *"di complemento"*, (reserve) as Olivi, who was only 22 years old, and this fact is proof of how highly he was regarded. Higher commands surely considered that his command was only temporary, and yet he remained commander of the 76ª until the day of his death.

Stoppani had left the squadron to become a test pilot, but there was another new fighter who was earning a name for himself, *serg.*

A double portrait of Luigi Olivi, mirrored in the rear view mirror attached to the windscreen of a Nieuport 11.

Not very serious pilots of the 76ª Squadriglia making derogatory gestures. Olivi is the second from right, sitting on the wheel of a Nieuport 11.

Giovanni Bartolomeo Arrigoni. The unit also had its first combat casualty on 14 May, when *ten.* Francesco Broili was shot down and killed, victim of *Zugsf.* Fejes and *Lt.* Tahy of Flik 19. Broili, when a Caproni pilot with the 7ª Squadriglia, had already been shot down, on 3 December 1916, a victim of pilot Adolf Heyrowsky, flying a Brandenburg of Flik 19.

The 76ª Squadriglia also received a few new Nieuport 17 fighters, which complemented the now obsolete Nieuport 11, which was just too slow and underpowered to effectively intercept Austrian two-seaters. Olivi first flew a Nieuport 17 110 hp on 3 April 1917, and since April 20 he flew exclusively on the new model. His combat introduction with the more powerful fighter was not a success, however. On 12, 13 and 19 April he had engagements with enemy airplanes with no result, then on 20 April he went after an enemy two-seater firing 50 bullets and forcing it back across the lines. The Austrian crew, however, still smarted for a fight, came back, got behind Olivi's tail, peppered him with bullets and then finally went away. The Austrian pilot possibly was Karl Benko of Flik 28, flying Brandenburg 29.65 with observer Franz Grof.

On 24 April Olivi attacked an enemy airplane over Ranziano, that got away diving at full speed back to its lines, while poor Olivi had an engine failure and smashed the landing gear and tail skid of its airplane on landing. The next day was just as unlucky, because returning from a Caproni escort flight he made another bad landing and put out of action another fighter.

As good weather returned, the Italian army went on the offensive again, and air activity increased as a consequence. Olivi flew no less than 34 missions in May, and had seven combats. Finally on 28 May 1917 Olivi attained his elusive fifth victory, which turned him into an ace. It was not an easy success: his enemy was a late model Brandenburg C.1 of Flik 32, which was engaged in a standing patrol in the area between Oslavia and Plava. Two top pilots attacked it over Mount S. Marco, Olivi flying his Nieuport 17 and *maggiore* Piccio flying a Spad 7.

The two Austrian airmen defended themselves very well and hit Olivi's Nieuport in its engine cowling, fuselage and horizontal planes, but after a long and hard dogfight the observer was mortally wounded and the pilot, also seriously wounded, managed a crash landing in Austrian-held territory destroying his airplane.

At that time, besides a few Fokker monoplanes received from Germany and which were based at Aisovizza airfield, the Austro-Hungarians used the excellent Hansa Brandenburg C.1 in all roles, including the one of fighter, and that plane was perfectly adequate to intercept such obsolete crates as the Italian Caudron, Voisin and Farman two-seaters. But things began to change for the Austrians too, and Olivi realized it on 30 May 1917, when he had his first dogfight with an enemy single-seat fighter, probably a Brandenburg KD, which managed to get behind the Italian's tail and to hit the right wing and the landing gear.

On 17 June 1917 Olivi scored his final victory. Some sources in the past wrongly credited him with two victories in his last day, actually another member of his squadron, *sottotenente* Bonavoglia, claimed as shot down the first Austrian airplane. Early in the morning, it was just 4.30 a.m., five Austrian airplanes were spotted as they headed toward Cormons. All efficient fighters of the 76ª took off, and *s.ten.* Bonavoglia attacked one of them and claimed it shot down over Ranziano. Olivi on his own intercepted an enemy just above Borgnano airfield, then the base of his squadron, pursued it all the way to Villa Fausta and managed to hit it and shoot it down across the Austrian trenches at Merna.

Olivi in the cockpit of Nieuport 11 serial 1651. The diagonal black bands indicated the flights within the squadron. *(Photo Lucchini).*

Olivi returned to base and landed. Two hours later he took off on Spad 7 140-hp serial 5403. According to his surviving logs, he seems never to have piloted a Spad before; probably he did it this time because Spads were not used just as fighters, but also for fast photoreconnaissance, and they could be equipped with a camera. So maybe he rushed to document the destruction of two enemy airplanes with the unquestionable proof of photographs before the Austrians could recover them, and this cost him his life.

He took off from Borgnano together with *s.ten.* Virgilio Appiani, who was flying a Nieuport. A few miles east of the airfield, near Moraro, as they were still at low level, the engine of his Spad failed and the propeller stopped. Olivi tried to turn back gliding to his airfield, but the characteristics of the Spad, heavier than the Nieuport to which he was accustomed, proved fatal for him. His airplane flew too flat, stalled and went into a spin. Olivi could not recover

Nieuport 11s of the 76ª Squadriglia: from the left, Ni 1623 carrying below its cockpit the name "Fringuello" (chaffinch); Ni 1700 with a Happy Hooligan character and a second protuction batch airplane, Ni 2271 with another cartoon character. The two pilots are *s.ten.* Bonavoglia and *s.ten.* Retinò.

Another line-up of the fighters of the 76ª Squadriglia. The frist one is 2271, its cartoon insignia appearing only on the right side, then two "second production batch" aircraft, the first of which is armed with Le Prieur rockers, then 1700 with its Happy Hooligan on both sides of its fusellage, one more and then 1623 "Fringuello" with its name on both sides. Only the last but one airplane is a French Nieuport 17.

Ni 1700

Ni 1700

and crashed to his death. His skull and backbone were broken, and death was immediate.

During his whole career as a fighter pilot, Olivi had flown 180 combat sorties, facing 48 air combats with the enemy, and scoring 6 confirmed victories. He posthumously received a second Silver Medal, issued in the field on the day of his death, which in its citation reports 102 combat missions, 31 air combats and two air victories in his final eight months of service.

Initially he was buried in the cemetery of Langoris, then his remains were moved to the memorial cemetery of Oslavia. The airport of Loreto, near his hometown Ancona, was named after him.

Right: One more picture of Olivi in the cockpit of another Niueport, serial 1636.

Below: Brandenburg C.I 29.54 of Flik 23, brought down by Olivi on 25 November 1916. (Photo Masajdek).

The pilots of the 76ª Squadriglia in front of a French-built Nieuport 17, serial N 3130, and armed with a single Lewis gun, in early 1917. From left: *sold.* Luigi Fanti, *cpr.* Edoardo Olivero, *serg.* Virgilio Appiani, *s.ten.* Giuseppe Retinò, *serg.* Otello Venchiarutti, *cap.* Ettore De Carolis, commander, *ten.* Luigi Olivi, *ten.* Ernesto Bonavoglia and *serg.* Giovanni B. Arrigoni. On the fuselage of the Nieuport there is a pet monkey.

## Luigi Olivi
## 6 confirmed victories

| Claim | Vict. # | Date | Sqn. | Aircraft | Hour | Place | Victim |
|---|---|---|---|---|---|---|---|
| 1 | 1 | 3.10.1916 | 76 | Ni 11 | 9h | Aisovizza | Two-seater |
| *Unknown pilot/Lt. i.d.R. Leopold Hirth DOW, Flik 2, Br. C.I 68.01* | | | | | | | |
| 2 | 2 | 11.10.1916 | 76 | Ni 11 | - | Biglia | Two-seater |
| *[with serg. Stoppani, 76ª Sq.] unknown pilot UNH/Korp. Gustav Weiser WIA, Flik 19, Br. C.I 61.72* | | | | | | | |
| 3 | 3 | 23.11.1916 | 76 | Ni 11 | 15h30 | Biglia | Two-seater |
| *Zugsf. titl. Fw. Josef Franke WIA/Lt. i.d.R. Béla Pecsely KIA, Flik 4, Br. C.I 68.10* | | | | | | | |
| 4 | 4 | 25.11.1916 | 76 | Ni 11 | - | Schönpass | Two-seater |
| *Oblt. Andreas von Kammerer UNH/Oblt. Böla Gantzstuckh von Hammersberg DOW, Flik 23, Br. C.I 29.54* | | | | | | | |
| 5 | 5 | 28.5.1917 | 76 | Ni 17 | - | Schönpass-Paskonicze | Two-seater |
| *[with magg. Piccio, 91ª Sq.] Korp. Paul Forgach WIA/Lt. i.d.R. Anton Boeck KIA, Flik 32, Br. C.I 229.01* | | | | | | | |
| 6 | 6 | 17.6.1917 | 76 | Ni 17 | 4h30 | Merna | Two-seater |

## Giuliano Parvis (Giorgio Pessi)

When the great European War began, Italy initially remained neutral, but many groups and forces pushed for its entry in the conflict on the side of the Allies, against the Central Powers, and one of their main arguments was the reunification in the Kingdom of Italy of the last Italians speaking provinces still part of the Austrian Empire. The name of the two main cities in the claimed provinces, "Trento e Trieste", was the rallying call of the "interventionists".

Trento was the most important city on the divide between the German and Italian speaking territories, while Trieste was then, and is still today, the border city between the Latin and Slavic languages and civilizations. In wartime propaganda the Austrian enemy was often portrayed with the physical characteristics of the arrogant German or else of the Moslem Slav, without discriminating among the many ethnic groups and religions that intermixed in the Balkans.

Trieste, in the upper Adriatic Sea, was then the crossroad and meeting point of these worlds. According to the languages, it was Trieste, Triest or Trst. It would not be historically correct to say that all the local Italian-speaking population earned to be united with Italy. It had been a city of the Austrian Empire for centuries, known in Latin as "Tergeste fedelissima", the very faithful Trieste, and it had a great importance in that mixture of races and nations. It was the biggest harbor of the Empire, and a first rate industrial city with shipyards and factories that bordered its shoreline and that ensured prosperity. Its population was 229,000 people. In the census poll of 1910 about 190,000 of them had indicated Italian as their main language, 53,000 Slovene, 12,000 the German language, 2,500 Serbo-Croat and another thousand spoke other languages. The rest of the population was made of foreigners, most of them Italians, "Regnicoli", subjects of the kingdom, as the people of Trieste called them disparagingly, immigrants coming from the bordering regions looking for work. The liberal-national party, which represented the Italian speaking middle class, ran the administration.

When the war broke out in August 1914 the Italians were still considered the allies of Austria and there were popular demonstrations against Serbia and the common "Slavic enemy" and there were riots between the opposing factions. The situation soon changed as Italy became more likely to join the war as an enemy, and not as an ally. Many militants of the Italian nationalist party felt threatened and escaped to Italy while they still could. Among them there was a young *triestino*, Giorgio Pessi.

He was born in Trieste on 17 November 1891, son of Edoardo and Maria Pitteri. His mother was the cousin of a famous poet of Trieste, Riccardo Pitteri, a close friend of Gabriele d'Annunzio, and the daughter of Ferdinando, mayor of the city. He came therefore from a very important family of strong pro-Italian feelings. He had an education fit for his social status, attending high school in Trieste and then the schools of engineering and architecture at the universities of Vienna and Munich. He spoke fluent German and French, and also some English.

An intense expression of Giorgio Pessi, a.k.a. Giuliano Parvis, in the cokpit of a Nieuport 11. (Photo Casirati).

Right: Parvis as an instructor at Malpensa at the controls of a Caproni bomber. The signed photo, dated October 1916, carries the words "Per Aspera ad Astra".

Escaping to Italy, Pessi volunteered in the army and was commissioned as a *sottotenente* in the Piemonte Reale Regiment, one of the most celebrated units, which gave to aviation its most famous pilot, Francesco Baracca. War of position in the rocky Carso hills was not fit for cavalry, however, and many officers decided to serve in more active arms. Pessi applied to be admitted to the Battaglione Aviatori and he began flying instruction in 1916 at Malpensa airfield. He got his pilot's license on 10 October 1916 flying a Nieuport 10, and then he remained there as an instructor until the spring of 1917.

He was then assigned to fighters, and before that he had to go to Pisa for the aerial gunnery instruction. There he picked up his cover name, Giuliano Parvis, as was required to an *irredento*, (literally: not redeemed) that is to say an ethnic Italian of Austrian nationality fighting in the Italian armed forces. Some of them in fact, captured in combat, were hanged by the Austrians, who considered them traitors. Even late in the war, after the defeat in the battle of mid-June 1918, the Austrian commands showed no mercy, venting their rage for the defeat hanging all the Czechoslovak legionnaires who fell in their hands during the offensive of the Piave. From then on, Giorgio Pessi was for everybody Giuliano Parvis.

For a brief period, Parvis was the commander of the 82ª Squadriglia a new unit that was formed in April 1917 and then deployed at S. Caterina, equipped with obsolete Nieuport 11s, but he had this command until 9 June, when *ten.* Fochessati replaced him, Parvis going to the 78ª Squadriglia at Istrana. This unit, with commander *cap.* Bolognesi, was also equipped mainly with the Nieuport 11, while its few modern Nieuport 17s were reserved for the more experienced pilots. That was a good period for the 78ª, tasked with interception missions against enemy aircraft over the Asiago plateau. On its strength there were some excellent sergeant pilots: Chiri, Fornagiari, Nardini and Magistrini, the last two future companions of Parvis at the 91ª Squadriglia.

He logged his first combat sortie on 14 June. That was a good day for the squadron, *serg.* Nardini scoring a double victory, but Parvis did not get a chance to meet the enemy until 25 June, when over Mount Zebio he found two enemy aircraft that refused combat and flew away. On 1 July he attacked an enemy fighter without results: there was not much that the underpowered Nieuport 80 hp, with a single Lewis gun above the top wing, could do.

On 15 July 1917 Parvis was reassigned to the 91ª Squadriglia at S. Caterina airfield, on the Isonzo front, getting there on the 22nd. The Squadron Diary reports: "Tenente Parvis has reached the squadron, this pilot is getting instruction on the Spad." This excellent fighter, that France was supplying to Italy since the spring, was quite a different airplane, with an in-line 140 hp engine, it was heavier and more demanding than the Nieuports. As there were not enough of them to fully equip a squadron, besides the 91ª, the idea was to supply a few of them to each squadron, to exploit their speed in interception missions and in photographic reconnaissance sor-

Parvis with a Nieuport 11 without armament. (Photo Casirati).

In the 91ª Squadriglia: Parvis next to his Spad with the personal insignia of the black crescent and a doubtful Giovanni Sabelli. (Photo Casirati).

ties, as they could carry a camera. The 91ª Squadriglia operated also as a conversion unit for pilots coming from different squadrons. It was further work for the unit, but it was a good way for Baracca to get to know new pilots with special qualities and when he picked one of them, he had no difficulty in getting him assigned to his élite squadron. Thus Parvis, who also originally came from the same Piemonte Reale Regiment as Baracca, never returned to the 78ª Squadriglia.

The 91ª then was a real crucible of aces; it was the only squadron formed only by officers. Its commander already had 13 victories, and there were other aces: Ruffo, also with 13 victories, Olivari with 12, Piccio with 6, and other talented pilots were Ranza, Sabelli, Di Rudinì and Costantini.

Parvis converted on the Spad very quickly and on 30 July he flew his first combat sortie, a standing patrol with *tenente* Sabelli. They spotted above Castagnevizza a two-seater of the type "Albatros", as the Italians then universally called the Hansa Brandenburg C.1, with two escort fighters and they engaged them, Sabelli attacking the two-seater and Parvis a fighter sporting a bright red tail, against which he fired 20 rounds without result. Opportunities would soon be abundant: the Italians were getting ready for a new big offensive, later to be called the Battle of the Bainsizza, and air activity was very intense, with frequent combats and successes for the squadron. On 2 August *magg.* Piccio began shooting down two enemies in a few minutes. Parvis was flying with ten. Ranza, they engaged an "Albatros" 2,600 meters over Castagnevizza and the Vippacco and after many attacks they shot it down near Ovcia Draga. It was the sixth victory for Ranza and the first one for Parvis, although after the war it was not confirmed. *Serg.* Sorrentino who witnessed the event reported that the two Spads showed personal insignia, an owlet and a black crescent.

As the offensive was unleashed Parvis flew all sort of missions, including the demanding escorts to Caproni bombers attacking Austrian positions. Since 18 August the pilots of the squadron had to fly this kind of missions every day. Combats with enemy airplanes, including fighters, were frequent, and whenever possible the pilots also went down to strafe the trenches. Austrian losses were considerable, but they still contested the Italian superiority in the air. On 28 August the whole squadron went in an escort mission, with seven fighters. At 6.30 p.m. Parvis saw two enemy fighters attacking an Italian SP pusher biplane over Gargaro. Immediately he went into a dive against one of the fighters and he fired 30 bullets. His intervention was too late, the SP, belonging to the 22ª Squadriglia, was on fire and it crashed on the Mount Sabotino, killing *sergente* Ernesti and *tenente* Piacentino. The enemy KD fighter had to get out of harm's way quickly so his pilot, the ace Godwin Brumowski, was not able to check the demise of the SP, and didn't get confirmation of this victory.

In September a new type of enemy fighter appeared, with V struts like the Nieuport. It was the Albatros D.III, license built in Austria by Oeffag. Parvis did not have much luck, and he had another combat only on 23 September. He lost contact with his wingman, *s.ten.* Ferreri and then a fighter bounced him. He managed to get free from its attack, fired 25 rounds and drove it away.

The next day together with Sabelli he met a two-seater and tangled with its escort fighter, again without results. The opportunity for victory was now ripe, however, as meanwhile Parvis had gained enough experience of air fighting. On 29 September he was flying with Sabelli, at 10.40 a.m. over Monfalcone, behind the Italian lines they saw a two-seater with its escort. Before take-off they had arranged their strategy, so Sabelli attacked the two-seater while Parvis engaged the fighter and drove it away. The two-seater defended itself well from Sabelli's attacks, maneuvering skillfully, until it succumbed, and trailing white smoke it crashed near the lake of Pietra Rossa, in Italian territory.

Combats raged until the end of the month, then bad weather set in, giving to airmen of both sides many days of rest, which they certainly appreciated. In this period there was a sad event, the death of Luigi Olivari, who crashed on take-off on 13 October.

The days to come were to be intense and eventful. The enemy was preparing its sudden offensive that broke through the Italian lines and pushed the front back to the Piave River, nullifying two years of terrible sacrifices of the Italian infantrymen on the Carso. The German troops came with a strong complement of squadrons, both fighters and reconnaissance aircraft, well equipped, more aggressive and better trained than Austrian units. The 91ª Squadriglia was ready to face them, and to make them pay a very heavy price.

Baracca was the first one on 22 October, shooting down two German DFW C.Vs in a few seconds. The offensive began on 24 October, but the weather cleared enough to allow flights only the next day, the 25th. Since the morning observers reported frantic activity of German aircraft in the area of Tolmino. All pilots took off but Parvis, going out with Ranza and Ferreri, was forced to return as his engine gave troubles. His two companions pursued a two-seater but German fighters bounced them, shooting down *s.ten.* Ferreri. Also Sabelli, his friend and wingman in many missions, was shot down and killed that day, but the other pilots of the squadron destroyed six enemy airplanes.

The next day, as news arrived that the front had collapsed and a retreat was possible, Parvis took off for an offensive patrol as wingman to *maggiore* Baracca, who was flying a Spad 13. The couple Baracca-Parvis in the following month produced excellent results: the two pilots co-operated perfectly, and the result was apparent since the first day. At 11.15 a.m. they attacked an "Aviatik" north of Cividale, and shot it down in flames. Soon after they met another one: the Germans defended themselves well, and twice they hit Baracca's fighter but at the end of a very long combat they crashed on flames over Mount Matajur. Baracca had to nurse his damaged airplane to a forced landing.

The following days were dramatic. On the 27th, under a pouring rain, the squadron moved to Pordenone, from where it launched strafing missions against the invaders. On 31 October it moved back to Arcade, beyond the Piave. Flying form Arcade, Parvis reached acedom. On 6 November he flew an offensive patrol with Baracca toward the Tagliamento River. At 10.30 a.m. they met two fighters of the renowned Flik 41/J, the squadron of Brumowski. Baracca engaged the leader and Parvis the wingman. The latter ran away, outdistancing Parvis who then returned to help Baracca, engaged in a frantic dogfight with the other Albatros. The Austrian tried to escape spinning down but the two Italians attacked it until it hit the ground. They saw the pilot motionless in the fuselage, marked with a black band. Hit in the chest, *Oblt.* Rudolf Szepessy Sokoll, who according to Austrian sources had become an ace the day before, shooting down two seaplanes, soon died for his serious wounds. The other enemy pilot, *Fw.* Radames Iskra, when alone exploited the situation and defected, landing in Italian territory. He was interrogated and, maybe trying to please his captors, he told a lot, and slandered his superiors.

The day was not over yet for the two Italian pilots. Soon after they found over Aviano two German two-seaters, resolute as usual, with some escorting fighters very far away. They engaged one of them that spun and crashed near Conegliano. Despite the chaotic situation of the retreat, Baracca and Parvis couldn't resist the temptation to pick up souvenirs, so they drove to the downed German machine. The other one, the Albatros was out of their reach, because Austrian vanguards had already reached the wreck.

The next day Baracca and Parvis flew together again to intercept enemy bombers attacking the bridges over the Piave used by the retreating Italians. At 11 a.m. they tangled with three Albatros D.III that had mauled a lonely Italian Nieuport 17, and then at 11.45 a.m. they met a two-seater coming to. They followed it trying to remain unseen, cutting off its escape route. When the Germans realized they were trapped, they turned and accepted combat, but the Spads were well coordinated and jointly brought it down to crash

Left: The Spad of Parvis taxies to take off at Santa Caterina. (Photo Casirati). Right: Italian soldiers examine the wreck of Brandenburg C.I 329.16 shot down at Pietra Rossa Lake on 29 September 1917 by Parvis with *ten.* Sabelli and *serg.* Rizzotto. (Photo Casirati).

in a field. Once again Baracca crossed the Piave to pick up parts of the downed airplane, risking a lot, because enemy vanguards were very close. He collected the machine guns, black crosses and rudder and then he set the wreck on fire and returned to Arcade. As he wrote to his family: "Thus I shot down my twenty-seventh and Parvis his seventh one."

A few days later, on 15 November, Parvis pointed out to Baracca a German reconnaissance airplane flying over Italian territory. The latter got very close and with a single burst from the guns of his Spad 13 he shot it down two miles from Istrana airfield. This time the victory was credited to Baracca alone.

Air activity was still very intense and in the second half of November the 91ª Squadriglia began to fly patrols with a large number of fighter, to contrast the German Jastas that sent across the frontline large formations of fighters. On 23 November the 91ª shot down three enemy aircraft, the last of them a victim of Parvis, who in this occasion was flying with *tenente* Guido Keller, artist, esthete and dreamer, surely the most eccentric pilot in the history of Italian aviation. Their mission was to escort a group of Capronis bombing Feltre. On the way back they met a two-seater and shot it down, forcing it to land on the slopes of Mount Moncader. It was a tough combat and the enemy observer damaged the Spad of Parvis, lodging three bullets in it. Soon after this combat some fighters attacked them, probably a patrol of the famous Flik 55/J formed by *Lt.* Lupfer, *Zugsf.* Kaska and *Korp.* Munczar. The two Italians managed to disengage and return to base, but their adversaries claimed one of the two "Nieuports" that they had attacked as shot down over Bassano.

Only after landing at Istrana did Parvis realize that the enemy observer was a good shot. His Spad was written off and was returned to an aviation park.

November had been a fruitful month, but it ended badly for Parvis. On the 27th failure of a rod put his Spad out of action and in the ensuing forced landing the fighter turned over, with no harm for the pilot. In the following days the 91ª Squadriglia operated mainly over the Asiago plateau, where the Austrians were striking the last blows against the still unstable Italian line of defense. The unit operated temporarily from Nove di Bassano, closer to the battlefield.

Parvis had his last air combat on that front, on 16 December. With *ten.* Novelli and the Argentinean Olivero he clashed repeatedly against Austrian and German fighters and two-seaters, but the deep cold kept freezing their guns and they didn't score any victory. A harsh winter then put an end to operations, both in the air and on the ground. While Baracca, Piccio and Ruffo went to Turin to evaluate new fighter models, the squadron saw limited activity, and only Ranza scored victories against two of the usual German photo reconnaissance airplanes checking from high altitude on movements around Treviso.

Besides the horrible weather, the 91ª Squadriglia had constant problems due to the new engines of Italian production. Night bombing raids on airports caused the loss of many Italian aircraft. Parvis went on leave and returned to the squadron on 12 March, but he didn't have time to fly more sorties, because on the 16th he was

Albatros (Oeffag) D.III 153.54 of Flik 41/J in which the ace Szepessy Sokoll was shot down by Parvis and Baracca, 6 November 1917. Baracca in his combat report described the fuselage band as black.

assigned to the Commissariato Generale d'Aeronautica. The reason for this shift was a general order to remove from the front all *irredenti* to move them away from too high a risk of death. Parvis returned to the large airfield of Malpensa, near Milan, where he flight-tested new Caproni bombers produced locally. In a note dated 4 March 1918 there is the proposal to send him to the United States as instructor of Caproni and Farman, instead of the deceased Pittaluga (killed in an accident on 26 February), in consideration of the fact that he was an instructor on Caproni at Malpensa and that he spoke many foreign languages. There was also a suggestion to send him to Foggia, to train American pilots getting instruction on the Caproni.

Parvis was sent to the Italian mission in the United States, reaching New York on 10 August. There until September he flew demonstration flight at Hazelhurst Field with the Caproni bomber that was to be license built in the USA. With *cap.* Ugo d'Annunzio, son of the poet, and Lt. Harold T. Harris he flew No. 42118, the prototype of the Caproni Ca.5 built by Standard in the USA, including a flight with ten newspapermen over New York City, over the Statue of Liberty and back to Hazelhurst. On 2 October one of the engines of 42118 backfired, starting a fire that burned through the wing. As a mechanic climbed out on the wing with a fire extinguisher, Parvis skillfully landed the damaged bomber that could be saved.

During the war Parvis claimed eight victories, of which the 5ª Sezione confirmed six. He was awarded two Silver Medals: first

Another picture of Parvis with his good friend Giovanni Sabelli, whose Spad with the black ladder insignia is parked behind the Spad S1544 with the crescent. (Photo Casirati).

one Medaglia d'Argento in the field, with a citation reporting four victories in the period from 2 August to 26 October 1917, and then another Argento postwar, for four more victories in the period from 6 to 23 November, with the specification that his victim on 23 November was a German airplane. As usual, the citations of medals, usually written soon after the events, are more generous than the strict criteria of the postwar "Bongiovanni Commission".

He returned to Italy, but in 1919 he was sent once more to the United States, where he flew a large trimotor bomber under the Brooklyn Bridge. Then he flew for the Caproni Company in many aviation events all around Europe: London, Berlin, Paris, and Brussels.

Returning to civilian life, he got into business with an old squadron mate, the eccentric Guido Keller. They created a flying school in Turkey, but it didn't last very long, and soon Parvis was back in Italy. In the following years he joined the "Società di Navigazione Aerea Aero Espresso Italiana", an airline from Genoa that flew services across the Mediterranean sea using the large Dornier Wal flying boats built in Italy by CMASA.

Pessi, as he now was called, back to his real name, was director of the line from Brindisi to Istanbul, via Athens and Rhodes. He was a manager, but often he piloted personally the seaplanes of Aero Espresso. On 18 July 1933 he took off with the Wal c/n 147 registered I-AZEE from Athens, heading for Rhodes. He never arrived there, and the seaplane and all its six occupants disappeared without traces.

The Dornier Wal transport flying boat of Aero Espresso Italiano which disappeared in the Aegean Sea ending the life of Giorgio Pessi.

## Giuliano Parvis
## 8 victory claims – 6 confirmed victories

| Claim | Vict. # | Date | Sqn. | Aircraft | Hour | Place | Victim |
|---|---|---|---|---|---|---|---|
| 1 | - | 2.8.1917 | 91 | Spad 7 | 10h13 | Ovcia Draga | Albatros |
| *[with ten. Ranza, 91ª Sq. & maybe serg. Sorrentino, 81ª Sq.]* | | | | | | | |
| 2 | 1 | 29.9.1917 | 91 | Spad 7 | 10h40 | Pietra Rossa Lake | Albatros |
| *[with ten. Sabelli, 91ª Sq. & serg. Rizzotto, 77ª Sq.] Zugsf. Andreas Kreidl KIA/Lt. Erich von Luerzer KIA, Flik 28, Br. C.I 329.16* | | | | | | | |
| 3 | 2 | 26.10.1917 | 91 | Spad 7 | 11h15 | Clabuzzaro | Aviatik |
| 4 | 3 | 26.10.1917 | 91 | Spad 7 | 12h | S. Pietro al Natisone | Aviatik |
| *[with magg. Baracca, 91ª Sq.] two out of the following German crews:*<br>*Vzfw. Bruno Fuchs KIA/Lt. Hans Joachim Schlüter KIA, FA 17*<br>*Uffz. Eduard Fricke KIA,/unknown observer*<br>*Vzfw. Kurt Berroth KIA/Lt. d. Res. Kurt Frommhold KIA, FA 39*<br>*Unknown pilot/Lt. d. Res. Gustav Krannig,*<br>*Unknown pilot/Lt. d. Res. Erich Lucas KIA, FA 19*<br>*Lt. d. Res. Erich Lucas/Lt. d. Res. Werner Schulze KIA, FA 39* | | | | | | | |
| 5 | 4 | 6.11.1917 | 91 | Spad 7 | 10h40 | Portogruaro-Fossalta | Berg fighter |
| *[with magg. Baracca, 91ª Sq.] Oblt. Rudolf Freiherr Szepessy Sokoll von Negyes et Reno KIA, Flik 41/J, Alb. DIII 153.54* | | | | | | | |
| 6 | 5 | 6.11.1917 | 91 | Spad 7 | 11h+ | Godega di Urbano | Aviatik |
| *[with magg. Baracca, 91ª Sq.] Vzfw. Werner Schröder KIA/Lt. d. Res. Albrecht Binder DOW, FA(A) 219* | | | | | | | |
| 7 | 6 | 7.11.17 | 91 | Spad 7 | 14h55 | Orsago | DFW 3955 |
| *[with magg. Baracca, 91ª Sq.] Gefr. Wilhelm Appelt KIA/Lt. Paul Wilkening KIA, FA (A) 204, DFW C.5 3955/17* | | | | | | | |
| - | - | 15.11.1917 | 91 | Spad 7 | 12h30 | Istrana | Aviatik |
| *[with magg. Baracca, 91ª Sq.] Lt. Alfred Müller KIA/Lt. d. Res. Erich Peucer KIA , FA 14* | | | | | | | |
| 8 | - | 23.11.1917 | 91 | Spad 7 | - | Mt. Moncader | Two-seater |
| *[with ten. Keller, 91ª Sq.]* | | | | | | | |

# Pier Ruggero Piccio

Pier Ruggero Piccio was an outstanding personality, a gallant war pilot and also the planner of a new doctrine of air warfare, and then the first Chief of Staff of the independent Italian air force, always admired and respected despite his harsh character. Tall, with a strong nose and dark penetrating eyes, his looks reflected his character, demanding with himself and with others, ready to face in violent discussions both his superiors and his subordinates, in the interest of his service and his pilots.

A very elegant captain Pier Ruggero Piccio, as commander of the 5° Squadriglia Nieuport, riding in a SPA truck. The letters SM of its registration SM 808 stand for Military Service.

Piccio was born in Rome on 27 September 1880 in Via dei Serpenti, a narrow typical street near the Coliseum. He studied in Rome up to the Classical High School and then at the age of 18 he joined the Military Academy of Modena. He was commissioned in 1900 as *sottotenente*, and was assigned to the 43° Reggimento di Fanteria, becoming a *tenente* three years later. The dull garrison life was not enough for this young officer. In 1902 Italy signed an agreement with Belgium, allowing officers to be placed in sabbatical leave to serve in the Congo Free State (then a personal property of King Leopold of Belgium) as post commanders. Many officers took up this opportunity until the Italian Government denounced the agreement, in 1907, after a strong press campaign by people like David Livingston and Roger Casement denouncing the cruelty of the Belgian administration. Piccio spent three years in Congo, as governor of a province, and then in 1906, as his contract expired, he tried to join the expedition of the Duke of the Abruzzi who was climbing the Ruwenzori. Piccio moved to the mountains of Uganda with native porters, but his expedition got lost in the "Region of lava" and had to turn back. In October Piccio left Africa, sailing from Mombassa, and returned to active service in Italy. He visited Paris, then at the peak of its splendor, and in a few days he spent his savings of three years. In 1908 he sailed to Crete, where a multinational force was based after a series of revolts and troubles by factions demanding union with Greece.

By 1909 his interest in aviation was fired. He mailed many requests, and even a petition to the king, asking to be assigned to aviation, but he got no answer, nor did he have enough money to attend the courses of private flying schools. At the end of 1911 he went to Libya, with the 37ª Fanteria Regiment. He was in command of a squad of machine gunners and he earned a Medaglia di Bronzo al Valor Militare during the battle of Homs.

After his return to Italy he finally managed to be assigned to the flying school at Cascina Malpensa. By now a captain, he got his pilot's license on 27 July 1913. Very soon his outstanding qualities became evident; on 20 December he flew a Nieuport monoplane to set the Italian record of altitude with a passenger, reaching 3,800 meters.

In early 1914 Piccio became commander of the 5ª Squadriglia Nieuport at Busto Arsizio, where he showed his capacities as a pilot and as a commander. When Italy entered World War I the 5ª Nieuport was based at Campoformido and it immediately went into action. On 25 May, second day of the conflict, Piccio bombed fac-

Right: Piccio in the cockpit of a Nieuport monoplane. The officer in front is *ten.* Calori, of the 6ª Squadriglia Nieuport, and at war's end commander of the VI Gruppo.

tories at Monfalcone, and then he attacked the power plant of Nabresina, which was feared to provide power to electrify the barbed wire across the Carso. The technical aspects of air warfare had to be invented and improvised in the field. 90-mm cannon shells were useless, thus Piccio got some 162-mm shells that didn't fit in the Nieuport, and so he devised a system of cables and a hole in the fuselage to drop those makeshift bombs.

On 27 June 1915 Piccio took off from Santa Maria La Longa, where his unit had moved, with *tenente* Messi and met the enemy in the air, an airplane that fired at them with a machine gun, while the observer returned fire with a Mauser P08 pistol, with Piccio filling the magazines holding the joystick between his knees. All they could manage was to hit their own propeller, but the Austrian quit the field. On 8 July the two officers also flew a night reconnaissance mission, but the Nieuport monoplane was outdated, and at the end of the month the 5ª Squadriglia was disbanded. For his service in this reconnaissance unit Piccio earned a Bronze Medal, his first decoration as an airman.

Piccio then went to Malpensa for conversion on the Caproni bomber and then he became commander of the 3ª Squadriglia Caproni, formed on 8 October 1915.

On 18 October there was a peculiar accident: Piccio was piloting a Caproni when he spotted an airplane, presumed to be enemy. He aimed the heavy bomber to attack it, but at the last moment he realized it was an Italian Caudron. Meanwhile the Italian artillery started firing on the Caudron, but there were no tragic results, and a round of drinks at the bar marked the end of this adventure.

On 17 January 1916 Piccio flew a long photoreconnaissance mission to Fiume, but it was so cold that he got pneumonia, so he could not take part in the raid to Lubjana of 18 February, in which *capitano* Oreste Salomone earned the first Gold Medal assigned to an airman. When he got well again Piccio finally managed to get himself assigned to fighters: on 1 March he passed command of the 3° Squadriglia to *cap.* Adolfo Resio and then he went to Plessis-Belville, France, for conversion on the Nieuport fighter.

In the late spring of 1916 Piccio returned to the War Zone, at Istrana, as commander of the new 77ª Squadriglia that was then

On 25 September 1916 Piccio reported a combat with an enemy twoseater, and that its observer was wounded. Brandenburg C.I 64.47 of Flik 12 force landed at Wippach after a combat with a Nieuport and a Voisin, its observer *Oblt.* Breitenecker died of his wounds. (Photo *Höfling*).

converting from the Nieuport 10 to the more agile Ni. 11. Two future aces were also serving in that unit, *serg.* Cosimo Rizzotto and *ten.* Ferruccio Ranza. During the offensive leading to the capture of Gorizia on 8 August 1916 the 77ª Squadriglia moved to Cascina Farello, near Aquileia. Piccio flew very much, but he never had the chance to meet Austrian aircraft, and as he returned he used to complain about his bad luck loudly and with a particularly strong vocabulary.

Kite balloons, or *drakens* as they were called then, were particularly important targets, because they directed artillery flights, but difficult to destroy with the plain bullets of a Lewis gun. Piccio asked in vain to the commands for the French Le Prieur rockets, only to discover that the first ones supplied to Italy had gone to a Farman squadron. He managed to retrieve them, fit them on a Nieuport and tested them against an old Austrian balloon inflated for tests at S. Caterina field. In the rain he found his target and fired the rockets. Returning he discovered that he was now under arrest because the balloon was moored on the ground, surrounded by people and by highly inflammable bottles of hydrogen. Finally on 18 October 1916 Piccio attacked a balloon near Mavinhie with Ranza and *serg.* Savio. While Ranza flew higher for protection, the two fighters placed themselves in the sun and dove on the balloon. The

A smiling Piccio next to a Nieuport 11 equipped with the then unusual and precious Le Prieur rockets for balloon busting.

*Cap.* Tacchini, commander of the 91ª Squadriglia at its inception. He partially hides a red heart painted on a Spad at Santa Caterina, which could be Piccio's own aircraft, although the same insignia remained as the unit emblem of the 77ª Squadriglia.

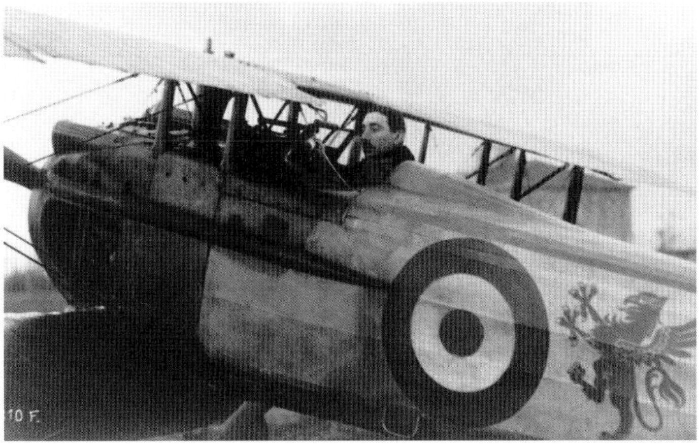
Pictures of Piccio with his own aircraft are very rare. He appears at the controls of a well worn Spad 7 carrying the griffon of the 91° Squadriglia.

On 14 September 1917 Piccio shot down Brandenburg C.1 229.31 of Flik 2/D, this is its wreck.

rockets of Savio failed, but the ones of Piccio hit the balloon and blew it up. This was Piccio's first air victory, which also earned him a Medaglia d'Argento al Valor Militare in the field. His feat had a large echo and it was specifically reported in the War Bulletin the next day: "Over Castel San Giovanni (Ivanigrad) west of Cominciano our airmen set fire to an enemy observation balloon." There was a small diplomatic problem, however, because the French fighter squadron at Venice had supplied the rockets for a joint action with two of their pilots but strangely that morning the Italian car that very punctually used to carry the Frenchmen to Cascina Farello was one hour late, and when the two French pilots arrived at the airfield it was all over, and the wily Piccio had used the rockets himself.

On 10 December 1916 Piccio was promoted to the rank of *maggiore*. On 26 January 1917 he left command of the 77ª to *cap.* Ettore Croce. His personal emblem, a red heart, was taken over as the unit insignia. The last wartime commander of the 77ª, *cap.* Filippo Serafini, later became president of the Aero Club of Bologna, which inherited the red heart and uses it to this day.

On 15 April 1917 Piccio became the first commander of the new X Gruppo. This group was directly assigned to the Supreme Command as a strategic force, and it received the new 91ª Squadriglia, equipped with the powerful Spad 7 fighter and manned by the best pilots. Piccio was not a young man, he was 37 then, and he had an important command, and yet he flew as much as possible, together with aces such as Ruffo, Baracca and Olivari, despite the pressures from above trying to force him to wage war behind an office desk.

His eagerness brought results. On 20 May, during the 10th Battle of the Isonzo, Piccio in the early morning was cruising between Gorizia and Plava. Following the signals of the observers he found a Brandenburg and promptly engaged it. The combat moved down, after the fifth attack of Piccio the enemy gunner stopped firing, the two-seater kept receiving bullets and then smoke started pouring out of its fuselage. Brandenburg C.1 129.52 of Flik 12 crashed in flames east of Plava, killing its crew, *Korporal* Michael Scholz and *Oberleutnant* Ludwig Varga. The Italian pilot celebrated his victory flying aerobatics above Santa Caterina airfield. Later he took care to have a message dropped behind the enemy lines reporting the death of the two aviators.

Eight days later, on 28 May, Piccio attacked a Brandenburg of Flik 32, serial 229.01, at an altitude of 3,500 meters near Mount Sabotino. After his second attack the observer *Lt.* Anton Böck was deadly hit and dropped his gun. The two-seater went into a dive to

The terrible, too often forgotten dark aspect of air warfare: the remains of *Rittmeister* Eugen *Graf* Somssich and *Oberleutenant* Bruno Kain, the unlucky crew of 229.31, laid out on stretchers before military burial.

A classic portrait of the three leaders of Italian fighters: Piccio, Baracca and Ruffo di Calabria, in the spring of 1918. Ruffo, being a prince, refuses to be throttled by a stiff collar.

Gorizia but Piccio, now joined by Olivi of the 76ª flying a Nieuport, didn't let it escape and pressed on. The Austrian pilot *Korp.* Paul Forgach was badly wounded but managed a crash landing near Schönpass, in which the plane was wrecked.

In the same area on 1 June Piccio forced down Brandenburg 29.63 of Flik 19, this time the crew was luckier, the airplane was destroyed but only its observer was wounded. On 6 June Piccio followed his friends of the 91ª Squadriglia as they moved to Istrana on the northern Asiago front to support the Italian attack to the Ortigara peak.

Back on the Isonzo front, in the morning of 29 June Piccio, flying from Campoformido airfield, spotted four enemies who from Ternova headed to Gorizia and he attacked the last one, claiming it shot down on the slopes of San Marco hill. He regained altitude and then he attacked another aircraft, forcing it down in the valley of the Vippacco River. This double victory was specifically reported in the daily Bulletin of the Supreme Command. In this period Piccio received another decoration, the Cross of the Ordine Militare di Savoia, an award for capacities of command.

In the afternoon of 28 July Piccio flew two escort flights for the Capronis bombing the mercury production plant at Idria. Coming back from the second sortie, in the afternoon, he attacked an enemy two-seater near Aidussina. He saw smoke coming out of its radiator but he couldn't follow it down, as six fighters attacked him. He managed to fight his way back to base, and there he complained with his superiors, as they credited him with a victory although he couldn't confirm it. He applied his strict criteria for claim verification first of all to himself, but actually an aircraft of Flik 19 was forced down near Selo, its crew unharmed, on that day.
Piccio himself in a letter to a friend, which a magazine published, described the events of 2 August 1917:

> "That morning I was ready on my airplane, the engine running, and I wondered: Should I take off now? Should I wait more? Then the horn sounded the alarm: 'enemy aircraft toward Mount Corada!'
>
> I give throttle, and I take off, to the Corada: it's in the direction of Tolmino. Go! Go! The engine is running too fast, but let's hope for the best. Over Globokak I see our anti-aircraft shots, and in fact I spot an airplane. Very good. This one is going to go down.
>
> I place myself with the sun on my back to be in the best conditions for aiming. With my eye at the telescope and my finger on the trigger I get behind it. I hear its machine gun singing, no trouble. Let's see who is the better shooter. I must, I want to aim well, and get it with a few bullets. So I wait to be real close before shooting. I fire about twenty rounds but I must stop because I am so close that I almost crash into it. I give rudder; my left wing literally scrapes its tailplane. I lose it from sight below my wings. I complete my turn banking, I straighten and the plane is again in front of me. Heck, it isn't gone yet, it's still here.

Piccio was not always terrible. He could even pet a puppy. The pilot to his right is *asp.* Amleto Degli Esposti.

> It is firing below my plane at very close range. I repeat my maneuver, another burst of twenty rounds. Oh, this time I see the red flame. The airplane is burning, its observer stops shooting, I see a tragedy in front of my eyes. The observer, a burning brazier in the sky, jumps from the airplane. I follow the plane down to the ground. It crashes in a huge flame near the barracks of Tolmino. I return to my airfield, and they welcome me with incredible enthusiasm. Good, good, two aircraft with a few rounds! I don't understand: why two aircraft? I shot down only one, flamed near the Tolmino barracks.
>
> No, no, look at the phonogram: you shot down two aircraft! Good, good, one flamed and one crashed to earth.
>
> That's the way it had been: the one I met and flamed was a different one. The first one had already gone down. That was another one, and it also had gone down. Two airplanes with 30 bullets, not bad at all. And yet to be convinced I had to go and see for myself."

The victims of Piccio's remarkable marksmanship were two airplanes of Flik 12, Brandenburg C.1 129.15, flying a photo reconnaissance mission and its escort Aviatik C.1 37.08. While the observer of the "Albatros" *Oblt.* Alexander Füredi jumped to his death, its pilot *Korp.* Franz Kröppl, although badly burned, managed to bring its plane down and to get out of the wreck. The Aviatik force landed behind enemy lines and was destroyed by Italian artillery, its crew was unharmed: they were the observer Johann von Herberstein and the great pilot Frank Linke-Crawford, the future "Falcon of Feltre".

A weird accident happened on 19 August, after an escort mission to Capronis over the Bainsizza plateau in which Baracca literally snatched a prey from Piccio's hands. Coming back Piccio suddenly lost his balance and felt nauseated. He retched violently, feeling shaken and exhausted, somehow managed to return to Santa Caterina, but he turned over in landing, damaging Spad 5408. Other pilots too felt a similar sickness, which was attributed to the effect of poison gas.

On 29 August Piccio, escorting Capronis, drove away a fighter that was threatening a bomber near Gorizia: it was one of the new Albatros D.III fighters, and the ace remarked its typical V struts, that were reported in the Diary of the 91ª Squadriglia.

In September the weather was good, the squadron flew every day but Piccio was grounded by his much disliked paperwork, and managed to fly only eight sorties, but he exploited them at best, claiming four victories, all of them perfectly confirmed by Austrian sources. The first one was in the morning of 7 September, when he shot down a two-seater of Flik 32/D that was operating between Britof and Gargaro. After his first burst the observer stopped firing, the bullets of the Vickers pierced the radiator producing a stream of vapor, then two more bursts marked the end of the Brandenburg that crashed vertically on a hill between Zagorje and Voglarji. The combat of 14 September was tougher, the "Albatros" over Globokak put up a desperate resistance trying to regain its lines, and before crashing it placed many bullets into the Spad. The Austrians credited a victory to the desperate crew who jumped to avoid being burned alive, but Piccio returned to his field, and only the right wing was badly damaged, and had to be replaced.

A similar posthumous but unfounded honor was attributed to another crew of that same unit, Flik 2/D, who perished on 23 September after another combat with Piccio in the same area. The Italian ace fired 150 bullets to flame the Brandenburg.

An unknown Austrian crew had better luck on 27 September, escaping thanks to a jammed gun on Piccio's Spad. On his return Piccio, whose character and attitude were not at all easy, made such a fuss to the ground personnel that cartoonist Muller, who was at the airfield, drew a funny cartoon. Everybody knew of Piccio's short temper, and *serg.* Aliperta, a good pilot of the 91ª, once having made a mistake had to hide behind his airplane fearing that Piccio would physically attack him. His fury against the armorers was useful, because on 29 September his gun worked well and with just 45 bullets Piccio shot down an airplane in the area of the Giulie Alps. It seems that no less than 30 of those bullets hit a two-seater of Flik 12 that force landed with its crew luckily unscathed.

The black period of Flik 2/D continued on 2 October, when Piccio shot down a third airplane near Auzza. The next day he scored his fifteenth victory, and also he was promoted to *tenente colonnello* for war merit. The Diary of the 91ª Squadriglia reports the event:

> "Maggiore PICCIO took off from Borgnano airfield at about 4 p.m. and met a group of enemy aircraft at 4,000 meters over the Bainsizza plateau. He attacked the first one, a two-seater. He followed it with such violence and insistence that he confused the Austrian airplane and forced it to land within our lines at Leupa near Auzza. Its pilots were slightly wounded and were made prisoner. A burst hit the fuselage of the Maggiore's airplane and a bullet struck the right front strut. He shot 150 bullets."

The two airmen in hospital had the visit of Piccio, who guaranteed he would drop a message about their fate in Austrian territory.

In the dark days of Caporetto Piccio, who now was *Ispettore della Caccia*, Inspector of fighter aviation, didn't spare himself. In the morning of 25 October he had to watch the death of Sabelli, as

The 91ª Squadriglia used in 1918 individual numbers in Roman numerals, repeated on top of the fuselage. This is the Spad 7 "X" of the Argentinean Edoardo Olivero, his persona insignia is an Indian chief's head on a pennant attached to the strut.

his gun was jammed, but by the end of the day he had shot down two airplanes.

As the Italian army retreated behind the Piave, Piccio untiringly organized the fighter units. New pilots arrived to the front and new aircraft replaced the ones lost in the retreat. His office work didn't stop him from flying whenever possible. In the morning of 23 November he attacked two reconnaissance airplanes over the Montello when an unidentified fighter bounced him and shot up badly his Spad, forcing him to land at Istrana airfield.

Before Christmas Piccio, having passed the command of the X Gruppo to *magg.* Gino Matteucci, went to Turin with Ruffo and Baracca, to the Direzione Tecnica Aviazione Militare to test new fighters like the Ansaldo A.1 "Balilla" and the Pomilio Gamma. Also in this case Piccio had the opportunity to show, besides his experience, an earnestness of opinions that disregarded the expectations of others. Namely, he found no reason to prefer the Balilla to the much-appreciated Spad. The officers returned to the frontline on 24 January 1918.

Until the spring he didn't fly very much, on account probably of bad weather and office work. Whenever possible, however, he flew his Spad looking for enemies. On 5 May he received Italy's highest decoration, the Medaglia d'Oro, the Gold Medal, with this citation:

> "Very bold fighter pilot, always ready for every daring deed, from 10 July to 30 November 1917 he shot down eleven enemy aircraft, bringing the total of his victories to seventeen. Wonderful personification of the best qualities of the Italian pilot, ideal leader of the sky fighters, heart burning of enthusiasm, soldier tested in the hardest air events, he is an example to all airmen and the pride of our army.– Sky of Isonzo and Carso, Aidussina, 28 July – Tolmino, 2 August – Volgarù, 7 September – Avscek, 14 September – Kal, 23 September – Ternova, 29 September – Auzza, 2 October – Mesmak, 3 October – Castelmonte, 25 October – Panovizza Wood, 25 October 1917."

Victory returned on 26 May 1918. Cruising along the Piave with *ten.* Keller and Novelli, Piccio found a patrol of nine enemies between the river and Treviso. He shook his wings to signal an attack to his wingmen and he got behind the tail of the last plane in the patrol, lagging behind. A hail of bullets struck the Austrian plane, which caught fire and crashed near Ronchi. In the wreck of Albatros D.III 153.230, Italian soldiers found the remains of *Zugsf.* Franz Hofstaedtler. Other Allied fighters somehow took part in this action and a British pilot, Capt. J.M. Mitchell of No. 28 Squadron also claimed Piccio's victory.

On 9 June Piccio took off early in the morning and at 7.30 a.m. he engaged an airplane over Moriago describing it as *"combat two-seater (new type)"*. The Austrian went in a dive and stopped firing after placing three bullets in his Spad but the Italian ace had to break off combat, because an enemy fighter attacked him. "The enemy aircraft was very fast. I believed I was attacking a fighter. His well-aimed fire hit my aircraft. The intervention of the enemy fighter after my second burst forced me to break off combat. I fired from a distance of 80 to 90 meters and my fire was accurate." Piccio's description would indicate an UFAG C.1 or a Phönix C.1, but the only known event corresponding to this claim happened to Brandenburg C.I 269.62 of Flik 44/D, forced down at Pramaggiore airfield with its crew unharmed.

As the information services warned of a coming Austrian offensive, the Italian Supreme Command organized for defense, and a large combined force of fighters, the *Massa da Caccia*, Fighter Mass, was formed, with Piccio as commander. When the offensive began, on 15 June, this Mass, including Squadriglie 70ª, 72ª, 75ª, 76ª, 79ª, 80ª, 82ª and 91ª, under the leadership of Piccio, acted as a deadly and agile instrument of war, causing very high losses to the Austrian air force, breaking its backbone and gaining immediately air supremacy. Then the fighters could concentrate on strafing and bombing attacks. All contemporary sources, including the Austrian ones, agree in considering Italian mastery of the air one of the determining elements in the failure of the Austrian attack.

On 19 July Piccio attacked at 10.45 a.m. an airplane described as *"Combat Biplace (German production)"* over Grave di Papadopoli. He reported other details about this aircraft in the Notes to his Combat Report:

> "The aircraft is fast = very large and short fuselage = Wings and fuselage with camouflage like the Spad = Very small crosses = Two struts on each side = Fire from the observer with incendiary bullets."

This description of the airplane and the details of this combat perfectly correspond to the loss of UFAG C.I 161.38 of Flik 47/F, forced to a crash landing between S. Michele di Conegliano and S. Polo di Piave. Piccio's bullets wounded its observer, but both crewmembers came out alive from the crashed wreck.

Ten days later, on July 29 Piccio claimed a victory over a fighter described as *"type D. 5"* that he attacked over Motta di Livenza without determining the place of its fall. Ground observers confirmed this victory and Austrian sources report the loss in that locality of Phönix 228.50 with the death of pilot *Zugsführer* Wilhelm Bott, although reporting the presence of three enemy fighters.

On 1 August Piccio together with Lieutenant C. McEvoy, of RAF No. 66 Squadron, shot down Albatros D.III 253.40, that crashed in flames killing *Oberleutnant* Vuk von Hreljanovic. This was his combat report:

> "I was cruising at 4,600 meters above the Piave, when I saw a patrol of five fighters about 400 meters higher than me. I followed them and attacked the last one from below and behind, above Lancenigo. After my first burst at close range and by surprise, my gun jammed. All five of them attacked me and I had to maneuver a lot to disengage as I was trying to clear my gun. Meanwhile a British patrol also engaged combat. As I cleared my gun I noticed a dogfight going on between an En-

glishman and one of the enemy fighters. Both were higher than me about 400 meters, more or less over Lovadina.

I could rejoin the combat over the Piave, and there I started firing again. The enemy fighter that was dogfighting to attack the Englishman started going down as soon as I joined the combat, and I was able to fire some accurate bursts. I had other jams. I followed the enemy aircraft down to 800 meters from the ground and I saw it landing in a field catching fire."

On 5 August Piccio, together with Adriano Bacula, shot down in flames near Conegliano Brandenburg C.I 369 of Flik 57/Rb. Just a few days later, on 8 August he met a new type of reconnaissance airplane that he described as very fast and failed to shoot down, but three days later he managed to bag this new kind of prey, a brand new Phönix C.1 of Flik 12/Rb:

"I took off for a patrol with tenente NOVELLI. I cruised at 5000 M. between Piave and Treviso. As we were in the whereabouts of Grave di Papadopoli I saw an aircraft higher than us coming from inside enemy territory toward Susegana. I turned southeast to Oderzo to let it come inside our lines. From its shape and speed I guessed it must be one of the new types.

The enemy aircraft when over Susegana turned to Treviso. When it realized that our patrol had surrounded it, it tried to escape using its speed, first toward S. Donà, and then to Venice. But maneuvering I was catching up so it began to dive (we were by now at about 5,000 meters) heading decidedly to the Grave. We engaged, it returned fire. Both of my guns jammed. I tried to force it to land in our territory just with maneuvers, but it was useless: it kept on shooting. Meanwhile I managed to clear one of the guns and I fired again. My last burst started a fire and probably killed the pilot, because the plane went down burning and crashed near Maserada within our lines."

By now the Austrian air force, that had suffered crippling losses in the battles of mid-June, was seldom to be seen in the air. Piccio, always looking for new experiences, on 20 August tried a night interception sortie. On 29 September Piccio attacked together with Ranza an airplane *"type Uffag"* that was returning to its lines from Montebelluna. He didn't claim it as a victory, but he was sure that some of his bullets had reached their target, because the gunner didn't return fire.

In the autumn Piccio was again in command of the new *Massa da Caccia*, now formed by Squadriglie 70ª, 71ª, 72ª, 77ª, 82ª and 91ª, and he was active fine-tuning it for the coming final offensive. When the Italians attacked, on 24 October, Piccio as usual for him was in flight at the head of his pilots, and on 27 October he shot down his final enemy near Sacile, possibly Albatros 153.162, its pilot *Oberleutnant* Richard Koderle, commander of Flik 101/G, surviving. Like in the battle of June, the ace did not hesitate to go down and strafe the enemy, facing its ground fire that could be deadlier than the fighters.

Wearing an Austrian overcoat, this is how Piccio reappeared at the 91ª Squadriglia mess after being shot down and evading. He liked the event so much that he gave this signed photograph as a souvenir to *cap.* Riva.

The Italian War Bulletin of 28 October gave special mention of Piccio, an honor so far bestowed only on Baracca, Scaroni and Baracchini (twice): "...eleven enemy aircraft and three kite balloons were shot down in combat. Lt. col. Piccio reached his 24th victory." In reality Piccio's acknowledged victories were 26 and the reason why toward the end of the war they were reduced to 24 is unknown, and above all we don't know which ones were deleted.

In the morning of 31 October he strafed the airfield of la Comina, and then he machine-gunned a train near the Tagliamento River, but some Austrian soldier placed a lucky bullet in the engine of his Spad serial 2962. Bacula from above saw his commander gliding with a still propeller, landing in a field, and then Piccio took off his overalls and waved. The anguished Bacula strafed the grounded Spad and tried to scatter the enemy soldiers to give Piccio a better chance to avoid capture. The loss of their commander was a blow to the morale of Italian fighters. Lacking news, they were fearing the worst, until in the evening of 4 November Piccio, wearing an Austrian overcoat and showing a long beard, simply appeared in the squadron mess, ordered a healthy meal and told his story. The Austrian had captured him bringing him to Villach, but he had escaped rather easily, as the whole enemy organization by now was falling apart. All these events had not tired the commander at all, nor dampened his sense of duty: at 2 o'clock at night Piccio asked for a car to be driven to Padua, to report to the Supreme Command.

The war was over. Later, Piccio was attributed another decoration, his second Medaglia d'Argento, celebrating his role as Inspector of fighter squadrons and his personal leaderhip during the Battle of the Piave, 15 to 25 June 1918.

On 31 December 1918 Piccio was assigned to the D.T.A.M., the technical command of the air service, and went to France, to the experimental establishment at Villacoublay, where he found, among others, the aces Avet and Masiero.

In July 1919 Piccio met at the fashionable "Fouquet's" restaurant of Paris the Belgian ace Willy Coppens. A shared drink on the Champs Elysées marked the beginning of a lasting friendship between the Italian and the Belgian, both of them sharp witted men who didn't mince words. Piccio had some commission in Italy, and then in April 1921 he returned to France as Air Attaché.

On 28 March 1923 the Regia Aeronautica was born, as the independent air force of Italy. Piccio was Commander General of the air force, and swore allegiance to its flag wearing the new blue uniform. In 1925 he was again in Paris, but then the rank of *Comandante Generale d'Aeronautica* was replaced by a new post, Chief of Staff, so Piccio returned to Italy and since 1 January 1926 he was the first Chief of Staff, with the rank of General. He relinquished this post in 1927, after a series of troubled events that included the end of his first marriage with an American woman, and in January 1928 he was back again in France.

On 17 September 1932 he was promoted to the rank of *Generale di Squadra Aerea* (3-stars General) and on 3 November 1933 king Victor Emmanuel III appointed him as Senator. Piccio took part with the Italian delegation in 1932 to the Conference on Disarmament at Geneva as an expert on aviation. Scaroni described his way of participating:

> "...the conference turned very soon in a dull verbal battle that dragged on for months. During one of the usual séances at the table of the aeronautical sub-commission a French delegate launched a tireless and endless discussion on the difference between offensive and defensive weapons. The chief of our sub-commission, the hot General Piccio, ace of aviation and chief of the Fighter Mass at the front faced him. Exasperated by his nitpicking counterpart, Piccio finally exploded. He jumped from his chair, grabbed a heavy glass inkpot from the table and raised it over the head of the speaker crying: "According to you what's this, an offensive or a defensive weapon?" The whole room gazed in silence at this unexpected reaction, and then everybody burst laughing, as a comment to such an impulsive but sincere gesture. Then talks went on, speakers followed one another and the conference ended as could be expected: accomplishing nothing."

General Piccio leads a group of generals hosting a foreign visitor in the late 1930s.The airplane behind is the S.74 airliner I-URBE used by Ala Littoria on the Rome-Paris route.

On permanent leave since 27 September 1935, Piccio remained in Paris acting as Special Counselor of the Regia Aeronautica, without pay, and he performed special tasks for his country, such as Royal Commissioner for the International Exhibition of Paris. Mature age didn't mellow his hot temper, and when he didn't like the international policy of the Italian government, he spoke his mind openly. Willy Coppens wrote that the two met at Geneva in 1940. After having embraced, the Belgian ace remarked that despite their friendship now they were officially enemies. Piccio replied showing a silver cigarette case that he had found on the body of a dead enemy, a squadron commander, with the signatures of the pilots. He opened up his mind with his usual lack of diplomacy: "From 1915 to 1918 Italy waged war to get rid of the spiked helmets, and now Mussolini brought them back to us."

During the war Piccio was in Switzerland, trying to help Italian refugees there during the German occupation, and he acted as a mediator between the French and Italian resistance movements. The third Italian ace and the first Chief of Staff of the Regia Aeronautica died in Rome on 31 July 1965. His son also served in the Air Force, rising to the rank of General and his grandson, with the same name, is now a Pilot with the Tornado strike fighters of the Aeronautica Militare Italiana.

## Pier Ruggero Piccio
## 26 victory claims - 24 confirmed victories

<table>
<tr><th>Claim</th><th>Vict. #</th><th>Date</th><th>Sqn.</th><th>Aircraft</th><th>Hour</th><th>Place</th><th>Victim</th></tr>
<tr><td>1</td><td>1</td><td>18.10.1916</td><td>77</td><td>Ni 11</td><td>-</td><td>Selo</td><td>Kite balloon</td></tr>
<tr><td>2</td><td>2</td><td>20.5.1917</td><td>91</td><td>Spad 7</td><td>9h30</td><td>Plava</td><td>Two-seater</td></tr>
<tr><td colspan="8">[with s.ten. Baracchini, 81ª Sq.] Zugsf. Ludwig Ferschl KIA/Lt. Eugen Csutka KIA, Flik 12, Br. C.1 229.10.</td></tr>
<tr><td>3</td><td>3</td><td>28.5.1917</td><td>91</td><td>Spad 7</td><td>-</td><td>East of San Marco</td><td>Two-seater</td></tr>
<tr><td colspan="8">[with ten.Olivi, 76ª Sq.] Korp. Paul Forgach WIA/Lt. i. d. R. Anton Boeck KIA, Flik 32, Br. C.1 229.01</td></tr>
<tr><td>4</td><td>4</td><td>1.6.1917</td><td>91</td><td>Spad 7</td><td>-</td><td>East di Staragora</td><td>Two-seater</td></tr>
<tr><td colspan="8">Unknown pilot/Oblt.i.d.R. Edmund Bernhardt WIA, Flik 19 a/c FTL</td></tr>
<tr><td>5</td><td>5</td><td>29.6.1917</td><td>91</td><td>Spad 7</td><td>morn</td><td>East di San Marco</td><td>Two-seater</td></tr>
<tr><td>6</td><td>6</td><td>29.6.1917</td><td>91</td><td>Spad 7</td><td>morn</td><td>Val Vippacco</td><td>Two-seater</td></tr>
<tr><td>7</td><td>7</td><td>28.7.1917</td><td>91</td><td>Spad 7</td><td>19h40</td><td>Aidussina</td><td>Two-seater</td></tr>
<tr><td colspan="8">Unknown crew UNH, Flik 19, Br. C.1 129.60 FTL</td></tr>
<tr><td>8</td><td>8</td><td>2.8.1917</td><td>91</td><td>Spad 7</td><td>morn</td><td>Tolmino (Volzano)</td><td>Two-seater</td></tr>
<tr><td colspan="8">Korp. Franz Kröppl WIA/Oblt. Alexander Füredi KIA, Flik 12, Br. C.1 129.15</td></tr>
<tr><td>9</td><td>9</td><td>2.8.1917</td><td>91</td><td>Spad 7</td><td>morn</td><td>Tolmino (Volzano)</td><td>Two-seater</td></tr>
<tr><td colspan="8">Oblt. Frank Linke-Crawford UNH/Oblt. Albert Graf von Herberstein, Freiherr zu Neuberg und Gutenhag UNH, Flik 12, Av. C.1 37.08 FTL</td></tr>
<tr><td>10</td><td>10</td><td>7.9.1917</td><td>91</td><td>Spad 7</td><td>morn</td><td>Zagorje/Voglaji</td><td>Two-seater</td></tr>
<tr><td colspan="8">Korp. Aladar Balasz KIA/Oblt. Arpad Wunder KIA, Flik 32/D, Br. C.1 29.18</td></tr>
<tr><td>11</td><td>11</td><td>14.9.1917</td><td>91</td><td>Spad 7</td><td>8h15</td><td>M. Verli</td><td>Two-seater</td></tr>
<tr><td colspan="8">Rittm. Eugen Graf Somssich KIA/Oblt. Bruno Kain KIA, Flik 2/D, Br. C.1 229.31</td></tr>
<tr><td>12</td><td>12</td><td>23.9.1917</td><td>91</td><td>Spad 7</td><td>19h</td><td>Chiapovano Valley</td><td>Two-seater</td></tr>
<tr><td colspan="8">Fw. Karl Pichler KIA/Lt. i. d. R. Alphons Kratochwill KIA, Flik 2/D, Br. C.1 69.44</td></tr>
<tr><td>13</td><td>13</td><td>29.9.1917</td><td>91</td><td>Spad 7</td><td>14h</td><td>Ternova</td><td>Two-seater</td></tr>
<tr><td colspan="8">Maybe: Korp. Alois Pauk UNH/Hptm. Eduard Vallo UNH, Flik 12, Br. C.1 129.44 FTL</td></tr>
<tr><td>14</td><td>14</td><td>2.10.1917</td><td>91</td><td>Spad 7</td><td>-</td><td>Bainsizza</td><td>Two-seater</td></tr>
<tr><td colspan="8">Fw. Johann Szakrànszky KIA/Lt. Georg Mücke KIA, Flik 2/D</td></tr>
<tr><td>15</td><td>15</td><td>3.10.1917</td><td>91</td><td>Spad 7</td><td>16+</td><td>Leupa/Auzza</td><td>Two-seater</td></tr>
<tr><td colspan="8">[with serg. Fornagiari, 78ª Sq.] Korp. Josef Czerip WIA, POW/Lt. Bruno Bèlohlàvek WIA, POW, Flik 53/D, Br. C.1 329.20</td></tr>
<tr><td>16</td><td>16</td><td>25.10.1917</td><td>91</td><td>Spad 7</td><td>11h</td><td>Panovizza</td><td>Two-seater</td></tr>
<tr><td colspan="8">[with cap. Baracca, 91ª Sq.]</td></tr>
<tr><td>17</td><td>17</td><td>25.10.1917</td><td>91</td><td>Spad 7</td><td>-</td><td>Castelmonte</td><td>"Aviatik"</td></tr>
<tr><td colspan="8">[with ten. Fochessati, 82ª Sq.] One of the following:<br>Lt. Paul Dotzenrodt KIA/Lt. Robert Lessing KIA, FA 17<br>Lt. Walter Conrad KIA/Lt. Albert Tawainer KIA, FA (A) 232<br>Vzfw. Friedrich Moritz KIA/Lt. Franz Neumarket KIA, FA 39</td></tr>
<tr><td>18</td><td>18</td><td>26.5.1918</td><td>91</td><td>Spad 7</td><td>8h40</td><td>Ronchi</td><td>Fighter</td></tr>
<tr><td colspan="8">[with ten. Novelli & Keller, 91ª Sq., Lieut. Mitchell, No.28 Sqn. RAF] Zugsf. Franz Hofstädter KIA, Flik 42/J, Alb. D.III 153.230</td></tr>
</table>

| Claim | Vict. # | Date | Sqn. | Aircraft | Hour | Place | Victim |
|---|---|---|---|---|---|---|---|
| 19 | 19 | 9.6.1918 | 91 | Spad 7 | 7h30 | Moriago | Two-seater |
| *Maybe: unknown pilot UNH/Lt. Alfons Benyovszky UNH, Flik 44/D, Br. C.I 269.62 FTL* | | | | | | | |
| 20 | 20 | 19.7.1918 | 91 | Spad 7 | 10h45 | S.Michele/S.Polo | Two-seater |
| *Fw. Robert Turek WIA/Oblt. Andreas Sitterli WIA, Flik 47/F, Ufag C.I 161.38* | | | | | | | |
| 21 | 21 | 29.7.1918 | 91 | Spad 7 | 10h10 | Motta di Livenza | "D5" fighter |
| *Zugsf. Wilhelm Bott KIA, Flik 37/P, Phönix D.I 228.50* | | | | | | | |
| 22 | 22 | 1.8.1918 | 91 | Spad 7 | 11h45 | S. Polo/Cimadolmo | Fighter |
| *[with Lieut. McEvoy, No.66 Sqn. RAF] Oblt. Vuk Edler von Hreljanovic KIA, Flik 56/J, Alb. D.III 253.40* | | | | | | | |
| 23 | 23 | 5.8.1918 | 91 | Spad 7 | 09h10 | Conegliano | Two-seater |
| *[with s.ten. Bacula, 91ª Sq.] Fw. Josef Richter KIA/Oblt. Josef Karner KIA, Flik 57/Rb, Br. C.I 369.38* | | | | | | | |
| 24 | 24 | 11.8.1918 | 91 | Spad 13 | - | Maserada | Two-seater |
| *[with ten. Novelli, 91ª Sq.] Oblt. Albert Graf von Herberstein KIA/Lt. Otto Krakowitzer KIA, Flik 12/Rb, Phönix C.I 121.26* | | | | | | | |
| 25 | -? | 29.9.1918 | 91 | Spad 7 | - | Maserada? | Two-seater |
| *[with ten. Ranza, 91ª Sq.] Zugsf. Alois Breier WIA/Oblt. i. d. R. Julius Karikas WIA, Flik 22/K, Ufag C.I 123.24 FTL* | | | | | | | |
| 26 | -? | 27.10.1918 | 91 | Spad 7 | - | Sacile? | EA |
| *Maybe: Oblt. Richard Koderle UNH, Flik 101/G, Alb. D.III 153.169* | | | | | | | |

# Orazio Pierozzi

The first Italian naval ace was born on 8 December 1889 in San Casciano Val di Pesa, a beautiful town in the hills of Tuscany, very close to Florence. Orazio Mario Pierozzi was the son of Gustavo, the local doctor, who could boast of having been one of the "1000", the volunteers with the national hero Giuseppe Garibaldi, who in 1860 had delivered the Kingdom of Naples to King Victor Emmanuel and brought the unification of Italy.

Rather short, with a sort of impish vitality, Orazio Pierozzi spent a comfortable and pleasant youth in his land of art and beauty, but instead of following his father's career in medicine, he joined the Naval Academy in 1908. This highly reputed institution, located in Livorno (Leghorn) turned him into an officer and a gentleman. In 1911 he became an *Aspirante* (Officer Cadet) and soon after he became *Guardiamarina*, ensign. Each summer of the years 1909, 1910 and 1911 he took cruises on the *Etna* training ship, and he visited the United States (where, in New York harbor, he witnessed and photographed a flight by Wilbur Wright). In October 1911 he was embarked on the battleship *Dandolo*. His classmates included the crown prince, Umberto, and many young men destined to become important naval aviators, like Giorgio Fiastri, Mario Primicerio, Carlo Daviso, Umberto Magaldi, and the future Atlantic flyer Francesco de Pinedo. A family tale that unfortunately cannot be checked out is that he met in Germany and befriended the young Manfred von Richthofen.

He became *Sottotenente di Vascello* in January 1914 and *Tenente di Vascello* on the last day of 1915. He had been embarked on the *R.N. Napoli* from September 1914 to October 1915, then on the *R.N. Coatit* until December and on the R.N. *Duilio* until 11 June 1916.

On 15 June 1916 he became a student pilot, at the military flying school of Sesto Calende, near the SIAI works, by the Lake Maggiore in Northern Italy and was quickly trained, on F.B.A. flying-boats, being licensed as a *Allievo Pilota* on 14 October and as seaplane military pilot on 11 November 1916.

When in flying training, Pierozzi was considered reckless, and at the same time with an excellent technical preparation. Disregarding the advice of his instructors, he banked his seaplane too much, until it went into a spin. Instead of panicking, Pierozzi coolly analyzed his experience as it was happening, and concluded that spinning was not a sudden but a controllable phenomenon, and that the best solution was not to try all kinds of hectic movements, but to throttle down the engine and leave the controls free, letting the plane re-stabilize itself. The fall caused him a broken leg, but when asked if he wanted to keep on flying, he answered:

> "It would be absurd to keep on flying after a fall whose causes are mysterious, but flying for me is now a well determined choice, because I have reconstructed the fall in every aspect, and I got from it precious information."

He was assigned on 16 October 1916 to the naval base of Brindisi, where an air station was located. In 1916, the Italian naval aviation forces at Brindisi were puny. Aircraft production was limited, and most new seaplanes went to the North Adriatic theater or to Valona, Albania. The strategic situation in that theater was complicated: the Italians, with relevant reinforcements from the Royal Navy and the French *Marine Nationale*, had to supply their army in Albania, through the port of Valona, and they also had to close the Adriatic at its narrowest point, the channel of Otranto, to preclude

Orazio Pierozzi in front of his first Macchi M.5 serial 7087 at Brindisi.

the raids of enemy submarines in the Mediterranean. The Austrians held the eastern shores of the Adriatic and the northern part of Albania, with the port of Durazzo. The two opposed naval air forces were engaged in patrols, reconnaissance and occasional raids against enemy bases but generally, throughout 1916, the *K.u.K. Seeflieger*, equipped with much better seaplanes, held the ascendancy.

On 30 December 1916 Pierozzi, with his observer Giorgio Fiastri, took off for a mission of reconnaissance over Durazzo, but he found himself in a bank of fog, and had to abort, while two other seaplanes with pilots Maddalena and Rigobello completed their mission. In the early months of 1917 activities at Brindisi were limited. In the spring, Italian naval aviation was completely reorganized, and Squadriglie were established. The aircraft of Brindisi naval air station thus became the 255ª Squadriglia, and Pierozzi was the commander of this unit. Under his command were twenty F.B.A. flying boats housed in a large hangar built for the three engined Bresciani bombers that never reached the frontline units. The naval aviators in Brindisi cooperated with the squadrons in Varano, on the Gargano, the "spur" of the Italian boot and in Valona, Albania, with the Royal Navy Shorts and Sopwith aircraft at Otranto and the French Navy seaplanes at Corfu, Greece.

On 4 May 1917 a raid was organized from Brindisi air station against the Austrians at Durazzo, Albania. Two flying boats were to raid the city and lure enemy airplanes, while four more escorted them, ready to pounce on the enemy. The crews of the four seaplanes were Pierozzi, with *STV* Carlo Daviso di Charvensod, Maddalena and Fiastri, Giannelli and Casagrande, De Sio and Magaldi. Two enemy planes took off to challenge the Italians, but they spotted the trap and eluded the eager fighters.

On 15 May the *K.u.K. Seestreitkräfte*, Royal Austrian Navy launched a big operation against the Allied convoys to Albania and against the line of anti-submarine drifters. This originated a whole day of confused fighting, in which cruisers traded gunfire, submarines launched torpedoes, mines wrecked ships and, generally, everything went well for the Austrians and bad luck persecuted the Allies, that lost a few merchantmen, two destroyers and many drifters, while HMS *Dartmouth* got a torpedo from a submarine and

Pierozzi, right, in the cockpit of an FBA flying boat at Brindisi.

limped home. The naval air forces of both sides took active part in this battle. During the fire exchange between the Allied group with *R.N. Mirabello* and four French destroyers and the Austrian force consisting of the *Saida*, *Novara* and *Helgoland* explorers, two Italian flying boats attacked the enemy ships. They were the F.B.A. 429 of Pierozzi and *STV* Luigi Lombardi di Lomborgo, together with *2° nocchiere* Aldo Giannelli and *STV* Eugenio Casagrande. Their bombs fell very near the target, and then the Italians tangled with two Austrian flying boats, K177 and K179, that had previously strafed the Allied ships and warned their own of another approaching naval group. In the mêlée, K179 of *Frglt.* Otto Freiherr von Klimburg and Heinrich Schramek got 10 bullet holes, but its return fire forced Pierozzi to alight on the sea in emergency. The two Italians found themselves between the artillery volleys of the

A clear view of an FBA at Brindisi, early 1917. The pilot in the cockpit is Umberto Maddalena, who became one of the most important Italian pilots.

An FBA floating among the warships in the harbor of Brindisi. Under its wing hangs a British bomb.

Pierozzi at the controls of a Sopwith Baby, N1034 of Blackburn production, at Brindisi.

two fleets, that further damaged the F.B.A. so they recovered the weapons, sank the boat and then the French destroyer *Bisson* picked them up.

Three weeks later, Pierozzi had his revenge. On 7 June 1917 three Austrian flying boats attacked a small anti-submarine airship, the D.E.5. The Italian airship, piloted by *STV* Catalano Gonzaga, managed to escape this attack, and regained its base with the envelope pierced by 30 bullets. Two Italians seaplanes took off to protect the airship, they were F.B.A. "14" of Pierozzi and gunner Ercole Bellingeri, and Macchi L.3 3440 of Giannelli and Daviso. They intercepted the Austrian seaplane K154 that maybe had mis-identified the two Italian planes as his two companions. The three seaplanes exchanged fire, so close that the Macchi's wing crashed against the hull of K154. It was Pierozzi who finished off the Austrian aircraft, with a burst that hit the crew. The enemy flying boat crashed in the sea, with the death of *Linienschiffleutnant* Alois Poljanek and *Stabsbootmann* Prauer. At the same time the Macchi crashed nearby. Pierozzi immediately alighted alongside to help the unharmed Daviso, he dived repeatedly to look for Giannelli but couldn't find him. The body of Giannelli was found later, with a bullet hole in his forehead. Pierozzi's first victory was dearly paid for, as the Adriatic Sea had taken the lives of three gallant airmen.

In July and August 1917 Pierozzi lead his squadron in several attacks and reconnaissance flights over the enemy bases of Durazzo and Antivari while the Austrians retaliated with similar raids against Valona, Otranto and Brindisi. Orazio Pierozzi remained in command of the Brindisi squadron until 17 March 1918. Meanwhile his unit had received Macchi L.3 flying boats, a few of the new Macchi M.5 single-seat fighters and one Sopwith Baby, serial 6849, that he liked to fly.

On 18 March 1918 Pierozzi was posted to Venice, to lead the Group of fighter squadrons, 260ª and 261ª Squadriglie, based at the "Miraglia" air station on Sant'Andrea Island.

The airmen at Venice had as their mission the protection of the frontline, at the time running along the Piave River, a few miles north of the Venice lagoon, and the contrast of the Austrian naval aviation based at Trieste and Pola. Trieste, the main port of the Austro-Hungarian Monarchy, was the goal of Italy's war effort while Pola, on the Istria Peninsula, was the main base of the Austrian Navy. Both cities and all the region became parts of Italy after World War I, but after World War II only Trieste remained, while Pola and the Istria became part of Yugoslavia, now of Slovenia.

Pierozzi was above all a natural leader, who imposed his superiority not only on account of his outstanding naval and military culture, but also for his quality as a teacher. He was always interested in flying new types of aircraft, claiming that a pilot must be

Right: Finally, a real naval fighter: Orazio Pierozzi in the cockpit of Macchi M.5 7087 at Brindisi.

The Macchi M.5 7087 was an early production aircraft with lateral floats directly attached to the wings. The ships in the background are the *Vittorio Emanuele II* and *Regina Elena* battleships.

Pierozzi in his Macchi M.5 with other fighter pilots of the Venice fighter group. The pilot sitting on the prow of the M.5 is Calvello.

able to control whatever airplane, and in practicing new maneuvers. Then he knew how to pass his abilities to all pilots, first cajoling them into trying apparently easy maneuvers, and then praising them for having succeeded in what actually was a very difficult feat. This way, he was very good at building the confidence and self-esteem of his new pilots.

Another typical trait of Pierozzi was his emphasis on the importance of formation flying and fighting, and his downplaying of individualism. When the King of Italy, during a visit to Venice, asked him which pilot had shot down a captured Austrian fighter, Pierozzi answered *"The whole squadron, Your Majesty."*

His first combat mission from Venice was an escort flight on 13 April 1918, followed by four more combat missions during the month. On 1 May 1918 the 261ª Squadriglia took off on a mission of escort to a camera-equipped Mach L.3 going to Trieste. Pierozzi was leading on Macchi M.5 "18" 7256 and his wingmen were *STV* Luigi De Riseis and *sottocapi* Vincenzo Burattini and Luigi Macchia. Two enemy fighters rose to challenge them, but then stayed clear of the larger Italian formation. On the way back, the Italian fighters spotted one enemy above Grado, and dived down to attack it. The single Austrian fighter had no chance, with the skillful maneuvering of the Italians to preclude its escape, and soon its engine was on fire, hit by Pierozzi's guns. The pilot Josef Niedermeyer had to force land his Hansa Brandenburg A67 but he had managed to put a few bullets into Pierozzi's fighter. Three days later, on 4 May 1918, a similar combat developed. The same Macchi L.3 was sent on another reconnaissance mission to Trieste, escorted by the 260ª Squadriglia with Pierozzi leading *TV* Martinengo, *STV* Calvello, chiefs Rivieri, Jannello and Pagliacci. Four Austrian fighters rose to the challenge: A91 again with Niedermeyer, A83 with *Fl. Meist.* Karl Schneider, A78 with *ZwW Egast* Franz Boros, soon to be joined by A82 piloted by the Austrian ace Gottfried Banfield. In the ensuing dogfight, above Punta Salvore and the estuary of the Tagliamento river the Italians claimed three enemy fighters: A91 and A78 were forced to land near Italian torpedo boats, they were captured together with their pilots and taken to Venice, while Banfield himself suffered damages, and force landed in friendly waters, returning to Trieste. Pierozzi, flying again Macchi "18" 7256, took part in this combat, (that is strangely reported to 3 May in his logbook, while the correct date, 4 May , is beyond question) but he did not claim personal victories although, as a participant and commander, he could easily have got the victories assigned.

In the next days, he flew again in patrol and escort missions, and one attack on a kite balloon. On 13 May he tested a SVA landplane. On 14 May 1918 the Italian special boat "Grillo" of Captain Mario Pellegrini unsuccessfully tried to attack the battleship *Viribus Unitis*. Austrian seaplanes attacked the ships of the Regia Marina that covered this operation, and the 261ª Squadriglia

Pierozzi atop his later Macchi M.7 serial 7256, individual number 18, showing the insignia of a dog holding in ita jaws a crushed Austrian seaplane.

gave battle. The formation lead by Pierozzi on his Macchi "18" consisted of *TV* Umberto Magaldi, *STV* Luigi De Riseis, *s.ten.* Giovanni Naccari, *2° capo* Edoardo Odierna and *marinaio* Renato Balleri. They tangled with two bombers, K216 and K217, that were about to attack the Italian ships escorted by four fighters, A65 of *Fl. Maat* Max Ott, A70 of *Fl. Maat.* Josef Gindl, A84 of *Frglt.* Stephan Grossschmied and A85 of *Skdt.* Franz Pichl. The Ks dropped their bombs in a hurry, and then K216 made its escape with a jammed gun. The dogfight went on for one hour, and the aircraft got down at wavetop height, until K217 got away, while A70 was hit in the engine and forced down, and A85 was shot down, with the death of *Seekadet* Pichl. After the war, Pierozzi found the photograph of Pichl, and kept it in his album, marked with words of sorrow and respect for his fallen enemy. Later books report of one Austrian victory, a Macchi M.5 forced down and captured intact, but actually this event, the loss of Macchi M.5 7280 of Magaldi, happened five days later, in a duel against Phoenix land fighters. The Italians, on their part, claimed three victories all against Ago fighters (Ago was the name commonly given by Italians to the Hansa Brandenburg A fighters, a name particularly appropriate, as *Ago* in Italian means needle) because, besides A70 and A85, they saw another seaplane fighter alight off Rovigno. Enemy bullets again hit Pierozzi's fighter, and the three victories were assigned to him. Austrian sources so far available confirm only two losses for the *K.u.K. Seeflieger.*

Pierozzi met again the enemy on 22 May, when flying on an exploration flight over Pola together with *2° nocchiere* Beniamino Piro, of 260° Squadriglia. In the sea off Rovigno they met a pair of Austrian landplanes, Phönix D.1 A 110 and A 115 piloted by *Frglt.* Lang and Wolleman. The Italian seaplanes dueled, and Pierozzi claimed an enemy fighter pursued until Rovigno and probably hit. He was underclaiming, because A 115 was indeed hit and force landed at sea, with its pilot *Fregattenleutnant* Stephan Wollemann wounded.

In the next days, Pierozzi flew more combat missions, and he also tested two new fighters, the Hanriot Hd.1 floatplane and the disappointing ISVA, the floatplane version of the successful SVA. His many missions over Pola were followed, after 15 June 1918, by strafing missions on the Piave, as the Austrian offensive developed.

He returned to victory on 2 July 1918. An Austrian formation of three seaplane fighters and one bomber, K394, was attacking Italian torpedo boats off Caorle, when it was bounced by a flight from 260ª Squadriglia, lead by Pierozzi and including *STV* Domenico Arcidiacono, chief Macchia and *STV* Vittorio Pomilio (brother of aircraft designer Ottorino Pomilio). The K-boat was shot down in flames and its crew, *Seefähnrich* Vinzenz Guglielmi, *Fliegermaat* Emil Modler and *Fliegermatrose* Niedoba were taken prisoners. On 6 July, Pierozzi escorted land bombers, and then went down to strafe an enemy truck and trenches. That was his final mission with his trusted Macchi M.5 "18", afterward he flew on a new M.5, "25".

On 13 July 1918, during an escort mission for a recce seaplane above the Piave, the Austrian flak hit Pierozzi's Macchi. In August, combat missions went on, and again his aircraft was hit by AA fire above the Piave on 19 August. In September, he started to fly training flights for night interceptions, and in one of these, on 15 September, he damaged his seaplane on alighting. Another mishap happened on 19 September when, escorting a raid of SIA 9b bombers of the Squadra San Marco over Pola, he ran out of fuel and had to alight 7 miles off Venice.

The portrait of *Seekadett* Franz Pichl that Pierozzi preserved as an homage to his fallen adversary.

His final missions took place in the last days of October, as the Austrian armed forces were collapsing. On 1 November he lead an escort mission above Trieste and strafed enemy barges in the Marano lagoon, the next day he returned over Trieste and on the 3rd he alighted there, as the Italian troops finally reached the liberated city.

By the end of the war, his logbook at Venice registered 96 combat missions, 146 hours of flight in war missions, six air combats and five victories (besides the first one from Brindisi). His decorations included no less than four Medaglie d'Argento and one Bronzo. His first Silver was for his general activities up to June 1917, the second one, assigned in the field, reported in its citation the luckless events of 15 May 1917. His third Medaglia d'Argento referred to his first air victory and to the missions of June 1917, and the last one, assigned posthumously, referred to his tragic death. His Medaglia di Bronzo was the award for a gallant mission against Durazzo on 16 July 1917.

On 3 November 1918 he took command of the enemy's nest, the Trieste Naval air station and during the rest of the year he flew liaison connections to Pola and Venice, above the now peaceful Adriatic.

In February 1919 Pierozzi, together with Ajmone di Savoia, visited the Caproni works in Milan and the SIAI factory in Sesto

Above: A general view of the wharf at the Miraglia Naval Air Station, with Macchi M.5s of the 260ª and 261ª Squadriglia and, in the forefront, the Austrian seaplane fighter A78 captured on 4 May 1918.

Below: Pilots at Venice: Pierozzi is the last one on the right, the third person is the tall prince Ajmone di Savoia Aosta, who was with Pierozzi when he had his fatal accident, after him there is commander Roberti di Castelvero.

Calende, where he witnessed the tests of the new S.13 flying boat, soon to be the winner of the Schneider Trophy race at Bournemouth, piloted by Guido Jannello. Young Ajmone di Savoia, son of the Duke of Aosta, commander of the Italian 3rd Army, had been a gallant airman with the 251ª Squadriglia. During World War II, he was named as the nominal king of the puppet kingdom of Croatia, but he didn't take that role too seriously, and never went there. Today his son Amedeo, Duke of Aosta, is an eager private pilot and defender of the heritage of the royal family.

On 17 March 1919 Pierozzi and Ajmone took off in a Macchi M.9 flying boat from Venice to Trieste, in order for the young prince to visit his father and brother. Weather conditions were good, but when they were ready to alight, the strong "bora" wind of Trieste started to blow, and overturned the aircraft, that fell in front of the harbor. Pierozzi hit his head violently and fainted, Ajmone was only bruised and kept afloat the body of his companion, holding to the wreck of the aircraft. They were soon rescued, but it was to no avail: Orazio Pierozzi died the next day in the Military Hospital.

The loss of the small and tough commander was deeply felt at the time. Among the flower wreaths at his funeral, there was one with a German inscription, offered by an unknown Austrian naval pilot, one of his former adversaries. It was a gesture that Banfield could easily have done. Later on, the Naval Air Station of Brindisi was named after him, and the main square of San Casciano Val di Pesa still carries his name.

## Orazio Pierozzi
## 6 confirmed victories

| Claim | Vict. # | Date | Sqn. | Aircraft | Hour | Place | Victim |
|---|---|---|---|---|---|---|---|
| 1 | 1 | 7.6.1917 | 255 | F.B.A.14 | - | Lower Adriatic | K154 |
| *[with Bellingieri, Giannelli & Daviso] Lslt. Alois Poljanec KIA, EF Stmw. Josef Prauer KIA, K154* | | | | | | | |
| 2 | 2 | 1.5.1918 | 261 | M.5 7256 | - | Trieste | A67 |
| *[with STV De Riseis, capo Burattini & Macchia, 261ª Sq.] Lt. Josef Niedermayer UNH, A67* | | | | | | | |
| 3 | 3 | 14.5.1918 | 261 | M.5 7256 | - | Pola | A70 |
| *[with TV Magaldi, STV De Riseis, s.ten. Naccari, 261ª Sq.] Fl.Maat Josef Gindl UNH, A70* | | | | | | | |
| 4 | 4 | 14.5.1918 | 261 | M.5 7256 | - | Pola | A85 |
| *[with STV De Riseis, s.ten. Naccari, capo Odierna, 261ª Sq.] Skdt.Franz Pichl KIA, A85* | | | | | | | |
| 5 | 5 | 14.5.1918 | 261 | M.5 7256 | - | Pola | Seaplane |
| - | - | 22.5.1918 | 261 | M.5 7256 | - | Rovigno | EA damaged |
| *Frglt. Stephan Wollemann WIA, A115* | | | | | | | |
| 6 | 6 | 2.7.1918 | 260 | M.5 7256 | - | Revedoli | K394 |
| *[with STV Pomilio, Arcidiacono & Macchia, 260ª Sq.] Fl. Vinzenz Guglielmi POW, Fl.Maat Emil Modler POW, Fl.Matrose Niedoba POW, K394* | | | | | | | |

# Ferruccio Ranza

Aviation enthusiasts studying photos of World War II have no problem in recognizing the profiles of two protagonists, Hermann Göring and Ferruccio Ranza. The chief of the Luftwaffe was nicknamed *"der Dick"*, the fat one, and also Ferruccio Ranza had considerable dimensions. It was rumored that his aircraft needed special oversize seats. There is a weird parallel development in the lives of these two men: both were fighter aces in World War I, both took the place of command of their nation's top ace, respectively von Richthofen and Baracca, leading the best squadron, both rose to the top of their air forces. But then this commonality branches out into radical differences. Göring turned into a corrupt man, a drug addict and an embezzler. Ranza was a man of clean morals, and a family man. Göring rose to the top thanks to politics, being first a servant and then an accomplice of dictatorship; Ranza kept his distance from fascism, and was the less politicized among Italian generals. Göring lost his courage, he didn't fly, he sent his pilots on mission from the safety of his well-protected and luxurious special train; Ranza was a man of reckless courage, who lead his men from the first airplane in a formation. These two different destinies epitomize how initially similar life stories can branch out in opposite directions, bringing out all the good or all the bad that is in the capability of man.

Ferruccio Ranza was born in Fiorenzuola d'Arda, a wealthy agricultural town in the valley of the Po River, on 9 September 1892. He went to school in Fiorenzuola and Piacenza, where in 1910 he got his high school diploma as accountant. He entered military life in January 1914 attending the first officers training course of the Battaglione Aviatori del Genio in Turin, being assigned to the 1° Reggimento Genio and on 11 November 1914 he was commissioned as *Sottotenente di Complemento*. That period formed Ranza, who had a strong passion for flying and a natural talent, so much that he was the best student in his class, but probably he didn't like military discipline applied with the hard formality derived from the old army of Piedmont, contrasting with his dashing character. Pilot training in Italy, using Bleriot monoplanes, was particularly slow and methodical, but it worked in producing well-trained pilots. Young men were impatient to get their wings, but the slow pace of the flying school had its advantages.

When Italy entered the great European War, on 24 May 1915, Ranza had obtained his brevet of military pilot, but he was sent to a unit only in September 1915. He transferred to Ghedi and joined the 4ª Squadriglia per l'Artiglieria, a new unit in formation with the Caudron G.3, a tractor biplane with tailbooms, slow, placid and unarmed, an apparently obsolescent aircraft that provided good service, particularly for its qualities of flight in mountain environment, so much that it remained in operations well into 1917.

The 4ª Squadriglia included among its personnel names destined to fame, like Ruffo di Calabria, Silvio Scaroni, Costantino Cattoi. It was based at Gonars, a village in the Friuli region, assigned to the Artillery Command of the 3ª Armata, on the front of the Carso. On 14 October Ranza was shifted to its twin unit, the 3ª Squadriglia, based not far away, at Medeuzza. Ranza distinguished himself for the first time on 1 April 1916, flying a daring low-level photoreconnaissance over the Carso, in which enemy artillery hit his airplane 11 times and he gained his first medal, a Medaglia di Bronzo al Valor Militare.

On 15 April 1916 the unit changed its name, becoming the 43ª Squadriglia. Ranza however was about to leave reconnaissance: his ability as a pilot was remarkable, and in June he was transferred to fighters. He was assigned to the new 77ª Squadriglia Caccia that was activated on 18 June 1916 at La Comina airfield, at the com-

Tenente Ferruccio Ranza at Quinto di Treviso, August 1918, with his Spad showing the black ladder insignia.

On 25 November 1916 Ranza claimed a victory, his likely victim was Brandenburg C.I 27.53 of Flik 28, which was damaged and its observer *Lt.* Schrimpl was wounded. The plane was repaired and put back into service, this photo shows it after a bad landing in 1917. *(Photo Cejka).*

mand of *cap.* Pier Ruggero Piccio, equipped with Nieuport 10 biplanes, soon replaced by the small and agile 80-hp Nieuport 11 Bebé.

The first encounter with the enemy was on 26 July when Ranza together with *s.ten.* Giovanni Fasoli had a combat against an Austrian "Albatros". The next day, 27 July, Ranza attacked another enemy two-seater together with *ten.* Carlo Savio: they hit the Austrian that dropped its machine gun and its camera in Italian territory, the observer was presumably hit and the plane was seen going down in Valsugana. It was Ranza's first victory, but it was not confirmed. The enemy airplane was Brandenburg 61.23 of Flik 24, with pilot *FP Zugsf.* Alois Jezek and the observer *Oblt.* Georg Kenzian Edler von Kenzianhausen, who were both wounded, their plane was badly damaged and landed in emergency behind its line. *Oblt.* Kenzian, then credited with two air victories against Italian airplanes, later became an Austrian fighter ace. This is the combat report:

> "I took off for my 3rd flight with Ten. Savio; as I was patrolling above Treviso-Bassano I saw some anti-aircraft shots and then an enemy Albatros type (reconnaissance) airplane coming from Val Sugana aiming for Treviso, at a height of about 4,200 meters. Before reaching Treviso it turned back toward Bassano, maybe because they had seen us.
>
> We reached it and first Ten. Savio attacked from the back, then dove and passed below it, probably to reload his gun. The enemy observer fired many times, I joined the attack shooting all the magazine of my Lewis (50 rounds). The enemy airplane went into a dive (hit, perhaps). I saw the back machine gun and some papers falling out of the plane. The enemy observer, possibly hit, disappeared within the fuselage. The pursued aircraft went down at full throttle.
>
> As I reloaded by gun and followed it, I saw the Newport of Ten. Savio attacking again and then banking. I attacked again at 1,000 meters firing a second magazine of rounds. The enemy plane went into the Val Sugana descending. While I reloaded my gun I lost it against the dark background of the terrain. I cannot say if it landed in our lines or in enemy territory."

Right: One of the victims of Ranza, probably Brandenburg C.I 69.92 shot down on 10 August 1917.

In August the 77ª Squadriglia moved to the airfield of Cascina Farello, with the task of providing protection for the seaplanes based at Grado. On 13 September the fighters escorted a large raid of Caproni land bombers and seaplanes against the arsenals of Trieste. The next day, 14 September, the Austrians attempted a reprisal, sending three flying boats against the Italian positions, while King Victor Emmanuel III and King Peter of Serbia were in a frontline observation post. The 77ª rose up to intercept, and Ranza, who flew a new Nieuport 17, French serial 2613 armed with a Vickers gun, Savio and Tesei claimed two victories against enemy seaplanes. The flying boats alighted in the gulf of Panzano and the gulf of Trieste, one flying irregularly, and the other one towed into harbor by Austrian ships, as the observers of the Navy attested. For this action Ranza received his second Bronze Medal, its citation credited him with one victory.

On 1 November Ranza was flying again a French-built Nieuport armed with a Vickers gun:

> "I took off with Ten. Fasoli and Serg. Tesei to escort two Voisin aircraft, going to bomb Kobilà-Glava and Phigcovitza. After the bombing, as we were returning, I saw an enemy Albatros type airplane in pursuit of a Voisin, and approaching it above the lines. I reached it and attacked it from behind, as it was banking, at a height of about 2,500 meters over the valley. I saw the enemy aircraft turning and diving toward the coast, getting down fast toward Prosecco. I didn't follow it, as I reached the Voisin. I returned to our airfield with my patrol having flown 75 minutes. I cannot determine if the aircraft crashed or force landed."

Ranza didn't claim a victory, but this event could be connected with the action of Brandenburg 20.08 of Flik 19, its crew *Hpt.* Adolf Heyrowsky and *Lt.* Othmar Biegler operated against an Italian bomber raid, had a combat with a Nieuport and force landed at Haidenschaft (Aidussina), both airmen being unhurt.

Ranza at Santa Caterina airfield in the summer of 1917, with his Spad showing his first personal insignia, the owlet. *(Photo Giorgio Apostolo).*

In November, after a few inconclusive combats, Ranza had his big day. On 25 November, still flying his Nieuport with a Vickers gun, Ranza took off at 9 a.m. with Tesei, intercepted an "Albatros" and shot it down over the Hermada. He returned to base, took off again at 11.45 and again he intercepted an enemy that, coming from Prosecco, was flying over Nad Logem. He attacked from behind and shot it down between Schönpass and the Ternova forest. The events of that day are confused, the Austrians lost a total of seven aircraft, so the victims of Ranza could be Brandenburgs 29.54 or 27.64 or 27.53.

At the end of January 1917 Ferruccio Ranza became commander of the 77ª Squadriglia. His logbook reports a combat on 4 April 1917 against an enemy that force landed at Oppachiasella.

Ranza remained with the 77ª until 1 May 1917, when he was assigned to the 91ª Squadriglia Caccia, formed on that day, mainly with forces from the 70ª, at Santa Caterina di Udine airfield. It was a crack unit, which included Francesco Baracca, Fulco Ruffo di Calabria, Luigi Olivari, the best pilots for the squadron tasked with the defense of Udine, base of the Supreme Command. The 91ª was

The Nieuport 11 fighters of the 77ª Squadriglia ready for action on 14 September 1916. Some of them appear to have a dark mottle over the basic finish.

equipped with Nieuports and with the French Spad 7, possibly the best fighter of the war, powered by an excellent Hispano Suiza engine and armed with a Vickers machine gun, a strong and maneuverable fighter.

In May Ranza returned to his old squadron, the 77ª at Aiello, leading a flight of the 91ª Squadriglia deployed closer to the frontline to provide air cover for the 10th Battle of the Isonzo. He remained there until 3 June, and then two days later he moved with many companions to Istrana, to support the Alpine troops in the battle of the Ortigara.

On 18 June, flying Spad 1307, he attacked an enemy two-seater above Borgo, but lost it after his gun jammed. He fixed the jam, saw an airplane that was attacking an Italian SP.2, engaged it and drove it away, probably hit. The first of these two enemies could be Brandenburg 129.65 of Flik 48 with pilot *Korp.* Adolf Habel and observer *Oblt.* Hermann Plenk, which had a combat and landed at Levico with damaged fuel tank.

On 23 June, again flying Spad 1307, Ranza scored his fifth confirmed victory between Mount Armentera and Cima Mandriolo and became an ace. He identified his victim as an "old type" Brandenburg two-seater, repeatedly hit and forced to land near Barco. This was also his first victory with the 91ª.

On 2 August 1917, now back on the Isonzo front, Ranza took off to test Spad 4691 and flew above Monfalcone:

> "Cruising at 10h15 toward Rubbia I saw an enemy airplane flying parallel to my direction beyond Castagnevizza. Higher than me another SPAD *(4690 of Parvis)* got above the enemy airplane and turning left tried to bring an attack from behind and below. The enemy aircraft got into a dive turning left, so it placed itself in front of me. I attacked from behind at a distance of about 100 m and immediately the enemy aircraft went into a spin for about 300 m, then it got out of the spin diving. I reached it and attacked it again from the side. It spun again and then dove. In a final attack at a height of 800-900 m I flew very close to it, then I lost it. I looked for it but couldn't see it, and as the combat from Castagnevizza we were now over the Vippacco Valley, I think the plane fell in that valley. From below I saw again the SPAD that had attacked with me. The enemy observer fired some 30 rounds. The enemy plane was a two-seater reconnaissance Albatros."

This victory finds no reference in Austrian data, and it was not credited to Parvis. Ground observers however confirmed that the Austrian landed at Ovcia Draga. Also *serg.* Ennio Sorrentino of the 81ª Squadriglia was there, on his Nieuport 17, who saw the event and described the insignia of the two Spads: an owl and a black crescent, Ranza and Parvis.

On 10 August, flying Spad 4789, Ranza at 8 p.m. was returning from an escort mission to Chiapovano when between Monte Santo and S. Michele he saw two Austrian airplanes, one white and one brown, followed by three more, stragglers from a formation of 16 aircraft on a big bombing raid. He attacked the second one per-

Ranza, with white collar, checks the gun alignment on a Spad 7.

fectly from behind, hit it making it dive, then as it recovered he attacked again, until the enemy fell vertically between Castagnevizza and Selo. Ranza then had to avoid the attack of two enemies, one of them a fighter. His victim, Brandenburg 69.92 of FlG.1, fell near Flondar and the Italians captured the airmen: *Lt.* Ebner was seriously wounded and did not survive, *Korporal* Kocher was questioned and described the action, to which also *aspirante* Amleto Degli Esposti, of the 80ª, participated with his Nieuport. In that same evening action FlugGeschwader 1 lost also Brandenburg 229.25, shot down by Sabelli and Allasia.

In the second decade of August the huge 11th offensive of the Isonzo was launched, the 91ª Squadriglia was active and escorted Caproni bomber raids. Ranza also flew many photoreconnaissance missions deep into enemy territory using camera-equipped Spads such as 4889. On 18 August he saw a fighter, which he initially mistook for a Spad, behind a Caproni. It was an Austrian fighter, but as his gun jammed Ranza could only maneuver, but he managed to protect the bomber. On 29 August Ranza was into action again in a Caproni escort over the Ternova forest, and he had a dogfight with Albatros D.IIIs that he described this way: "...they shot at me from afar using tracing bullets. The craft were Nieuport type with V struts painted whitish yellow, with inline engine."

On 15 September 1917, again escorting Capronis, Ranza made the mistake of not watching his back, and he was bounced. At 11 a.m.:

> "I attacked an enemy fighter that was below me trying to approach the Capronis but I didn't realize that there were two other fighters behind me, and as I was approaching the first one I heard shooting behind me. I turned and saw a fighter shooting me from far away and as I maneuvered to attack it the first one and another one from above also engaged me. I got away diving toward S. Gabriele at full throttle, while the enemies cruised above Ternova. The enemies had V struts and were painted dark khaki, with fuselage insignias; highly agile and better climbers than the SPAD. I had the impression that the wingtips were canted. They shot tracer bullets."

His full speed dive mislead the Austrians, and ace Julius Arigi of Flik 55J, flying Albatros 153.15, claimed one Spad shot down.

On 23 September Ranza flying with Sabelli over Castagnevizza met a Brandenburg with one escorting fighter. The latter was engaged by Ranza and got away, so the two Italians could easily shoot down the two-seater, Br. C.1 129.48 of Flik 35/D, that fell on the S. Michele, at Cotici. The two Austrian airmen, *Zugsf.* Josef Baier and *Oblt. i.d.R.* Eduard Hafner were killed.

On 13 October Ranza witnessed the tragic incident that killed his companion Luigi Olivari, one of the most successful Italian aces. On 23 October, together with *aspirante* Antonio Pagliari, Ranza escorted a Pomilio PE over the Valley of the Sava, but he found a total cloud cover (which hid most of the preparation of the coming offensive). So they flew to the coast and reached Fiume, where Ranza discovered that his camera was not working.

On 24 October 1917 the Austrians and Germans pierced the front at Caporetto. Air activity was very intense, on 25 October the 91ª Squadriglia put up 17 fighter sorties and lost two pilots, Sabelli and Ferreri, while scoring six victories. Among them was Ranza's ninth claim, a German two-seater shot down near Lom Tolmino at

Fighters at Santa Caterina, spring 1917: first is Spad 7 serial 4691, flown by Ferruccio Ranza, then seven Nieuport 11s and at the end of the line a silver Nieuport 17.

9.15 a.m. Ranza took off again at 10.20 together with *tenenti* Parvis and Ferreri, and clashed with several aircraft, but he witnessed the demise of his wingman:

> "I took off at 10h20 with tenenti Parvis and Ferreri heading for Gorizia, then I flew up the Isonzo. I spotted an airplane near Tricorno and I followed it together with another Spad, I engaged, I fired a burst then my gun jammed. The other Spad following me dove as the enemy plane dove toward Tricorno. Then I spotted nearby a separate group of enemy aircraft. I was at the level of the two lower ones (4500 m.). Then I attacked the lower one that banked, while above me I saw three airplanes that were after the other Spad that was also pursuing an enemy fighter. I saw tracer bullets aimed at the Spad, and then I saw the latter diving and spinning low across the Isonzo, toward Tolmino. I didn't see it hit the ground, but I had the impression that the plane was not under control. The first combat took place at 13h18, the second one at 13h25."

Ranza's evaluation unfortunately was right, *ten.* Ferreri did not come back, shot down by German fighters of Jasta 1.

The next day Ranza was sent to La Comina, to prepare a base in the rear, while the front had broken down, but later in the day he flew a mission, had a duel over Cividale, but his gun jammed and his German adversary was saved. In the following days the 91ª left S. Caterina di Udine, burning unserviceable equipment, moved to La Comina and then to Padua. The efforts of its pilots should not be measured only in terms of victories, but also of anonymous patrol

Ferruccio Ranza in the cockpit of a Spad. An unusual central rear view mirror is attached below the feeder tank.

and escort flights. On 17 November Ranza again attacked a two-seater, but a jammed gun deprived him of a possible victory. On 21 November he took off in a thick fog for an offensive patrol. Over Bassano he found a clear sky, and a German two-seater, that he shot down over Casoni after a brief combat.

On 24 November Ranza flew another mission in the fog, and when he returned he failed to find Padua airfield, so he landed at Istrana. On 28 November, together with *aspirante* Edoardo Olivero, he attacked two two-seaters over the Montello, but a whole squadron of twelve German fighters bounced him, and he barely managed to evade them, his plane getting some bullets. On 30 Novem-

ber, with *ten.* Novelli and *serg.* Magistrini he shot down another German reconnaissance plane near Moriago, behind Italian lines. The return fire of the German was just as accurate, however, hitting the engine of the Spad, so Ranza force landed between Arcade and Istrana, the Spad was completely wrecked but the pilot was unhurt, thrown out of the cockpit at the moment of crash. Literature may have described dogfights like romantic duels of winged knights, but actually combat flying in World War I was a grueling business, that physically wore down the pilots.

On 6 December Ranza aborted a patrol flight with his mates. He took off alone, after having fixed the trouble, and met a German two-seater, but his attack failed for one more gun jam. The next day, 7 December, the 91ª Squadriglia moved to Nove di Bassano, to support the battle to stop the final Austrian offensive on the Asiago plateau. There were many combats, with one victory for commander Baracca and one more victory for Ranza, the 11th according to the Squadron Diary: a Brandenburg of Flik 39/D shot down with *serg.* Magistrini:

> "At 12h15 from my height of 4100 m. I saw an Albatros coming from the plateau toward Cittadella flanked by a SPAD that I initially mistook for an escort fighter. I approached to attack but the SPAD that was on my left attacked slightly behind, and then it moved away. I then attacked from behind but I had to stop shooting because the SPAD got into my gunsight as it attacked again the Albatros. The enemy airplane shot only one burst, it got into a dive, turned left and then right. I attacked again together with the SPAD that forced me to stop shooting many times, getting in my way. The enemy Albatros went down out of control breaking up in its fall and it burned near S. Pietro (between Nove and Cittadella) I saw the SPAD following it in its fall while a patrol of British fighters surrounded me. The attacking SPAD was the one of Sergente Magistrini. People who went to look at the aircraft found it destroyed. Its pilot was burned, the observer who had jumped has not been found yet."

Bad weather then curtailed flying activities until 30 December when, escorting a two-seater, Ranza found and attacked an enemy airplane near Fonzaso. This was his combat report:

> "I took off at 11h45 from S. Pietro in Gu airfield with tenente Novelli and tenente Pagliari, for SAML escort; at 12h30 I saw an enemy airplane flying very high toward the Grappa. I reached it at a height of 4400 m and attacked it from behind. After a few rounds my gun jammed ... The Albatros turned right then left, then it started diving toward the Seren valley. I reached it and attacked it again from behind, but again my gun soon jammed. I cleared it and attacked a third time at a height of 4000 m. The enemy aircraft went down immediately, then it pulled back and dove again turning toward Mount Tomatico, in the valley between Fonzaso and Feltre. I followed it in its dive to a level of 1600 m. The enemy plane was painted in patches of different colors, standing out well against the snow, and then it disappeared over a pine forest, at 12h45 in front of Mount Aurin. When I return enemy anti-aircraft artillery targeted me. The enemy airplane was an Albatros two-seater with sloping struts, it looked to me smaller than the usual ones and rather fast. The enemy observer shot me several bursts during my second and third attack, then he stopped shooting."

From this report, the victory doesn't seem certain, and Austrian records don't supply a confirmation.

The year 1917 ended on a lighter note. A photoreconnaissance had to go to the airports of Aviano and La Comina, now in enemy hands. This mission was so important that Guido Keller, flying the camera-equipped Spad, had no less than five escort fighters: Ranza, *ten.* Adriano Bacula, *s.ten.* Antonio Pagliari, *asp.* Olivero and *serg.* Magistrini. Weather was beautiful and the Italian formation performed its task without any opposition. It could be one of their many successful missions, of the 91ª Squadriglia, were it not for the fact that somebody forgot to load the photographic plates in the camera. While they were over La Comina, an enemy two-seater was spotted. Ranza and Magistrini attacked it forcing it to land in a field next to the airport.

On 12 January 1918 Ranza scored one more victory against a German reconnaissance airplane, that he felled near Campo San Pietro. Ranza was alone on a standing patrol above Padua when he spotted the enemy flying at 5600 meters. He approached, but his Spad couldn't climb above 5400 meters. Ranza tried to put up the nose and shoot, while the German observer fired from above. The German pilot made the mistake of getting down to dogfight with the Spad, losing the advantage of height:

> "After many bursts the enemy aircraft began to go down in spirals, slightly canted to the left, and I saw the enemy observer raising his hands. I stopped shooting and followed the plane down, but then it got back straight and level trying to run away toward the Piave. Then I attacked again at close range and it caught fire. The observer jumped and the plane broke up in the air falling near Camposampiero a 13h30. The bodies of both pilot and observer were found out of the wreck. The pilot was a sergeant decorated with the Iron Cross first class and the observer was a Lieutenant of the Ulahns, both of German nationality."

This aircraft was one of the high-flying Rumplers belonging to Reihenbild Zug 1. The Germans, asking for the fate of their comrades, wrote a message indicating even the model and serial number of the missing plane: Rumpler 8266/17.

There was not much air activity in January, and this was the only victory of the 91ª Squadriglia.

On 3 February Ranza flew in patrol with Baracca, and had an inconclusive duel. Then he returned to base, refueled and took off again with Parvis and Keller. He found an enemy two-seater, but an enemy fighter was quicker, and bounced him, but his two wingmen managed to extricate him. The next day Ranza met another fighter and drove it off toward Cima Mandriolo, possibly damaged. According to the War Diary of the 91ª this plane was forced to land.

On 10 February Ranza flying with Bacula surprised a patrol of two two-seater and a fighter, and immediately shot down in flames one of them, while Bacula kept the fighter a bay, then he reached the second one and forced it to land in Val d'Astico:

> "I took off at 10h with tenente Bacula, I cruised until 11 between Col Santo and Cima Mandriolo Val d'Adige. At 11h15 a few shells of anti-aircraft fire reported the presence of enemy airplanes between Arsiero and Schio. I immediately saw a fighter above and a two-seater below. With Tenente Bacula I approached the two-seater and attacked it at close range while Tenente Bacula engaged the fighter that was trying to get on my back. After a few rounds the Albatros fell down in flames in Val dei Signori. Hearing shooting on my back I turned and saw Tenente Bacula dueling with the fighter and another Albatros two-seater that was trying to cross the lines. I approached it and after a burst I saw it going down steeply in Val d'Astico near the opening of the Val d'Assa and disappearing against the background. This second Albatros reconnaissance plane went down in our lines at Mosson south of Arsiero. The combat took place at 4600 meters."

*Tenente* Ferruccio Ranza with his commander, *maggiore* Francesco Baracca.

Austrian data allow the identification of only one of the claims of the Italian fighters (identified as "five British Sopwiths") in this action.

Nothing much happened during the spring: the front was stationary along the Piave, the two armies were building up their strength and air activity was limited. Ranza was able to go on leave, returning on 25 May. Then he resumed patrol and escort flights. On 14 June he strafed enemy trenches with Baracca.

When the Austrian offensive began, he was in the air, and on 15 June together with *s.ten.* Olivero he shot down in flames a Brandenburg that was bombing the trenches. The plane crashed on a sand bank near Grave di Papadopoli. That day Ranza flew three combat missions, and then he flew many strafes, facing the intense ground fire. This dangerous activity claimed the life of *maggiore* Baracca on 19 June, and Ranza, together with *ten.* Osnago and the journalist Garinei had the sad task of recovering his body and bringing it back to the airport. *Capitano* Ruffo di Calabria took command of the 91ª Squadriglia. At the end of the month Ranza was sent to Lugo di Romagna for the funerals of Baracca in his hometown.

On 8 August Piccio, Ranza and *serg.* De Corato found over the Piave a new model of enemy aircraft that was faster than the Spads and outdistanced them. Ranza scored another victory on 16 August, shooting down an enemy plane near S. Polo after a very low combat, that went on from altitude down to a few meters from the ground:

> "Quinto 16 August 1918
>
> Airplane Spad 180, Wickers machine gun, 150 bullets fired.
>
> I took off on a patrol with Ten.ti Keller, Osnago, Sott. Olivero and sergente Aliperta, as patrol leader at 18h20, for a Caproni escort. After half an hour of flight I saw 2 Caproni, which on account of the thick fog were returning toward Mestre and Padova. Then I flew to the Piave and cruised at a level of 1200-1500 meters until late, between Grave-Fagaré di Piave. I was now alone with sott. Ten. Oliviero, at about 19h35 I noticed from ground signals that an enemy Albatros type airplane (reconnaissance) was approaching and I attacked it above the Piave at a height of about 1000 meters between Cimadolmo and San Polo.
>
> After my first attack the enemy turned and started a dive across the Piave, giving me the impression that it wanted to land, then it tried to fly on and escape, but I attacked again firing all the way to a level of 50-30 meters and I saw it land in a field behind the village, turning over.
>
> As I returned flying low above the ground (about 20 m.) strong rifle and machine gun fire targeted me and as I crossed the Piave near Fagaré di Piave, a bullet struck my fuel tank and smashed my gunsight. I turned on the gravity feeder tank and landed at Aiello airfield.
>
> No other pilot took part in this combat.
>
> Ten. Ranza"

Ranza carrying the casket of Francesco Baracca after the recovery of his body on the Montello.

After the war, Ranza showing his Spad and the griffon of the 91ª Squadriglia to the old folks back home.

Missions were many, but fruitless. Ranza flew patrols and escort to bombers, that mainly dropped leaflets, an important weapon to sap the morale of the composite Austro-Hungarian army. On 19 September 1918 Ruffo di Calabria became commander of the new XVII Gruppo, and thus Ferruccio Ranza took command of the 91ª, the *Squadriglia degli Assi*, now officially named "Squadriglia Baracca." On 29 September Piccio and Ranza attacked an enemy. Austrian archives reveal the identity of their victim: UFAG C.1 123.24 of Flik 22/K that landed in emergency at Corbolone airfield, its crewmembers *Zugsf.* Alois Breier and *Oblt.* Julius Karikas both wounded in action.

Ferruccio Ranza claimed his last two victories on 29 October. The Italian final offensive was launched on 24 October, but it was not an easy victory. Despite the political turmoil at home, the Austrian army was still united and combative, and losses were high on both sides. Aviation took part with constant harassing of the battlefield, and lost a large quantity of airplanes. On 29 October Ranza was cruising near Oderzo and met a patrol of fighters, he surprised them and shot one of them down in flames between Ponte di Piave and Oderzo and then brought another one to a forced landing near the Livenza River with its propeller stopped.

Italian documents report that after the Austrian retreat an unknown pilot was found dead near his airplane, Aviatik D.1 338.06 at Casa Gregoletto, who was identified as one of Ranza's victims, but they might also have been two Albatros D.III of Flik 51/J shot down that day in the same area, one of them flown by the famous ace Eugen Bönsch.

On 30 October Ranza lead a strafe in which his flight claimed two locomotives and five airplanes destroyed on the ground, but coming back his Spad 2482 was hit in the tank and force landed at S. Fior airfield, abandoned by the Austrians. The ace could easily refill his plane with Austrian gasoline and returned to base. It was the third time that he was shot down. His total war roll is impressing: three years of uninterrupted warfare, 465 combat missions, 43 strafes, 57 air combats with 22 victories claimed, and 17 confirmed, according to the official list. He was awarded three Medaglie d'Argento al Valor Militare and two Bronzo.

His first Silver reported in its citation the victories of 25 November 1916 and 4 April 1917. The second medal, in the field, concerned his victories on the summer of 1917, reporting victories on 23 June, 2 and 14 August (the last victory surely an error for 10 August). The third medal for his role as squadron commander had a rousing citation:

> "At the command of the squadron named after Francesco Baracca, in remarkable air combats, shooting down enemy airplanes, he imposed our mastery of the air. At the head of his squadrons, in all missions of strafing during our offensive he spread slaughter, hardy and gallant in the fight to the end. Sky of Piave and Tagliamento Rivers, June-October 1918."

Ranza remained as commander of the 91ª Squadriglia until March 1919. It was the difficult and confused postwar period. Many airmen were dissatisfied with peace and detached from the feelings of the majority of the population, but Ranza was not the kind of man to entangle himself in the seditious activities of d'Annunzio. His remarkable courage didn't mean that he looked for adventure for its own sake, and his allegiance to the King made him steer clear of the Fiume adventure. But he was not the kind of man to rest on his laurels, so he joined a different sort of senseless but fascinating feat, the formation flight from Rome to Tokyo. D'Annunzio conceived this raid, and the government immediately supported it, as a way to keep the poet out of mischief. The Direzione Generale d'Aeronautica spared no expenses, sending liaison officers, mechanics and spares all along the route, in the Middle East and Asia.

After many of the pilots of the 87ª Squadriglia, the unit that had flown to Vienna, quit the project for different reasons, new pilots were chosen among those who did not pose a risk of diverting to Fiume. First four Caproni bombers left in January 1920, one managing to reach Syria and the other three Turkey, before being stopped by sundry accidents. On 14 February 1920 Arturo Ferrarin and Guido Masiero took off with two SVAs, and they were the only ones who eventually reached Tokyo. Finally on 11 March 1920 the main squadron of five SVA 10 two-seaters took off from Rome.

Ferruccio Ranza in a big airplane, a Fiat BR bomber named "Ferruccio Marzari" after a fallen pilot, during the Baracca Cup air race of 1922. The man in the back seat il *col.* Moizo, who was the first chief of the Italian air force in pioneer days.

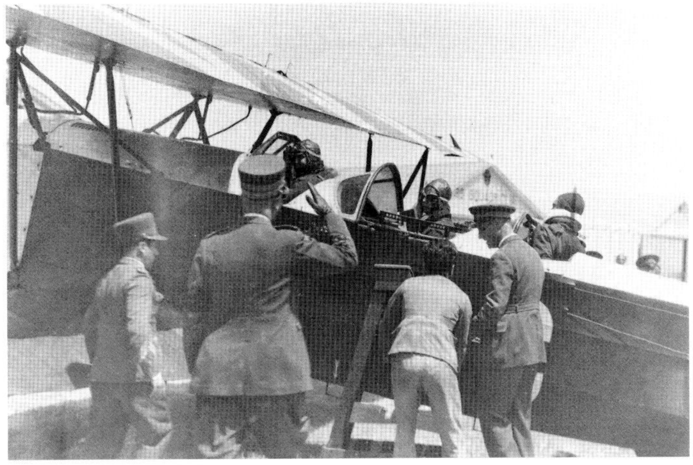

Ranza in Libya, about 1930, in the cockpit of a Romeo Ro.1 of the 89ª Squadriglia with two special passengers, Crown Prince Umberto and his wife Princess Maria José. The general in the forefront saluting is no less than Pietro Badoglio, Chief of the General Staff and at the time also Governor of Libya.

*Capitano* Ferruccio Ranza piloted SVA serial 13197 with fitter Ricciotti Brigidi, the other four SVAs carried *cap.* Gordesco and *ten.* Grassa, *cap.* Re with *ten.* Alberini, *ten.* Mecozzi with *ten.* Bilisco, *ten.* Marzari with motorist Diamonte.

Bad weather immediately stopped them at Brindisi. They took off again on 20 March and four reached Athens, as *cap.* Re quit at Valona, Albania, for mechanical troubles. On 26 March they took off to Adalia, Turkey, but on landing the SVAs of Ranza and Marzari collided and were wrecked. Ranza, with the help of fitter Brigidi, cannibalized one airplane from the two and took off again, after a few days, now together with Marzari, following Gordesco and Mecozzi who had already flown to Aleppo, Syria. Their engine failed them and the two airmen force landed near a desolate Syrian village.

The series of events that developed looks like a film of Rudolf Valentino or Gary Cooper. The two Italians were captured by the rebels:

> "a rabble of some four hundred bandits, some mounted, some on foot, of whom a dozen looked like Turks, the other ones were a mixture of Arabs and Kurds, armed with Mauser rifles, hand grenades, three machine guns and other weapons captured from a French force a few days before, having killed the officers and completely undressed the captured soldiers."

They were to be shot, they were imprisoned but during the night a cavalry troop of Sheriff militia freed them. Ranza was not the kind of person to give up. "Two days later, disguised as an Arab, I went with an escort of twenty men in five carts headed for Aleppo". In that city he found his companions and witnessed the crash of the SVA of Mecozzi, probably victim of sabotage, and the departure of Grassa and Gordesco, then he returned to pick up his airplane.

"The area is festering with Kurds and armed bands. Nobody can pass there after sunset without being assaulted. The car driver had lost his head and his self-control, the Sheriff's officers were worried, told us to wear Arab dress and forbid us to speak in any language. We were all armed with rifles." Ranza found his companion and his SVA, fixed somewhat the plane and took off to Aleppo. There Ranza and Marzari fixed the plane properly and on 14 April they departed to Baghdad. The next day Ranza flew from Baghdad to Basra, on 16 April from Basra to Bouchir, where three days earlier Gordesco and Grassa had crashed to their death on landing. Dangers, bandits, mechanical failures, a horrible weather, the death of their friends, nothing stopped Ranza who on 17 April flew to Bandar Abbas, on the 18th to Ciau Bar, on the 21th to Karachi, on the 23rd to Nasdirahbad, on the 24th to Delhi, on the 25th to Allahbad, where on landing he found a hole and broke his plane's tail.

"I was disheartened and started crying like a child…" But he didn't give up. He was told that in Calcutta there was a SVA fuselage left there by Ferrarin, got there by railroad, picked it up, brought it back to Allahbad, reassembled an aircraft and on 6 May 1920 he flew to Calcutta. He stopped there a few days for a proper assembly of the airplane (if that jerry-rigged contraption could still be called a plane) and took off on 12 May headed to Rangoon, Burma. The engine failed: "I remained cool until the last moment; two meters from the ground with a quick bank I crashed with my left wings, avoiding a fuselage first crash that could have killed me and my motorist. Motorist Zecchini was just bruised, I badly wounded in my legs, with painful consequences for my shins."

The adventure was over, and on 25 June Ranza returned to Italy by ship. His raid may now look like a marginal footnote in the history of aviation, but actually he had flown 8,500 kilometers, along a route that would be demanding even today, showing not only his courage but also his competence. Quotations are taken from his story "*Da Roma a Calcutta in volo*" that he wrote for the magazine "L'Aeronautica" in 1921, one of the very few published works by Ranza, who showed a lively talent as a writer.

In the following years he pursued his career as an officer of the new Regia Aeronautica. Since January 1925 he was commander of the XIII Gruppo Caccia. From 25 April 1927 to 1 October 1928, now a *tenente colonnello*, he was commander of the 2° Stormo

Caccia in Turin. In July 1928 he took part in a fast raid from Rome to London, Amsterdam and Berlin, using a squadron of Ansaldo A.120 and Fiat R.22 reconnaissance airplanes, lead by Italo Balbo, the young minister of aviation. Ranza flew without any trouble, but for the rest of the formation the adventure went badly: the airplanes had accidents all along the route, Balbo was furious and swore that all his future flight were going to be with flying boats, and this raid was talked about as little as possible, then and ever since.

In 1928 Ranza began a new adventure: he became commander of the Aviazione della Tripolitania. It was not an easy command in some sort of tropical paradise, it was a war command, because a tough colonial war was going on to wrest control of Libya from the native insurgents, and the air force was the main weapon against rebel groups operating in that huge desert territory. A commander could and should remain behind his desk organizing and giving orders; he was not actually expected to fly and fight, but not Ranza. For him, to command meant flying in front of everybody else.

This is how Ranza operated (time is April 1929, and this quotation is drawn from "L'Italia in Africa - L'opera dell'Aeronautica" by Vincenzo Lioy, a semi-official history):

> "The band of Khalifa Zaui, overwhelmed by the forces of Sef en Nasser, was in a dangerous situation. The intervention of the air force was requested: at that time at Mizda, command base of Graziani there was the commander of the Aviazione della Tripolitania ten. col. Ferruccio Ranza, who put on his flight overalls, loaded with bombs the only Romeo available at Mizda and took off for the assigned objective. He found the enemy, dropped his lethal load causing losses and panic and strafed. Khalifa, exploiting the temporary panic of the enemy, moved to a better position, but as soon as the airplane flew away toward Ghariat the attack resumed.
>
> The airplane loaded bombs at Ghariat and twice returned over the enemy to bomb. Then in the afternoon four more airplanes went into action to complete the rout. Later on, to show his gratitude, Khalifa Zaui told to ten. col. Ranza: "You are my father and my mother. After God and general Graziani, for me there is you!""

During the conquest of the Fezzan mountain range, in January 1930, Ranza returned from a combat mission with two cylinders shot off his engine. In September 1930 col. Ranza personally lead a patrol of airplanes deep into the desert, opening the road for ground troops to the remote oasis of Tazerbo, and then he flew a reconnaissance to the oasis of Rebiana and El Giof, dropping leaflets. That long forgotten war may seem minor compared to the combats of the Great War, and the Libyan rebels less dangerous than Austro-German fighters. One must consider, however, that every flight accident could mean death, either by the enemy or by the desert itself.

*Col.* Ferruccio Ranza remained in Libya from 1928 to 1931. Our contemporary awareness of politically correct causes a different judgment of colonial wars: yet the man and the fighter had remarkable courage, earnestness, love for adventure, straight behavior. He was more at home under a tent than in the palaces of Rome.

In 1933 Ranza was back in Italy as Chief of Staff of the 1st Territorial Air Region. In 1934 he had the command of the 2ª Brigata Aerea at Mirafiori, Turin. But shortly after he was back in Africa, surely on his own request. On 21 January 1935 *col.* Ranza arrived in Eritrea and became High Commissioner of the new Comando Aeronautica dell'Africa Orientale, (air command of East Africa) that combined the two small air forces of Eritrea and Somalia, that were to be much reinforced for the forthcoming invasion of Ethiopia.

Ranza was in charge of the huge growth of the Italian Aeronautica, until 5 September 1935, when suddenly command was given to General Mario Ajmone Cat. The reason why Ranza was dumped was the opposition of Minister of Colonies Lessona and above all of the Chief of the JCS General Badoglio, a man accustomed to the formality of rigid discipline, who didn't appreciate Ranza's rogue ways. Badoglio wrote in a report to Mussolini after an inspection tour in East Africa in October 1935:

> "Initially the commander of aviation was Colonel, now General Ranza. A top class element in terms of gallantry and will, but wavering in the disciplinary field, and not always well oriented in the delicate use of the air force. I knew him well in Tripolitania where he served me very well. But always and very carefully I had to guide him in order to correct his defects and to exploit his brilliant qualities."

Ranza surely preferred to be flying in action than to waste time on discipline of its own sake and formalities. Promoted to *Generale di Brigata Aerea* (Brigadier General) on 21 March 1935, Ranza on 5 September became commander of the new III Brigata Aerea, a command formed with the tactical squadrons in Eritrea. Still, the situation was unpleasant, Gen. Ajmone Cat was tactful but there was an obvious rivalry, which had to be solved sending Ranza somewhere else. There was another problem, a strong contrast between Ranza and the Ministry of Aviation on where to build the air bases for the attack to Ethiopia. Ranza was in favor of locating them on the Eritrean plateau, where weather was mild; the Ministry wanted them along the coast, fearing that airplanes would have difficulties taking off from runways located above 1,000 meters of altitude in the African heat. The air force had it its own way, but later experience proved that Ranza had been right: the airfields on the plateau operated without any limitation, while the air bases built near the Red Sea, Otumlo, Zula, Assab, expensively built with infrastructures, generators, barracks and refrigerating systems were just too hot, and were hardly used.

Ranza personally guided the first combat mission of the war, an attack by the 118ª Squadriglia with light bombs at Mai Barat, on 3 October 1935. On 28 November 1935 he had another close call, when the notoriously weak landing gear of his Romeo 37 collapsed during a landing at Gura. On 12 January 1936 Ranza lead a bomber formation to Mai Timchet.

At the end of January 1936 Ranza quit the difficult coexistence with Ajmone Cat in Eritrea, and took command of the Aeronautica della Somalia, and in the southern colony he found again the atmosphere of Libya: the war in the desert, the hard but romantic life under the tents.

In every air action on the southern front Ranza was always in the lead aircraft. He was the only general in the history of Italian aviation to lead his units from the cockpit.

On 6 March 1936 Ranza flew a mission that was dangerous, and not just for the reaction of the enemy: flying a Caproni 133 with a crew formed by *cap.* Porru Locci and *cap.* Zanchetta he took off from Neghelli and flew above Addis Ababa, remaining a long time over the Ethiopian capital taking photographs. This feat was much appreciated by the journalists in Somalia, but the command in Eritrea was not amused, and General Ajmone Cat notified the press that a Siai Marchetti S.81 from Eritrea had flown over Addis Ababa before. General Valle, Chief of Staff of the air force in Rome, had to settle the question and his decision was that the S.81 had only reached the outskirts of the capital, and the first real overflight was the one by Ranza. For this feat, he received his fourth Medaglia d'Argento.

In late March 1936 Ranza led large formations of bombers attacking Ethiopian strongpoints along the route of Italian advance. On 29 March the city of Harar was bombed. On the way back a Ro.37 suffered engine troubles and force landed at Uarandab, in enemy territory. With no hesitation, Gen. Ranza landed his Caproni Ca.101E next to the Romeo, in half an hour the problem was fixed and the airplanes took off together. On 24 April 1936, during the battle of Gunu Gadu, Ranza remained in flight above the battlefield four hours, strafing enemy positions and reporting by radio the progress of the battle to Gen. Graziani, as seven bullets hit his airplane.

After the conquest of Ethiopia Ranza, surely not a prolific writer, wrote a short essay, "*From Ual Ual to Dire Dawa*" that is just a chronicle of events, with a peculiarity, that he never names either Mussolini or the régime, and uses the word "fascist" only once, quoting the words of somebody else. Considering that in those years when a General wrote something, it had to be embellished with flattering homage to the dictator "who is always right", this article shows his nature of an apolitical soldier, but probably it didn't make him any more popular.

In October 1936 Ranza received the command of the Divisione Aerea "Centauro" based at Bologna. Years passed and a new world war began. *Generale di Squadra Aerea* (three star General) Ranza fully deserved a ministerial position in Rome, but that was not his style. He assumed command of the Air Force in Albania, and carried the burden of the war in Greece, that began on 28 October 1940. Greek resistance humiliated the Italian Army, and the sacrifices of the Air Force were fundamental to stop a difficult situation from turning into a rout.

Ranza with his group commanders in Somalia, 1936. The air force officer on the right is Guido Masiero.

Giuseppe D'Avanzo, who wrote a masterly history of the Italian Air Force using a pen dipped in poison, so described him:

> "The leathery Ferruccio Ranza, a three stars general who, although he could claim a higher, and safer, position, asked to keep the command of the forces in Albania in order to remain in a zone of operations."

In that time, a secret informer sent to Mussolini a report on Ranza:

> "A brilliant soldier, in love with his profession. Gallant. 3 extraordinary promotions for war merit. 3 Medaglie d'argento etc. Impulsive character – not much balanced – he has particular fondness for certain people, so in his judgment of subordinates sometimes he is not objective. But all his subordinates like him very much, and he is open with them to the point of empathy.
>
> He ignores discipline – his office is a mess and it's only thanks to his collaborator Gen. Sozzani that things work in Albania in the fields of organization and administration. His courageous character makes him usually behave with superficiality, jeopardizing the orderliness that should be the rule in a high Command."

Right: Ranza, chief of the Italian air force in Albania, with a Breda 44, registered I-ORIO, similar to I-AURA on which he was shot down by an Italian fighter on 15 April 1941.

This anonymous informer writing to the dictator wanted to slander Ranza, but what comes out is the portrait of a true man, who dislikes red tape, but takes care of his subordinates, who discounts discipline but who does his job of an airman with passion.

As for danger, in Albania Ranza managed to be shot down once more. Having braved the bullets of Austrians, Germans, Libyans and Ethiopians, this time he was shot down by Italians. On 15 April 1941, as he was flying a Breda 44 (a small twin engines transport biplane) to test the practicability of a small airfield, a Fiat CR.42 fighter mistook the Breda for an enemy and attacked it, soon followed by the fire of ground guns. The Breda was badly hit; Ranza crash-landed but survived with all his passengers.

For his role as a commander in Libya Ranza was decorated as *Cavaliere dell'Ordine Militare di Savoia* (Knight of the Military Order of Savoy); for his action as commander of the air force during the war with Greece he received the higher decoration of *Commendatore dell'Ordine Militare di Savoia*.

Ranza, center, commander of the Italian Air Force in Albania, talking with General Pricolo, chief of staff of the Air Force in 1941.

In late 1941 Ranza was shifted to the command of the 4ª Squadra Aerea (4th air force) at Bari. On 8 September 1943 when Italy surrendered Ranza was there with General Enrico Grande, another remarkable commander. While many other generals ran away leaving their troops without orders, Ranza remained at his place and with his presence and his leadership he kept the unity of the troops, and saved the airports from German destruction. Thus in October Ranza was able to deliver to the Allies the airfields of Lecce, Gioia del Colle, Manduria, Bari, Grottaglie, San Vito dei Normanni, Foggia, Scanzano, Brindisi, San Pancrazio and Taranto and their infrastructures, and to organize battalions of thousands of airmen who cooperated with the Americans.

The role of Ranza in those days was fundamental: dealing directly with Air Commodore Foster, who had the task of liquidating what little remained of the Regia Aeronautica, Ranza convinced him that Italians could and should fight for the liberation of their country from Nazi occupation. The Allies had totally forbidden any kind of flight since 15 September 1943, Italian airplanes instead were still operating against the Germans, but it was Ranza who talked to Air Marshals Tedder and Cunningham and managed to convince the British that Italy had the force and the right to contribute to its own liberation. Maybe the military honor of Ferruccio Ranza and his allegiance to his king were most nobly expressed in his now forgotten decisions of command in the fall of 1943 that had so much indirect importance for the interests of his country.

The years passed, and in 1952 it was time for Ranza to retire. On his final years he generously helped historians of aviation, always kind, available and good-natured. Ferruccio Ranza died in Bologna on 25 April 1973. Italian airmen have not forgotten him.

## Ferruccio Ranza
## 22/24 victory claims – 18 confirmed victories

| Claim | Vict. # | Date | Sqn. | Aircraft | Hour | Place | Victim |
|---|---|---|---|---|---|---|---|
| - | - | 27.7.1917 | 77 | Nieuport | - | Martel Valsugana | Two-seater |
| *[with ten. Savio, 77ª Sq.] Zugsf. Alois Jezek WIA/Oblt. Georg Kenzian Edler vom Kenzianhausen WIA, Flik 24, Br. C.1 61.23* | | | | | | | |
| 1 | 1 | 14.9.1916 | 77 | Ni 17 2613 | - | Gulf of Panzano | Seaplane |
| 2 | - | 14.9.1916 | 77 | Ni 17 2613 | - | Gulf of Trieste | Seaplane |
| *[with ten. Savio, 77ª Sq.]* | | | | | | | |
| - | - | 1.11.1916 | 77 | Nieuport | - | Hermada | Albatros |
| *Maybe: Hptm. Adolf Heyrowsky WIA/Lt .i .d .R Othmar Biegler UNH, Flik 19, Br. C.1 20.08* | | | | | | | |
| 3 | 2 | 25.11.1916 | 77 | Nieuport | 9h30 | Schönpass | Albatros |
| *[with ten. Olivi, 76ª Sq.] Maybe: Oblt. Andreas von Kammerer UNH/Oblt. Béla Gantzstuckh von Hammersberg KIA, Flik 23, Br. C.1 29.54* | | | | | | | |
| 4 | 3 | 25.11.1916 | 77 | Nieuport | 11h30 | Hermada | Albatros |
| *Maybe: Korp. Josef Magyar UNH/Lt. i.d. R. Otto Schrimpl WIA, Flik 28, Br. C.1 27.53* | | | | | | | |

| Claim | Vict. # | Date | Sqn. | Aircraft | Hour | Place | Victim |
|---|---|---|---|---|---|---|---|
| 5 | 4 | 4.4.1917 | 77 | Nieuport | - | Prosecco | EA |
| 6 | 5? | 23.6.1917 | 91 | Spad 1307 | - | Barco | "Old" Albatros |
| 7 | -? | 2.8.1917 | 91 | Spad 4691 | 10h13 | Ovcia | Draga Albatros |
| *[with ten. Parvis, 91ª Sq.]* | | | | | | | |
| 8 | 6 | 10.8.1917 | 91 | Spad 4789 | 20h | Castagnevizza | Albatros |
| *[with asp . Degli Esposti, 80ª Sq.] Korp. Michael Kocher WIA, POW/Oblt. i. d. R. Franz Ebner DOW, FlG.1, Br. C.1 69.92* | | | | | | | |
| 9 | 7? | 23.9.1917 | 91 | Spad | 9h30 | Kotici | Albatros |
| *[with ten. Sabelli, 91ª Sq.] Zugsf. Josef Baier KIA/Lt. i.d. R. Eduard Hafner KIA, Flik 35/D, Br. C.1 129.48* | | | | | | | |
| 10 | 8? | 25.10.1917 | 91 | Spad | 9h15 | Lom | Aviatik |
| *Lt. Paul Dotzenrodtf KIA/Lt. Robert Lessing KIA, FA 17 or: Lt. Walter Conrad KIA/Lt. Albert Tavernier KIA, FA (A) 232 or: Vzfw. Friedrich Moritz/Lt. Franz Neumarker, FA 39* | | | | | | | |
| 11 | 9? | 21.11.1917 | 91 | Spad 6367 | - | Casoni | German two-seater |
| *Vzfw. Konrad Strelocke KIA/Lt. d. Res. Detlev Lamp KIA, FA 39* | | | | | | | |
| 12 | 10? | 30.11.1917 | 91 | Spad 7 | 13h40 | Moriago | Aviatik |
| *[with ten. Novelli & serg. Magistrini, 91ª Sq.] Uffz. Kurt Kieling KIA/Lt. d. Res. Hugo Huck KIA, FA (A) 232* | | | | | | | |
| 13 | 11? | 7.12.1917 | 91 | Spad 7 | 12h15 | S.Pietro | Albatros |
| *[with serg. Magistrini, 91ª Sq.] Korp. Johann Frick KIA/Lt. Johann Knezevic KIA, Flik 39/D, Br. C.1 369.21* | | | | | | | |
| 14 | 12? | 30.12.1917 | 91 | Spad 7 | 12h45 | Fonzaso | Albatros |
| 15 | - | 31.12.1917 | 91 | Spad | - | La Comina | EA |
| 16 | 13? | 12.1.1918 | 91 | Spad | 13h30 | Campo S.Pietro | Rumpler |
| *Vzfw. Werner Hochheim MIA/Lt. Heinrich von Ahlefeld MIA, RHBZ 1, Rumpler 8266/17* | | | | | | | |
| 17 | 14? | 4.2.1918 | 91 | Spad | 8h45 | Cima Mandriolo | Fighter |
| 18 | 15? | 10.2.1918 | 91 | Spad | 11h40 | Val dei Signori | Albatros |
| 19 | 16? | 10.2.1918 | 91 | Spad | 11h40 | Val d'Astico | Albatros |
| *[with ten. Bacula, 91ª Sq.] Fähnr. i. d. R. Guido Prodam POW/Oblt. i. d. R. Heinrich Marsalek KIA; Flik 17/D, Br. C.1 29.68* | | | | | | | |
| 20 | 17? | 15.6.1918 | 91 | Spad | 9h10 | Fagarè di Piave | Two-seater |
| *[with s.ten. Olivero] Korp. Aladar Kerschbaum MIA/Lt.i.d.R. Franz Petzny MIA, Flik 44/D, Br. C.1369.23* | | | | | | | |
| 21 | 18? | 16.8.1918 | 91 | Spad 7 | 19h35 | S.Polo | Albatros |
| 22 | 18? | 29.9.1918 | 91 | Spad | - | Maserada? | Two-seater |
| *[with t.col. Piccio, 91ª Sq.] Fw. Alois Breier WIA/Oblt. i.d.R. Julius Karikas WIA, Flik 2/K2, UFAG C.1 123.24* | | | | | | | |
| 23 | -? | 29.10.1918 | 91 | Spad | 12h10 | P.te Piave-Oderzo | Fighter |
| 24 | -? | 29.10.1918 | 91 | Spad | 12h10 | Oderzo-Livenza | Fighter |
| *One of the following: unknown pilot KIA, Av. D.1 338.06, or: Zugsf. Eugen Bönsch UNH, Flik 51/J, Alb. D.III 253, or: Fw. Josef Siegel KIA, Flik 51/J, Alb. D.III 253* | | | | | | | |

# Antonio Reali

At the turn of the 20th century Italy was a country of deep social and economic troubles. For many Italians of the lower classes, living close to poverty, the only hope for a better future was emigration and up to 600,000 of them were doing just that every year. While people from Southern Italy usually went to America, those from the North usually went to Europe, especially if they had some sort of industrial training. Such was the choice of Antonio Reali, born in Ozegna, a small town in the Canavese near Turin on 31 March 1891.

Information on his early years is scarce, and often misleading: an article that appeared on the magazine "Il Cielo" doesn't describe him as an immigrant, but credits him with obtaining a pilot's license in Germany. Reali as a young worker surely lived in Switzerland, where he married very young, having two children, a girl, born in St. Gallen in 1912, called Carmen Beatrice Lucia, and a boy, Giuseppe, born in Zurich in 1915.

He didn't have time to enjoy his family, because he was called back to Italy for the draft and, as he had some technical expertise, he was posted as a soldier in the *Corpo del Genio* (Corps of Engineers). Very soon he volunteered for the air service in the Battaglione Aviatori.

The first flight marked down in his logbook is dated 15 November 1915, on a Maurice Farman with the instructor Nardini. Reali got his pilot's licenses and remained at the flying school until 3 September 1916, then he was assigned to Cascina Costa for a further period of training on Nieuport fighters. Then, now with the rank of *caporale*, Reali was placed in force with the newly created 79ª Squadriglia da caccia at the Centro Formazione Squadriglie (squadron formation center) at Arcade. On 13 January 1917 the 79ª Squadriglia flew to its operative base, the airfield of Istrana, a few miles west of Treviso. From that base it used its Nieuport 80 hp for standing patrols, escorts and offensive patrols over the Asiago plateau, reporting to the Comando Truppe Altipiani.

Reali flew his first combat mission on 20 January 1917, a standing patrol above Asiago, but his first encounter with the enemy in the air took place only on 4 April. An enemy airplane was spotted over Feltre; Reali took off on alarm and attacked it near Grigno, forcing it to return to its own lines despite the fact that his gun was jammed.

In June the 79ª Squadriglia was engaged supporting the offensive of the Ortigara and during these operations Reali had his second air duel. In the morning of 2 June, during an escort mission for two Caudrons over Mount Verena, Reali attacked an enemy airplane and after a brief exchange of gunfire drew it away.

The two-seater that he attacked the next day, 3 June, was less lucky. During a cruising flight over the Plateau Reali, flying Nieuport 11 serial 2110, clearly saw after his attack, "...the observer slumping over the left side of the fuselage, as the machine-gun was unmanned and the engine had failed ", as he wrote in his combat report. Reali followed his adversary down to a height of 2,500 meters,

Antonio Reali, next to a Savoia Pomilio. On his sleeve the winged propeller, emplem of enlisted pilots.

and then he disengaged, as the mountains were too close. In his opinion the enemy possibly force landed between Borgo and Roncegno, but as there were no witnesses this victory was not confirmed. Another pilot of his squadron, *cpr.* Cerutti, participated in this combat.

In the afternoon of 17 June Reali had another duel, in which he fired the whole double drum magazine of the Lewis on his Nieuport Ni 2363:

> "At 6 p.m. as I was cruising over the area between Borgo and Tresche at a height of 4,500 meters I saw an enemy airplane type Albatros heading toward Borgo a few hundred meters lower than me. I was able to reach it and immediately I attacked firing 97 rounds in several bursts. Then I saw the air-

Right: Antonio Reali in the cockpit of a Nieuport 17 emblazoned with the red "fez" or bonnet of the Bersaglieri troops, probably not his own airplane.

Pilots of the 79ª Squadriglia in 1917: from the left, *serg.* Attilio Imolesi, *serg.* Antonio Reali, *ten.* Federico Comirato, *ten.* Umberto Mazzini, *cap.* Cesare Bertoletti, squadron commander, *s.ten.* Vittorio Melloni, *serg.* Marziale Cerutti, *serg.* Giovanni Nicelli, *sold.* Remo Bozzetti (?) *serg.* Giuseppe Ciotti.

plane going down in the clouds; I tried to follow it down to be sure of where it was heading, as it was diving toward Levico, but when I got down to 3,000 meters I was targeted by strong anti-aircraft fire so I turned back and I couldn't see where such airplane had landed."

The squadron commander *tenente* Cesare Bertoletti noticed the qualities of Reali as a fighter pilot and he praised his eagerness in a report dated 16 August 1917. On 18 August Reali had the usual encounter for all pilots of the 79ª Squadriglia patrolling the Plateau: an Austrian two-seater with its escort fighter. The combat was sharp and furious, but there were no winners. In this period Reali was often detached for the protection of Padua, and he flew an incredibly large number of sorties (40 in September, 24 in October) without claims.

After the Caporetto retreat, as the frontline shifted west, the sector of the plateaus became the keystone of the Italian defensive system, and now air combats raged in that area. The 79ª Squadriglia, moved to the airfield of Nove di Bassano, was heavily engaged. On 20 December Reali claimed his first victory: he found two enemy aircraft over Asiago as they were dropping leaflets, and he claimed one of them shot down in Val d'Assa.

In the first two months of 1918 the 79ª Squadriglia supported Italian operations for the conquest of the Tre Monti. On 10 January 1918 Reali took part in an escort mission for a SAML of the 115ª Squadriglia. Near Mount Lisser a patrol of Flik 60/J lead by ace Frank Linke-Crawford engaged the Italian formation. The Austrian ace got on the tail of the SAML but then Reali attacked him, so Linke-Crawford went into a dive. Reali followed him, but he had to watch his tail from the other Phönix, while Marziale Cerutti took part in the combat despite a jammed gun. The fighter had a furious dogfight until the two-seater could get away safely, then the Italians turned for home and got away from the combat. Luck had been on their side, because the SAML was riddled with bullets but its crew was unharmed, while Reali back on his airfield counted twenty incendiary bullet holes in his fighter. A bit optimistically, Reali claimed a fighter shot down while Linke and Gruber were credited with a fighter and a reconnaissance airplane shot down.

On 14 January Reali finally had his first confirmed victory. In the early afternoon the alarm sounded, warning of Austrian airplanes flying from Asiago to the Grappa. Reali took off with *sergente* Attilio Imolesi, another gallant pilot and ace, and near Mount Rubbio he discovered a two-seater with a fighter as escort. The latter, seeing the Italians, left the field, leaving the two-seater as an easy prey. Reali was the first one to attack, but after a first pass his gun jammed so it was Imolesi who finished off the enemy aircraft, that crashed near Valstagna.

The combat that Reali had on 25 January is proof of how the criteria for the confirmation of victories in the confused development of an air combat were different among the various air forces. That morning the Italian pilot returned to base declaring that with his Nieuport Ni.27 5845 he had attacked an enemy aircraft near Mount Verena and to have followed it down until it landed. It was Albatros D.III 153.70 with pilot Franz Lahner, who was hit, made a forced landing and got out of it unscathed. Despite the fact that his squadron commander, *tenente* Umberto Mazzini, and Group commander *tenente colonnello* Glauco Visconti, both asked for this victory to be confirmed, stressing Reali's reliability, the higher commands did not accept this claim, because the artillery observers could only certify that they saw an enemy plane going down, but not actually reaching the ground.

Reali got his revenge in the afternoon of 28 January claiming a double victory that received all the official confirmations. Flying in a standing patrol over the plateau with a Nieuport 27, he spotted an enemy airplane low near Bertiaga and attacked it diving upon it. After a close combat with the participation of Cerutti, the Austrian plane went down trailing smoke toward Valle Vecchia. Reali, as he was climbing and returning to Cima Echar, saw another enemy, attacked it also and sent it down to crash near Mount Melago. Of this second victory of Reali, we now know that it was Hansa Brandenburg C.1 129.31 of Flik 21, its pilot *Zugsführer* Max Gaderer became prisoner of war while the observer *Leutnant* Max Seewald was killed in the crash. As usual, a message was dropped to the Austrians informing them of the death of their officer.

Two days later, 30 January, Reali was escorting a Savoia Pomilio over Sasso Rosso together with another outstanding pilot of his unit, *sergente* Giovanni Nicelli. At about 10 in the morning the two sergeants saw a pair of Austrian airplanes, as usual a two-seater with a fighter as escort, and engaged them resolutely. Very soon the two-seater was hit and went down in a spin, but the fighter managed to get out of this predicament, when the two Italians had to return to the SP that was itself in danger. Their victim was Hansa Brandenburg C.1 29.07 of Flik 45, its airmen *Zugsführer* Ernest Winkler and *Oberleutnant* Oskar Knoll were posted as missing in action. For this combat the Comando Truppe Altipiani issued a written commendation of both Reali and Nicelli.

Just thirty hours later, in the afternoon of 1 February Reali officially became an ace, shooting down another Austrian airplane in the area of Asiago.

On 4 February 1918 while flying Nieuport 27 5845 in an escort mission for a SP, Reali claimed another victory. This is his Combat Report:

> "I took off for an S.P. escort at 2.50 p.m., having completed the escort back to the airfield, I returned to the lines (Val Frenzela). After a long time of standing patrol I saw lower than me four fighters that were coming in my direction from Asiago. One of them went down separating from the other ones, I turned to guess his intentions and I realized that he was about to attack one of my mates. I jumped unto him and when I had aimed well I fired while he was turning back to his lines. I followed him down until he disappeared near Valstagna. I landed back at 5 p.m."

The career of this Italian pilot peculiarly had long periods without victories, and then several successes concentrated in a few days. He had to wait one year of service at the front to score, and then he became an ace in 19 days. Then he had to wait until June before another victory was credited to him, although his activity up to that time was very intense, and he didn't lack in personal claims. On 5 February, once again escorting an SP, he had a combat with a fighter that he saw going down near Valstagna. On 11 February he had another duel against two Austrians: first he attacked from a distance of 200 m, then as the two enemies challenged him again, Reali fired another 200 rounds, but his engine failed and he force landed.

On 8 April his logbook reports:

> "Shot down an enemy aircraft in a patrol flight: at 5.30 p.m. he saw an enemy reconnaissance airplane returning to its lines near Conegliano, he pursued it and fired a long burst after which the enemy crashed to the ground near S. Lucia di Piave. There is no official confirmation of the combat at the 79ª".

This Albatros D.III 153.181 was damaged on 5 July 1918 in a combat with two Italian fighter, its pilot *Lt.* Skoczdopole was wounded. One of its winners maybe was *serg.* Reali. *(Photo Csizek).*

On 12 April Reali shot down an enemy kite balloon near the Piave, with repeated attacks. Ground observers confirmed that the balloon was deflated and came down. A victory over a kite balloon could be allowed only if the "draken" caught fire, according to the instructions of *t.col.* Piccio, and this is the reason why this claim was not mentioned later on.

He was flying so much (very often three or four missions in a single day) that he had many opportunities to meet the enemy. On 1 May 1918 Reali met five fighters flying from Belluno to the Montello. He attacked from below one of them, separated from the group, and it crashed near Pieve di Soligo. The Italian pilot then managed to avoid the attacks of the four other fighters and returned home. Once again he got no confirmation for this claim. On 17 May Reali, flying Nieuport 27 5865, participated in the confused combat in which the Austrian ace Franz Gräser was killed. Although he is reported among the many pilots who claimed credit for this victory (which was confirmed only to *ten.* Novelli and *serg.* Magistrini and Nardini of the 91ª) there is no claim in his logbook, despite the claim in his combat report:

> "I took off on alarm over Treviso, when I reached an altitude of 3000 m I could attack one of the many enemy aircraft flying over Maserada. The plane I attacked was a two-seater and right after a burst I saw it falling down and catching fire even before reaching the ground. In the place where the burning airplane fell, also an Italian airplane landed with a still propeller. This place is near Breda di Piave. I returned to the airfield at 10.40."

Actually the enemy plane that crashed in flames was a fighter, not a two-seater, and this is probably the reason why he received no credit. Other pilots forced the two-seater he had attacked to land in Italian territory.

On 31 May 1918 Reali had another combat, and the note in his logbook just says: "Air combat for which there is no pilot's report but there is the official confirmation, Report No. 35, Command 420". No Austrian data has been found to match this paucity of Italian records.

On 10 June 1918 Reali met an airplane near Falzè di Piave and shot it down at very low altitude near S. Lucia, seeing it crash in a wood. Ground observers confirmed this victory.

During the offensive of mid-June the activity of Reali increased even further, to a peak of five combat missions for a total of 5 hours 25 minutes of combat flying on 22 June, and his claims were just as abundant. On 15 June flying with another airplane of his squadron (*ten.* Toffoletti) Reali fired at short range against an enemy airplane that went down trailing black smoke. In the official list of victories he is credited with another victory at Valdobbiadene - Mosnigo, which however is reported in his logbook without any information. On 18 June the logbook reads:

> "Enemy a/c shot down. Attacked at 10.15 at an altitude of 500 m over the Piave during an alarm flight. At the first burst the enemy went into a steep dive without recovering. He couldn't attack a second enemy a/c, which like the first one had come out of the clouds because his engine was working badly. The command of the 79ª Squad. has no official confirmation of this victory."

On 20 June 1918 Reali got a treble victory. The *Diario Storico della 79ª*, the squadron war record, reports:

> "Serg.te Reali takes off for patrol at 15h20 – strafing troops left bank of the Piave - returns to airfield at 17h. Height 1,500 m. Takes off on alarm at 18h5. Reaching the level of 1,500 m over the Montello he sees 8 Albatros fighters in formation from Susegana toward the Montello. Near Colle della Sotta he attacked one of them that had strayed from its group and shot it down. He got at firing distance from a second one and fired several bursts causing it to crash while Ten. Toffoletti (his wingman) shot down a third one after it had attacked him from the back. Places of crashes: one near Canareggio a second one north of Susegana and the third one nearby. Height 1500 – time 18h40."

Still spoiling for a fight, Reali returned, changed airplane and took off again for another strafing mission.

In the late afternoon of the next day, 21 June, with Toffoletti and Cerutti, Reali claimed another treble victory. First they met two enemy aircraft, reportedly two-seaters that were dropping food to the Austrian troops holding on the Montello and shot them down, then on the way back two more, one of which they forced to land near Susegana. That was not all. On 22 June Reali claimed four victories. First he met a large squadron of Austrian fighters, and together with Baracchini, of the 81° Squadriglia, he claimed two of them shot down, one crashing in flames across the Piave. Then he met four two-seaters over the Montello, engaged them and brought two of them down, on the left bank of the Piave, in front of Nervesa. 11 victories in three days were certainly a very remarkable claim, which at least shows the aggressiveness of this born fighter.

*Sergenti piloti* during a ceremony of formal delivery of awards: proudly wearing a Medaglia d'Argento are, from the left, Fornagiari, Chiri, Cerutti and Reali.

The end of the offensive didn't mean the end of action. 5 July:

> "Shot down enemy a/c in a fighter flight. At 8.55 a.m. between Farè and Collalbrigo he saw an enemy a/c at an altitude of 50 m. He shot at it repeatedly and it sideslipped and crashed. During the combat the attacked aircraft and another one that arrived shot at him. Coming back he faced intensive ground fire. He was wounded in his right hand by an explosive bullet."

The wound was not serious, because Reali was back in the air the next day, claiming the attack to another Austrian airplane that he followed down to low altitude and then he lost from sight.

The notes to his logbook for 31 July read:

> "Combat and destruction of an enemy aircraft in an alarm flight. He saw an enemy aircraft at 5.15 p.m. at an altitude of 4,500 m that was returning to its lines toward Conegliano. He

Antonio Reali in the mid 1930s getting refresher training and climbing into a Fiat CR.20 biplane fighter, not that much of an improvement over the Nieuports of 20 years before.

VM
N
19798

A line-up of Nieuport 27s of the 79ª Squadriglia late in the war, showing the black she-wolf, emblem of the unit. One of the wolves is white. The Ni. 27 marked VM is the one of *ten.* Vittorio Melloni, the one marked FPF is the one of *ten.* Franco Pagnani Fusconi.

engaged above Arcade and fird a first time. The enemy went into a steep dive trying to escape but serg. Reali reached it again firing repeatedly until he saw the aircraft going down first in a spin and then in a steep dive, disappearing among the hills nort-west of Conegliano (The Command of the 79ª Squadriglia has no official confirmation of this combat)."

A similar event was reported the next day, 1 August:

> "Enemy aircraft shot down in a patrol flight. At 9.15 a.m. he saw an enemy aircraft that was flying from Mandre toward Susegana. He engaged it and followed it down to low altitude until it disappeared in the vegetation. (The Command of the 79° Squadriglia has no official confirmation of this combat)."

On 12 August he had a combat with a "small two-seater" over Valdobbiadene, in which his bullets struck, but Reali couldn't determine its fate.

On 11 September, flying on alarm together with *capitano* Mario Omizzolo, commander of the 79ª Squadriglia, he claimed an Austrian airplane shot down north of Susegana. Austrian air activity was reduced in the summer of 1918, but still Reali flew all the time. On 20 September he met an enemy aircraft that was strafing the troops, he had a long combat in which he fired 212 rounds with no results. Then in September and October he had more combats, and forced kite balloons to be winched down with his attacks. The last confirmed victory of the war for Reali, his eleventh, arrived in the morning of 4 October 1918, with his wingman. "Serg.ti Reali and Lucentini attacked and shot down an enemy airplane that fell south of Moriago." The victim of the two Italian pilots was an Albatros D.III, 253.51 of Flik 56/J, it pilot *Korporal* Otto Berggold jumped out but his parachute was already on fire, so he lost his life. It is likely that also *ten.* Eleuteri of the 70ª Squadriglia took part in this combat. In the next two days he had two more combats with uncertain result, then he doesn't seem to have flown any longer, after 8 October. If, as it seems, he had been removed from combat, it was a well-deserved rest for a pilot who had given so much.

At the end of the year, on 28 December 1918, Antonio Reali was put on permanent leave for the benefit of industry, and he went to Como, the city of his family, working for the Pessina company. All his hectic war activity earned him only one Silver Medal in the field, for his combats of 28 and 30 January 1918. There is no doubt that Reali was somewhat optimistic in claiming victories, but the 5ª Sezione was merciless in cutting down the number of his confirmed victories. What one can say is that Reali certainly didn't lack courage, as he always attacked the enemy, whatever its numbers, and that the quantity of his combat flying is phenomenal: in 20 months, with no period of rest, he flew almost 500 combat missions!

Even the aircraft he usually flew, the Nieuport 27, was surely not a top fighter in 1918: on the Western front by the end of 1917 it was considered obsolete and used only on secondary sectors or for training. The 79ª Squadriglia was equipped with second-hand aircraft of this type from the end of 1917, and it flew them until the end of the war with remarkable success.

Reali, without mustaches, in Ethiopia next to a Romeo Ro.1 of the 38ª Squadriglia, which he briefly commanded.

Very modestly, Reali left the limelight and the press, that had talked about him and published his portrait, forgot him forever. When the Regia Aeronautica was formed as an independent weapon, Reali was assigned to the Reserve, and he took refresher courses flying a Fiat CR.20 fighter. When Italy invaded Ethiopia, Reali was called back to service and flew in that war with the 38ª Squadriglia, a reconnaissance unit equipped with Romeo Ro.1. He briefly became commander of that squadron, then based at Dessiè, on 24 October 1936, for just a few days, because in mid November *capitano* Ferruccio Vosilla replaced him. In 1940 the forgotten ace was promoted to the rank of *capitano*.

After World War II Reali lived in Como, but his marriage by then had ended. He had a workshop for car electrical repairs. He had a very serious car accident, so his children brought him to Fano, the city where they lived, but his wounds were too bad, and Antonio Reali died on 19 January 1975.

## Antonio Reali
## 27-41? victory claims – 11 confirmed victories

| Claim | Vict. # | Date | Sqn. | Aircraft | Hour | Place | Victim |
|---|---|---|---|---|---|---|---|
| - | - | 3.6.1917 | 79 | Ni 2110 | 9h30 | Borgo | Two-seater |
| *[with cpr. Cerutti, 79ª Sq.] EA damaged, observer seen slumping in the cockpit* | | | | | | | |
| - | - | 14.7.1917 | 79 | - | - | Forni | EA |
| 1 | - | 20.12.1917 | 79 | - | - | Val d'Assa/Verena | Two-seater |
| *Maybe: Korp. Tit. Zgsf. Karl Linner UNH/Oblt. Ludwig Draxler WIA, Flik 24/F. Br. C.I* | | | | | | | |
| 2 | - | 10.1.1918 | 79 | Ni.27 | - | Primolano | Fighter |
| 3 | 1 | 14.1.1918 | 79 | Ni.27 | 15h05 | Valstagna | Two-seater |
| *[with serg. Imolesi, 79ª Sq.]* | | | | | | | |
| 4 | | 25.1.1918 | 79 | Ni.27 5845 | 10h45 | Monte Zebio | EA |
| *Zugsf. Franz Lahner UNH, Flik 55/J, Alb. D.III 153.70 FTL.* | | | | | | | |
| 5 | 2 | 28.1.1918 | 79 | Ni 27 | 14h10 | Valle Ronchi | EA |
| *[with serg, Cerutti, 79ª Sq.]* | | | | | | | |
| 6 | 3 | 28.1.1918 | 79 | Ni 27 | 15h | Cima Echele/M. Melago | Two-seater |
| *Zugsf. Max Gaderer POW/Lt. Max Seewald KIA, Flik 21/D, Br. C.I 129.31* | | | | | | | |
| 7 | 4 | 30.1.1918 | 79 | - | 10h | Col Echele/Costalunga | Two-seater |
| *[with serg. Nicelli, 79ª Sq.] Zugsf. Ernst Winkler MIA/Oblt. Oskar Knoll MIA, Flik 45/D, Br. C.I 29.07.* | | | | | | | |
| 8 | 5 | 1.2.1918 | 79 | - | 15h | Cismon | EA |
| 9 | - | 4.2.1918 | 79 | Ni.27 5845 | 11h | Valstagna | EA |
| 10 | - | 4.2.1918 | 79 | Ni.27 5845 | 11h | Valstagna | Fighter |
| 11 | - | 11.3.1918 | 79 | - | 14h | M. Erio/M. Civello | EA |
| 12 | - | 8.4.1918 | 79 | - | 17h30 | S. Lucia di Piave | Two-seater |
| 13 | - | 12.4.1918 | 79 | - | 9h50 | Piave | Kite balloon |
| 14 | - | 1.5.1918 | 79 | - | 16h30 | Pieve di Soligo | EA |
| - | - | 17.5.1918 | 79 | Ni.27 5865 | - | Pero | Fighter |
| *[confirmed only to Magistrini, Nardini, Novelli, 91ª Sq.] Lt. Franz Gräser KIA, Flik 61/J, Alb.D.III 153.221.* | | | | | | | |
| 15 | - | 31.5.1918 | 79 | - | - | S. Salvatore | EA |
| 16 | - | 10.6.1918 | 79 | - | 7h35 | Falzè di Piave | EA |
| 17 | - | 15.6.1918 | 79 | Ni.27 | 7h40 | Farra di Soligo | Albatros |
| *[with ten. Toffoletti & serg, Ciotti, 79ª Sq.] Possibly one of these:<br>Fw. Rudolf Berger KIA/Oblt. Ernst Petric KIA, Flik 26/DS, Br. C.I 329.48<br>Stfw. Franz Koudela UNH/Oblt. Anton Chrzascz WIA,Flik 26/DS, Br. C.I 329.22<br>Unknown pilot/Lt. i.d.R. Johann Reinstaller WIA, Flik 52/D, Br. C.I<br>Unknown pilot/Lt. Ernst Winkler WIA, Flik 52/D, Br. C.I* | | | | | | | |
| 18 | - | 15.6.1918 | 79 | Ni.27 | 12? | Valdobbiadene/Mosnigo | EA |

| Claim | Vict. # | Date | Sqn. | Aircraft | Hour | Place | Victim |
|---|---|---|---|---|---|---|---|
| 19 | - | 18.6.1918 | 79 | - | 10h15 | Ponte della Priula | EA |
| 20 | 6 | 20.6.1918 | 79 | - | 18h40 | Canareggio | Fighter |
| 21 | 7 | 20.6.1918 | 79 | - | 18h40 | Lusignana? | Fighter |
| 22 | - | 20.6.1918 | 79 | - | 18h40 | Lusignana? | Fighter |
| *[with ten. Toffoletti, 79ª Sq.]* | | | | | | | |
| 23 | 8 | 21.6.1918 | 79 | - | 19h45 | Susegana | Two-seater |
| 24 | 9 | 21.6.1918 | 79 | - | 19h45 | Susegana | Two-seater |
| 25 | - | 21.6.1918 | 79 | - | 19h45 | Susegana | Two-seater |
| *[with serg. Cerutti & ten. Toffoletti, 79ª Sq.] Korp. Raimund Praschl WIA/Oblt. Jakob von Jaskiewicz WIA, Flik 65/DS, Br. C.I 269.22* | | | | | | | |
| 26 | - | 22.6.1918 | 79 | - | 19h45 | Nervesa | Fighter |
| 27 | - | 22.6.1918 | 79 | - | 19h45 | Nervesa | Fighter |
| *[with ten. Baracchini, 81ª Sq] Maybe: Fw. Hubert Graf, Flik 42/J, Alb. D.III FTL* | | | | | | | |
| 28 | - | 22.6.1918 | 79 | - | 19h45 | Nervesa | Two-seater |
| 29 | - | 22.6.1918 | 79 | - | 19h45 | Nervesa | Two-seater |
| 30 | - | 5.7.1918 | 79 | - | 8h55 | Collabrigo | EA |
| *[Maybe with serg. Soliani & sold. Rustici, 79ª Sq.] Maybe: Lt. Heinrich Skoczdopole WIA, Flik 61/J, Alb. D.III 153.181* | | | | | | | |
| 31 | - | 31.7.1918 | 79 | - | - | Conegliano | EA |
| 32 | - | 1.8.1918 | 79 | - | 9h15 | Susegana | EA |
| - | - | 12.8.1918 | 79 | - | - | Valdobbiadene | Small two-seater |
| 33 | 10 | 11.9.1918 | 79 | - | - | North of Susegana | EA |
| *[with cap. Omizzoli, 79ª Sq.]* | | | | | | | |
| 34 | - | 23.9.1918 | 79 | - | - | Susegana/Conegliano | EA |
| 35 | 11 | 4.10.1918 | 79 | - | - | Fontigo | EA |
| *[with serg. Lucentini, 79ª Sq.& ten. Eleuteri, 70ª] Korp. Otto Berggold KIA, Flik 56/J, Alb. D.III 252.51* | | | | | | | |
| - | - | 5.10.1918 | 79 | - | - | Conegliano | EA |
| - | - | 6.10.1918 | 79 | - | - | Parè | EA |
| *Maybe: Fw. Karl Kallasch KIA, Oblt. Johann Dopita MIA, Flik 57/Rb, Br. C.I 121.61* | | | | | | | |

There is much confusion about Reali's claims in the list issued in 1919 by the Bongiovanni Commission, that gives 27 claims for Reali, some of them with a wrong date (17 June 1918, 1 July 1918 and 30 July 1918 instead of 18 June, 5 July and 31 July respectively). According to his logbook he claimed 14 further victories. Probably they were disallowed immediately and thus they were not mentioned in the later list, but we cannot be sure of this.

# Cosimo Rennella

Secondigliano, a suburb north of Naples, was the birthplace of Cosimo Rennella, born on 15 February 1890, son of Salvatore and Andretta Barbatto. Little Cosimo didn't get to know Naples, because when he was two years old his father decided to immigrate to America. Unlike many Neapolitans who went to the United States, he decided to try his lot in Ecuador, which was then a young nation, expected to have a great future. The Rennellas opened a small commercial business in Guayaquil. A typical Neapolitan, young Cosimo was described as cheerful and a great lover of sports. At the age of 17 he was already in a football team and he won many sport competitions. As his family had gained a certain wealth, he was able to study, and attended with good results the Colegio Nacional Vicente Rocafuerte, liked by his school mates and appreciated as a mechanic, mending in his father's workshop the bicycles of all the kids in the neighborhood.

From the beginning, his life seemed marked by sport and risk. In 1909 he enrolled in a battalion formed by the students of his college, called "Patria I" and he took part in the brief war between Ecuador and Peru. He then practiced with success the sport of walking, and again he won many races, and boxing, also in this case with success, winning a championship in 1910, and beating boxers of international repute. It seems that no sport was unattractive to him, as he took part in competitions of gymnastics, wrestling, boxing and shooting.

His life changed on 20 April 1912, when he signed a contract with the Club Guayas de Tiro y Aviacion. This so-called club actually had a para-military character. It wanted to send one of its members to Europe, to learn to fly and then spread aviation in Ecuador. The chosen one was young Rennella who received a scholarship for Europe. He went first to the famous Bleriot school at Pau, France, and then to Chiribiri of Turin, Italy. In the report of the aviation meeting at Cuneo, 11 and 12 August 1912, attended by the king of Italy, among the guests also *"the Captain from Ecuador Rennella"* was quoted. Meantime in his adopted country the progress of their first airman were followed in great trepidation.

Rennella got the civil pilot license number 166 on 28 August 1912 in Mirafiori. Soon after he sailed to South America bringing two Nieuport airplanes, with which he flew above Panama City. What he really wanted to do was to fly over the Canal, but first lack of money, and then the veto of the United States forced him to fall back on a simpler feat. He returned to Ecuador welcomed as a hero. The following year he returned to Italy to buy an airplane with the money that Ecuador allocated to set up a flying school. He used this opportunity to get his military pilot license again in Mirafiori, on 28 July 1913 and he returned to Ecuador with a Chiribiri monoplane.

On 8 October 1913 Rennella was the first national pilot to fly an airplane in Ecuador, taking off from the Jockey Club of Guayaquil field. He flew again on the 19th and the crowd was so excited that it poured on the flying ground before Rennella had fully landed, forcing him to turn sharply not to hit a woman, damaging beyond repair his precious airplane.

Cosimo Rennella in a Chiribiri monoplane, 1913.

In 1914 Rennella traveled all around Latin America, trying to offer his capacity as a pilot, but with little success. He was in Costa Rica, Peru, from where he was expelled on account of an affair with the French mistress of the local dictator, and in Chile, ending up as a military pilot in the Mexican air force, fighting against the revolutionaries. He was supposed to set up the first flying school in Mexico, but the European war had started, so he returned to Italy.

Initially he probably had problems with his nationality, because he had first to join as an infantryman, although he volunteered in the Battaglione Aviatori. On 15 September 1915 he was admitted to the flying school of Pisa, and in a short time he confirmed his qualities, and got once more a military pilot license on 15 February 1916.

After a brief stay at Busto Arsizio, base of a training flight for Farman airplanes, he was finally posted to a squadron at the front. Almost one year after Italy's entry in the war, on 14 April 1916 he reached the 31ª Squadriglia, a very experienced unit. Based at Verona-Tombetta airfield, the 31ª Squadriglia flew Maurice Farman MF.14 biplanes and its commander was *capitano* Bonazzi. It operated on the northern front of the Trentino, less important initially than the eastern Isonzo front, but still a theater of intensive action, and there had already been air combats with enemy airplanes. The Farman, carrying a machine gun, was even supposed to act as an interceptor and stop the occasional Austrian raids against Verona that caused minimal damages but had a great impact on the civil population. Rennella didn't stay long with the 31ª, just the time for five sorties, two of them "fighter" missions, and then on 1 May 1916 he was posted to the 48ª Squadriglia, that operated on the same front, in the narrow valley that holds the city of Belluno. This unit consisted in one flight with Caudrons and two flights with Farmans, and it performed reconnaissance flights supporting the troops that garrisoned the scenic Dolomites, taking off from an airfield located at 380 meters a.s.l. in a valley and climbing over 6,000 feet mountains. Not an easy task, and Rennella flew about 40 sorties, many of them test flights and "attempted reconnaissance", meaning aborted flights for the prohibitive weather conditions.

Rennella remained in contact with Ecuador, and the newspaper of Quito, "El Comercio" published one of his articles, quite realistic in describing the technical aspects of flying, despite any censorship.

> "....I hear a sound of death getting closer: it is the Shrapnel with their buff colored clouds that welcome me: I must climb higher. The observer turns to look at the altimeter: 1800 meters, and without worrying he turns again and gets ready to enter in communication with our batteries at the right moment. The area that we must observe is marked on the map but I easily recognize every spot, because we flew above and we took photographs countless times. There are trenches, strongpoints and forts well hidden in the woods and presently our 305-mm batteries rain upon them volleys of shells.
>
> The cannon shots of enemy artillery are intermittent; it seems that our heavy artillery is making its deadly effect felt. The tempest of shots that they vomit sounds like a deep and wailing cry. As it passes, every grain of earth is mixed up, suffocated and covered in a curtain of iron. The curtain of civilization!"

For some reason, photographs of Rennella are very rare, and he doesn't appear in known group photos of the pilots of the 78ª Squadriglia. Here he appears in a souvenir photo of the best sergeants of the XXIII Gruppo (78ª and 79ª Squadriglia): from the left, Fornagiari, Reali, Cerutti, Rennella, Aime and, standing behind, Chiri. *(Photo Selinger).*

As the 48ª Squadriglia converted to the twin-engine Caudron G.4, more appropriate for that high mountain sector, Rennella, now an experienced pilot, on 27 July 1916 was shifted with other pilots to the Farman-equipped 45ª Squadriglia, based at Oleis, on the Isonzo front, with *cap.* Giovine as commander.

The air force of Austria-Hungary in this period had only a handful of Fokker fighters. Their deadly reputation spread among Italian units, but their actual combat successes were rather limited, and combats against Italian airplanes were rare. Rennella experienced his first air combat on 6 August 1916, when a Fokker monoplane attacked his airplane, that carried *capitano* Giovine. They got away undamaged. Two days later there was another attack, but again Rennella and his observer, *ten.* Monti, avoided the bullets of the enemy fighter. On 31 August his war experience was acknowledged and he rose to the rank of *sergente*.

On 31 October 1916 as he was flying with *ten.* Allegra, who became one of his favorite observers, Rennella clashed with two enemy "Albatros", as the Italians then called the Brandenburg. Once again the combat brought no result. Anti-aircraft guns were the major threat, however. During 1916 at the 45ª Squadriglia Rennella flew 45 sorties, and then 10 in January, 11 in February and 7 in March 1917. In April his missions were 14, in one of which, 11 April, he had a combat supposedly with a Fokker (not likely to be still in frontline service at that time). Ground fire had more telling effects: both on 15 May and on 19 June 1917, both times with *ten.* Allegra, he returned to base with his aircraft damaged by ground fire. The latter was his final reconnaissance mission. He was finally assigned to fighter conversion, having performed a total of 125 flights at the 45ª Squadriglia.

Posted to the Comando Aeronautica of Turin, on 26 June he was at Mirafiori airfield to convert on the Nieuport, which he did successfully, and then moved to aerial gunnery training at Pisa. In

Right: A signed portrait of a handsome Cosme Rennella.

The first victim of Cosme Rennella, Aviatik Berg D.I 38.17 shot down on 24 September 1917 with the death of *Fw.* Kotzian. *(Photo Csizek)*.

Another valiant sergeant of the 78ª Squadriglia was Oreste Codeghini. This is his Hanriot in flight.

just two months he was back at the front, now assigned to the 78° Squadriglia that he reached on 24 August 1917.

The 78ª, with commander *cap.* Bolognesi, and since mid October *ten.* Antonio Riva, still had to become one of the best Italian fighter units. Formerly based at Istrana and operating on the Trentino front, it had recently moved to Borgnano, on the Isonzo front. It had no aces yet, but its pilots had scored about a dozen victories. Among them were the future aces Chiri, Fornagiari and Magistrini, but the most successful pilot was Guido Nardini, then credited with four victories. Its material, Nieuport 11s and a few Nieuport 17, was obsolete and this fact hampered the squadron. In this period it began receiving the excellent Hanriot Hd.1 as the first new fighters were arriving from France.

Papers don't report with what kind of fighter Rennella began operations as, after a few familiarization flights, he flew his first mission on 1 September, a patrol above the city of Gorizia.

As was rather usual, conversion from reconnaissance to fighters was not painless, and in a few days everything happened: on 5 and 11 September he broke the *bequille* (skid) on landing, on 6 September during an escort to Capronis bombing Chiapovano he was hit by anti-aircraft fire, force landed at Oligna and turned over. His first combats brought no results, either for gun jams or for the escape of the enemy.

He seemed to be jinxed: on 16 September he took off for an usual Caproni escort, but after two minutes he had to force-land with the engine stopped by a magneto failure, he hit some telephone cables and smashed the left wing of his plane. Later on he took off again, but he had to land again for one more engine failure. At least the commander noticed his eagerness, and on the 20th he wrote a proposal for the Medaglia d'Argento al Valor Militare.

This trust was well placed, as on 24 September Rennella scored his first victory. Escorting a Savoia Pomilio to Aidussina he faced the attack of three fighters, saved the two-seater and shot one of them down west of Zagorje. His victim was one of the new Austrian fighters, an Aviatik D.1 of Flik 32/D that crashed in its own lines killing pilot *Fw.* Karl Kotzian. Inspecting the wreck the Austrians found its engine riddled with bullets.

Two days later, on the 26th, during another Caproni escort flight over Lom he had a dogfight with an enemy fighter and claimed it shot down, while the enemy put a bullet through his top wing spar but this victory was not credited to him. On the 28th he had no less than three combats in a day, but each time he had to quit because of gun jams. Once more the engine failed him on 29 September, and landing back on his airfield he turned over and totally wrecked his fighter. During the month he flew a total of 36 flights.

October went by relatively quiet, for Rennella, who had only one combat and many escorts to the SP and SAML of the II Gruppo going on reconnaissance to S. Lucia, Monte Nero, Ternova and Chiapovano, where the Austro-German offensive was about to spring. When it arrived, on 24 October, the 78ª Squadriglia, based close to the first line, had to retreat. On the 27th it moved to la Comina, and under the pressure of the enemy advance, with horrible weather, twelve precious Hanriots had to be burned and the unit crossed the Piave without airplanes. By this time Rennella had flown over 50 missions, had 8 combats and two victory claims.

By now he was an experienced fighter pilot, but his squadron, back at Istrana, had only two Hanriots and Rennella couldn't fly until 14 November, when he ferried a brand new Hd from Verona.

On 21 November he returned to action and to victory, shooting down another fighter. Escorting a two-seater over Vidor, Rennella spotted five attacking fighters and shot down one of them at Mosnigo firing 136 rounds of his machine gun. Till the end of the year 1917 he flew 35 more missions from Istrana, meeting the enemy only twice and without success.

In January 1918 air activities were curtailed by bad weather but Rennella was busy escorting Capronis in bombing missions. Those were difficult flights, that included transfers to staging airfields and long tiring flights with the engine throttled down to keep pace with the slow bombers, always watchful for possible fighter attacks.

On 14 January, now in offensive patrol, he attacked an enemy two-seater over the Grappa and claimed it going down out of control in the direction of Val di Pez. The press of Ecuador published the description of this combat written by their national hero:

"..In the clear sky of dawn stars vanish and all the pilots of the squadron are ready for take off, on alarm. Not much to wait: the telephone rings telling of enemy aircraft arriving from Northeast. We take off and climb into space almost simultaneously. I check my altimeter, 1500...2000...2600 meters, as my Macchi Nieuport gains height a strange restlessness fill me. Maybe it is a foreboding that something important is about to happen, or the desire that the unknown, the inevitable happen as soon as possible? Four thousand meters… I see among the clouds of anti-aircraft fire three black spots coming in: it's the enemy! I remove the lock of my gun and test it: today it works properly. I turn and go in their direction, picking up an enemy that I aim at a speed of 200 km. I aim directly at the nose of the airplane and I pass below it at a few meters. The enemy opens fire, the battle has started, I turn and gain height getting above the enemy plane, as I distinctly see its occupants. The pilot understands my maneuver and tries to get away banking. I dive upon them and fire two bursts of hundreds of bullets in three seconds; twice I clearly hear the noise of ripped canvas.

Two of our planes come to help me, maneuvering skillfully. I see one of ours going down wavering, from the insignia painted on the rudder I understand it is my commander: obviously his engine is out of action due to enemy fire. Seeing this I return to the dogfight with more determination, full of boldness. The dogfight gets more and more violent, the beauty of combat and hatred for the enemies of my Fatherland give me an aggressive fury that fills me as I get ready to attack.

Will he get away once again? No, I will attack, pursue until I have one bullet in my gun and a bit of fuel in my tank. No reason to doubt: the enemy airplane, harassed by our guns flies here and there looking for escape. Finally I have it fully in my gunsight and by instinct, before my own will, I realize I shot. One more burst, I hit it fully. Twice it banks sharply, then it goes down vertically in a steep flight to disaster.

There are no maneuvers; the pilot is not attempting anything to avoid the crash. Did I wound him or is he dead? Quick, like a lightning, compassion fills me: I am here as he was, up in the air, at over 3,000 meters. I am full of life, dreams and hopes, as he also was a few moments ago, I think of my loved ones, who wait for my return.

My mates are still in the sky, watchful sentinels, as I start my descent, impatient to land. When I land the rumble of guns from the Alps nearby quiets me, as I think that also up there, there are millions of men who fight and die, all for the same cause and defending the same ideal."

Unidentified pilots of the 78ª Squadriglia: Cosme Rennella could be the one on the extreme right.

One more colorful Hanriot of the 78ª Squadriglia, the one of *cpr.* Rinaldo Bulgheroni.

The next day, 15 January, in another Caproni escort Rennella attacked an intercepting enemy fighter and shot it down near Arsiè. Both these victories were credited to him after the war.

In February he was out of action for two weeks, probably on leave. When he returned to his unit preys were scarce, and only on 17 March, as he was simply flying from Nove to S. Luca, he had a combat with an enemy reconnaissance plane, shooting it down in Val Seren: this victory had no confirmation. He did destroy a plane on 21 March, Hanriot 509, wrecked in a forced landing, possibly for the usual engine failure.

Rennella scored two more victories on 17 April shooting down two airplanes over Conegliano and Valdobbiadene. Unfortunately there are no details of this event, but since it was a morning flight it probably was the same combat in which six Hanriots of the 70ª Squadriglia took part, even if the Italian War Bulletin reported them as two separate combats. That day two flights of Flik 42/J consisting of nine Albatros D.III escorting a Brandenburg two-seater of Flik 52/D clashed with what they described as 6 or 8 "Sopwith", losing in the combat the leader, *Oblt.* Michael Gassner-Nordnon whose fighter shed its wings in flight and the two-seater that, despite the protection of the two fighters of *Fw.* Oberst and *Lt.* Ujvary, was hit by three fighters and fell, also losing its wings. Both the commander and the two crewmen of the Brandenburg were killed.

This hard combat generated an amount of overclaiming: the pilots of the 70ª claimed two fighters and a two-seater; Rennella two unidentified aircraft, probably the same ones. The Austrians claimed three fighters certainly shot down, including the one that had shot down their leader, and another one probable, but there were no Italian losses. On that day British pilots claimed no less than eleven Albatros fighters shot down, but the details and localities that they reported don't correspond to the ones of this combat, and no other Austrian losses are known for that day.

Two weeks later, on 1 May, Rennella had another lucky day that brought him a double claim. In a first sortie with his squadron commander, *capitano* Riva, he flamed an enemy airplane above Fontanelle, while in a second sortie he claimed one shot down over Cimadolmo – S. Lucia di Piave. The first one was one of the new Phönix C.1 reconnaissance planes that had just entered service, with as pilot the commander of Flik 52/D who was wounded in the crash as was his observer, while the plane was totally destroyed. These two victories of Rennella (other sources quote three) were not confirmed in the final list of aces issued after the war, although we now know that at least one of them is certain.

Once again two weeks passed before he met enemy aircraft, and again he was victorious. It was 17 May, and also in this case data are contradictory: his logbook reports two enemy planes shot down, a fighter that fell in the Piave and then a two-seater at Maserada, but his Combat Report shows only one plane attacked and destroyed. The mystery can be explained by the fact that Rennella didn't see the end of the fighter and didn't mention it in his report, while his commander in a later document claimed for his subordinate also a fighter shot down, getting a strong reproach from the higher commands, (*ten. col.* Piccio) that reminded him that one couldn't claim victories when there was no verification if the fall of an enemy was really a crash and not just a dive.

That was the combat that cost the life of the Austrian ace Gräser, with the participation of many Italian pilots, who later disputed the merit if his demise. Rennella attacked first a fighter and then the two-seater with the escort of the formation of the unlucky Gräser. That one, Brandenburg 229.30, was his only victim that fell within Italian lines. A later investigation gave most of the credit for this victory to him together with *tenenti* Fucini and Avet. The enemy pilot declared that three fighters had shot him down, but it seems that after the war this victory was credited to Avet alone. This is how Rennella described the events in his brief Combat Report:

"Time of take-off 9h58
Time of landing 10h27
Height reached 3400 meters
Type of plane HD
Machine gun Wickers
Bullets fired 155

I took off on alarm at 9h58 at a height of 3400 meters I saw a group of enemy airplanes one of them, a large two-seater probably escorted by the other ones, was flying to Maserada, where I reached it and engaged combat. After a few gun bursts I saw it going down in tight spirals and following it close I shot at it again each time it tried to reach its lines, so I followed it down to 100 meters and I saw it hitting a wing in a small field near Maserada. I returned to my field and landed at 10h27. Sergente Cosme Rennella."

One should notice that he always signed himself "Cosme", the Spanish version of his Italian name Cosimo.

Three days later he had the honor of receiving the Medaglia d'Argento during a ceremony headed by General Pennella, commander of the 2ª Armata. Its citation quoted 15 combats and four victories in the period up to 15 January 1918. His portrait after that event was published on the cover of the magazine "Nel Cielo" together with other successful sergeants, Fornagiari, Reali, Cerutti, Chiri and Aime.

On 30 May during an offensive patrol to Susegana Rennella spotted five enemies, attacked one of them and shot it down near Lovadina shooting 47 rounds. June brought hard chores for Rennella, often tasked with the escort of two-seaters spotting for the artillery that was firing at concentration of enemy troops, preparing for a big offensive. When this started, on 15 June, he was always in the air, looking for airplanes or strafing enemy forces across the Piave.

On 20 June, tasked with the escort of a SAML of the 118ª Squadriglia he saw an enemy squadron about to attack the two-seater, and he rushed to its defense with his wingman. Once again the excitement and the tension of hard combat caused pilots on both sides to overclaim, with no confirmation from official documents. In the mid morning of that day over the Montello at least two patrols of Hanriots of the 78ª Squadriglia and two RAF Sopwith Camels clashed with a big force of fighters from the best Austrian fighter squadrons, Flik 41/J and 42/J that were attacking some SIAs over Nervesa. A hard combat ensued and ground observers declared that they saw two enemy airplanes crashing. Rennella, rather optimistically, claimed three shot down on the eastern slope of the Montello, the first one in flames, probably the same ones claimed by other Italian and British pilots. The Austrians again claimed several "Sopwiths" shot down, but both the pilots of the 78ª and the British didn't suffer losses, according to surviving data. The greatest Austrian ace, Brumowski, returned to base with his plane riddled with bullets and Hefty, the ace of Flik 42/J landed at Pianzano with a badly damaged plane.

On June 23 as he was in patrol over the Montello at 11.20 a.m. Rennella saw five Albatros D.III coming out of a cloud, he attacked and saw one crashing between Falzè and Barbisano. After the storm of the mid June battle, he flew only once in July, while in August he returned to the usual level of activity, but enemy aviation was absent, and he had to wait until 31 August to score again. That day Rennella flew as wingman for two outstanding pilots, *tenente* Fucini and *sergente* Chiri. They found an enemy two-seater and attacked

A gallant Austrian ace, *Stfw.* Friedrich Hefty, whose Albatros 153.169, appearing here, was damaged in combat on 20 June 1918, possibly at the hands of Cosme Rennella. *(Photo Höfling).*

it in perfect coordination, and shot it down across the Piave, watching it break up against a row of trees.

During September Rennella didn't fly very much, he had only one unsuccessful combat. During the final month of the war he was again active flying many offensive patrols, but he found enemy aircraft only during the last final offensive. Italian pilots generally didn't like very much alarm service, which meant sitting at the airfield waiting for the unlikely arrival of enemy aircraft; when an enemy was spotted, interception was a rare event. Rennella however was often lucky in this service. On 27 October, taking off after the alarm, he found an enemy reconnaissance airplane and shot it down north of S. Polo, while other fighters engaged an enemy formation. Then he attacked another one north of Susegana, and finally he ended his sortie by dropping six bombs on Conegliano. Once again, this victory was not confirmed after the war. His companion in this combat probably was *sergente* Chiri of the 78ª who claimed a two-seater in the same area.

In the following days Rennella flew many strafing and bombing missions against the retreating enemy army. His final war mission took place on 2 November 1918, a peaceful cruise over Belluno, the city where his war adventure had begun two years before, but now there were no more enemy aircraft to face him.

In 1919, when the official list of victories was issued, Rennella and generally all of his squadron, the 78ª, were the ones that most suffered the cuts by General Bongiovanni, and possibly rightly so. According to unofficial data he had claimed from 17 to 20 enemy aircraft, but he was credited only with 7. Beyond his actual number of victories, the contribution of Rennella is expressed by his logbook: 414 flights, including 65 reconnaissance missions, 150 fighter patrols, 86 escorts without ever losing a two-seater or bomber, and 39 air combats. After the war he was awarded a second Medaglia d' Argento al V.M., which in its citation credited him with 14 victories at 23 June 1918. With peace Rennella remained at the 78ª Squadriglia until March 1919, then he went to Montecelio, the new air force test establishment near Rome, where in July he gave the baptism of flight to a dubious Consul of Ecuador. His heart had remained in Latin America, and he returned there as soon as he was discharged.

The Italian government gave as a gift a surplus fighter to each ace, and Rennella got a Hanriot. He also bought a Caudron G.3 for the cheap price of 500 Lire and in February 1920 he reached Venezuela with his small air fleet and a contract for air shows. The local public was enthusiastic of his aerobatics and the press praised him. He also experimented in the fast transport of mail and newspapers. One of his flights is considered the first air mail flight of that country.

General Gomez, Chief of State, whom Rennella had turned into an enthusiast of aviation, decided to create an air force buying the two airplanes of the Italian pilot for a national flying school. It was during his period in Venezuela that Rennella learnt of his second Silver Medal.

He returned briefly to Italy, then in 1922 he was back in Venezuela, where Gomez charged him with touring Europe to get all information about the development of civil aviation in order to form a Venezuelan air line. He hoped in vain to be appointed as chief of the Aviation School at Maracay.

At the end of 1924, after some contrast with Gen. Gomez, Rennella returned to Ecuador, called there by the new President of the Republic Gonzalo S. Còrdova: his fame as a pilot was still high, and he was appointed as instructor at the Military Aviation School "El Còndor" of Duràn, with the rank of captain. Other enthusiastic Italian pilots, Elia Liut, Ferruccio Guicciardi and Attilio Canzini had founded this school but its conditions were terrible. Despite its great enthusiasm, the country was poor and there were problems keeping efficient their few airplanes, of which the more modern ones were an old Hanriot brought by Liut, called "Telegrafo 1", protagonist of several raids, a SVA, a Sopwith 1 1/2 Strutter and a Caudron G.3. The other ones were Gabardini training monoplanes. Rennella's task was to teach aerobatics to the first pilots of Ecuador, a subject in which the ace was peerless.

His great fame however brought him enemies, particularly within the flying schools. Pupils, believing to be more expert than they really were, didn't like his style of teaching, his promiscuous behavior in civil life was criticized as unfit for a member of the army, and finally he didn't go along with Attilio Canzini, the good Italian pilot who was primary instructor. As a result, the students officially asked the expulsion of Rennella from the school and from the Army.

It was an unpleasant situation, but the authorities and public opinion defended him, and the price was paid not by the unruly students but by innocent Canzini, whom the press portrayed as a foreign agitator who had instigated his students.

When a revolution in 1925 overthrew the constitutional government of Còrdova, Rennella was briefly dismissed, but he was soon recalled, and organized many long distance flights, using whatever he had: in 1926 an old Sopwith 1 1/2 Strutter, in 1930 a Travel Air. The press often published essays by him on how to organize the Fuerza Aérea Ecuadoreña: since the country had limited financial resources, he proposed to use military personnel also for civil

Cosme Rennella in a late production camouflaged Hanriot, serial 20960, that here appears without its machine gun. *(Photo Longhi).*

flights, updating their training and at the same time contributing to the development of commercial aviation. All this remained on paper, due to lack of funds. In 1932, piloting a Ryan called "Ecuador" he flew with Luis Mantilla in a tour all around the republic, that lasted 13 days. Rennella's career continued with raids and photographic surveys of the dangerous Andes with the limited means of the air force of Ecuador.

In 1934 he was teacher of Air Photography for a new pilots course. He was promoted to Major and he had the satisfaction of his son Ernesto following his footsteps and becoming a military pilot. In 1936 he was appointed director of the new school of civil aviation, but once again financial problems scuttled this enterprise. In that year he also visited the United States and attended an Aviation Meeting at Los Angels, 5 to 7 September.

Rennella died when what he considered his new country was establishing a real air force, as he had always wished. He returned from Dayton, Ohio, where he had attended a meeting of World War I Aces, but he fell sick with pneumonia. He was hospitalized in the military hospital of Quito, but despite all efforts he died on 3 May 1937. His death caused great commotion in Ecuador, where his deeds were the stuff of legend. The famous poet Pablo Hannibal Vela wrote a beautiful poem. At his funeral the most beautiful homage was the flight and aerobatics performed by major Fritz W. Hammer, a German pilot who shared with Rennella the common love for flying. After his death Rennella was promoted to the rank of Lieut. Colonel.

One of the airplanes that he flew above the Andes, the "Telegrafo 1" that he had smashed in a forced landing was restored and preserved, and is now exhibited in the Museo Aerónautico y del Espacio of Quito. Nobody now remembers Cosimo Rennella in Italy, but in Ecuador he is not forgotten, and is considered one of the fathers of the Fuerza Aérea Ecuadoreña.

## Cosimo Rennella
## 18 victory claims – 7 confirmed victories

| Claim | Vict. # | Date | Sqn. | Aircraft | Hour | Place | Victim |
|---|---|---|---|---|---|---|---|
| 1 | 1 | 24.9.1917 | 78 | Hd? | - | Zagorje | Fighter |
| *Fw. Karl Kotzian KIA, Flik 32/D, Av .D.I 38.17* | | | | | | | |
| 2 | - | 26.9.1917 | 78 | Hd? | - | Lom | Fighter |
| 3 | 2 | 21.11.1917 | 78 | Hd | - | Mosnigo | Fighter |
| 4 | 3 | 14.1.1918 | 78 | Hd | - | Val di Pez | Two-seater |
| 5 | 4 | 15.1.1918 | 78 | Hd | - | Arsiè | Fighter |
| 6 | - | 17.3.1918 | 78 | Hd | 9h+ | Val Seren | Two-seater |
| 7 | - | 17.4.1918 | 78 | Hd | 7h30+ | Conegliano | Fighter |
| *[with ten. Avet, Eleuteri, Resch and serg. Bocchese,70ª Sq.] Maybe: Oblt. Michael Gassner-Nordon Von Laudon KIA, Flik 42/J, Alb. D.III 153.152* | | | | | | | |
| 8 | - | 17.4.1918 | 78 | Hd | 7h30+ | Valdobbiadene | Fighter |
| *[with ten. Avet, Eleuteri, Resch and serg. Bocchese,70ª Sq.] Maybe: Oblt. Richard Fitz,UNH, Flik 42/J, Albatros D.III, returns to own airfield with damages.* | | | | | | | |
| 9 | - | 1.5.1918 | 78 | Hd | 10h20 | S. Lucia/Cimadolmo | EA |
| *[with cap. Riva, 78ª Sq.] Rittm. Géza Csenkey WIA/Oblt. Johann Ruzek WIA, Flik 52/D, Ph. C.I 121.08* | | | | | | | |
| ? | ? | 1.5.1918 | 78 | Hd | - | Cimadolmo | EA |
| 10 | - | 1.5.1918 | 78 | Hd | 12h | Fontanelle | EA |
| ? | - | 17.5.1918 | 78 | Hd | 10h | Beyond Piave | Fighter |
| *[with ten. Novelli, serg. Magistrini & Nardini, 91ª Sq.] Lt. i.d.R. Franz Gräser KIA, Flik 61/J, Alb. D.III 153.221* | | | | | | | |
| 11 | - | 17.5.1918 | 78 | Hd | 10h | Maserada | Two-seater |
| *[with ten. Fucini, 78ª Sq. & ten. Avet, 70ª Sq.] Fw. Johann Valenta WIA, POW/Lt. Viktor Herzog POW, Flik12/Rb, Br. C.I 229.30* | | | | | | | |
| 12 | - | 30.5.1918 | 78 | Hd | - | Lovadina | EA |
| *[Maybe with Lieut. Boysen, 2nd Lieut. Row & Bartlett, No. 66 Sqn., RAF] Maybe : Offstv. Karl Gebhardt KIA, Flik 41/J, Alb. D.III 153.219* | | | | | | | |
| 13 | 5? | 20.6.1918 | 78 | Hd | 10h45 | Montello | Fighter |
| *Maybe: Stfw. Friedrich Hefty, Flik 42/J, Alb. DIII 153.169 damaged* | | | | | | | |
| 14 | 6? | 20.6.1918 | 78 | Hd | 10h45 | Campagnole | Fighter |
| *Maybe: Hptm. Godwin Brumowski, Flik 41/J, Alb. D.III 153.209 damaged* | | | | | | | |
| 15 | 6? | 20.6.1918 | 78 | Hd | 10h45 | Fontana Boera | Fighter |
| 16 | 6? | 23.6.1918 | 78 | Hd | 11h20 | Falzè/Barbisano | Alb. D.III |
| 17 | 7 | 31.8.1918 | 78 | Hd | 13h30 | Mandre | Two-seater |
| *[with ten. Fucini & serg. Chiri 78ª Sq.] Maybe: Stfw. Franz Prochàzka UNH/Lt. i.d.R. Emil Bauer DOW, Flik 22/D, Ufag C.I 161.10* | | | | | | | |
| 18 | - | 27.10.1918 | 78 | Hd | 14h45 | North S. Polo | Two-seater |
| *[maybe with serg. Chiri, 78ª Sq.]* | | | | | | | |

The logbook of Rennella reports only six confirmed victories, but the official list of Aces credits him with seven. The missing victory is probably one of the three claims of 20 June 1918, of which only one was credited, according to his logbook. The medal citation for his second Medaglia d'Argento in fact mentions two victories in the same day, crediting him with 14 victories, and they can be only the ones claimed on 20 June.

# Alessandro Resch

Alessandro Resch was born in Avezzano, in the province of Aquila, region of Abruzzo, on 19 November 1892 from a family of small businessmen, and like so many young men at the end of the 19th century mechanics strongly attracted him. As a boy he helped his father and brothers running the small family businesses, particularly the workshops, and he led a sporting life, hunting, horse riding and bicycling. He was conscripted in September 1912, assigned to the 7° Battaglione Bersaglieri Ciclisti, (bicycle-mounted Bersaglieri battalion) based in the Goito Barracks at Brescia. He eagerly collaborated with the technical officer maintaining the bicycles and motorcycles assigned for testing to the battalion.

On 13 January 1915 he suffered a heavy blow, as the terrible earthquake that struck the region of Avezzano, causing over 30,000 dead, also killed his mother and his five sisters.

On 18 May 1915, one week before Italy's entry in the war, Resch was recalled and sent to the front with the rank of *caporalmaggiore*, and he soon distinguished himself. Although assigned to a technical job, tending motor trucks, on 19 July on his own initiative he reached his battalion engaged in combat, volunteered as a messenger and brought to safety a wounded officer, braving the fire of the enemy. *Maggiore* Leoncini, his battalion commander, proposed him for a Silver Medal for Bravery, which was assigned, although scaled down to Bronze. The officer whose life Resch saved in the Fifties, now a General, met in Milan one of the sons of the ace, and expressed his gratitude for that selfless deed.

Resch took part with his unit in the bloody combats at San Canziano, Cave di Selz and Monte Sei Busi, and was raised to the rank of *sergente* for special qualities. He then volunteered for flying training and in May 1916 he was admitted to the pilot courses at Mirafiori flight school, near Turin. His flying training was remarkably brief, and in just 45 days he got his second, military pilot, brevet.

After a further period of operational training the pilot from Abruzzo, now with the rank of *aspirante ufficiale*, was posted on 8 July 1916 to the 26ª Squadriglia based at S. Maria La Longa, a reconnaissance unit equipped with the strong and efficient Voisin pusher biplane, operating in reconnaissance, standing patrol and bombing missions on the southern part of the Carso front, in which other future aces, Baracchini and Masiero, also served.

It was common for many fighter pilots to start their career in reconnaissance units, but Resch was unusual because he claimed his first victory on a two-seater, hardly one month after his arrival at the front. On 15 August 1916 four Voisins of the 26ª Squadriglia went out to bomb the railroad station at Reifenberg (now called Branik in Slovenia). Resch was flying a SIT-built Voisin with the observer *sottoten.* Vincenzo Lioy, later a renowned aviation historian, and two Austrian fighters intercepted them, attacking from the side and the back. The Voisin returned fire, the observer emptied two magazines of the Fiat machine gun, then Resch went into a dive, got in front of a fighter, fired more bullets and saw the enemy monoplane shedding part of its wing and falling down, while the remaining fighter got away toward Aidussina. The Austrians had

*Tenente* Alessandro Resch in front of the Ansaldo A.1 Balilla of his 70ª Squadriglia, summer 1918.

been good shooters, and when back at their airfield the Italians counted 30 bullet holes in their sturdy Voisin, including some that had chipped the propeller, causing it to vibrate violently. They didn't know, however, that both Austrian fighters, two Fokker A.III, had crashed killing their pilots. Some Austrian sources motivate these losses to a mid air collision for unknown causes.

Just one week later, Tuesday 22 August, the Austrians retaliated with their anti-aircraft artillery. Resch was flying over the Carso in a reconnaissance mission when a shrapnel explosion riddled his Voisin. The Italian pilot was wounded but he brought back the damaged airplane, almost blinded by blood running down his face. For this action Resch received his first Medaglia d'Argento al Valor Militare in the field. The observer expressed his gratitude with a gift and with words that mattered as much as the citation of a medal: "My dear Resch, please accept this modest watch that I send you as a souvenir of the hectic moments that we lived together. Once more I thank you and I wish all the deserved fortunes for your qualities of soldier and man. Dearly yours, Ten. U. d'Ippolito."

The 26ª Squadriglia was disbanded on 20 February 1917 and its components were scattered. Resch was assigned to the outstanding and reputed 25ª Squadriglia, then the last frontline unit still equipped with the Voisin, at Pozzuolo del Friuli. On 25 April 1917 Resch flew with the squadron commander, *cap.* Giuseppe Gabbin, later to die in action leading his squadron on 25 October. They returned from a mission in the area of Castagnevizza with their Voisin badly hit. The weather was not good, and Resch flew just a few missions with this unit, then he was assigned to fighter training.

Now with his officer commission and the rank of *sottotenente*, Resch was licensed on Nieuport fighters in July 1917, but he didn't return to the frontline, as he was sent back to train on Savoia Pomilio reconnaissance aircraft in late September.

Archive data concerning the assignment of Resch are lacking and confused. He was rushed back to the front after the Caporetto retreat, some documents report him serving with the 70ª Squadriglia since 27 October, other documents report a period of service with the 82ª Squadriglia, from 1 November to 15 December. Whatever happened in those troubled days when fighting was more important than filling forms, Resch flew his first recorded fighter combat mission on 6 December 1917 with the 70ª Squadriglia, that will be his unit till the end of the war. The squadron, equipped with new Hanriot biplanes, was based at Istrana, assigned to the Comando Truppe Altipiani and operated in an area where the Austrians were applying their last and most dangerous pressure.

When on 26 December the Germans launched their ill-fated air attack against Istrana, Resch was on alarm service and he was one of the first to jump in the cockpit and go, but before he could take off the Germans hit his Hanriot with machine gun fire and put it out of action.

On 25 January 1918 *ten.* Flaminio Avet, the squadron commander, lead an offensive patrol formed by Resch, *ten.* Aldo Bognetti and *serg.* Aldo Bocchese over the Asiago plateau. Bognetti felt sick and had to turn back, Bocchese got lost, so the two officers cruised between Gallio and San Marino, targeted by accurate ground fire. Near 10.20 a.m. they spotted three enemy aircraft north of Gallio so they gained altitude to engage them. As Avet went in a dive to attack, Resch noticed an Austrian fighter getting behind the tail of his leader, so he went after them, placed himself several times behind the tail of the Austrian but each time his gun fired a single bullet and then jammed. All they could accomplish was to watch one of the enemies get away in a dive. Then they landed at San Pietro in Gu, then the base of their squadron.

In early March 1918 Resch got his commission as *tenente*. On 10 March his Hanriot had an abrupt engine failure, but the pilot used to advantage his altitude and brought it to a safe landing at Casoni. On 3 April an Italian flight went to Conegliano to attack by surprise the usual kite balloon floating there. While Avet and *ten.* Enrico Guglielmotti remained high to provide protection from fighters, Resch and *ten.* Leopoldo Eleuteri went down at low level, Resch managed to fire 270 rounds, but the Austrian ground personnel hoisted down the balloon.

On 17 April the Austrian fighters of Flik 42/J escorting a two-seater of Flik 52/D were engaged in a tough dogfight above the Piave. Avet was leading between Quero and Ponte alla Priula a flight of 70ª Squadriglia, including Resch. At an altitude of 4,000 meters above Quero they spotted, some ten miles away, an Austrian formation which, according to Italian reports, included eight fighters and two reconnaissance airplanes. The Italian leader banked and aimed at the enemy. An Austrian fighter engaged Resch frontally, the Italian returned fire while climbing, but his Hanriot serial 6252 stalled and spun. Resch regained control 500 meters lower, finding himself in front of an Austrian reconnaissance airplane under the

Alessandro Resch at the controls of a Salmson Canton Unnè-powered Voisin of the 26ª Squadriglia as a soldier loads a 162-mm bomb. The plane is armed with a Fiat and a Villar Perosa machine guns.

Fokker monoplane 03.52 crashed on 15 August 1916 after a duel with the Voisin of Alessandro Resch and *s.ten.* Vincenzo Lioy. *(Photo Csizek-Lytai).*

Austrian soldiers surround the wreck of Fokker A.III 03.44, another victim of the air combat against the Voisin of Resch. *(Photo Csizek-Lytai).*

attack of Avet. As soon as it was at range Resch opened fire, the two-seater caught fire, lost a wing and crashed near Bigolino. Another airplane crashed on the road from Bigolino to Valdobbiadene, described by the officer commanding Observation Post 28 with poetic words: "…it began to go down and to fall, similar to a leaf detached from its branch." There was little poetry in the reality of air warfare: the Italians claimed three victories, two of them now with the proof of enemy records, and three gallant Austro-Hungarian airmen perished.

In his combat report Avet credited these victories to all the components of his flight, each of them somehow participating, either actually shooting or just protecting their leader's back.

On 22 April a freak accident almost cost Resch his life. The front fuel tank of his Hanriot broke and gasoline immediately soaked the cockpit and the pilot's clothes. He turned off the engine and tried to glide back to his airfield, but he faced a strong head wind and he touched down in a field, short of his base. The fighter turned over but luckily no spark ignited the fuel and Resch came out of this misadventure unhurt.

On 3 May he took part in a dogfight in which a flight of the 70ª Squadriglia protected a SAML two-seater of the 117ª Squadriglia from the threat of Austrian fighters, which brought a confirmed victory for Avet.

After six months of uninterrupted service at the front, Resch was rotated, on 14 May he was detached to the Defense Flight of Padua, and then ten days later he went on leave. He came back to his squadron on 14 June, just in time to join the actions following the large Austrian mid-June offensive. He flew many strafing missions, reported in the citation to his Bronze Medal, his only award for his service as a fighter pilot, besides the Silver Medal he received in 1916.

In the morning of 12 July Resch was flying in a group patrolling between the Grappa and the Montello. He spotted enemy aircraft above and he climbed at full throttle to position himself higher and behind them, while his companions lagged behind. He engaged, and after his first bullets an enemy airplane went away in a dive, while another one, according to his Combat Report, "…stood its ground, but it received many bursts and then it went spiraling down to 2.500 m. and then it went down in a flat spin, occasionally trailing smoke…" The Italian pilot couldn't follow it further, because he had to defend himself from two other fighters, but he presumed that the enemy, identified as a "Halberstad", may have crashed in

Resch with a beautifully emblazoned Spad 7 of the 70ª Squadriglia. There is no contemporary data to give a cue at what the colors of the stripes may be.

the area of *"Boffat-Col dell'Orso"*. It was possibly an airplane of Flik 30/J, Albatros D.III 153.259, damaged, while its unknown pilot was not harmed. One has to remark that this single Austrian loss corresponds to no less than three Italian claims (Resch, Scaroni, Scaroni again with Magistrini and Keller) and seven British claims (five of them for Lt. C. E. Howell of No. 45 Sqn.) in a large and confused combat in which Silvio Scaroni, Italy's second ace was shot down and badly wounded.

The serious losses of June and lack of supplies reduced Austrian air activities in the summer of 1918, and Italian fighter pilots had few opportunities to increase their score. In the hectic actions during the final offensive, Resch flew ground attack missions. On 29 October he dropped Bettica light bombs and strafed carts near Sacile. On 2 November he flew his last combat sortie, leading a flight armed with Bettica bombs against transports and troops in the area of Latisana. On 4 November the armistice on the Italian front went into effect, and on that same day Resch was promoted to the rank of captain.

Among the papers concerning this pilot preserved in the Historical Archive of the Italian Aeronautica Militare there is the letter of the so-called Bongiovanni Commission for the confirmation of victory claims, one of the few that is known to survive. That document does not consider his first victory, while piloting a Voisin with the 26° Squadriglia. In the folder there is no later addition on this

HD 6252, the personal Hanriot of Alessandro Resch surrounded by pilots of the 70ª Squadriglia, including Flaminio Avet, center. The insignia on the fin is the collar tab of the Bersaglieri, the corps to which Resch belonged.

Resch used individual number 13 like two other aces, Cabruna and Leonardi. The pilot leaning over the roundel is *ten.* Enrico Guglielmotti.

subject, but clearly Resch must have reported in the required way and time his further claim, with valid proofs for confirmation, because he was admitted in the "five down and glory" roll of aces.

On 8 December 1918 he was assigned to the 91ª Squadriglia "Baracca", with which he flew on 2 February 1919 a formation flight from Rome to Treviso and back for the ceremony of naming that airfield after the great ace. Together with other outstanding pilots, in June 1919 Resch went to France as member of the Italian Military Mission. During that period he was ordered to fly from Paris to Amsterdam for the ELTA aeronautical exhibition. His report shows how long distance flights with visual navigation were still a problem then, but simple ingenuity could get a pilot out of troubles. "I had lost my bearing, so I landed in a field and I asked a farmer which way was Amsterdam, and I arrived there smoothly."

Resch then returned from France and he was put on leave on 30 October 1919. Like many other veterans, he had trouble returning to civilian life, it was not a happy period and it was not easy for him to find a decent job. He joined the Fascist movement, and in 1922 he participated in the "March to Rome" which brought Mussolini to power. When the Milizia Volontaria Sicurezza Nazionale, the militia of the Fascist party was founded, he joined it, with the rank of *Capomanipolo* in February 1923 and he served in that body until 1 October 1927, when he was called up for refresher training in the Regia Aeronautica. His commission in the air force lasted three months, and would be renewed for similar periods, not enough to provide security to Resch, who was now a widower with a small son, so the ace in 1928 left the air force and became a pilot with Aviolinee Italiane, ALI, an airline belonging to the Fiat group. This was probably due to the intervention of his good friend Arturo Ferrarin, the famous record pilot, who at that time was working for Fiat.

Resch pursued his career as an airline pilot, becoming Chief Pilot in the early Thirties and operating also as a test pilot both for the airline and for Fiat. During the preparation of the mass flight of Siai S.55 seaplanes from Italy to America, the *Crociera del Decennale* of 1933, he flew a series of weather flights over the Alps, earning the special appreciation of Italo Balbo and the medal commemorating that air feat.

Alessandro Resch with Guido Masiero at Villacoublay, France. Although a captain, he was still wearing the lapel marks of a *tenente*, two stars.

Resch and Masiero, in the background, with two Ansaldo SVAs that they flew to Villacoublay. Masiero is holding a camera, with which he maybe took pictures of the captured German aircraft in the hangar, like the Fokker D-VII on the right.

The Fiat APR.2 that Resch flew in 1938: only one example was built of this elegant small airliner.

As a commander pilot of ALI he was not called back to service during the mobilization of 1935 for the war on Ethiopia. In 1938 he flew a record flight for passenger aircraft from Rome to Paris, at the controls of the small Fiat APR.2 airliner. In that period he was *Direttore di Rete* (director of operations) but he still flew normal service routes. When Italy joined World War II Resch flew many transport missions to the Balkans and Greece, and he was manager of personnel of Aviolinee, that had been militarized, with the name Nucleo Comunicazioni Aviolinee of the Comando Servizi Aerei Speciali, which controlled the assets of the militarized airlines. He got back into a uniform only in late 1941, now with the rank of *tenente colonnello* for extraordinary merit. He kept flying on the routes to the Balkans assigned to ALI and on 1 July 1942 he became commander of the Nucleo Comunicazioni ALI.

When Italy asked for an armistice Resch was in Milan, and he decided to stay at his job, and so he served with the air force of the Repubblica Sociale Italiana, the state which Mussolini set up in Northern Italy under German occupation. He was put on leave in August 1944, as there was nothing for him to do. In the same period also Aviolinee Italiane, which had stopped all flight operations, also terminated his employment. He was hired again only in 1949, and he remained with that company until it was liquidated.

Alessandro Resch died of a tumor at Cerchio, in his native Abruzzi region, on 8 January 1966. His son Arturo, an officer in the Italian Aeronautica Militare, was a pilot of the "Lancieri Neri" aerobatic patrol equipped with F-86E Sabres in the late Fifties.

## Alessandro Resch
## 5 confirmed victories

| Claim | Vict. # | Date | Sqn. | Aircraft | Hour | Place | Victim |
|---|---|---|---|---|---|---|---|
| 1 | 1 | 15.8.1916 | 26 | Voisin | - | Komen | Fighter |
| *Fw. Stefan Szücs KIA, Flik 19, Fokker A.III 03.44 or: Fw. Franz Gregl KIA, Flik 28, Fokker A.III 03.52* | | | | | | | |
| 2 | 2 | 17.4.1918 | 70 | Hd | - | Valdobbiadene | Two-seater |
| *[with ten. Avet, Eleuteri & serg. Bocchese, 70ª Sq.] Korp. Viktor Stibor KIA/Lt. Ladislaus Rajcsics KIA, Flik 52/D, Br. C.I 169.35* | | | | | | | |
| 3 | 3 | 17.4.1918 | 70 | Hd | - | Valdobbiadene | Fighter |
| *[with ten. Avet, Eleuteri & serg. Bocchese, 70ª Sq.] Oblt. Michael Gassner-Nordon von Laudon KIA, Flik 42/J, Alb. D.III* | | | | | | | |
| 3 | 4 | 17.04.18 | 70 | Hd 523 | - | Valdobbiadene | Fighter |
| *[with ten. Avet, Eleuteri & serg. Bocchese, 70ª Sq.] Maybe: Oblt. Richard Fitz, Flik 42/J, Alb. D.III damaged* | | | | | | | |
| 5 | 5 | 12.7.1918 | 70 | Hd | - | Boffat/Col dell'Orso | Fighter |
| *Maybe: unknown pilot UNH, Flik 30/J, Albatros D.III 153.259* | | | | | | | |

# Antonio Riva

If at the turn of the 20th Century many Italians left the poverty of the country and went abroad as immigrants, there were other expatriate Italians who were running prospering commercial enterprises in all continents of the world. One of them was Achille Riva, who in Shanghai ran a very important business in the commerce of silk, as China witnessed the decay of its Heavenly Empire. Achille kept in strict contact with Italy and he returned to the mother country to get married. His wife Teresa Barbaran Capra belonged to one of the most outstanding families of the nobility of Veneto, which among other things had a famous round villa built by the great architect Palladio. Antonio Riva was born in Shanghai on 8 April 1896. We don't know much about his youth, besides the fact that he studied in Florence, in the Collegio La Querce, a highly reputed Catholic boarding school which also Baracca had attended, a few years before. Before the war Riva was attending the Law School of the University of Florence, and officially he resided in that city in Via di Poggio Imperiale, a wonderful hillside avenue near the Piazzale Michelangelo.

In the last day of 1914 the young man volunteered as an officer trainee in the reserve, and was assigned to the 84° Reggimento of infantry. When Riva at the end of the course received his commission as a *sottotenente* in July 1915, ranked ninth in his course out of 3,000 applicants, Italy was already at war, so he was immediately posted to the front, wearing on his collar the yellow-black tabs of the 70ª Fanteria of the Brigata Ancona. He remained with that unit in the sector of Mount Croce until October. In early November the Italians attacked on the Isonzo, in what was called the Fourth Battle. The fiercest combats raged in the area of Oslavia, lost and regained. In that little town Riva, who was major adjutant of the 2nd Battalion, was wounded in combat on 12 November and he was brought to a hospital with a wounded right shoulder and the proposal for a Medaglia d'Argento al Valor Militare. His wound was serious, as only in March he was dismissed from the hospital and returned to the war zone on 15 June 1916, now assigned to the 201° Reggimento. His stay at the front was short; just thirteen days later during a combat on Mount Maggio he broke his right forefinger and was again forced to rest. Antonio Riva probably had had his fill of the trenches, because after two months of healing and one of rest, on 23 September he was attached to the Battaglione Squadriglie Aviatori and three days later he had his first flying lesson on a Maurice Farman with 80 hp engine. He obtained his pilot's licenses at San Giusto, near Pisa, and after a further period of instruction on Farman 100 hp at the Foggia North flying school, on 11 April 1917 he, now with the rank of *tenente*, reached the 29ª Squadriglia, a reconnaissance unit based at the airfield of Cavazzo Carnico, in the high valley of the Tagliamento River, that operated in the difficult mountain region of the Carnia.

On 3 May 1917, as he was flying with the observer *sottotenente* Roberto Lordi, an enemy airplane attacked their Farman above Mount Rombon but without success. Combats in that area, where mainly against the Austrian squadrons based at Villach, like Flik 16, operated, were so rare that it is usually possible to identify the

Capitano Antonio Riva.

opposing aircraft: in this case Brandenburg C.1 68.59 with pilot Josef Friedrichs, who together with Josef Siegel on a KD, claimed a Farman shot down, and with this "victory" became an ace.

Riva had logged a total of eight cruise and seven reconnaissance missions when he left the front and went to Malpensa on 14 June for conversion on the Nieuport fighter. Then he was sent to S. Giusto, for aerial gunnery training. This period was very brief, and soon Riva was back at the front, assigned to the 71ª Squadriglia, based at Villaverla, a Nieuport equipped fighter unit. He flew his first mission, a drop of leaflets on the Pasubio on 30 June.

Riva had his first contact with the enemy on 6 August 1917 during a standing patrol between the Pasubio and Cengio mountains, when together with *sergente* Montalto Ercoli he engaged two enemy aircraft. Ercoli saw one of them "going down in irregular gliding flight toward Arsiero", but Riva had problems with his gun, was targeted by one of the Austrians and had to disengage from the combat. Likely their opponents were two aircraft of Flik 48, a Brandenburg C.1 and an escorting Albatros D.III, one of the new fighters that had just reached the Trentino front. While the two-seater was badly damaged in a forced landing, *Zugsführer* Meissner misunderstood the quick departure of the Italian fighter and claimed one Nieuport shot down, getting also confirmation for it.

Tenente Riva in front of a Nieuport 17 of the 71ª Squadriglia.

Riva tries his hand with a Spad 7 supplied in the summer of 1917 to the 71ª Squadriglia.

Riva's first victory was not slow in coming. In the morning of 24 August Riva, together with *sergenti* Tola and Amantea forced to land Albatros D.III 53.33 north of the Fort of Luserna. Its pilot, *Stabsfeldwebel* Julius Kowalczic, one of the best pilots of Flik 24, got out unharmed from this combat.

Riva had some more air combats, all of them inconclusive, and he logged a total of 26 war missions until, on 13 October he was assigned as the new commander of the 78ª Squadriglia, replacing *capitano* Domenico Bolognesi. Reporting the events of this unit is particularly difficult, because most of its records were lost during the Caporetto retreat. In the late fall of 1917 Riva was promoted to *capitano*. At the time of Riva's arrival the 78ª Squadriglia was based at Borgnano, an advanced field on the Isonzo front. After the Austro-German breakthrough it moved back to Istrana. There, on 26 December, the so-called "Air battle of Istrana" marked the turn of the tide, as Italian fighters took a heavy toll of an attacking German formation. Riva took part in that slaughter, and described it in his combat report:

> "...I took off with ten other pilots of the 78 and 76 squadriglia, and engaged three combats. In the first one I was alone against a DFW which crash landed in a little field next to the Church of Signoressa, (Treviso- square 9001, Map of Italy 25000). The combat had started at a height of 700 meters. I followed closely the enemy in its dive, and after three bursts it went down. I climbed back very quickly and then I saw and attacked a second DFW and, in collaboration with three other Italian fighters and a British one I shot it down near Camalò. I had a third combat but it was unsuccessful as I had ran out of bullets, having fired 350 shots."

Such was the confusion in sorting out these victories and crediting them that eight months later Riva complained to the higher commands that only the victory at Camalò had been credited to him, and not the one at Signoressa, despite the declaration of the shot down German observer, lieutenant Pallasch, bedridden in the English hospital at Istrana, who declared that only one fighter had shot him down, "white and with V dihedral wings (Hanriot)". Other witnesses, including three artillery officers, declared that the victorious fighter carried on its fuselage a red pennant, the personal insignia of Riva.

The use of a pennant as the squadron badge of the 78ª Squadriglia started in summer 1917, but it was under Riva's leadership in 1918 that a growing number of pilots of that unit painted their fighters with pennants of different colors, and also in the post-war years the Italian air force maintained the use of squadron pennants derived from the one of Riva.

Right: Antonio Riva in leather flying jacket in front of an unidentified Nieuport 27 that appears devoid of camouflage.

The Command of the 4ª Armata gave him the Medaglia d'Argento al Valor Militare reporting in its citation fourteen air combats, three victories and many strafes. It was, however, the only decoration of this gallant officer. Possibly, as a squadron commander, he used to write applications for medals for his pilots, but not for himself.

Riva claimed two more victories on 27 January, in a tough combat that had him opposed to a truly outstanding Austrian pilot, the ace Josef Kiss, of Flik 55/J from Pergine. In the past credit for the wounding of Kiss was assigned to Silvio Scaroni, but this has to be excluded, because Scaroni had his presumed combat on 25 January, while Austrian sources unfailingly date it 27 January, and on that day Scaroni didn't fly. Other authors credit this feat to 2nd Lieutenant Matthew Brown "Bunty" Frew, but the Scottish pilot claimed a victory in a totally different part of the front, at far-away Conegliano, while Flik 55/J operated in the Asiago sector, an area where in that time only Italian fighter squadrons operated.

That day Riva took off at 14.40 leading a patrol formed by *sergenti* Fornagiari and Germino in an escort flight for a SIA 7b of the 32ª Squadriglia. Near Sasso Rosso, where Italian troops were on the offensive, a gaggle of enemy fighters, eight or nine, attacked the Italian formation. Although smaller in number, the Hanriots of the 78ª defended themselves well and protected the two-seater. Three times Fornagiari saw a squadron mate under attack and set him free from the danger. At the end of this combat the Italians claimed two enemies shot down, a fighter and, strangely, a two-seater. Riva was credited with a fighter and Fornagiari with a two-seater. Germino, who had lost contact with his two more expert squadron companions, was badly wounded and landed behind the Italian lines.

Some books reported that Kiss was flying alone, but authoritative Austrian sources disprove this. What is certain is that Kiss landed in emergency at Pergine airfield near sunset, which in a close valley like Valsugana in January, comes very early. Also another pilot of the Austrian squadron, *Korporal* Gottlieb Munczar, was forced to make an emergency landing with a damaged airplane. As there were no other combats in this sector of the front, the claim for Kiss' misadventure for the 78ª Squadriglia seems well grounded, although there is no final proof. The Austrian ace Gruber of Flik 60/J claimed credit for shooting down Germino, and this fact may explain why the Italians reported a large number of enemies, as they probably had faced the joined forces of two of the enemy's best squadrons, Flik 55/J and 60/J. Later on Riva complained that the commands credited him with only one victory, instead of two, in collaboration with Fornagiari. What is not explained is the identity of the two-seater, although it may have been a wrong transcript on the papers.

By the end of 1917 several French and British squadrons, both reconnaissance and fighters, were sent to the Italian front. There was a common respect for the ability and professionalism of the allies, but contacts were rather scarce, besides official ceremonies, particularly with the British, and the language barrier was a factor, because in those years if an Italian knew a foreign language, it was French. Riva was one of the few Italian commanders who had excellent relations with the British allies who at the end of 1917 were sharing the same airfield with his squadron. As Fucini recalled in his book, "mister Riva", a gentleman and a man of the world, spoke perfect English and was an excellent host, and often a guest of the British for dinner and for shows.

Riva in the cockpit of his Hanriot of the 78ª Squadriglia showing his personal insignia, a red pennant.

As the last winter of war was ending, the 78ª left Istrana, too much targeted by enemy night bombers, and on 16 February it moved to Nove di Bassano and then, on 15 March, to San Luca, a hamlet close to Istrana, together with the 79ª Squadriglia, both units now belonging to the XV Gruppo.

On 1 May Riva took off with *serg.* Cosimo Rennella for a standing patrol between Ponte alla Priula and Cimadolmo, in which they also had to drop psy-warfare leaflets. Near Cimadolmo they met an enemy airplane and attacked it, Riva firing 300 rounds. It may have been Phönix C.1 121.08 of Flik 52/D with as pilot *Rittmeister* Géza

Captain Riva receives a decoration from General Pennella, commander of the XII Army Corps.

Riva, center, with the pilots of the 78ª Squadriglia. The first three on the left are sergeants Chiri, Codeghini and Capparucci, third from right is *ten.* Amedeo Mecozzi, the other ones are not identified.

Csenkey, commander of that unit, and as observer *Oberleutnant* Johann Ruzek. The Phönix was wrecked in a forced landing near Santa Maria and the two Austrian airmen were wounded.

Other victories for Riva arrived in the hard combats that broke the strength of the Austro-Hungarian air force during the great offensive of mid June, the period that the imperial pilots called with bitterness "the Black Weeks". In the morning of 15 June, first day of the offensive, Riva and *ten.* Mecozzi spotted three enemy aircraft flying to the Italian lines from Grave di Papadopoli. Near Maserada the two fighters attacked in a dive an airplane, described as a "Brandenburg" in the combat report, which immediately tried to return to its own territory. Riva fired a long burst, but his gun jammed, so his wingman stepped in, as the commander tried to clear his gun. Mecozzi also had troubles with his Vickers, but the two pilots took turns against an observer who was fighting back very effectively, until the enemy airplane crashed in the Piave River, in front of Cimadolmo. For this victory both pilots received a prize of 500 Lire.

The next day Riva, flying his Hanriot decorated with a long red pennant, lead a patrol with *ten.* Fucini and *soldato* Venier in a standing patrol from the Montello to Nervesa. As they were flying above Busa delle Radici at a level of 600 meters, they saw a Brandenburg low down and attacked it. Riva once more had his gun jammed, so Fucini completed the attack, while Venier protected the back of his commander. When he had cleared his gun Riva returned to the attack of the Austrian plane that was trying to run away to the Piave. The enemy crossed the river a few feet from the ground and attempted an emergency landing, ending up upturned against a hedge between Malborghetto and Fontigo. In this case also the winning fighter pilots got their prize.

In the following days Riva, like all pilots of the Italian air force, from the top aces to the inexperienced rookies, was in action strafing enemy ground forces, and often returned with bullet holes in his wings.

In the morning of 12 August Riva, flying French-built Hanriot 606, together with *cap.* Bortolo Costantini of the 91ª Squadriglia shot down at S. Lucia di Piave the Phönix D.I 128.11 photoreconnaissance airplane of Flik 40/P with the death of its pilot, *Oberleutnant* Ludwig Stillmungus. These are the *Note Complementari* (Additional Notes) to his combat report:

> "I was somewhat lower than my adversary that was climbing away from me. I tried to get closer to it, both in height and in distance, now climbing and then flying straight and level. I realized that I was not gaining on it, so I fired a first burst of about 110 rounds, and as a result the enemy set into straight flight. After we crossed the Piave, a Spad 200 HP arrived from the direction of Candelù and attacked the enemy at close range. The latter did a wing looping and then got into a steep dive coming toward me and passing below, with the Spad still on its tail. I also got into a dive and attacked it frontally with a second burst of 40 rounds. The enemy had disappeared below, so I banked and saw it upside down, gliding in spirals, always with the Spad following it."

French-built Hanriot Hd.1 serial 606 of *cap.* Antonio Riva.

A handwritten pencil note on that original document, not totally clear, seems to indicate that Riva did not claim a victory and left all the merit to Costantini.

During the summer of 1918 Riva flew almost every day, and he also had a night flying training sortie on 22 August, but there were no relevant events.

He had a tough combat in the afternoon of 6 October:

> "At 15h50 in patrol with tenente Mecozzi and sergente Gandini of my squadron, I saw anti-aircraft fire at our level in the direction of Asolo – We headed there from the Montello and we saw a reconnaissance airplane with an escort of seven fighters. Tenente Mecozzi was the first to attack at about 15h55. The escort fighters reacted violently while the D.F.W. outraced us diving toward the Piave. – In this very fierce combat I believe that two enemy fighters were shot down."

This dogfights was surely extremely violent, because Riva, remarking that the place of the crash was behind Col Fosco, added in his Additional Notes:

> "Out of the escorting D.5, two of them were particularly aggressive and bold and they had on their fuselages as insignia the capital letters K and Y. The former targeted tenente Mecozzi

Antonio Riva, center, with pilots of the 78ª Squadriglia in front of Antonio Chiri's black tailed Hanriot. Chiri himself is fourth from rigth.

*Sergente* Emanuele Morfino was the valuable wingman of *cap.* Riva. Here he is pictured in his camouflaged Hanriot with a dark tail of unknown color.

with a burst and hit his airplane in 5 parts. I believe that the D.5 marked with the letter K was one of the two shot down, because I saw it going down banking to the left."

In this case the opposed claims balanced each other because the Austrian fighters of Flik 56/J suffered no losses and claimed the non-existent destruction of two Italian airplanes. There is a difference, however, in the confirmation of claims, which the Italian commands denied, while the Austrians admitted them.

Riva had another combat in the afternoon of 27 October, when with *cpr.* Morfino he attacked a two-seater over Moriago and saw it "falling vertically at Pieve di Soligo". He claimed one more victory on 28 October, when together with three British Sopwith Camels of No. 28 Squadron – of which he noticed the insignia – he shot down at Cascina Gregoletti, near Vazzola, Hansa Brandenburg C.1 369.70 of Flik 35/K.

The last victim of the Italian ace fell the next day, 29 October:

"In offensive patrol with Caporale Morfino, over Ormelle, we saw among the clouds at our level two enemy aircraft that ran away towards Oderzo. Over the town of Oderzo one of the two turned left getting down in a dive to the interior. The other one did the same maneuver to the right, in a steep dive. We reached it near C. Colussi and shot it down."

That aircraft, Brandenburg C.1 369.175, crashed against a row of poplars and was found after the Austrian retreat in the place that the Italian pilots had indicated. The local townsfolk told that its

Speaking excellent English, Riva had many friends among RAF pilots who gave him souvenir photographs. This is Lt. C. M. McEwen, No. 28 Squadron with his colorful Sopwith Camel D8239.

pilot was a lance corporal with a broken arm, who was helped by Hungarian soldiers. Those were the last, exalting and frantic last days of the war, and Riva claimed three victories in three days.

At the end of the war Antonio Riva had logged 257 combat missions in 18 months of combat service. Not too many, but one must consider that Riva, as squadron commander, had many more tasks and chores than a normal pilot.

On 8 December 1918 Riva was assigned as commander to the SVA-equipped 90ª Squadriglia, ready to depart for Libya, but on 10 January 1919 he got sick and was hospitalized for one month in Vicenza. He was then posted to the Comando Scuole for conversion on the Ansaldo biplane.

The higher commands, in consideration of his knowledge of land and languages, sent him to the Far East to assist the forthcoming raid of a formation of SVAs from Rome to Tokyo, and specifically to arrange the base at Seoul. In the Korean city Riva met Arturo Ferrarin and Guido Masiero, the latter a former pilot in his 78ª Squadriglia.

Riva then returned to Italy and on 31 December 1920 he was discharged. Not much is known of the remaining years of his life. In 1927, when he was shifted from the ranks of the Army to those of the Regia Aeronautica he was back in China. The great Asian country in those years was torn by civil strife, threatened by foreign powers, beset by secret societies, warlords, members of the clandestine Communist party, agents of Russia, Japan and many other nations, and the "big Game" of politics mixed with the sordid enterprises of adventurers, "Gentlemen of fortune", soldiers of lost wars, arm dealers and impoverished noblemen. Antonio Riva tried to sell aircraft to the Chinese generals who exploited the civil war to increase their power and wealth. He did not accomplish much, as Daniele Varè, who was then chief of the Italian military mission to China wrote later. The fighter ace was just too honest and lacked the financial resources necessary to succeed in that shady and surly environment. There Riva met Roberto Lordi, his old companion at the 29ª Squadriglia, who was chief of the Italian Aeronautical Mission. Lordi, who had been a gallant officer in World War I and in the reconquest of Libya, was shot by the Nazis because of his participation in the Resistance during the Fosse Ardeatine massacre on 24 March 1944.

Riva, center, in white uniform, leads a group of European well-wishers who examine the Ansaldo SVAs of Ferrarin and Masiero on their way to Tokyo. Guido Masiero is the pilot in overcoat, below the propeller.

Riva took up his father's business in the commerce of silk, but he was also a journalist, writing dispatches from China for the daily "La Stampa" of Turin. When Mao Tse Tung's communists took power in China, many Westerners left the country, fearing for their safety. Riva had married Catherine Lum, an American girl and daughter of a painter who was in China (to get inspiration from Oriental art), and sister of the writer Peter Lum. They had four children, the older of them was sent back to Italy. Riva remained with his family in China, fearing no evil, as he was well known and considered China as his country to all effects. Unfortunately his trust was badly placed. In his house, near the Western Gate of the Forbidden City of Beijing someone found (or placed) a mortar, maybe a wartime souvenir. The ace was arrested together with two other foreign citizens, a German and a Japanese, called Gentner and Yamaguchi, in the afternoon of 27 September 1950. He was put on trial by a military tribunal for "counter-revolutionary activities", was found guilty and shot in Beijing on 17 August 1951.

Where his remains now rest is not known, because the Catholic cemetery where he was buried was later turned into a park.

Antonio Riva, Guido Masiero and Arturo Ferrarin with two Japanese officers in Seoul, Korea.

Antonio Riva with his beautiful and courageous American wife, Catherine Lum.

## Antonio Riva
## 14 victory claims – 7 confirmed victories

| Claim | Vict. # | Date | Sqn. | Aircraft | Hour | Place | Victim |
|---|---|---|---|---|---|---|---|
| 1 | 1 | 24.8.1917 | 71 | Ni.17 | 10h | Fort Luserna | Fighter |
| *[with serg. Tola & Amantea, 71ª Sq.] Fw. Julius Kowalczik UNH, Flik 24, Alb. D.III 53.33 FTL* | | | | | | | |
| 2 | - | 26.12.1917 | 78 | Hd | - | Signoressa | German two-seater |
| *Uffz. Hedessinski POW/Lt. Pallasch WIA, POW, FA 2, DFW C.5* | | | | | | | |
| 3 | 2 | 26.12.1917 | 78 | Hd | - | Camalò | German two-seater |
| *[with s.ten. Scaroni & Michetti, 76ª Sq., serg. D'Urso, Benvenuti & Teobaldi, 82ª Sq., 2nd Lieuts. Jarvis, Frayne & Mulholland, No.28 Sqn. RAF and possibly ten. Avet & Comandone, 82ª Sq., s.ten. Fucini, 76ª Sq.] Possible victims:*<br>*Lt. Kessler WIA/Lt .d. Res. Edebohls KIA, FA 2, DFW C.5*<br>*Uffz. Wili Petzold KIA/Lt. d. Res. Wilhelm Voigt KIA, FA 2, DFW C.5*<br>*Vzfw. Karl Uecker KIA/Lt. d. Res. Heinrich Pfeiffer WIA, DOW, FA 2, DFW C.5*<br>*Vzfw. Pollman WIA, POW/Lt. Schlamm WIA, POW, FA (A) 219, DFW C.5* | | | | | | | |
| 4 | 3 | 27.1.1918 | 78 | Hd | 15h+ | Lisser-Val Brenta | Fighter |
| 5 | - | 27.1.1918 | 78 | Hd | 15h+ | S. Gaetano | Two-seater |
| *[with serg.m. Fornagiari, 78ª Sq.] Offst. Josef Kiss WIA, Flik 55/J*<br>*Korp. Gottlieb Munczar UNH, Flik 55/J* | | | | | | | |
| 6 | - | 1.5.1918 | 78 | Hd | - | Cimadolmo | EA |
| *[with serg. Rennella, 78ª Sq.] Rittm. Géza Csenkey WIA/Oblt. Johann Ruzek WIA, Flik 52/D, Ph. C.I 121.08* | | | | | | | |
| 7 | 4 | 15.6.1918 | 78 | Hd | 9h55 | Cimadolmo/Maserada | Two-seater |
| *[with ten. Mecozzi, 78ª Sq.] One of these:*<br>*Korp. Franz Hyks KIA/Oblt. Alexander Gherardini WIA, Flik 62/D, Br. C.1 69.61 or:*<br>*Fw. Alfred Brandt/Oblt. Karl Hiemesch WIA, Flik 19/D, Br. C.1 369.69 or:*<br>*Zugsf. Wilhelm Watzl MIA/Lt. Sylvester Walitza MIA, Flik 5/F, Br. C.1 369.126* | | | | | | | |
| 8 | 5 | 16.6.1918 | 78 | Hd | 10h30 | Malborghetto/Fontigo | Brandenburg |
| *[with ten. Mecozzi & sold. Venier, 78ª Sq.] Korp. Bruno Koziarski UNH/Oblt. i .d. R. Julius Tiefbrunner UNH, Flik 38/D, Br. C.1 369.61* | | | | | | | |
| 9 | - | 12.8.1918 | 78 | Hd 606 | 11h | S.Lucia di Piave | Fighter |
| *[with cap. Costantini, 91ª Sq.] Oblt. Ludwig Stillmungus KIA, Flik 40, Phönix D.1 128.11* | | | | | | | |
| 10 | - | 6.10.1918 | 78 | Hd | 15h50 | M. Solder-Susegana | Fighter |
| 11 | - | 6.10.1918 | 78 | Hd | 15h50 | M. Solder-Susegana | Fighter |
| *[with ten. Mecozzi & serg. Gandini, 78ª Sq.]* | | | | | | | |
| 12 | - | 27.10.1918 | 78 | Hd | 16h | Pieve di Soligo | Two-seater |
| *[with cpr. Morfino, 78ª Sq.]* | | | | | | | |
| 13 | 6 | 28.10.1918 | 78 | Hd | - | Vazzole | Two-seater |
| *[with pilots of No. 28 Sqn. RAF] Zugsf. Johann Matousek WIA/Lt. Stefan Burazovic UNH, Flik 35/K, Br. C.1 369.70* | | | | | | | |
| 14 | 7 | 29.10.1918 | 78 | Hd | - | Cascina Colussi | Two-seater |
| *[with cpr. Morfino, 78ª Sq.] crew and unit unknown, Br. C.1 369.175* | | | | | | | |

# Cosimo Damiano Rizzotto

Cosimo Rizzotto is certainly one of the lesser-known Italian aces. Apparently one of those reliable, humble and hard-working sergeant pilots, who were the backbone of the air force of the Italian Army, actually he had an adventurous and fascinating life that suggests a much more intriguing personality than the little data one can surmise from surviving official records. Rizzotto was born on 6 June 1893 in Cologna Veneta, (not Colognola ai Colli as reported by some documents) a small agricultural town in the plain between Padua and Verona. Christened Cosma Damiano, after the two healer saints, he had his unusual name changed to the more common Cosimo by the draft-board employees. He was brown-haired, of medium height, with gray eyes. He was drafted as a soldier in September 1913, and he was assigned to the Battaglione Aviatori, probably because of his mechanical specialization, attested in his *"foglio matricolare"* (military résumé). On 24 May 1915, the day of Italy's entry in World War I, Rizzotto was admitted to pilot training at San Giusto airfield, Pisa. On 19 December he was moved to Malpensa training field, near Milan, and on 1 January 1916 he was licensed as a pilot. On 31 January 1916 he rose to the rank of corporal, and on 8 March 1916 he was posted to the Nieuport flight at Cascina Costa training field. On 30 March Rizzotto was sent to France for fighter training, and returned exactly two months later.

Rizzotto was assigned to the new 77ª Squadriglia, which was formed on 18 June 1916 at Istrana airfield, near Treviso. This unit was initially equipped with the Nieuport 10, but soon it received the nimble Nieuport 11, or Nieuport 80 hp, as it was called in Italy, with which it began combat missions on 30 July 1916. Rizzotto, who had been promoted to *sergente* on 31 July 1916, took part in the intense activity of this unit, which in the summer moved to Cascina Farello, on the southernmost part of the Isonzo front. He flew 9 uneventful fighter missions in August, 15 in September, including on the 17th a remarkable combat against two Austrian airplanes in defense of a Caproni, 21 in October, 31 in November. On 1 December he had another inconclusive combat with the enemy, an "Albatros", that is to say a Brandenburg, that escaped in a steep dive.

In the new year, 1917, Rizzotto kept on high rate of missions: 7 in January, 22 in February, and in the last one of these, on 28 February 1917 Rizzotto scored his first victory, shooting down an aircraft at Quota 40 (Hill 40), near Redipuglia, and then watched as Italian artillery fire wrecked the grounded airplane. His victim was Hansa-Brandenburg C.1 27.60 of Flik 34 that, according to Austrian data, claimed a Nieuport after a long duel south of Monfalcone (in the operating area of 77ª Squadriglia) and was then forced down by anti-aircraft fire, its crew of *Korp.* Alfons Langer and *Oblt.* Ludwig Ruppenthal escaping unharmed.

In March Rizzotto's unit moved to Aiello, still in the sector of the Italian 3rd Army, the one that operated closest to the sea, and received its first Spad 7s. During the spring and summer 1917 the 77ª Squadriglia flew a mix of Nieuport 11s, Nieuport 17s and Spad 7s. Rizzotto had a duel on 25 April, facing three enemies and forcing them to return to their own lines. On 1 May, together with *serg.* Piaggio and *serg.* Tesci, he met a patrol of enemy fighters, that they engaged in combat, until one of them landed back at Prosecco airfield, one dove away behind the Hermada plateau and the third one returned to its own lines. On 2 May Rizzotto flew his first photographic reconnaissance, surely with a Spad 7 equipped with cameras. On 31 May he had a combat with two enemy aircraft that were forced to land at Aisovizza. His logbook registers 10 combat flights in March, 23 in April, 29 in May, 20 in June.

Sergente Cosimo Rizzotto, ace of the 77ª Squadriglia.

On 7 July 1917 in the late evening, Cosimo Rizzotto claimed a victory over Mount Stol, with the help of *ten.* Giulio Sambonet. The enemy airplane burned and crashed near Castagnevizza, on Hill 219. His combat report was:

> "Today at 20h10 flying on an alarm mission on the Stol Mountain I had a combat with an enemy airplane. The enemy was jumped and didn't have the time to defend itself. After my first burst I saw it immediately stumble and go down toward Castagnevizza. I didn't leave it until I finished my ammunitions and saw it turn on its wing, spin and go down leaving a long stream of fire and smoke. I could not pinpoint the place where it fell because of the haze. I here enclose copy of a report of the observation point of Coti, *(re: Cotici)* signed by captain Bugni."

Later the command of the 14th Infantry Division confirmed that pieces of the airplane, its machine gun, part of the fuselage carrying the number 112-2 and a strut were recovered from the crash site, but not the remains of the crew. This number corresponds to a Lohner C.II that was never used on the Italian front; thus there is no confirmation for this victory in Austrian sources.

Four days later, on 11 July, Rizzotto had another victory over Voiscizza, at 5.50 in the morning. "This morning, leaving on a cruise at 4.45, after about one hour (time 5:50) I saw an enemy fighter plane near Doberdò that was flying to Gradisca. Immediately I rushed at it and attacked it on the side, firing my first burst and the enemy answered shooting with its gun. So we started a very hard dogfight that went on until my enemy stopped the fight diving at very high speed near Voiscica (between Temnica and Voiscica). Getting down to 1500 meters on that spot where it had fallen, I could clearly see the shape of the airplane." Again the observers at Cotici confirmed this victory.

On 29 July again Rizzotto met an enemy aircraft and, according to his logbook, forced it to land by San Daniele, but no other records of this event have been found.

29 combat flights in July, 36 in August, 40 in September 1917 are reported in his logbook, a very high level of activity, showing that the 77ª was considered a "workhorse" squadron.

On 29 September 1917 an enemy reconnaissance airplane with red wheels was attacked over the Isonzo by a Spad with a ladder symbol, flown by *ten.* Sabelli of the 91ª Squadriglia. Other pilots watched this action, and other Spads joined in, although the reports state that only the first fighter was actually determinant in this victory. One of them was Rizzotto, who shared this claim over an enemy identified as Brandenburg 329.16 of Flik 28 that crashed by Pietrarossa Lake with the loss of its crew, *Zugsf.* Kreidl and *Lt.* von Luerzer. The diary of the 91ª Squadriglia, however, ignores the involvement of any other squadron's pilots in this action. This victory was later confirmed only to Rizzotto and Sabelli, not to *ten.* Parvis of the 91ª. The two fallen Austrians were buried with full military honors.

On 9 October Rizzotto at 5 p.m. attacked an enemy fighter all painted red over Punta Sdobba: it may have been the top ace Brumowski, who had just shot down a captive balloon, the one of observer *ten.* Venditelli, of the 20ª Sezione Aerostieri, who jumped with his parachute. Rizzotto went after him down to low level, toward the sea, shooting 150 rounds but his engine started coughing, so the tables turned, the hunter became the hunted one until his engine failed completely and his fighter splashed down in a marsh at Matarussi. This event could legitimately justify a victory claim, but it is not reported by Austrian data. If that red fighter was indeed Brumowski's, it surely had attracted the attention of the Italian air force and of all the anti-aircraft guns this side of the Isonzo. A few minutes before, flying in a formation with other fighters, it had been spotted by Ruffo di Calabria, who bounced it, but had to retreat because his gun jammed. Having fixed his gun, Ruffo attacked again the red fighter above Duino, but could not hit it.

After the Caporetto breakthrough, in later October 1917 the 77ª Squadriglia retreated from Aiello, burning some of its Spads, moving first to La Comina, then to Arcade and finally, on 10 November, to Marcon, in the mainland near Venice, that will remain its base for the final year of the war.

Meanwhile on 6 November, flying from Arcade, Rizzotto shot down Brandenburg C.1 229.24 of Flik 12 at San Michele di Conegliano, in collaboration with *serg.* Leonardi of the 80ª Squadriglia. This is his combat report:

A Nieuport 11 of the 77ª Squadriglia with an unusual finish: polished metal cowling and camouflage color dapple all over the fuselage and above the wings. This Nieuport, serial 1611, was also flown by Ranza.

Rizzotto scored a victory, on 7 July 1917, with *ten.* Giulio Sambonet, who here appears on a Nieuport 11 of the 80ª Squadriglia with *ten.* Guido Keller, another pilot and, in the cockpit, *cap.* Gordesco.

"This morning at 11:30 taking off from Arcade airfield for an alarm above San Donà di Piave together with another Nieuport 110 HP of the 80a squadriglia having reached the level of 3,000 meters I spotted an enemy airplane in the direction of Conegliano. Trying without hesitation to cut its path, I managed to reach it and engaged combat while the other Nieuport, also maneuvering, had also reached the correct distance for the combat. After exact and violent bursts of machine-gun, the enemy airplane was hit and soon started a steep dive. After other bursts, the airplane crashed burning near S. Michele di Conegliano. Two Spads were present at this com bat, piloted by major Baracca and lieut. Parvis".

The observer *Lt.* Schwarzenbach was dying, while the unfortunate pilot *Zugsf.* Josef Feiler was already dead.

After the hectic actions while the Italian air force tried to contest the brief mastery of the battlefield gained by the Austrians and their German allies, Rizzotto didn't score any more, and remained in the unglamorous, bur demanding daily missions of a sergeant pilot. He was assigned his first Medaglia d'Argento al Valor Militare, with this citation:

"Airplane pilot, bold and able, he flew very many war flights, showing his value in many combats. On 28 February 1917 with repeated attacks he hit an enemy airplane, which, in the sky above Monfalcone, crashed at Hill 40."

In May 1918 he received an unexpected letter: ingegner Alessandro Marchetti, who had been his instructor in aero engine technology during the course in Pisa, and who was now designer with the Vickers Terni armament factory in La Spezia, having heard of his prowess as a pilot, asked him to test-flight at Coltano, near Pisa, a new fighter, built by Vickers Terni, powered by a 200 hp SPA 6a engine, and very fast. This was to be the MVT fighter. Rizzotto, for unknown reasons did not pick up this offer, and the test pilot of the MVT was Sergeant Elia Liut, also of the 77ª Squadriglia, who with the MVT established an unofficial world speed record.

Cosimo Rizzotto returned to victory during the frantic combats of 15 June 1918, when the Austrians launched their final offensive across the Piave, and the Italian fighters wrought havoc first against the enemy air units, and then against the troops, strafing boats, bridges and footbridges on the Piave. Rizzotto was assigned a victory over an enemy airplane a Grassaga, on 15 June 1918. In the veritable "turkey shoot" that was the action on that day, it is impossible to match Rizzotto's claim with the many Austrian losses. At the war's end, he was assigned his second silver medal, with this citation:

"Daring fighter pilot, for seventeen months he did his duty uninterruptedly, with strong will and dash, giving a good proof of energy and uncommon courage. In many and difficult combats he shot down five enemy aircraft. Sky of the Carso and Conegliano, July-November 1917."

On 19 September serg. Rizzotto was transferred to the Ispettorato Squadriglie da Caccia (that is to say, Fighter Command) at Padua where the next day, in a big ceremony with dozens of airplanes lining the airfield, King Victor Emmanuel decorated many airmen. During the war he had flown in combat for almost 500 hours in about 350 combat missions and had six confirmed victories.

After the war, Cosimo Rizzotto was demobilized and placed on leave on 14 September 1919. Unexpectedly, this pilot with no planes to fly did not return to his town and folkspeople, but embarked on a new and adventurous life. He moved to Argentina. In the summer of 1920 he was at Castelar, working for the Escuela Italo-Argentina de Aviación, a new flying school that was organized by Lieut. Ernani Mazzoleni and Eduardo Olivero, an Argentinean who had been a valiant pilot in the 91ª Squadriglia "Baracca" and who later flew daring raids for his country. The other instructors, besides Rizzotto, were Nicola Bo, formerly a sergeant pilot in the 2ª Sezione SVA, with three victory claims during the war, and Angelo Pescarmona, formerly a pilot with the 131ª reconnaissance squadron. In August 1922 Rizzotto offered his services to the government of Paraguay that was fighting against the insurgency of a group of officers. Rizzotto arrived in Paraguay with a Breguet XIV two-seater, and flew many reconnaissance and bombing missions against rebel positions. On August 25, while on a reconnaissance mission, the propeller of his Breguet splintered, and the engine caught fire. Rizzotto crash-landed the airplane, that was destroyed, and quit the country, returning to Argentina where until May 1923 he worked as instructor with the Escuela Civil de Aviacion Italo-Argentina at Castelar. He remained in South America for fifteen years, until 1935. Until 1926 he directed the construction of an Italian agricultural establishment for war veterans in Salto, Uruguay, then he moved to Brazil, as assistant director of an agricultural company. He also had an appointment as a consular attaché, with the Italian Embassy in Brazil. He then returned to Italy, and was called back in service, in the Royal Italian Air Force, for the invasion of Ethiopia. He was shifted from Italy to Mogadishu, Somalia, in the summer of 1936, when the war was already won, and the Italian Empire in Africa had been instituted. He remained in Mogadishu one year, and then was assigned to the Dire Dawa airport. In 1937 he was placed on leave, but he remained in Africa. He obtained the concession of a large tract of virgin land in the south-eastern region of Arussi, where he built a rich and prosperous farm, spending all the money he had saved in South America. When Italy

Cosimo Rizzotto in the cockpit of a Nieuport 17 showing the red heart of the 77ª Squadriglia. *(Photo Ing. Maurizio Longoni).*

Mixed equipment of the 77ª Squadriglia in mid 1917: first there is a Spad with a red heart, then another Spad with a fuselage roundel, and earlier type of radiator, then a Nieuport 17 with an individual insignia (not a red heart) and then a mix of Nieuport 11s and 17s, all with the red heart. The airfield is Aiello.

Brandenburg C.I 329.16, *serg.* Rizzotto cooperated in its destruction together with *tenenti* Parvis and Sabelli of the 91ª Squadriglia on 29 September 1917.

The business end of the Spad 7, the centrally located synchronized Vickers machine gun.

Cosimo Rizzotto is flying this Breguet 14 during the civil war in Paraguay, August 1922. (Photo *Museo de Historia Militar, Asuncion via Sapienza*).

entered World War II, in June 1940, Rizzotto was mobilized again. In April 1941 he was by his farm, where he organized the defense of the Italians against the attacks of Ethiopian rebels. They organized a relief column that was attacked by the rebels, with heavy losses, and then on 11 April 1941 he became a prisoner of the British. He remained in a prisoner camp in Kenya for five years, and returned to Italy in 1946. As a war veteran and a refugee from East Africa, he got a job with the city administration of Milan, a far cry from his previous restless life. He spent much time and effort trying to get from the state some compensation for his farm, and he finally obtained Lire 15,900 that, at the time, was about 25 US dollars. In 1958, having turned 65, he retired. He died in Milan on 18 February 1963.

## Cosimo Rizzotto
## 7 victory claims - 6 confirmed victories

| Claim | Vict. # | Date | Sqn. | Aircraft | Hour | Place | Victim |
|---|---|---|---|---|---|---|---|
| 1 | 1 | 28.2.17 | 77 | Nieuport | - | Monfalcone | Two-seater |
| *Korp. Alfons Langer UNH/Oblt. i.d.R. Ludwig Ruppenthal UNH, Flik 34, HB C.1 27.60* | | | | | | | |
| 2 | 2 | 7.7.1917 | 77 | Nieuport? | 20h10 | Mount Stol | EA "112.2" |
| *[with ten. Sambonet, 80ª Sq.]* | | | | | | | |
| 3 | 3 | 11.7.1917 | 77 | Nieuport | 5h50 | Voiscica | Fighter |
| *Budzinsky KIA?* | | | | | | | |
| - | - | 29.7.1917 | 77 | ? | - | S.Daniele | EA FTL |
| 4 | 4 | 29.9.1917 | 77 | Spad | 10h40 | Pietrarossa | EA |
| *[with ten. Parvis & ten. Sabelli, 91ª Sq.] Zugsf. Andreas Kreidl KIA/Lt. Erich von Luerzer KIA, Flik 28, Br. C.1 329.16* | | | | | | | |
| 5 | 5 | 6.11.1917 | 77 | Ni.17 | 12 | Conegliano | EA |
| *[with serg. Leonardi, 80ª Sq.] Zugsf. Josef Feiler KIA/Lt. i. d. R. Othmar Schwarzenbach POW, DOW, Flik 12, Br. C.1 229.24* | | | | | | | |
| 6 | 6 | 15.6.1918 | 77 | Spad | - | Grassaga | "Small Brandenburg" |

## Fulco Ruffo di Calabria

In collective imagination, the typical World War One fighter pilot has a specific look: tall, elegant, noble, with the rogue spirit of an adventurer, and a dashing and ironical look in his eyes. This is just the portrait of Fulco Ruffo di Calabria, emphasized in his handsome Latin face by his musketeer's mustaches and goatee. The family of Fulco, Duke of Guardia Lombarda, Count of Sinopoli, Neapolitan Patrician and Scion of the Princes of Sicily, was an ancient noble house of Southern Italy, that produced many gallant soldiers, among them Marquis Paolo, who fought at Waterloo with the English Dragoons.

Fulco was born in Naples, in Villa Palazzolo, on 12 August 1884. His father, Don Beniamino, had long been mayor of the city; his mother, Laura Messelman du Chenoy was a Belgian. The connection of the Ruffo di Calabria family with Belgium was destined to be long and important.

The young man was raised according to the standards of nobility, and after high school he volunteered, of course in cavalry, to serve in the Cavalleggeri di Foggia Regiment. He was put on leave on 30 November 1905, and then he got a commission as *sottotenente di complemento*, 1st lieutenant of the reserve, on 18 February 1906 and served three months. After the army, the restless young duke scorned comfortable life and went to Africa, then the final frontier. He worked first as an agent and then as assistant manager of Wegimont, a Belgian-Italian trading company that operated ships on the Giuba River, in Somalia. Ruffo remained a long time in Africa, with periods of work and alternatively hunting expeditions and explorations to places where no white man had ever been. He briefly came back to Italy in a trip that looked like a novel: a duel, a meeting with his family, and off again. His second departure from Africa was longer: he came back to find financing to establish a company, but war had broken out. Italy was still neutral, but in September 1914 he was recalled to arms to get seniority for promotion. On 20 December he returned to the active service, and he volunteered for the Battaglione Squadriglie Aviatori, the air service. He got flying training at Mirafiori, Turin, and then, now with the rank of tenente, he completed his training at Pisa.

Silvio Scaroni described his personality:

> "We were about twenty students, among them Ruffo and Stoppani. Ruffo, officer in the reserve, came from Africa and he had a black servant. He was a great lord. He had a penchant for writing letters, he wrote about ten every day and, considering his rank, all addressed to ladies of high society. We were a little bit intimidated by his aristocratic and elegant personality, and we were also a bit envious because he was loaded with money."

On 29 September 1915 he was considered ready for service and was sent to the front, at the 4ª Squadriglia per l'Artiglieria,

Fulco Ruffo di Calabria in training for Nieuport conversion, sitting next to Luigi Olivi, Cascina Costa, June 1916

where also Scaroni was. That unit, based at Gonars, near Palmanova, equipped with Caudron G.3, operated for the guns of the VII Army Corps. On 12 November Ruffo earned his first mention in a dispatch for a reconnaissance over the lower Isonzo. On 26 January 1916 Ruffo was shifted to the 2ª Squadriglia d'Artiglieria, based at Chiasottis. In this period he earned two Medaglie di Bronzo al Valor Militare in less than three months, one for a daring reconnaissance over an enemy battery at Lokvica, in which his airplane was hit eleven times, on 11 February 1916, another one for similar feats in April. His qualities now apparent, the Neapolitan officer was assigned to the training field at Cascina Costa for conversion on the Nieuport fighter.

On 28 July 1916 Ruffo returned to the front, assigned to the 70ª Squadriglia at S. Caterina di Udine airfield, the unit that was the spearhead of Italian fighters. On 9 August he flew his first combat mission, escorting the Capronis raiding Dornberg together with Baracca, Caselli and Olivari.

He scored his first victory less than one month after his arrival at the squadron, on 23 August. He took off with Baracca in the

Right: Ruffo sits in the cockpit of a Nieuport, Adriano Bacula and Ferruccio Ranza, on the right, look on, whole Luigi Olivi on the left touches his private parts as a good luck gesture.

morning, to intercept an enemy that they found between Gradisca and Gorizia. Baracca attacked first but he couldn't shoot, Ruffo had ice in the carburetor and the engine went off, but he managed to glide to 25 meters from the enemy and fire. The Austrian airplane, Brandenburg C.1 61.61, among the bullets of the Nieuport, dove and landed behind its lines, between Bukowitza and Ranziano, as the observers of the 37° Reggimento d'Artiglieria da Campagna noticed. Ruffo also was forced to land near Corona, so Italian cannons opened fire on the Brandenburg but the Austrian ones targeted the Nieuport and hit its wing. Ruffo got the help of a mechanic from a Farman that also landed nearby, fixed the engine and took off for S. Caterina.

His second victory, on 16 September, was described in a letter by Baracca:

> "Another fine victory for my squadron, and partly mine also, as I was leading the combat against the shot down enemy.
>
> We knew about an infantry attack in the depression of Plezzo next Saturday, and a friend of mine down there told me that airplanes often operated there among the mountains, so three of us took off and we got there at about 9 a.m. at an altitude of 4,000 meters. I was in the lead, behind me 100 meters there were two mates, tenente Ruffo and aspirante Olivari, each of them has already shot down an airplane, and for half an hour we kept waiting cruising among peaks and gorges. There were clouds that often covered the ground and a strong wind. I was in the mind to return, but I was enthralled by the battle of the Rombon, with countless shells exploding at 2,200 meters.
>
> Finally I saw a black spot coming from Villach and getting bigger, the enemy. I gave a hand signal to my partners, flew a wide turn to remain unseen, retreating toward Caporetto and the Austrian unaware of the danger, crossed our lines near Mont Stol. He was flying at 3,700 meters, and we from 4,000 pounced upon him. He turned immediately toward Plezzo, firing with the machine gun and the repeating rifle, but we surrounded it and unloaded our guns from a distance of 50 meters. Then Ruffo, a very daring pilot, flew with his airplane at full speed very close to mine, and we almost collided. The enemy fell out of control, rearing up, banking, then balancing again, then it went down 500 meters trailing smoke, as we followed it down.."

The three pilots went by car to the crash site, and saw that pilot Franz Morozko was dead, with two bullets in his head. His observer, *Leutnant* Franz von Czàby was still alive, badly wounded in the abdomen. Italian surgeons operated him, but the unfortunate officer died on the 19th. He and his pilot were buried with military honors in the cemetery at Staro Selo. Italian airmen then dropped a message to the Austrians reporting the sad event.

The Nieuport 11 serial 1685 of Ruffo showing the first version of his black skull insignia. This airplane carries a Colt machine gun.

Ruffo stands in front of a line-up of Nieuport 11s. These airplane show an improvised from of camouflage, with a darker color applied on the field over the natural fabric. *(Photo Cejka).*

For this combat Ruffo received a Silver Medal, which in its citation reported also his victory of 23 August. Strangely, for the same event, he also received a Bronze Medal three months later.

On 3 October he had a crash landing, smashing his airplane. During that month the squadron was very busy escorting raids of Caproni bombers, and it received Le Prieur rockets for attacking captive balloons, first testing them against an old balloon raised on the field of the 33ª Squadriglia. On 21 October the 70ª Squadriglia received its first Nieuport 17, or "15 mq" as the Italians called it referring to its wing surface. It was serial 2614, and Baracca got it. Probably in this time Ruffo introduced his personal insignia, a black skull. He flew Nieuport 11 serial 1685, which is shown, armed with a 1914 model Colt machine gun, generally believed much inferior to the Lewis.

In the afternoon of 1 January 1917 Ruffo had a combat that developed from Sagrado to Monfalcone. This may have been one of the victories that were credited to him only one year after the facts. On that day there were three Italian claims in that area, while four Austrian airplanes were damaged to varying degrees, two of them forced to a crash landing.

11 February 1917 was a day of intense Austrian air activity all along the front, from the Carnia Mountains to the sea. After noon two airplanes were spotted in the direction of Pavia d'Udine. Ruffo took off with *sergente* Anselmo Caselli, soon followed by all the other pilots. The sergeant was the first one to engage the enemy, but after a few rounds he stalled and had to quit the combat, with his engine starved. Then Baracca attacked, followed by Ruffo and the other pilots. The Austrians defended themselves well, but finally they were beaten and force landed, crashing against a row of trees near Premariacco. The victim, Brandenburg C.1 27.74 of Flik 35 carried pilot *Korporal* Ludwig Fleck and observer *Leutnant* Wilhelm Graf Siemienki, both wounded, who became prisoners. In June Count Siemienski, from Galicia, still hospitalized, had the visit of Baracca and Ruffo, who offered him a splinter from his propeller engraved with a dedication. He was later repatriated on account of the gravity of his wounds. For this victory Ruffo was awarded his second Medaglia di Bronzo.

In the morning of Wednesday 28 February Ruffo, flying Nieuport 17 3139, attacked above Gradisca a two-seater of Flik 34 in a photo-reconnaissance mission. After having fired 70 bullets, Ruffo saw "a flat and large object colored yellow with a black mark in its middle" falling off the enemy airplane, which returned to its own line almost out of control. The Austrian crew managed to nurse the airplane back to its own airfield and to perform a successful emergency landing. The combat was so close that not only the two Austrians could relate that the fighter attacking them was emblazoned with a black skull, but they even stated that they had lost the rudder in a collision. Ruffo and Baracca picked up the rudder, and then presented it to the Duke of Aosta, Commander of the 3rd Army.

On 22 March Ruffo was officially assigned to the *Servizio Permanente Effettivo* (permanent active service). On 1 May the 91ª

Ruffo and Baracca standing next to the wreck of Brandenbrug 27.74 of Flik 35 which they shot down on 11 February 1917.

Above: Ruffo di Calabria, left, leaning on his Nieuport 17, with the second version of his black skull. On the right stands Francesco Baracca.

Right: Ruffo and his Nieuport 17. Besides the black skull, there is another personal touch, a small baby doll attached to the inverted V strut next to the windscreen.

Squadriglia was formed on the field of S. Caterina di Udine with Spad fighters and with the best pilots, and he was obviously one of them. On 3 May 1917 Ruffo attacked over Sagrado a two-seater that had successfully evaded a Nieuport: its pilot was particularly able, because he fended off the Spad for a long time. Ruffo fired 120 bullets, 12 of them hit the Austrian but then his gun jammed and he let the gallant enemy go. His outstanding adversary was *Fw.* Josef Franke of Flik 4, who brought his Brandenburg to a forced landing, unseen by Ruffo, who was thus deprived of a victory.

On 5 May Ruffo had another victory, as he shot down in the valley of the Vippacco River Br. C.1 69.20, with pilot *Hauptmann* Hübner, commander of Flik 32, and observer *Leutenant* Emil von Weinmann. He jumped that airplane, already limping for an artillery shell that had pierced its fuel tank, over Sagrado. His first bullets wounded the observer, who was dead for loss of blood before Hübner could land at his own base. The airplane was then sent to Flep 1 at Marburg for repairs.

In the next days, as the Tenth Battle of the Isonzo began, the Italian pilot continued in his winning streak and in a few days he claimed four victories, exploiting the qualities of the Spad 7, a fighter much superior to anything the Austrians could oppose it. The first one was on 10 May, when above Dolegna he spotted a fighter that was after an Italian two-seater; Ruffo attacked it firing 125 rounds and saw it getting down over Gorizia and then landing close to Biglia, behind enemy lines.

Two days later, 12 May, flying over Gorizia he saw a Farman of the 30ª Squadriglia under attack by a fighter. Ruffo stepped in immediately, fired one single burst before his gun jammed, but it was enough to drive the fighter away. The Farman managed to return to its airfield with 16 holes and was declared out of action. Its crew, the pilot *aspirante* Nicola Cena and *sottotenente* Gino Graziani, declared that Ruffo, whom they could recognize as his insignia was now well known to friend and foe, had saved them from certain death. His adversary probably was Brumowski who claimed a Farman shot down that day. The Austrian ace flew with Flik 41/J, one of the new wholly fighter units of the Austrian air force, equipped with the Hansa Brandenburg D.1 fighter, also called KD, a very fast single-seater but not maneuverable and difficult to fly, on account of its instability. That afternoon Ruffo had another combat:

> "I took off a second time on standing patrol at 7 p.m. I saw three enemy aircraft over Vertoiba, seven or eight hundred meters higher than me. I retreated pretending to run away in order to gain altitude. When I reached the level of my enemies I realized that two of them were about to attack one of our Caudron bimotors that was between Podgora and Gorizia, but as those two were getting closer to our reconnaissance airplane, I had got unnoticed behind one of them and I began to fire. My target went down, the other one ran away and as I went after the falling one diving steeply, the third enemy aircraft that had stayed high was now after me and tried to hit me with its frontal gun, but it missed completely. The airplane I shot down fell in the direction of the S. Marco".

The next day Ruffo claimed an "Albatros", (that is to say a Brandenburg two-seater), seen going down behind the Bainsizza at 10 a.m. Ground observers confirmed this victory, stating that the aircraft touched down at Felenih.

May 14, after his second combat of the day, Ruffo felt strong vibrations and saw sparks in his cockpit. He switched off contact and glided down to Volzano. On the ground he realized that an enemy incendiary bullet had hit his propeller and burned off its hub.

May 20 was another unlucky day for Ruffo. He had a first combat in the morning, at 9.20 a.m., attacking an "Albatros" over Zagora but letting it go because if a gun jam. This event corresponds in time and place to a victory claim for an artillery-spotting Brandenburg of Flik 2 with *Feldwebel* Stephan Wognar, slightly wounded in the back, and *Leutnant* Franz Gräser, two future Austrian aces. One year later Gräser was shot down in flames by three pilots also of the 91ª Squadriglia. Things went from bad to worse once Ruffo fixed his gun and attacked another "Albatros": a fighter bounced him and he had to quit the combat and land at Medeol field, his Spad repeatedly hit, with bullets in its fuel tank, radiator and wings. Actually Spad 4703 was so shot up that it was written off. Another future Austrian ace, *Zugsführer* Stefan Fejes, flying a Brandenburg of Flik 19 was credited with this victory.

Ruffo in the cockpit of his Nieuport 17. In this image, the Vickers gun over the cowling is not fitted, and the fighter is armed just with a Lewis gun attached over the top wing with an unusual French tripod.

Ruffo struck back on 26 May, with his ninth victory. He flamed and shot down near Britof Br. C.1 129.53 of Flik 32, its two airmen, *Zugsführer* Franz Zaratin and observer *Leutnant i.d.R.* Josef Eder were killed.

His ninth victory also earned him his second Medaglia d'Argento in the field with a citation referring to five air victories in the period from 5 to 26 May 1917.

In the morning of May 30 Ruffo saw three airplanes in the valley of Chiapovano flying to Gorizia and engaged them. He drove one away, attacked another one but lost it for the intervention of the third one. Back to base he saw that explosive bullets had torn his left aileron and damaged a strut. In May 1917 the 91ª had proven its worth, shooting down 14 out of 27 airplanes claimed by Italian aviation, and Ruffo had played his part, with five victories in 18 combats, three times returning with a damaged airplane.

On 6 June the squadron was shifted to Istrana, on the Asiago plateau front, to support the Italian attack on the Ortigara. On that front Austrian air activity was limited, and also bad weather curtailed flights. Ruffo met again the enemy only on 19 June:

> "I was flying today over the Valley of Assa and at 10.15 a.m. I saw an albatros attacking an S.P. of ours. I reached it as it placed itself behind our aircraft, I opened fire and it immediately dove down and retreated to its lines at very low altitude. I followed the albatros a bit more firing a few more bullets, and then I returned near our S.P. that was able to go on with its reconnaissance. The S.P. carried serial 4438.
>
> My second combat also took place in the sky of Val d'Assa at about 10.45 a.m. I was at an altitude of 4000 meters and I saw a Caproni returning from its bombing mission, followed far behind by a completely green airplane, with no crosses. Initially I thought that it was another Italian plane, but just in

case I got closer. The suspect plane in fact was an Austrian old type albatros with green painted wings and fuselage. Also the top part of the bottom wings was painted green and from a distance it could be confused with an Italian aircraft. It was not fast, and for some minutes it had been trying to reach the Caproni that was heading home in a slight dive. In a few more seconds it would have reached it but I fired a burst at it and it banked and the observer fired at me with his gun..

Our combat lasted some time, during which we fired at each other repeatedly, then the enemy went down in a dive toward Levico, and I had the opportunity to fire at it again. And then a thick smoke as I had seen other times came out of the enemy plane and it lasted so long that I believed it was on fire, so I stopped shooting. However I didn't see that airplane catching fire, nor crashing, and I lost it in the clouds as it got lower."

The Austrians lost three Brandenburgs that day, and Ruffo's enemy was probably an aircraft of Flik 24, its pilot, *Korporal* Stefan Volosin, who had gallantly attacked the Caproni four times despite the escort of the 91ª Squadriglia, force landed at Pergine airfield and survived.

On 26 June Ruffo fought over Asiago two "Albatros", one of which went down in a dive, but the ace couldn't determine the result of this action because the second one engaged him. Maybe the adversary of the prince was an unlucky crew of Flik 21, flying a standing patrol protecting a visit of the Austrian Sovereign Karl I to Pergine airfield from Italian intrusions. The bodies of the two airmen, badly mauled, were found at Malga Belen. Nobody from the ground had noticed the air combat, but the observer had fired 62 bullets from his gun and in the wreck there was an open roll of gauze, showing that somebody had been wounded. Different hours are reported for the Italian claim and the Austrian loss, but Ruffo's engagement is the only known combat on that day and in that part of the front.

After the tragic failure of the Italian offensive, in early July, the 91ª Squadriglia returned to its usual eastern front. On 14 July Ruffo shot down in flames with just 30 bullets near Comen a two-seater, according to the squadron diary. Actually he shot down a single seat KD fighter of Flik 35, its pilot *Feldwebel* Karl Sailer falling down from his burning plane at 4,000 meters, as reported by *cap.* Costantini. The KD crashed over a village, but Ruffo couldn't identify it, as he was sick after a long dive after the enemy. Austrian sources report the place of the fall as near Gorjansko.

On 17 July Ruffo shot down another fighter *"on a grassy knoll on the left bank of the Isonzo south of Selo"*. He remarked that the airplane had stuck its engine in the dirt and lay on its left side, with the right wing upright. Austrian airplanes were now flying in large formations, but that did not deter the fighter prince.

Friday 20 July Ruffo scored a double victory flying Spad 5408. This was his Combat report:

Brandenburg 69.20 forced down by Ruffo on 5 May 1917.This airplane was eventually repaired and put back into service. *(Photo Höfling).*

"I took off from Bolzano airfield, where I had landed the day before for engine failure. At about 9.15 a.m. also thanks to ground signals I spotted near the Valley a group of five enemy aircraft flying inland. Two of them were fighters and three were two-seaters. I was at an altitude of 3,000 and I engaged first a fighter and a two-seater that were above me; they returned my fire and then they turned back to their lines. Meanwhile I immediately engaged a second fighter escorting a two-seater. After a brief combat the fighter went in a steep dive, I had the impression that it had been hit and was trying to land; I then turned at the two-seater and after a short burst at very close range (possibly less than 30 meters) it went down catching fire and crashed near Oppachiesella.

Right: Next version of the black skull, grinning and applied over the fuselage roundel of a Spad 7.

A French squadron equipped with Sopwith 1 1/2 Strutters was sent to Italy in the summer of 1917 and the worldly Ruffo here chats with his French colleagues.

The survivors ran away. During the whole combat I absolutely didn't see any other Italian airplane."

His two-seater victim was a Brandenburg of Flik 23, while the two escort fighters were expert pilots, *Feldwebel* Magerl and Ristics, of Flik 42/J, who were not hurt, according to Austrian sources, although existing data don't allow us to know whether their airplanes were damaged or shot down.

This combat, bringing his total score to 13, was widely reported in the press, Ruffo was proposed for the Medaglia d'Oro al Valor Militare and on 2 August he was promoted to *capitano.* He had shown in the field that he deserved both the medal and the rank, as he was now close behind Baracca as Italy's top ace.

The citation for his Medaglia d'Oro, officially issued on 5 May 1918, was:

> "Having outstanding military virtues, fighter pilot of peerless courage, tested in fifty-three air combats, with a spirit of sacrifice equal to his valor, he kept on looking for victory wherever he may find it. In two months he shot down four enemy airplanes with his accurate fire. On 20 July 1917, with incredible daring, he attacked alone a tight squadron of five enemy aircraft, shot down two of them and drove away the other ones. Wonderful example to the bold. –Sky of Castagnevizza, 14 July –Sky of Tolmino, 17 July –Sky of Nova Vas, 20 July 1917."

This was his final award, bringing his total to one Gold, two Silver and two Bronze medals.

A few days later again he attacked alone a group of five enemy aircraft, but his dash was made void by a gun jam. His health however was suffering from the constant strain, and he was so exhausted that at the end of August he was sent on leave. It was to be a brief period of rest, but he remained away from the front a long time. While the newspapers were praising his deeds, Ruffo was in the hospital of Naples, and he was there under treatment for two months. From his bed he wrote to his good friend Baracca, including sarcastic comments about other fighter pilots who were too cavalier in claiming victories.

He returned to his unit only on 5 October, long after the end of the battle of the Bainsizza, the last Italian offensive. Weather was horrible, but on 9 October he was again in the air, alone as usual. He spotted a two-seater with four fighters escorting it, and despite the odds he attacked the fighters, but he had to get away, as his gun had jammed. Later on, he had another fruitless combat.

In the darkest moment of the war, on 25 October, the valor and ability of the prince were particularly evident. Ruffo flew three fighter missions from 8 a.m. to 3 p.m. and in each one he shot down an enemy two-seater, the last one in flames. One of them was an able and determined German crew. This was his Combat Report:

> "1s t Flight: I took off for a fighter flight at 7.50 and I attacked at about 8.20 an enemy reconnaissance airplane above Lom di Tolmino. The enemy noticed my attack and before I could aim properly he started maneuvering all around, with outstanding ability, but after I had fired three times he gave up

One of the many instances in which Ruffo couldn't bring his battle-damaged airplane back to base. Although the ace is trying to smile, the stress and strain of combat flying, especially with the intensity of Ruffo, must have taken a terrible physical toll.

the dogfight and dove toward home. As I fired my fourth burst he went down in spirals and now and then heading to Canale. The combat began at an altitude of 3,500 meters and at 2,000 meters the descent of the enemy that I still harassed and shot at, turned into a fall and I had the impression that the enemy aircraft was now out of control. Still I kept shooting at it. I expected it to catch fire or to crash down, but as I had to turn at an altitude of 1,000 meters I lost it. I looked for it below me but I failed to see it again. Meanwhile anti-aircraft guns were firing at me. Considering that I had lost it at very low level, I am strongly convinced that I had shot it down. I returned to the airfield at 8.47 a.m.

2nd Flight: I took off for a fighter flight at 10.40 a.m. together with Ten. Costantini and at 12.20 I attacked an enemy reconnaissance plane at an altitude of 3,700 m. I fired a few bursts and the enemy caught fire and near 3,000 meters it was completely engulfed in flames and it crashed. During this combat Tenente Costantini, as we had agreed upon, was always close to me. The wreck of the enemy plane fell down near Lom di Tolmino. I landed at 12.43.

3rd Flight. I took off again for a fighter flight at 2.05 p.m. and at 2.30 I attacked and I flamed with a few bursts an enemy two-seater which broke apart burning at an altitude of 3,500 meters. The remains of my adversary fell down near Gorizia. High above me there were three enemy aircraft, which fired at me and then got away. I landed at 2.53 p.m."

A more elaborate version of the black skull on a Spad 7. As he shows the rank of *capitano*, this picture can be dated after August 1917.

It was a remarkable treble victory, which was followed by a long empty period. As the front moved back to the Piave, only on 18 November he had another combat with an enemy plane, together with two other Spads of the 91ª Squadriglia, but all of them had gun jams and had to give up. It was the beginning of a black period for Ruffo: while his squadron mates had plenty of combats, he never happened to meet the enemy. On 27 November he had an engine failure (a broken rod) and force landed turning over his fighter, but thanks to the strong structure of the Spad he got out of it unhurt.

A few days later he had another turn-over with Spad 1526, at Nove di Bassano airfield.

On 23 December, as the front was by now stabilized, Ruffo together with Baracca went to Turin, where they tested with Piccio the new models of fighters, and he returned to Padua on 24 January 1918. On 6 February the kings Victor Emmanuel III and Albert of

Ruffo di Calabria in a signed photograph with his Spad 7.

Ruffo in the late summer of 1918, with his Spad 13 showing the final variation of the black skull. Second from left is Conelli de' Prosperi, in the cockpit Mario De Bernardi, sitting astride the fuselage Gastone Novelli.

Belgium visited Padua airfield. The Belgian sovereign pinned on Ruffo's breast the Cross of Leopold I.

On 9 March, as winter had brought operations almost to a standstill, Ruffo went on leave and returned on 6 April, a few days before the movement of his 91ª Squadriglia to Quinto di Treviso. His feelings then were expressed in a letter that Baracca wrote on 15 May: "Piccio and Ruffo are furious because for many months they haven't been able to find any Austrian airplane to attack."

Finally five days later his eagerness was satisfied. On 20 May Ruffo went in an offensive patrol over the Grappa with *serg.* D'Urso and found an enemy flight. He realized that one of them was trailing behind and he attacked it, but after a few rounds his gun jammed and the Austrian got away diving toward the Montello. Ground observers reported that this airplane might have crashed toward Nervesa. This victory earned Ruffo a prize of 1,000 Lire, a good sum at that time.

On 7 and 9 June Ruffo flew as escort to the Capronis bombing Godega and Pramaggiore. In the first day of the final Austrian offensive, 15 June 1918, Ruffo, with D'Urso again as his trusted wingman, shot down an airplane near the Grave di Papadopoli. His victim was described as an *"old type Albatros"*, probably a Brandenburg, and it lost pieces before hitting the ground. The squadron diary marked down this one as his 20th victory, which proves that some of his earlier claims were validated only long after the events had happened.

In the evening of 17 June he flew for the last time together with Baracca, in a strafing mission. Out of ten fighters, five returned with damages, an omen for the fatal mission of Baracca two days later. When the ace of aces died, Ruffo became commander of the 91ª Squadriglia, and he was a pallbearer when the body of Baracca was recovered.

During the summer he flew routine escort missions and tested two Ansaldo Balillas which De Bernardi and Nardini had ferried to the unit. Like other expert pilots, Ruffo was not at all impressed by the new Italian fighter.

On 19 September Ranza replaced Ruffo as squadron commander at the 91ª, and the prince became commander of the new XVII Gruppo, formed on that day with the 91ª and the 71ª Squadriglia which moved to Quinto. His new post didn't stop him from flying. During the final offensive of Vittorio Veneto, on 29 October, the prince strafed the road from Conegliano to Vittorio Veneto; his Spad was hit in the fuel tank and the pilot force landed near Marano. He managed to escape capture and returned to the Italian lines. Piccio, with less luck, in similar circumstances was shot down and captured, so Ruffo flew to la Comina carrying a bicycle in his Spad and went riding around looking for information about the missing commander.

On 12 February 1919 Ruffo was assigned to the Centro Formazione Squadriglie of Turin, and a few days later to the Ispettorato Aeroporti. His many years of war service, however, had taken their toll, and in 1920 he was removed from service because he was diagnosed suffering from fatigue and anemia. He was called back to the active service in 1923 in a Cavalry regiment. In 1925 he was dismissed, and he spent his following years tending his family properties, particularly the beautiful estate of Paliano, near Frosinone, Latium.

On 6 April 1934 he was made a Senator of the Kingdom. In spite of his official role, he did not refrain from criticizing the fascist regime. When racial discrimination was introduced in Italy, in 1938, Ruffo, then the Whip of the Senate, refused to remove Senator Mayer, who was a Jew. After the armistice of September 1943 his sons fought against the Nazis, and one of them died in combat. Ruffo himself collected money for the Resistance and using the help of a Belgian priest he reported to the Allies the presence of German troops in his estate of Paliano. Despite his behavior, which proved his democratic feelings, he was deprived of his place in the Senate on 27 September 1945 and he had to defend himself from groundless indictments.

Prince Fulco Ruffo di Calabria died at Ronchi d'Apuania on 23 August 1946. His beautiful daughter Paola married on 2 July 1959 Alberto of Lieges and she is now Queen of Belgium.

*Cap.* Ruffo next to a Spad 13, which shows camouflaged cowling and wings, but plain varnished cloth fuselage.

## Fulco Ruffo di Calabria
## 20 confirmed victories

| Claim | Vict. # | Date | Sqn. | Aircraft | Hour | Place | Victim |
|---|---|---|---|---|---|---|---|
| 1 | 1 | 23.8.1916 | 70 | Ni .11 | - | Ranziano | Two-seater |
| *Zugsf. Anton Himmer UNH/unknown obs. UNH, Flik 12, Br. C.1 61.61* | | | | | | | |
| 2 | 2 | 16.9.1916 | 70 | Ni .11 | - | Mount Stol | "Lohner 160 hp" |
| *[with ten. Baracca & asp. Olivari, 70ª Sq.] Zugsf. Franz Morozko KIA/Lt. Anton von Csaby DOW, Flik 16, Lloyd C III (WKF) 43.74* | | | | | | | |
| - | 3? | 1.1.1917 | 70 | Ni.17? | 15h15 | Duino | Two-seater |
| *Possibly: Fw. Josef Franke UNH/Lt.i.d.R. Wilhelm Schandlbauer UNH, Flik 4, Br. C.1 29.79 or: Korp. Eugen Bönsch UNH/unknown observer, Flik 12, Br. C.1 68.24 damaged* | | | | | | | |
| 3 | 4 | 11.2.1917 | 70 | Ni.17? | - | Romansacco | Two-seater |
| *[with ten. Baracca, serg. Poli & Gorini, 70ª Sq.] Korp. Ludwig Fleck POW/Lt. i. d. R. Wilhelm Graf Siemienski WIA, POW, Flik 35, Br. C.1 27.74* | | | | | | | |
| 4 | 5 | 28.2.1917 | 70 | Ni.17 3139 | - | Gradisca | Two-seater |
| *Korp. Stefan Morth UNH/Lt. Nikolaus von Szebenyi UNH, Flik 34, Br. C.1 27.56 FTL* | | | | | | | |
| - | - | 3.5.1917 | 91 | Spad 7 | - | Sagrado | Two-seater |
| *Fw. Josef Franke UNH/Lt. i. d. R. Arpad Pindter von Pindtershofen UNH, Flik 4, Br. C.1 129.41 FTL* | | | | | | | |
| 5 | 6 | 5.5.1917 | 91 | Spad 7? | 10h | Sagrado | Two-seater |
| *Hptm. Richard Hübner UNH/Lt. Emil von Weinmann KIA, Flik 32, Br. C.1 69.20* | | | | | | | |
| 6 | 7 | 10.5.1917 | 91 | Spad 7 4703? | 11h | Biglia | Fighter |
| 7 | 8 | 12.5.1917 | 91 | Spad 7 4703? | 19h | S. Marco | EA |
| 8 | 9 | 13.5.1917 | 91 | Spad 7 4703? | 10h | Jelenik | Two-seater |
| 9 | 10 | 26.05.17 | 91 | Spad 7 5408? | 11h45 | Britof | Two-seater |
| *Zugsf. Franz Zaratin KIA/Lt. i. d. R. Josef Eder KIA, Flik 32, Br. C.1 129.53* | | | | | | | |
| - | 11? | 19.6.1917 | 91 | Spad 7 | 10h45 | Near Levico | Two-seater |
| *Korp. Stefan Volosin UNH/unknow observer UNH, Flik 24, Br. C.1 29.60* | | | | | | | |

| Claim | Vict. # | Date | Sqn. | Aircraft | Hour | Place | Victim |
|---|---|---|---|---|---|---|---|
| - | - | 26.6.1917 | 91 | Spad 7 | 8h30 | Asiago | Two-seater |
| *Fw. Adolf Kind KIA/Lt. Franz Nagy KIA, Flik 21, Br. C.1 129.27* | | | | | | | |
| 10 | 12 | 14.7.1917 | 91 | Spad 7 | 9h15 | Comen | EA |
| *Fw. Karl Sailer KIA, Flik 35, Br. KD 28.43* | | | | | | | |
| 11 | 13 | 17.7.1917 | 91 | Spad 7 | - | S.Lucia di Tolmino | Fighter |
| 12 | 14 | 20.7.1917 | 91 | Spad 7 5408 | 9h30 | Castagnevizza | Fighter |
| *Either Fw. Magerl, KD 28.18 or Fw. Ristics, KD 28.58, Flik 42/J, who were not shot down* | | | | | | | |
| 13 | 15 | 20.7.1917 | 91 | Spad 7 5407 | 9h30 | Nova Vas | Two-seater |
| *Korp. Rudolf Blaszko KIA/Oblt. Alexius Celle KIA, Flik 23, Br. C.1 69.84* | | | | | | | |
| 14 | 16 | 25.10.1917 | 91 | Spad 7 | 8h20 | Lom | Two-seater |
| *Korp. Andreas Kulcsar WIA/Lt.i.d.R. Julius Thassy von Miske et Monostor WIA, DOW, Flik 4/D, Br. C.1 129.61* | | | | | | | |
| 15 | 17 | 25.10.17 | 91 | Spad 7 | 12h20 | Tolmino | Two-seater |
| *[with ten. Costantini, 91ª Sq.] Maybe: Lt. Paul Dotzenrodt KIA/Lt. Robert Lessing KIA, FA 17 or:*<br>*Lt. Walter Conrad KIA/Lt. Albert Tawainer KIA, FA (A) 232*<br>*Vzfw. Friedrich Moritz KIA/Lt. Franz Neumarket KIA, FA 39* | | | | | | | |
| 16 | 18 | 25.10.1917 | 91 | Spad 7 | 14h30 | S. Marco | Two-seater |
| *Fw. Anton Lambert KIA/Oblt. Jan Pappius KIA, Flik 19/D, Br. C.1 29.63* | | | | | | | |
| 17 | 19 | 20.5.1918 | 91 | Spad 7 | 9h15 | Nervesa | Fighter |
| 18 | 20 | 15.6.1918 | 91 | Spad 7 | 7h10 | Grave di Papadopoli | Two-seater |
| *[with serg. D'Urso, 91ª Sq.] One of the following:*<br>*Korp. Karl Szabo WIA/Oblt. i. d. R. Fritz Rosenstingl KIA, Flik 69/S, Br. C.1 169.111*<br>*Fw. Alfred Brandt UNH/Oblt. i. d. R. Karl Hiemesch WIA, Flik 19/D, Br. C.1 369.69*<br>*Korp. Wilhelm Watzl MIA/Lt. i. d. R. Sylvester Walitza MIA, Flik 5/F, Br. C.1 369.126.* | | | | | | | |

## Giovanni Sabelli

In the official list of confirmed victories in World War I, there are 42 Italian army pilots who qualified as aces, but, differently from other nations, only two of them were killed in action, Baracca and Sabelli, five lost their life in accidents and two were badly wounded. The percentage of losses for fighter pilots was very low, compared to the losses of Caproni bomber crews and, of course, of infantry. The circumstances of the death of the greatest Italian ace, Baracca, are still a source of discussion but another member of his squadron, Giovanni Sabelli, had the sad privilege of being the only ace certainly shot down and killed in air combat.

Giovanni Sabelli was born in Naples on 23 September 1886 from a wealthy family. Little is known of his youth, other that he studied engineering in New York City. At the age of 26 he was in England, where he got his civil pilot license on 30 January 1912 at Brooklands. He took part in many air meetings flying a Deperdussin, and he had a certain amount of fame when he volunteered as a military pilot in the army of Bulgaria, which was at war with Turkey. Sabelli wrote his impressions of the war in a small notebook that has survived, and that provides an amount of scarce information.

Giovanni Sabelli at Istrana airfield, commenting on his combat on 19 June 1917.

Sabelli liked his period in Bulgaria. The airfield was located at Mustapha Pasha, base of the general staff of the Bulgarian Second Army that was besieging Hadrianopolis. Some of the pilots were Bulgarians, but most of them were foreigners: Russians, Frenchmen and one Englishman. The airplanes also were international: there were Russian-built Farman biplanes, an Albatros biplane, a Voisin, a Kalinoff, and a side-by-side Bleriot (or "cote a cote" as Sabelli called him in French, then the language of aviation) with a 70 hp engine and a speed of 115 km/hour, his personal aircraft. He briefly piloted also a Bristol monoplane but he smashed it in a forced landing due to an engine failure:

> "...With the tandem Bleriot I have flown the most brilliant reconnaissance of the war. Weather was very cloudy and there was a strong wind, but it was necessary to fly a reconnaissance. Three of us took off. A Russian monoplane immediately returned because its pilot related that it was wholly impossible to fly, an Albatros biplane which carried as passenger a British reporter, Mr. Gore who wanted to take pictures of Hadrianopolis: it could only arrive near the city. I took off with the commander of the flotilla Major Zlataroff as passenger, who had to drop some powerful bombs and leaflets. We soon climbed into the clouds. When I reached 1800 meters we were flying between two huge layers of clouds. Every now and then we crossed cloudbanks dark like the night. But changing our altitude we managed a most successful reconnaissance of great strategic importance over Hadrianopolis flying about 250 km. We dropped big bombs that exploded with great results and generated much panic. We also dropped big bundles of leaflets and we returned to the airfield. Major Zlataroff is a wonderful

Right: Giovanni Sabelli at Hendon, England, standing next to his Anzani-powered Deperdussin monoplane.

A smiling Sabelli with a Nieuport monoplane at Hendon. (Photo Selinger).

The London Aerodrome Trophy presented to G. Sabelli to commemorate his flight in a high wind from Brooklands to Hendon September 7, 1912.

man, quiet and cold as an Englishman, with the courage of a lion, and with a heart of gold. I send you his photograph, where you can see my officer's epaulets. I also send another picture showing the General commander of engineers, Colonel Vassilief and Major Zlataroff. General of the engineers Santeff studied in Italy, in Turin, and he speaks Italian well."

Sabelli described to his mother the kind of life he found in Bulgaria, remarking the backwardness of certain aspects of that country:

"...our life is very methodical. We get up each morning at 6. At 7 we have breakfast at the Officer's Club then we reach the airfield by coach. What I call coach is a very primitive vehicle without springs, drawn by two horses. You can imagine the jolts you get with such a vehicle on these roads. In certain places we have to cross streams with water reaching the breast of the horses. At noon we return to Mustapha and the same in the evening. Meals are always very jolly occasions. Bulgarian officers are very nice people, happy and warm-hearted. They have an extraordinary talent and passion for music. They sing in choir with different voices so well that you would say that they just rehearsed their part. About 7 we go home. We are lodged here and there in the houses of the town. Now for two nights we could sleep soundly, because generally every evening and night the Turks tried desperate sorties that the Bulgarians stopped with a terrible bombing that lasted for hours and shook the glass panes of the windows. During those bombings we could see the shells exploding. Each time the Turks saw us arriving over their forts and the city they greeted us with countless balls and shells that never hit us. Shrapnel makes a peculiar sound, like a soft knocking on a door that stands out over the engine noise. Also you could see little puffs of smoke.

The activity of airplanes was very useful and gave excellent results. Our only problem is the weather. Nowadays the real enemy of the airplane is not wind, because a good airplane can fly even with a very strong wind, but clouds. Particularly in winter the sky is always cloudy. When it's cloudy we can fly, but it is impossible to reach an altitude of 1500 meters where one can challenge the shells and perform a reconnaissance..."

Sabelli, like many gentlemen in those years, spoke excellent French and he had a dispute with a French reporter, Mr. Rebaud, who had criticized Bulgarian aviation and its organization in order to defend another Frenchman, a certain Henry Olivier, dismissed by the Bulgarians and whom Sabelli also considered unfit: "... He was so obviously inexperienced that they didn't allow him to fly anything else and after a few days they sent him back to Sofia."

The war had a brief respite, then when hostilities flared up again Sabelli flew reconnaissance missions to Gallipoli, where the Bulgarians deployed their 4th Army to preclude sorties from 60,000 besieged Turks and keep them pinned down in the peninsula. Sabelli was quite pleased with his Bleriot: what mattered then in an airplane, besides performance, was strength of the airframe and landing gear, because forced landings in the open country were the rule:

"The Bleriot touched down almost like a bird, after having glided ten meters. As I got out I wondered how was it possible to land on such a terrain. The Bleriot surely has its defects, but to my knowledge there is no better landing gear. I have landed very often on very bad fields with shrubs and ditches but the only difference is that instead of running 50 or 60 meters it stops easily in 10 or 15. But it is absolutely necessary to touch down at a very low speed."

The Italian magazine "Lega aerea nazionale" in May 1914 reported that Sabelli had been hired as chief pilot instructor of the Bulgarian Military Aviation School. That nation had ordered twelve airplanes from Germany to form two squadrons. European public opinion and the military were interested in the first applications of the new weapon. After the first experiments of the war of 1911 between Italy and Turkey, Sabelli was the first Italian to experiment in practice what was just a theory, so he was interviewed and quoted by the press. This appeared in "Lega aerea nazionale":

"Let's consider vulnerability of airplanes to enemy action. Sapelli (sic) tells that his airplane throughout the war was fired upon unceasingly first by the Turks, then, in the second part of the conflict, by the Greeks and Serbians, and sometimes even by the Bulgarians themselves, who mistook his airplane for a Turkish monoplane. Once, his passenger counted up to 65 Shrapnel shells exploding around the airplane and always, in reconnaissance over Hadrianopolis, Gallipoli, Bulair, Ciatalgia and the coast of Asia Minor, always and everywhere the airplane with the Italian pilot got out unscathed from the exploding cannon shells."

Sabelli returned to Italy after the end of the war. When Italy entered the European war, he was assigned to the Battaglione Aviatori and on 22 August 1915 he was commissioned with the rank of *sottotenente*. As he had only a civil pilot license, despite his war experience he had to repeat the tests. On 30 August he received his military pilot license on Aviatik airplanes, ands he was assigned to the Aviatik defense flight at Aviano.

He remained there briefly, as on 15 October 1915 he was sent to Le Bourget, France, for flying training on what was then the most modern airplane, the Nieuport biplane. The best and boldest pilots were chosen to convert on the first real fighter. Sabelli in Paris lived in the Hotel Suisse in Rue Lafayette, and on 15 November he wrote to his friend Arrigo Abba showing his enthusiasm for the new airplane:

"...I have flown a lot the nieuportino that is a wonderful airplane and not the least bit dangerous. When I return we must see that you get assigned to my squadron so that you can fly with me as gunner..."

Training was not as fast as Sabelli hoped, possibly winter weather curtailed flights, and in January 1916 he was still in France, from where he wrote:

Sabelli at the controls of a Bleriot during his service with the Bulgarian aviation in the war against Turkey.

"...Lucky you who had the satisfaction of flying over the enemy. I haven't had this pleasure yet, but I will make up for lost time. We remain here a bit longer to pilot the latest type of Nieuport, which is smaller than the one of the Squadriglia Tacchini, and much faster, as it does exactly 178 km/hour. A few days ago at Le Bourget I looped the loop twice, and I am the one and only in this squadron. When I come to Aviano I'll do it again. With the Nieuport it is more impressive than with the Bleriot, because you lose more altitude. I am impatient to return to Italy and try to shoot down some Taube..."

The new airplane that he praised was the Nieuport 11 single-seater that soon would be equipping Italian fighter squadrons. A few days later Sabelli returned to Italy, and on 1 February 1916 he was assigned to the 2ª Squadriglia da caccia Nieuport. It was a new unit that had to apply on the Italian front the new tactical doctrines of air fighting developed by the French. Its equipment was first rate for that time, and justified the enthusiasm of Sabelli: two French-built Nieuport 11 80 hp, one Nieuport 10 "18 square meters" single-seater and five two-seaters. The new fighter pilot had been assigned to a élite unit, although its commander *cap.* Chiaperotti had a small number of pilots, just five.

Sabelli at Le Bourget with sold. Silvio Amico, who was killed at Cascina Farello in an accident with Nieuport 11 serial 1452 of the 2ª Squadriglia Caccia on 22 March 1916.

Actual combat experiences arrived right away, with the escort to the Caproni bombers facing the threat of the Austrian Fokker monoplanes based at Aisovizza. On 18 February 1916 Sabelli, with his friend Abba, who had indeed succeeded in getting attached to his squadron, flew a patrol over the enemy airfields protecting the Capronis that were flying to bomb the city of Lubjana. His Combat Report describes the first fighter missions:

"2ª Sq. Nieuport da caccia
Aircraft 1263
Pilot S.Ten. Sabelli - Obs. gunner. S. Ten. Abba

Mission – protection of the 300 hp Caproni aircraft returning from the action on Lubjana from attacks of the aircraft from the enemy airfields of Aisovizza and Aidussina.
Take-off from Comina airfield hour 9.10
Route: Comina-Codroipo-Mortegliano-Cormons-Gorizia-Aisovizza-Aidussina-Zoll-Schwarzenberg-Aidussina-Aisovizza-Gorizia-Aisovizza-Gorizia-Cormons-Gonars-Codroipo-Comina.
Flight altitude 2500-3000.
Hour 9,45: from Aidussina I see a returning 300 Caproni airplane at a lower height. Over Zoll that airplane changes its direction and goes over the Tarnova range. We go on without seeing any other 300 Ca up to Schwarzenberg. Returning over the airfields of Aidussina and Aisovizza an enemy airplane of Morane Saulnier type passes below us about 400 meters, turns and then turns again and lands at Aisovizza. As we return a third time over Aisovizza, an enemy airplane of the same type, maybe the same one, takes off and heads for the Vippacco (hour 10). We return to Comina airfield that we reach at hour 10,40 with 5 liters of gasoline."

In mid April the 2ª Squadriglia got the new designation 71ª Squadriglia, but equipment and personnel remained unchanged. The first air combat, that Sabelli so much longed for, arrived on 6 April when he engaged a flying boat that was bombing torpedo boats, and drove it off. On 29 April five Nieuports intercepted a two-seater above Ruda, attacked it with no apparent result but forcing it to run. In those times just forcing an enemy to clear the field was an accomplishment and the components of that patrol were praised, but the real winner was the Austrian gunner who had placed a few bullets in Sabelli's airplane and slightly wounded in the shoulder another pilot, Guicciardi.

Sabelli's first experiences as a fighter pilot were limited to brief combats with no result. Meanwhile the Nieuport 10 two-seater were retired as front line fighters, replaced by the smaller Nieuport 11. On 29 May the 71ª Squadriglia was moved to the Trentino front. Sabelli had two more combats on 25 July and 24 September against enemy airplanes attempting raids against Vicenza. Then he was briefly posted at Malpensa airfield, near Milan, as test pilot for the new Nieuports now produced by the Nieuport Macchi Company of Varese, under French license. It was a difficult task, showing his

Sabelli at Taliedo airfield, Milan, September 1915. The airplane is a Salmson-powered Aviatik, the first airplane used as a fighter by the Italian air force.

high reputation but obviously Sabelli wanted to remain in the first line, and he didn't stay long there. His efforts up to that time were recognized, and he was decorated with a Medaglia d'Argento, that credited him with ten air combats (but no victory) up to September 1916.

Promoted to the rank of *tenente* on 9 September 1916, Sabelli was posted as commander of the Sezione Nieuport based in Albania on 6 October. To be charged with the command of a unit was certainly good news, but the destination was not. The far-away Albanian front was not the place to find the air victories that he longed for, and the flight had only three pilots, himself, and *sergenti* Lovadina and Palpacelli. Bad weather made the airfield at Tahiraqua unusable, and air activity there was to all effects non-existent.

Sabelli remained in Albania until April 1917, when he was recalled to Italy and assigned again to the 71ª Squadriglia that was still on the Trentino front. He probably didn't even reach it, as a few days later, on 9 May, he was assigned to the 70ª Squadriglia, and five days later to the 91ª. The reason for these contradictory orders is probably due to the fact that on 1 May the best pilots of the 70ª Squadriglia were shifted to the newly formed 91ª so that Baracca could lead a crack unit equipped with the best fighter then available in Italy, the Spad 7, a few of which had just arrived. Besides its best pilots, the 91ª Squadriglia received from the 70ª all its Spads, leaving it with the Nieuports. In that period the best pilots from all squadrons were posted to the units equipped with the Spad to convert on this model, with flight characteristics radically different from the Nieuport, and as the 70ª no longer had it, Sabelli was shifted to the 91ª, but his stay in the Squadriglia of Baracca was to be a lasting one.

Austrian aviation in that time still used the Hanša Brandeburg C.1 as a fighter, together with a small number of Brandenburg KD single-seaters. The latter fighter was fast, but its flying characteristics were much inferior to the Spad, and it was quite dangerous, causing many accidents, a lot of them fatal. The Spad was fast enough to intercept easily any enemy airplane, it was extremely strong and it could perform any maneuver without fear of structural failure. Since it had entered service the pilots of the Squadriglia, first among them Baracca and Ruffo, had amassed a long string of victories, bleeding the gallant and able Austrian units facing them.

At the 91ª Squadriglia each pilot had his own personal insignia, and Sabelli introduced a peculiar ladder design, of unknown meaning. His companions, when they heard of his unusual prewar combat experiences, nicknamed him *"the Bulgarian"*.

The Squadron Diary named Sabelli for the first time on 24 May 1917, when he flew a Caproni escort mission together with Ruffo, but while the latter had five combats against Austrian aircraft that tried to attack the slow bombers, no combat is reported for him.

He didn't have much to wait: on 31 May at 7.30 p.m. as he was flying alone at an altitude of 3,000 meters he spotted between Plava and Canale three enemy airplanes, attacked one firing 300 rounds and forced it lower and lower to escape his attacks. He had to quit, however, as another enemy joined the dogfight, and all he got was the vision of the enemies crossing back their lines.

On 6 June the 91ª Squadriglia was shifted from the Isonzo to the Trentino front, where the Italians were about to launch an offensive. Sabelli went with Piccio, Baracca, Ruffo, Ranza, Di Rudinì, Costantini, Olivari and *serg.* Gorini. It was a good band of brothers for a pilot who one and a half years before had written to his friend: "I think just like you, my dear Abba; nothing is worth like a bit of glory and nothing I wished more than a beautiful death facing the enemy, for a just cause..."

On the new front he flew above the peaks around the Asiago plateau and the beautiful Valsugana. The fighters of the 91ª flew Caproni escorts and standing patrols behind the Nieuport units, ready to intercept some fast reconnaissance airplane that had penetrated the ring of the slower fighters. On 19 June finally he had another combat, again during a bomber escort, together with three other pilots of the 91ª Squadriglia. Over the scenic Lake of Levico in Valsugana he saw a Brandenburg C.1, clearly coming from the enemy airfield of Pergine. He attacked it from behind, followed it down to 2,700 meters and then he lost it. The enemy two-seater defended itself well, placing bullets in his engine cowling and left

Right: Giovanni Sabelli checking on the Nieuport. This French Ni.11 has a dark finish and a polished metal cowling but it shows Italian colors below its wings.

Below: Sabelli, right, getting ready for take-off on a Nieuport 10 in February 1916.

The Spad of Sabelli with the insignia of the black ladder.

wing, damaging two ribs. The airplane that Sabelli attacked maybe was one of the two Brandenburg C.1s that on that day had to force land at Pergine with serious damages, but with the crews unhurt.

*Capitano* Baracca noticed the qualities of Sabelli, who on 23 June had the honor of a permanent attachment to the 91ª Squadriglia. On 26 June Sabelli had a combat with a fighter over Mount Pasubio, but the enemy got away in a spiraling dive. On 2 July, as the offensive was over, the unit was ordered back to Santa Caterina, and two days later it returned to its base. Sabelli was back to the usual activity of Caproni escort and offensive patrols, but while the other pilots of his unit kept collecting victories, he had only two combats, on 28 and 30 July. In that period he also began to fly photo reconnaissance missions with a camera-equipped Spad. He flew to Idria, where there was an electric powerplant, shooting 14 photos from an altitude of 3,200 meters. On 2 and 6 August he went out to take photographs of an enemy airfield near the Lake of Seebach.

The month of August 1917, while a major Italian offensive was in preparation, brought many new combats for Sabelli. Both on 7 and 8 August he returned to base with his airplane damaged by enemy fire, but he had his revenge on the 10th. Flying along the lines late in the evening he saw an enemy formation and engaged a Brandenburg that carried long ribbons attached to its struts. He attacked from an altitude of 2,200 meters laterally from behind, shooting 75 rounds. The plane went down trailing a long plume of smoke, but he couldn't follow its fall. Ground observers declared that it had force landed behind enemy lines in the sector of Mount Stol. Once again his aircraft had been hit, but Sabelli now had his first confirmed victory.

As the Italian offensive developed, he was busy providing support to the infantry. Each day the 91ª Squadriglia flew to escort the bombing raids of the Capronis and then went down to strafe enemy trenches. In order to meet again the enemy in the air, Sabelli had to wait until 6 September. He was the wingman of his commander, *capitano* Baracca. At 6.30 p.m. they spotted at an altitude of 2,000 meters a Brandenburg over the Sabotino, dropping bombs. Sabelli was the first one to attack, and then both pilots attacked together and drew the enemy down, trailing a long column of white smoke. Baracca wrote that Sabelli was so rash in his attack that they often risked a collision, and it was to avoid crashing into each other that they lost their enemy, which force landed behind the Austrian lines. Still, their fire had been mortally accurate: from the aircraft, a Brandenburg of Flik 34, the Austrian rescuers took out the pilot who was seriously wounded, while the observer was dead in his combat position. Sabelli's aggressiveness in the attack much impressed Baracca, who wrote about it to his mother.

On 17 September, alone this time, he had two combats against the enemy. The first one was cut short by a gun jam, but fifteen minutes later Sabelli managed to place his bullets into an enemy two-seater, but again his gun jammed, so he had to quit the combat. The enemy airplane got away at full throttle, and ground observers saw it force landing in its own territory. Both Austrian crew members were badly wounded, and they were credited with the destruction of the Spad that had attacked them, together with a fighter pilot of Flik 41/J, *Fw.* Radames Iskra. Probably Sabelli's quick departure due to the gun jam made the enemies believe that he had been hit, but he returned unharmed to base, with his third confirmed victory.

Sabelli scored once more on 23 September together with *ten.* Ranza who described this way the dramatic air combat:

> "In a patrol flight on the front of the third Army at 9.35 a.m. I saw two airplanes coming from Aidussina toward the Vallone, at an altitude of about 4200 meters. I was above Costagnevica, with Ten. Sabelli (SPAD with a black ladder) when a reconnaissance airplane passed below us. It was an Albatros, with a fighter following it far away.
>
> While Ten. Sabelli attacked the Albatros from behind; the other enemy plane attacked Ten. Sabelli from behind, and then

A gathering of fighters at S. Caterina airfield, about mid 1917. On the right there are the Spads of the 91ª Squadriglia, including the ones with a crescent (Parvis) pennant (Costantini) and ladder (Sabelli). On the left the Nieuport 11s of the 82ª Squadriglia, and behind Ni.11s and Ni.17s of the 70ª and 82ª Squadriglie.

server, was
was written
were not rep
was shot do
self, after th
Moreover, n
that day. Lef
ably faced t
Spads were
*Vzfw.* Münni

His loss
body apprec

**Giovanni**
**5 confirm**

| Claim | Vi |
|---|---|
| - | - |
| *Korp. Stefan V* | |
| *Stfw. Richard* | |
| 1 | 1 |
| *[with serg. All* | |
| 2 | 2 |
| *[with cap. Bar* | |
| 3 | 3 |
| *Zugsf. Rudolf* | |
| 4 | 4 |
| *[with ten. Ran* | |
| 5 | 5 |
| *[with ten. Par* | |

haunted him all though his life. "The memory of that adventure never left me. Sure, it was war, but the look of certain dead men stays in the eyes, and return every night: even now," is how he recalled that event in an interview a few years before his death.

Scaroni scored his fifth victory and became an ace on 10 December when, together with Masiero he shot down a two-seater that after a hard combat caught fire and shed its wings, crashing near Noventa di Piave. He received a second Medaglia d'Argento al Valor Militare that in its citation reported his victories as on 14, 17 and 18 November 1917.

After his sixth victory near Conegliano on 19 December Christmas came, and the next day there was the big combat that was called the Battle of Istrana, of which Scaroni was the main protagonist, shooting down three planes. On Christmas day Capt. Barker, of Squadron 28, RFC, attacked on his own initiative a German airfield strafing barracks and wounding and killing ground and flying personnel. That action, that Scaroni described as *"one of his usual eccentric pranks"* had offended the Germans, maybe rightly so, as there was a sort of tacit truce on Christmas day. So they launched a reprisal attacking the airfield of Trevignano, called after Istrana by the Italians, believing it the base of British squadrons. Unfortunately for them, also many Italian Squadriglie, all equipped with Hanriots were there, $70^a$, $76^a$, $78^a$, $81^a$ and $82^a$. Scaroni described the events in a letter to his family:

> "About 9 yesterday morning the spotters of the field gave the alarm, as two large groups of enemy aircraft were coming toward the airport from the Montello. I watched that way with binoculars and I saw the aircraft in close formation at a level of about 3,000 meters, and higher another large group of small fighters, for a total of about forty planes. I dropped the binoculars, jumped in my fighter, had the propeller turned and took off. I had not yet completed a circuit of the airfield and I was at a level of about 500 meters when I saw three enemy planes zooming below me, while bombs of all sizes were exploding around the hangars. I was surprised, as I didn't expect them so soon: with a single dive they had zoomed against the hangars to exploit the surprise effect. I didn't waste time thinking and didn't worry about other aircraft; I attacked the closer one, which was engaged strafing the airplanes of my squadron lined on the field. With two short, well-aimed bursts I brought it down just at the border of the airfield. It crashed and caught fire.
>
> Then another large aircraft passed above me 2 o 300 meters higher. I gave full throttle and I quickly got at it. Keeping a bit lower and behind to remain hidden I started firing, aiming more or less at the cockpit section. At my first rounds the pilot tried to maneuver and get free of my attack so that the observer could use his machine gun against me, but he couldn't do it so he went in a dive and I was above him. Then another Hanriot came next to mine, and from the insignia I saw it was Michetti. The enemy was trying to regain his lines but it was a long trip and we were determined not to let him go. Michetti and I yoyed attacking him with short bursts. The observer thanks to the excellent maneuvering of the pilot fought us back with admirable ability and tenacity making our attacks difficult and ineffective. We had harassed him down to 50 meters above the ground and its maneuvers were getting slower and more difficult, while we were sure of victory and got more and more aggressive. The observer stopped firing and I understood why: I saw him quickly removing a new cartridge belt tied around his waist to replace the empty belt on his gun. I exploited that moment and from close range I fired one last burst. This time the pilot went down, hoping to escape with a good landing, but when it touched the ground the airplane turned over. Now we saw it, all white, with its huge crosses, defeated. As we flew around it we saw the passenger crouching and getting out, then with a special explosive device he torched the wreck without worrying about the pilot who died miserably in the fire. The explosion of the fuel tanks splashed with burning fuel also the passenger, who started thrashing and rolling in the ground. Some artillery men reached him and helped him removing his burning clothes, then they captured him."

Scaroni at the controls of his Hanriot. This picture was autographed by the ace himself on 30 May 1970, and dedicated to aviation historian Franz Selinger, brother of *Oblt.* August Selinger, of Flik 42/J, shot down and killed on 28 August 1918. *(Photo Selinger).*

Despite the failure of their first raid the Germans attacked again at 12h30 and Scaroni scored a third victory.

> "I was cruising above the Grappa when I noticed intense anti-aircraft fire in the direction of the Montello. I went that way and I found a large group of enemy airplanes dueling with our fighters. As I was getting ready for combat I saw a large twin-engine bomber coming toward me with a fighter on its tail shooting at close distance. From the insignia I recognized Tenente Brenta. I got on his side and we both fired a long burst at the enemy at the same time. The monster went down on a wing, fell for hundreds of meters, broke apart and crashed, catching fire. When I turned back, there were no more planes, either friend or foe, so I returned to the airfield."

Scaroni shared two victories with *sergente* (later *aspirante*) Raimondo Di Loreto, on 11 February and on 7 July 1918. Di Loreto flew Hanriots with checkers markings, like this 6206. *(Photo Selinger).*

The balance sheet of the day was positive for the Italians, who lost no fighter in combat, despite three German claims. Damages at Istrana were minor, a few Hanriots were wrecked or damaged and six men were killed and more wounded among the ground personnel. When the remains of the aircraft were picked up they were found to be all Germans and Scaroni, who didn't have a high opinion of Austro-Hungarian crews, commented: "This explained their aggressiveness and their daring." He ended the day quarreling with the furious German officer of his second victory, who denied having had Italians as adversaries and claimed that it was the British who had shot him down. Scaroni and Michetti thought this was Prussian arrogance, but besides the fact that many pilots took part in the battles, some of them British, the fact was that it was not easy for the enemy to tell apart a Sopwith Camel from an Hanriot, a new and little known model. All through the war Italian Hanriots were unfailingly identified as Sopwiths in the enemy battle reports. Concerning losses, the Italian War Bulletin reports eight planes shot down in the first raid and three in the second one, eight of them within Italian lines.

Beyond the accuracy of those figures, it was a brilliant victory that marked the turning point in the fight for air supremacy that returned firmly in the grasp of the Italians and their allies. The apprenticeship of Scaroni as a fighter was brilliant, having scored nine victories in just two months.

At the beginning of 1918 bad weather curtailed all flying activities, but by mid January things changed. On 12 January together with Michetti and *serg.* Razzi Scaroni attacked and shot down over Mount Cismon an enemy two-seater that was seen going down, but was lost in the mist of the valley. This victory was not assigned to him after the war, but the enemy airplane, an Aviatik C.1 of Flik 16/D indeed had crashed at Arsié breaking apart, its crew miraculously unharmed.

Two days later, 14 January, Scaroni flew on offensive patrol over Mount Asolone with two companions, but soon he was alone, as one returned for engine trouble and the other one got lost. At 11 a.m. he noticed an enemy airplane coming from the Brenta Valley that first lingered over the lines and then headed to Bassano. He let it go, then attacked. The enemy tried to escape but Scaroni got closer, firing short bursts, until he noticed tracer bullets going through his wings. He had been bounced; it was an unexpected KD fighter. They had a dogfight for ten minutes, until the enemy tried to break away to his lines but Scaroni used the opportunity to get in a long burst. The fighter went into a spin trying to glide to its lines with its propeller stopped. Scaroni reached it and gestured to turn toward the Italian side, the enemy flew some spirals pretending surrender, and then again he tried to turn to his lines, and after a further burst from Scaroni's gun he crash-landed in the snow in no man's land, turning over the plane.

The pilot hid in a shell hole, and then an Italian patrol captured him, keeping at bay an enemy patrol as Scaroni flew overhead. In the night the Italians tied ropes to the fighter and pulled it in, while the Austrians had also used ropes and were pulling in the opposite directions, the struggle ending with a brief skirmish. Scaroni met a smiling enemy pilot, Karl Cislaghi, from Vienna, and had a pleasant talk with him, and an exchange of gifts. The Austrian, who claimed four Italian and one British plane shot down, wore his decorations, and gave him his pilot's badge, while Scaroni gave him his signed portrait. Strangely a ground witness, Colonel Biancardi of the Brigata Regina, described Scaroni's plane as "a fighter with black triangle tail".

After this combat, bad weather grounded the airplanes for ten days. Scaroni had asked for a leave, but the first Italian counter-offensive after Caporetto was in preparation on the Grappa, and the 76ª Squadriglia was tasked with protecting that sector, the keystone of Italian defense, so he had to postpone. On 25 January weather improved and Scaroni in the morning flew a photoreconnaissance to the airfields of Arsié and Feltre. After a second test flight, he took off in the afternoon for an offensive patrol over the Asiago plateau, leading a patrol. The combat that followed was to be the hardest and the most celebrated of his career:

> "..In the afternoon of 25 January, as I was cruising over the Asiago plateau with Capitano Sequi and sergenti Razzi and Di Loreto, we met near Gallio a flight of three enemy fighters. I waited, hoping that they would turn to our lines, but, as they were undecided I went toward them with my wingmen. Their obvious intention was to draw us into their territory but giving full throttle we reached them, I immediately engaged the patrol leader and my companions took care of the other two. From the beginning I realized he was a pilot of great gallantry and exceptional ability. Whatever I did using the agility of my plane, I could never aim and fire effectively, he always cheated me, going through all the range of aerobatics in a perfect style and then trying to attack me. By maneuvering and keeping almost always a slight ascendancy I had managed to bring him down to 1,500 meters, but I had strayed away from the lines. We were now over the Valsugana, not far from Grigno, and I didn't see my companions, probably still dueling the other fighters. We were down to 100 meters level, one close against the other, when following one of my bursts he turned sharply, passed below me and although I followed him I lost him. It was getting dark and the dark background of the valley precluded my search for him, also on account of the fact that his fighter was painted all black. I was far away from the lines and it was late, so I climbed to return to base. The huge mountains surrounding me made me shiver, thinking of a possible engine failure… as God wished, I managed to get in the Brenta Valley, to Cismon where I got my usual ration of anti-aircraft shells from Col del Chior and finally I reached the airfield, where my companions who had landed before me were worried and waiting for me."

Later Scaroni added that the black enemy fighter was marked with a big white K. From the declarations of a prisoner and the confirmation after the war of an Austrian squadron leader his gallant adversary was identified as the great Hungarian ace Josef Kiss, of Flik 55/J, based at Pergine. It was believed that Scaroni had wounded him in this combat, but we now know that Kiss was wounded on 27 January, not on the 25th, and our data confirm that it was two other Italian aces, *capitano* Riva and *sergente* Fornagiari, of the 78ª Squadriglia, who wounded the great enemy pilot.

On 28 January, taking off on alarm, he reached the Montello and saw an enemy two-seater and an Hanriot attacking it. He recognized his friend Fucini, joined him in the attack and shot down the enemy near Biadene, within the Italian lines. The two German airmen of FA (A) 219 were killed.

On 1 February Scaroni claimed a two-seater shot down at Col S. Martino and ten days later two more: cruising with three squadron mates, Razzi, Sequi and Di Loreto, he met a two-seater with an escort of five fighters. They let them cross the lines, then they attacked and before the enemies could react Scaroni attacked one from the side, saw it rolling and going down. Thinking it was a diversionary maneuver, Scaroni fired again at close range, then he turned to the other ones, but only one enemy had remained. He reached it as it was dogfighting with the other Hanriots and as soon as it got below him, he fired and saw it going down in a steep dive, shedding a wing and crashing near Mount Fontanasecca, behind the enemy lines on the Grappa Massif. The pilot, *Offstv.* Alfons Schuller of Flik 8/D was badly wounded and later died in a hospital. Ground observers confirmed also the other victory, but it was not assigned to him. Actually the Austrians suffered another casualty on that day and in that area, but they attribute it to ground fire.

Those planes did not belong to fighter squadrons: in the Austrian air force, until mid 1918 also reconnaissance units had on charge a

Silvio Scaroni at Casoni airfield, with his armorer. Hanriot Hd.1 7517 shows his personal markings, a white, black borderd square, and its armament of two Vickers machine guns.

few fighters tasked with providing direct escort to the two-seaters, but this doesn't seem to have been a wise choice. Without specific fighter training, those pilots were easy prey for Italian and British fighters, as in this case. With these two victories Scaroni had reached 15 claims in three months, and he was now part of the cream of Italian aces. Only Baracca dominated with 30 victories, but the other ones, Piccio, Ruffo and Ranza at that time had only one or two more than he did. His name was leading in the second competition of "hunters of the sky" sponsored by the Pirelli company, in front of Cerutti and Reali, but the final results of the competition were made public only in the summer, when he was hospitalized. He also received a second Medaglia d'Argento, for his victories in November 1917.

By day now the Italians had supremacy in the air, but by night the Germans kept on bombing the airports, often with very effective results. The snow facilitated the attacks of the Austrians and Germans, making airport structures more visible, and also the 76ª Squadriglia suffered from their attacks. In the night of 31 January/1 February 1918 two bombs fell close to the hangar of the squadron damaging seven of its fighters, and grounding half of the unit. It was necessary on 2 February to move from Istrana, by now too much exposed, to Isola di Carturo, and then, on 17 February, to Casoni, that was to remain its base throughout the war.

On 18 February Scaroni was flying alone over the Grappa when he saw a two-seater passing below him headed for Possagno. He attacked immediately and saw that the observer, concentrated on his mission, had not noticed his fighter. After firing from just 50 meters he saw the enemy aircraft standing up on its nose, banking and then falling down in a spin. Scaroni went to the location of its crash and saw the final honors to the two dead airmen. Strangely the identity of these two victims, killed behind Italian lines, is unknown to this day, as Austrian data don't report losses corresponding to this victory. Then for almost one month the ace didn't meet any more enemy airplane, as if there were no flying activity by day. By night, however, raids went on, and the affected the nervous system of the pilots, who were forced every night to sleep in dugouts to avoid the bombs falling over Casoni from sunset to daybreak. The situation seemed hopeless, so the order came to try a night interception, which was assigned to the best pilot of the squadron, Scaroni himself. In the night of 17 February he took off hoping to intercept an enemy bomber. And intercept he did: he saw the shape of an airplane in the night sky, got close to it, fired his tracing bullets, attacked twice and saw it going down steeply and landing on the airfield. Scaroni landed, approached cautiously the enemy plane and the pilot standing besides it, but to his surprise he realized that his prey was another Hanriot, and its pilots was *capitano* Zoboli, the commander of the 81ª Squadriglia, the companion unit to the 76ª! Before departure they had arranged for no other airplane to take off, to avoid mistakes, but *cap.* Zoboli had insisted on taking off anyway, explaining to the worried Michetti: *"Scaroni is not going to see me!"*. Luckily in the night his aim was not as deadly as by day, and the captain was unscathed. The real enemy could keep on bombing Casoni with impunity, and in the night between 26 and 27

*Sergente* Raimondo Di Loreto, who shared with Scaroni the destruction of Phönix D.I 128.19 of *Zugsf.* Josef Laczko of Flik 14/J. Despite the total destruction of his fighter, the Hungarian pilot survived unharmed.

February they destroyed many buildings and wrecked ten Hanriots of the 76ª.

An apparently weird fact is that a few pilots scored so many successes, while most fighter pilots didn't accomplish much. Scaroni considered himself lucky, but this is only partially true. Actually he exploited at best his few opportunities, and he had a professional approach to his job. First of all, he flew much more than what his tasks required, he practiced gun aiming all the time with the reconnaissance planes that he escorted; he kept his gun properly maintained and he personally tested every single bullet in the belt. After two or three landings he returned to the firing range to check the alignment of his gun, spending much time but ensuring good results. He could trust on the friendship and collaboration of two excellent specialist, chief gunsmith "Bigio" and mechanic "Toni", who kept his fighter in top efficiency. Bigio in the dialect of their native Brescia explained to him all the *"secrets"* of the Vickers gun and with Scaroni he designed an aiming sight without lenses, simple and effective, because the standard telescope gunsight used to fog after a prolonged dive. The two men supported by squadron commander *cap.* De Bernardi tested the installation of a second gun on his fighter. We don't know when this exactly happened, photos show a second Vickers since June 1918 also on many other fighters of the VI Gruppo, particularly the ones of Lega, Baracchini and Fucini.

He was methodical in his pursuit of the enemy, reading all the notices of ground observers in order to understand where did the Austrian planes usually cross the lines and personally asking for information from people at the front. Thanks to a special informer, his brother Annibale, he finally had the opportunity of one more victory. His brother told him that every morning an enemy two-seater came very early over his position and then turned back. Scaroni was bent on catching it, and every day he took off before sunrise together with Michetti, who didn't share his eagerness.

Scaroni, almost unconscious, fell down with his fighter from 4800 to 800 meters, and then he banged his head against the central strut, partially regained consciousness and managed to straighten the Hanriot. He saw the Piave, and got ready to cross the lines near Mount Tomba, but then the engine gave up and exploded, throwing away its cowling. Scaroni turned off the contacts and glided toward a stream flowing down from Mount Grappa, where he attempted a forced landing near Villa Pasini. He touched down properly, but then the landing gear struck a hole in the ground and the Hanriot turned over, throwing out the pilot, who fainted again.

*Caporalmaggiore* Domenico Bertello and soldiers of the 143ª Rgt. infantry came to the rescue of the unconscious pilot and rushed him to the 10th Field Hospital at Oné di Fonte. The doctor *Tenente* Loiacono first attended him and realized how serious his condition was. A bullet had hit him in the buttocks and had splintered against the spine. Pieces of his flying clothing were embedded in his flesh, causing danger of infection. The spine itself, luckily, was not damaged but it was a serious wound nonetheless, and Scaroni was out of action for the rest of the conflict.

He regained consciousness in the evening and the first person he saw was his friend Michetti. The next day he was shown the Daily War Bulletin that reported: "...Despite bad weather conditions air activity was quite intense. Ten enemy aircraft were shot down. Tenente Silvio Scaroni scored his thirtieth air victory."

The final air combat of Scaroni had two other protagonists, Lieutenants Rice Oxley and Howell of RAF No. 45 Squadron who, in typical British style, claimed no less than seven Austrian fighters shot down, two for Rice Oxley and five for Howell. Other Italian pilots, whom Scaroni didn't see, also put forth their claims: *ten.* Resch of the 70ª Squadriglia, and *ten.* Keller with *serg.* Magistini of the 91ª Squadriglia. The two pilots of the 91ª were credited with one victory shared with Scaroni. In their combat report the two pilots of the Baracca Squadron wrote that they witnessed an Italian fighter shot down during that hectic engagement. The British pilots also confirmed the two victory claims by Scaroni, so the ace was officially credited with them. The war Bulletin reported ten victories, but as there was no other combat on that day, one can surmise that they all belong to that single combat. Austrian data, however, report only one single fighter of Flik 30/J badly damaged in combat over Mount Tomba, while their pilots claimed, and were credited with three victories, two *"Sopwith"* and one two-seater, while the Hanriot of Scaroni is the only known Allied loss.

This event was one of the most blatant examples of how difficult it is to reconstruct past events determining the truth in a maze of overclaims. Even the direct reports of protagonists, with their apparent accuracy, must be taken with a grain of salt: in the excitement of combat, almost every participant had a tendency to exaggerate and to claim personally any enemy airplane seen going into a dive. Beyond this single event, however, the victory claims of Scaroni almost always found confirmation in enemy documents, giving proof that this Italian ace was particularly honest and reliable.

The following months brought his slow recovery, but when he was finally ready for action, the war was over. In 1919, when the official list of Italian aces was issued, Scaroni took second place, after Baracca himself, with 26 confirmed victories out of 30 claims, while Baracchini had only 21 victories credited to him.

Right after the war Scaroni moved to Rome together with his good friend Michetti, who had lots of connections in the capital, through his father, the famous painter. They were two famous and glorious war heroes, so they were readily accepted in the Roman high society. They became constant guests of the exclusive parties of *donna* Lilia Patamia Curioni, companion of Rep. Pasquale Masciantonio, under-secretary to the Treasury, and there they met all the VIPs of postwar Rome, such as the writer Matilde Serao and even Guglielmo Marconi. For a short time Scaroni and Michetti were busy canvassing for Masciantonio and the presence of the two young war heroes was useful, as he was re-elected. Scaroni however came from humble origins and he never felt at ease in that environment, so he was probably happy to return to active service and to join the aeronautical mission to Argentina, that took many famous pilots, such as the reconnaissance ace Locatelli, to sponsor the sale of Italian airplanes in Latin America.

The mission sailed from Genoa on 25 February 1919 bringing many airplanes, most of them Ansaldo SVA and A.1 Balilla. The local population, that included many of Italian ancestry, welcomed them with incredible warmth. Based at Palomar, the Italians flew many propaganda raids all over that nation. Scaroni and Locatelli with two Balillas flew a raid from Buenos Aires to Rosario, Scaroni alone flew from Buenos Aires to Cordoba, always to an enthusiastic welcome. The most daring feat was the one of the expert Locatelli, who flew from Buenos Aires, close to the Atlantic Ocean, to Santiago, Chile, near the Pacific, flying above the Andes, close to the Aconcagua mountain, that peaked above 7,000 meters. A huge crowd welcomed Locatelli when he returned to Buenos Aires.

Meanwhile the third Silver Medal of Scaroni, awarded on 12 June 1919 had been changed one month later into a Gold Medal, the Medaglia d'Oro al Valor Militare, the highest Italian decoration, mostly assigned posthumously. This was its citation:

> "Fighter pilot, master of valor for the valiant, thirty times winner in splendid air duels, he matched his incomparable ability with equal valor, giving to his heroism only one goal: victory. To those who worked more and gave more, he was always an example to work and give even more. He gave to his fatherland all the energies of his bold youth, and his last victory over the enemy was even more brilliant because it was stained with his own generous blood. Sky of the Piave and Brenta, 5 December 1917 - 12 July 1918."

Scaroni returned to Italy in late 1919 and he was placed on leave on 3 January 1920. In 1922 he published a book called *"Impressioni e ricordi di guerra aerea"* (Impressions and memories of air warfare): his memories were exposed in a simple way, using mainly the letters written to his family. The book is reliable and pleasant, and most revealing for its sincerity and simplicity, without any of the bombastic jingoism of other war books. In 1923, when the Regia Aeronautica was formed as an independent arm,

## Mario Stoppani

Mario Stoppani was one of the élite of really great Italian pilots: test pilots and record-breakers who marked the history of aviation not just in Italy, but in all the world: men like Mario de Bernardi, Arturo Ferrarin, Francesco De Pinedo, Francesco Brach Papa. He was also a World War I fighter ace, and a remarkable one, having claimed seven victories (six of them confirmed to him by the Italian command) in just five months in 1916, when air duels were rare.

Mario Stoppani was born on 24 May 1895 in Lovere, a small town in the province of Bergamo, at the feet of the Alps, an area renowned for its hard working men, who in those years used to emigrate in all northern Italy and abroad, working as masons. It was an area that also produced a large quantity of remarkable aviators.

Gifted for mechanical works, as a teen-ager he worked in a mechanics shop, until, at the age of eighteen, he joined the Battaglione Aviatori as a volunteer. Initially he flew Bleriots at Mirafiori, then he attended the first course of the flying school at San Giusto, Pisa, where he was taught to fly on Bleriots, in a small group of men that included plenty of future aces: Ruffo di Calabria, Scaroni, Michetti. He got his first license on 11 June 1915, then he moved on to Somma Lombardo, got his second license on 15 July and, as Italy had entered the war, on 24 May 1915, Stoppani was quickly assigned, like most of the Pisa graduates, to the new Squadriglie per l'Artiglieria (artillery squadrons).

On 25 July 1915 as a simple private, *soldato*, he was assigned to the 2ª Squadriglia, based at Medeuzza and commanded by *cap.* Mario Santi, and then on 7 October he passed at the 3ª Squadriglia, commander *cap.* Giulio Maurel, also based at Medeuzza. These units were very efficient in artillery spotting, thanks to an effective cooperation established with the artillery, but they were severely hampered by the dangerous flying qualities of their mount, the Macchi Parasol, or PM, an airplane that was highly unstable, and that was quickly replaced by the much more liked Caudron G.3, in November 1915.

Stoppani flew 76 sorties from 26 July to 19 November 1915, using several Parasols, although his first combat mission was unauspicious, as he turned over landing with PM 397. On 26 August the pellets from an enemy shell damaged his PM and bruised him. The dangerous qualities of the Parasol are illustrated by the motivation of the Silver Medal that was assigned to Stoppani: "Despite several flying accidents, he kept on flying a type of airplane of difficult handling, disregarding the bad weather conditions and the lively enemy fire, that once hit his airplane seven times, and himself too, although without wounding him." It is remarkable that Stoppani, just a soldier, got a "Silver" while, for instance, *cap.* Piccio, a career officer, for a comparable service that included some daring bombing raids got the humbler "Bronze", and it shows that Stoppani enjoyed a really high reputation.

After this first period at the front, when the PMs were withdrawn from use, on 19 November Stoppani was assigned to the flying school at Malpensa as an instructor, a common fate for particularly good pilots. Very soon, however, he asked to be sent back to an operational squadron. It was the 76ª Squadriglia, formed on 25 May 1916 at La Comina airport, equipped with Nieuport 11s serialled 1042, 1043, 1805 and 1806. Commanded by *cap.* Ettore De Carolis, it had as pilots *sottotenente* Giuseppe Rigoni, *sergente* Mario Stoppani and the volunteer soldiers Giovanni Menegoni and Otello Venchiaruti.

Mario Stoppani

On 29 May, the squadron moved to the airfield of S. Maria la Longa, near the fortress-city of Palmanova, passing under the command of the I Gruppo, assigned to the southern 3ª Armata. Its first contact with the enemy happened on the first day of June, during an escort flight to two Voisins. Stoppani spotted a flying boat south of the gulf of Sistiana and attacked it, but had to desist when his gun jammed, after having fired seven rounds. However, he had tasted combat, and he quickly proved that he was good at it. On 8 June he had a combat with an "Aviatik" but he could fire only 2 bullets, the next day he threatened a kite balloon that was hoisted down, on 14 June he had another combat with an "Aviatik", fired a whole drum magazine from his gun and saw his adversary landing, and he had another combat near the end of the month. Overall in June 1916 Stoppani flew no less than 45 combat missions.

Stoppani, now with the rank of *sergente*, got his first victory on 9 July 1916. Over Salcano, a suburb of the contested town of Gorizia, he forced down an enemy airplane, together with other pilots of his squadron, *s.ten.* Rigoni and *sold.* Venchiaruti.

On 18 July Italian observers spotted a formation of seven Austrian airplanes approaching. Four Nieuports took off to intercept them, and formed a standing patrol cruise between Gorizia and

Monfalcone. Stoppani discovered an enemy aircraft and attacked it, emptying a whole drum of his Lewis gun. While he was recharging the gun, he lost sight of the enemy. He had not failed, however: the aircraft was a Lloyd C.III of Flik 4, engaged in a reconnaissance mission over Italian gun batteries, which force-landed with an unharmed crew. Stoppani had two more combats on 26 and 27 July, in the second of which he forced the enemy to land, according to his logbook.

On 28 July four Nieuports of the squadron took off in an escort flight for Voisins of the 25ª and 26ª Squadriglia going on a bombing mission. Austrian fighters attacked them, and in the ensuing combat *ten.* Giuseppe Grassa of the 76ª Squadriglia shot down an airplane described as an *"Albatros with machine gun"*. Actually it was a Fokker monoplane, 03.51 of *StFw.* Stefan Huzjan, who had just riddled with bullets the Voisins of two future aces, Mario Fucini of the 25ª and Guido Masiero of the 26ª, and then, hit by the Nieuport, force-landed unhurt. Stoppani in this event failed to engage the enemy. In July he had flown 26 combat missions, plus a very large number of test flights.

On 6 August, while escorting Caproni bombers that attacked the Opicina railway station, Stoppani saw some Austrian seaplanes that were going to attack Italian warships in the High Adriatic Sea. They were flying boats L 114, L 126 and L 98. He attacked them, forcing them to retreat to Muggia, and then returned to his escort mission. Meanwhile, unfortunately, the Austrian naval ace Gottfried Banfield had shot down the Caproni serial 1134 of the 4ª Squadriglia.

In August 1916, the Italians launched an offensive that allowed them to conquer Gorizia, and air combats were more frequent. Stoppani had two more combats on the 13th, near Ranziano, behind Gorizia. He fired two magazines and logged that one of the enemies flew out of control. This event, however, is reported as a full victory in the "official list". He had another similar combat three days later, but in this case his engine failed and Stoppani force landed at Medeuzza, his old airfield when flying the dreaded Parasol.

Pilots of the 76ª Squadriglia: Luigi Olivi, unknown and Mario Stoppani

In August he flew 25 combat missions, and in September 17 more, including several escorts to Caproni bombers, often he pursued enemy aircraft, but the Nieuport was just too slow to engage, unless in favorable conditions.

In October the 76ª Squadriglia was assigned to the II Group, supporting the 2ª Armata, facing the Austrians along the Isonzo River but it still remained based at S. Maria la Longa. From 9 to 12 October 1916, the 8th Battle of the Isonzo took place. The Italian Army conquered the irrelevant Sober hill, for the loss of 23,000 casualties. On 11 October Stoppani shared with *ten.* Luigi Olivi, of

Nieuport 11s of the 76ª Squadriglia in 1916. Visible serials are 1630, 16?3 and 1635.

his 76ª Squadriglia, one more victory, as they forced down Brandenburg C.1 61.72 of Flik 19, wounding its gunner *Korp.* Gustav Wieser. On the 18th he had three separate combats in two different missions, shooting three magazines. For some reasons, while other pilots were plagued by gun jams, the Lewis gun of Stoppani never seems to have failed him.

At the end of the month, the 9th Battle was launched, with another futile attack that led to the capture of the Faiti hill and 9,000 prisoners, against high losses. On 27 October Stoppani had another combat, and during the month he also twice tested rockets, surely the Le Prieur French devices, in 10' flights. On 31 October a swarm of Italian fighters again attacked a Hansa Brandenburg C.1 of the valiant Flik 12 flying as escort to two reconnaissance aircraft, which had a series of combats with them before its destruction. Its pilot, *Oblt.* Franz Cik, was killed and the airplane crashed, while the observer, *Zugsf.* Gustav Resch, was wounded and Italian soldiers brought him to a military hospital (unusually, this crew had an officer pilot and an NCO observer). The winners were Stoppani, together with *serg.* Giuseppe Tesei of the 77ª and *serg.* Anselmo Caselli and Arrigo Rossi of the 70ª Squadriglia.

As the winter stopped ground operations, air operations continued. Stoppani had a combat on 1 November, escorting Capronis, and another on the 25th, and this time, unusually, his gun jammed after firing ten rounds.

An earnest and deadly fighter pilot, Stoppani was also a healthy young man. The priest of the local parish had a habit of warning girls against the dangerous airmen, particularly "an envoy of the Devil, sergente Stoppani" who lured one of them to the sins of the flesh. Stoppani tried to clear his reputation with that priest but the self-righteous man of the church warned him that he had to stop

Right: Sergente Mario Stoppani during his service with the 76ª Squadriglia.

sinning "particularly considering that he was likely to die suddenly any moment". Having knocked on wood, Stoppani asked the priest to stop it, or else he would tear off the lightning rod of his church. The priest went on with his firebrand sermons and Stoppani retaliated, ripping off the lightning rod from the steeple with the landing gear of his fighter.

A final victory claim of Stoppani is dated 1 December 1916 over San Marco Hill. Actually he only logged an enemy airplane flying irregularly after a combat. He had three more combats on 8, 11 and 26 December.

In early 1917, Stoppani first flew with the new French Nieuport 17 that was being delivered to the 76ª Squadriglia, and on the 11th he had his last combat, attacking two enemy aircraft and driving them away. Overall with the 76ª Squadriglia he flew 280 flights, 163 of them combat missions in just nine months.

On 14 February he left the Front, as he was shifted to the Direzione Tecnica dell'Aviazione Militare, DTAM, as a test pilot. Ing. Brezzi, designer of the Ansaldo company, had contacted him since November 1916, offering him to become test pilot of the new fast fighter that was under construction. The huge Ansaldo industrial organization at Genoa had just entered aircraft production, building the Sopwith Baby floatplane, and was assigned the task of mass producing the SV, the revolutionary new design of Umberto Savoia and Rodolfo Verduzio, designers of the DTAM. Later called the SVA, it became one of the best airplanes of the War, used as a fighter, reconnaissance and bomber, both as a single- and two-seater, getting world renown for the flight to Vienna of Gabriele d'Annunzio and remaining in service until the early 1930s.

A second Silver Medal was assigned to Stoppani in 1917:

> "Fighter pilot of great courage, eagerness and value, in the short period of six months he flew many war missions, always putting to flight enemy aircraft that he pursued often at low level behind their lines. In brilliant air combats, he shot down two enemy Albatros: Biglia, 11 October 1916; Nad Logem, 31 October 1916."

As can be seen, only two victories were then confirmed to Stoppani, only one in the famed letter of the 5ª Sezione in January 1919 inquiring to establish the final and confirmed list of aces. Yet, later, Stoppani was correctly admitted to the roster of Aces.

Mario Stoppani first flew the SVA in late February 1917, at Grosseto airfield, and then tested all the production SVAs at Borzoli airfield, near Genoa. In the summer of 1917, he flew from Turin to Udine and back, 1,100 km in five hours, the first of countless fast flights by Ansaldo SVA airplanes for propaganda purposes. The Ansaldo Company publicized his record flight from Turin to Rome on 28 September 1917, with little concern for the secrecy of their new fast airplane.

He never returned to the front, but gave just the same an important contribution to Italy's victory. On 11 November 1918 he flew an A.1 Balilla fighter over Rome, giving an aerobatics display and dropping leaflets to celebrate the victory. On 12 May 1919 he flew a SVA from Turin to Barcelona in a single flight and in June 1919 he demonstrated the SVA in Spain, with other remarkable flights. He returned to Italy flying from Madrid to Rome, 1,900 km without stops, in less than 10 hours of flight. His mount was the Ansaldo A.5 that carried a message of King Alfonso XIII to King Victor Emmanuel III.

In September-October 1919, with designer ing. Brezzi as a passenger, Stoppani flew the Ansaldo A.300, a new reconnaissance two-seater, from Turin to the capitals of Europe, stopping along the way in Lyons, Paris, Brussels, Amsterdam, Dijon and back to Turin,

Mario Stoppani in the cockpit of Nieuport 11 serial 1651.

Right: A bad landing for this Nieuport 11, but whose fault? Mario Stoppani stands in the center, with white shirt.

Pilots of the 76ª Squadriglia: standing, second *cap.* Calori, sitting down *sold.* Venchiaruti and *serg.* Stoppani.

and then to Udine, Vienna, Krakow, Warsaw, Posen, Berlin, Prague, Vienna again. This trip was successful, and both the Air Force of Poland and the Red Air Force of Soviet Russia adopted the Ansaldo A.300. In 1920 he participated with success in the aviation competitions that went together with the Olympic Games in Antwerp, Belgium.

In 1922, after having left Ansaldo, Stoppani established his own flying school at Foggia, where American pilots used to be trained during the war. The company was called Società Garutti & Stoppani, and in the spring of 1923 it registered a bunch of war surplus airplanes: five Aviatik trainers, a Caudron G.3, two Hanriot Hd.1s (registered I-FEDE and I-FIFA, meaning Faith and Fear) and even a Fokker powered by a 110 hp Oberursel, probably a war-booty D.VIII. His initiative lasted less than three years, but no government training contract arrived, nor was there yet a public for such an enterprise. Stoppani had to quit, and in 1925 he became instructor at the SAI company of ing. Ambrosini, that ran a seaplane flying school at Passignano, on the Trasimeno Lake in central Italy that trained pilots for the Air Force.

In 1927, Stoppani was hired by SISA; the airline and flying school belonging to the Cosulich group that also owned the Cantiere Navale Triestino yard in Monfalcone. This was the beginning of a long cooperation that joined Stoppani's destiny with the one of Cant aircraft. In 1934 he became chief test pilot of the company, that was now called Cantieri Riuniti dell'Adriatico, CRDA, producer of Cant aircraft, and got to work with its gifted designer, Filippo Zappata.

On 18 October 1934 Stoppani, with second pilot *cap.* Corradino Corrado, flew a Cant Z.501 single engine flying boat, prototype I-AGIL, from Monfalcone to Massawa, Eritrea, a distance of 4,130 km that set the new world record of distance in a straight line for seaplanes. That record was beaten in June 1935 by the French, so on 16 July 1935 Stoppani, with second pilot *cap.* Casimiro Babbi, flew again the Cant Z.501 I-AGIL, this time with a new Piaggio variable pitch propeller, on a 4,929 km flight from Monfalcone to Berbera, Gulf of Aden, that recaptured the world record to Italy.

Mario Stoppani in the cockpit of the Ansaldo SVA prototype.

In 1936, Stoppani piloted the Cant Z.506 three engine floatplane, the most successful and long lasting of all of Zappata's designs, to capture ten world records: on 7 July he flew the prototype I-CANT (a most inappropriate registration) on a circuit setting eight world records for speed with loads of up to 2,000 kg over distances of 1,000 and 2,000 km. On 29 November he flew the same seaplane to set the height record with 2,000-kg load, reaching 7,831 m., and on 1 December he took a 5,000-kg load to a height of 6,727 m.

On 13 April 1937, using the large Cant Z.508, a three-engine parasol wing flying boat, Stoppani climbed to 4,863 meters carrying a payload of 10,000 kg.

On 27 May 1937 Stoppani flew the Cant Z.506 I-LERO setting ten further world records for speed over 1,000 and 2,000 km with 1,000-kg load, and also setting a closed circuit record of 5,200 km at an average speed of 308 km/hour. In November, he piloted on three occasions a military Cant Z.506B to the conquest of the world records of height with loads of 2,000, 5,000 and 500 and 1,000 kg, reaching respectively 8,951, 7,410 and 10,388 meters.

Many historians later described the search for world aviation records in the late Thirties as a futile example of the bombastic propaganda of the Fascist régime, but the fact remains that the Cant Z seaplanes were excellent aircraft, and Stoppani was an excellent pilot.

The record with the highest value for propaganda, absolute distance on a straight line, set by the Cant Z.501, meanwhile had been beaten twice, lastly by the French Latècoére "Lt. De V. Paris" with a 5,780 km flight. A Cant Z.506B, I-LAMA, received additional tanks for a Trans-Atlantic flight. The crew of Stoppani included second pilot *cap.* Enrico Comani, warrant officer Demetrio Jaria, wireless operator, and motorist Renato Pogliani. I-LAMA took off from Cadiz, Spain, on 28 December 1937 and reached Caravelas, Brazil, 400 miles north of Rio de Janeiro, 26 hours later, having flown 7,000 km. Stoppani exhibited the Cant Z.506B in South America, establishing the bases for a future Italian transatlantic commercial service, while an aerobatics team of Fiat CR.32 fighters was touring Argentina, Brazil and Uruguay. On 2 February 1938 he took off for the return flight, carrying *cap.* Mario Viola, of the CR.32 team, but an engine fire forced the Cant down on the sea. The crew launched a SOS, but an explosion of the wrecked seaplane killed them all. Stoppani alone survived, swimming in the sea covered by burning gasoline, and teeming with dangerous sharks, until a Dornier seaplane of Lufthansa that had heard the SOS saved him. For this record flight he was assigned the *Medaglia d'Oro al*

Ingegner Brezzi, designer with Ansaldo, and Mario Stoppani . Behind them an Ansaldo A.1 Balilla.

One of the Hanriots used by Stoppani in his Garutti & Stoppani private flying training company at Foggia carried the registration I-FIFA.

Mario Stoppani with the prototype of the SVA.

Stoppani, left in a dark overcoat, with journalist Pietro Negro, ready to depart for Poland with an Ansaldo A.300.

*Valor Aeronautico* (Gold Medal for Aeronautical merit): "With a fast flight of about 7,000 km without stop, he conquered for Italy the international distance record for seaplanes."

He gained his final international records on 30 March 1938 flying the Cant Z.509, a larger, Ocean crossing version of the 506 that improved the records for speed with loads of 1,000 and 2,000 kg flying in a circuit at a speed of about 400 km/hour.

During World War II, Stoppani tested the Cant Z.1007 bomber in its various versions, the beautiful Cant Z.511 four engine floatplane transport and the Cant Z.1018 twin engine bomber.

After the war, Stoppani followed Zappata, by now a close friend, at Breda, in Milan, where the gifted designer prepared the Breda BZ.308, a very beautiful four engine transport landplane, powered by four Bristol Centaurus, that first flew in 1948, but was not series produced.

Wearing his characteristic bonnet, which gave him a working-class look, having lost his youthful thin looks, Stoppani was a well-liked character: modest, never boasting, a man of great integrity and capacity. He looked like one of those grumpy and able test pilots played in American movies by Spencer Tracy. He had a difficult family life, but he was an adoring father, and an example to all around him. One of his sons also became an airman, and was a fighter pilot in the Spanish Civil War and in World War II.

In 1952, after Breda left the aviation business, the aging Stoppani became test pilot of Nardi in Milan, where he test flew the F.N.333 "Riviera" tourist amphibian. The next year he was hired by Siai Marchetti where he test flew overhauled Douglas C-47s for the Italian airlines. His last flight was a demonstration of the F.N.333 during an air show in Vergiate, on 13 September 1959. He died suddenly on 20 September 1959.

Left: Stoppani in the cockpit of Cant Z.506B I-LAMA ready for departure for his record-breaking trans-Atlantic flight, December 1937. Right: The old test pilot, Mario Stoppani with the Nardi FN-333 amphibian.

## Mario Stoppani
## 6 confirmed victories

| Claim | Vict. # | Date | Sqn. | Aircraft | Hour | Place | Victim |
|---|---|---|---|---|---|---|---|
| - | - | 14.6.1916 | 76 | Ni.11 | 8h45 | Isonzo | Aviatik |
| *Korp. Michael Giesz UNH/Lt. i. d. R. Oswald Kanitz UNH, Flik 4, Br. C.1 61.16 FTL* | | | | | | | |
| 1 | 1 | 9.7.1916 | 76 | Ni.11 | - | Aisovizza | EA |
| *[with s.ten. Rigoni & sold. Venchiaruti, 76ª Sq.]* | | | | | | | |
| 2 | 2 | 18.7.1916 | 76 | Ni.11 | - | S. Marco | EA |
| *Zugsf. Josef Franke UNH/Lt. Johann Kiss UNH, Flik 4, Lloyd C.III (WKF) 43.83* | | | | | | | |
| - | - | 27.7.1917 | 76 | Ni.11 | - | ? | EA FTL |
| - | 3 | 13.8.1916 | 76 | Ni.11 | - | Ranziano | EA OOC |
| 3 | 4 | 11.10.1916 | 76 | Ni.11 | - | Aisovizza | EA |
| *[with ten. Olivi, 76ª Sq.] unknow pilot UNH/Korp. Gustav Weiser WIA, Flik 19, Br. C.1 61.72* | | | | | | | |
| 4 | 5 | 31.10.1916 | 76 | Ni.11 | - | Nad Logem | EA |
| *[with serg. Tesei, 77ª Sq . & Caselli, Rossi, 70ª Sq.] Oblt. Frank Cik KIA/Zugsf. Gustav Resch WIA POW,Flik 12, Br. C.1 68.25* | | | | | | | |
| - | 6 | 1.12.1916 | 76 | Ni.11 | - | S.Marco | EA |

Mario Stoppani is listed as a six-victories ace, although in his logbook he specifically claimed only four victories. Originally the 5° Sezione confirmed only one victory to him. In the list of Italian fighter pilots published in February 1919 he erroneously appears twice, as a six-victories and one-victory ace.

# Romolo Ticconi

Acuto today is a small and pretty village a few miles from Fiuggi, and its economy benefits from the tourism in that nearby spa, but one can easily picture it 100 years ago, a rural town of less than 3,000 people, surviving from a less than generous land, where electricity first arrived – to the church – only in 1910. There in the night of 25 March 1893 Concetta, the young bride of Maurizio Ticconi, a peasant, gave birth to a boy who was christened Romolo Gabriele. There were other children, then the family left their farm and moved to the village, living in a small house at Via del Colle 4, a lane behind the castle of the Counts of Acuto. Probably to escape the poverty of the fields, the Ticconi family moved to Rome on 26 January 1911. In the capital Romolo, the oldest brother, helped the family working. Military papers report his profession as mason at the time of conscription, but Scaroni later wrote that Ticconi was a cart driver.

At least, despite his low social standing, he was able to read and write, no mean feat for that period. He was drafted and assigned to the 81st Infantry Regiment on 24 August 1915. There is no information about his service in infantry, but he had a good progression in the ranks, being promoted corporal on 16 December 1916 and sergeant on 20 September 1917. Meanwhile, on 21 August 1917, he had been admitted to the air service, posted to the Deposito Aviatori for pilot training. We find again *sergente* Romolo Ticconi in January 1918, assigned to the 76ª Squadriglia da caccia, then based at Istrana. Equipped with the excellent Hanriot Hd.1, it was an outstanding unit, second only to the renowned 91ª. Its most famous ace, Baracchini, was still recuperating from his bad wound, but other pilots had taken his place, first among them Silvio Scaroni, who by the time of Ticconi's arrival had already shot down 9 aircraft in just six weeks.

Flying schools were churning out pilots at a quick rate, but the level of training was less than adequate, and many commanders wanted to test newcomers before committing them to battle. Ticconi had to submit to this precaution: despite the lack of pilots, his first month at the front was spent exclusively in training and familiarization flights. Meanwhile the 76ª Squadriglia changed its bases in order to avoid night bombing raids, and it moved to Casoni.

On 5 February 1918 Ticconi finally flew his first operational mission, an offensive patrol over Carmignano di Brenta. He then flew several escorts to the SIA 7b two-seaters and more offensive patrols, but he didn't have combats with enemy aircraft. On 13 March after an escort flight he crashed on landing, wrecking Hanriot 6200 and risking cutting short his career. He had a peculiar mission on 1 May, when he was ordered to fly to the enemy airfield at Feltre and to drop a message, most likely information about shot down Austrian airmen.

On 3 May 1918 Ticconi finally claimed his first victory when, together with other pilots of the 76ª Squadriglia, *ten.* Alessandro Buzio, *ten.* Ludovico Censi, *s.ten.* Renato Donati and *ten.* Dante Nannini, despite the intervention of escort fighters he shot down a Brandenburg C.1 of Flik 16/D after a long combat that began above Col Moschin and ended when the two-seater crashed on Mount

Romolo Ticconi in the cockpit of a Caudron G.3

Cismon. The enemy crew, *Korp.* Sigl and *Oblt.* Weintritt, declared that they had been attacked by seven "Sopwiths", the usual name applied by the Austrians to the Hanriot, an airplane that strongly resembled the Sopwith Camel.

After this good start, Ticconi had a misadventure on 24 May, during an escort mission to Caproni bombers attacking Feltre. As he was flying 4,000 m above the target, he felt a sudden heat in the cockpit and he saw fire, probably due to a short circuit, coming out of the engine. He went down in a glide and managed an emergency landing at San Martino di Fonte, near Asolo, in friendly territory, before flames completely engulfed his airplane. His right leg was badly burned, but in a few days he recovered and was back in action.

On 15 June 1918 the Austrians launched their huge offensive along the whole front, from the mountains to the sea, and Ticconi was active, as all Italian pilots: suffice to say that on 19 June alone he flew five combat sorties. Still, despite his frantic activity, he had to wait until June 25 for another air combat. That day, during a badly coordinated escort mission to Caproni bombers, he and a few other pilots of his squadron, Scaroni as formation leader, cap. Lega and maybe *serg.* Pietro Mondini had to protect the second wave of bombers. Several big formations of enemy fighters, probably from Flik 56/J and Flik 42/J challenged them, but the Italians managed to thwart their threat to the bombers. At the end of the combat ground observers confirmed the fall of two enemy fighters, that were seen going down near Mareno di Piave and that were credited in collaboration to all the members of the patrol. Scaroni described this combat in his second book of war memories, written in the 1930s:

> "Ticconi seems to be possessed by all devils of hell. I see him everywhere, at my side, then behind the tail of one or two enemies, then above me, than again next to me. He dogfights

like an elf around my airplane, distracting the enemies as they shuttle behind me to hit me... What a wonderful fighter this boy is! When he was a coachman he clearly was in the wrong job. War has revealed many of these lions with an insignificant look."

Just a few days passed then, on 7 July, Ticconi returned to victory. That day he was cruising with Scaroni over the Asiago plateau when they spotted a two-seater and three enemy fighters heading to the Italian lines. With a clever surrounding maneuver they positioned themselves behind the enemy, ready to intercept them on the way back. Scaroni described this combat:

> "They were far from expecting our ambush, and such was their surprise that they all got into a dive at full throttle in the hope of evading us. We let them all pass below us then we quickly turned back and approached them. The enemy patrol was set up with the two-seater leading; behind it, at close distance and higher there were two fighters, the third one, higher still, trailed behind. Exploiting our higher speed from the dive I reached the last one and fired a terrible burst, before it could attempt any maneuver: then keeping my direction I got on the side of the fighter on the left and I briefly fired at it twice, then I kicked the rudder and headed for the two-seater. I went down in a little dive, positioning myself below its tail, I aimed at it exactly pulling up the nose and I fired a long burst and I kept on firing until my Hanriot lost all speed and stalled, losing over 100 meters of altitude. As soon as I recovered the enemy plane falling down passed next to me like a meteor, trailing a huge column of black smoke. Almost at the same time, much lower, one of the fighters, possibly the first one that I attacked, was trying to reach back its lines. I rushed toward it and immediately I fired two close bursts ... a huge blaze engulfed the entire airplane. I was so close that I could feel the heat of that terrible fire.
>
> I saw two hands sticking out of the fire and desperately grabbing the edge of the top wing; I saw the head of the pilot coming out, covered by a yellow leather helmet... he left the controls and stood up on the seat as if trying to escape the fire that was now engulfing him. He briefly turned toward me, looked straight as me as if asking for help... I felt the look of that man at me, and I was forced to watch that hopeless fight, losing no detail... The airplane suddenly banked and he fell out. Ticconi, on his part, had not been idle. He attacked another fighter and after a long dogfight had shot it down. Unfortunately he was not credited with this victory for lack of confirmations; later on, instead, after his death, a letter from General Pantano determined in a unequivocal way even the spot where this fighter had fallen."

Actually Ticconi claimed two fighters shot down, one over Mount Valbella and one over Mount Spitz, but he was credited with one only, obviously the one flamed by Scaroni, that fell within the Italian lines. The other one, with pilot *Korp.* Franz Pensl, was not

*Sergente* Romolo Ticconi, the trusted wingman of Silvio Scaroni.

*Ten.* Dante Nannini, who shared with Ticconi the victory of 3 May 1918. (Photo Molteni).

The Hanriot of *s.ten.* Renato Donati, which had a shoe as personal insignia. The pilot is sitting astride the fuselage and on the right there is *cap.* Lega. *(Photo Selinger).*

credited to him, not unfairly, considering that the Austrian pilot, wounded in a shoulder, managed an emergency landing at Ospedaletto airfield.

These victories earned him a Medaglia d'Argento al Valor Militare that in its citation credited him with five victories from 3 May to 7 July.

After Silvio Scaroni was badly wounded, on 12 July, Ticconi kept on fighting with his usual aggressiveness and he shot down two more airplanes, the first one on 14 August, that he attacked together with Michetti above Cismon and that was seen crashing at Rasai. His last victory came the next day, when he spotted a lonely fighter above Treviso, and he pursued it across the enemy lines. It was a photographic reconnaissance Phoenix D.I single-seater of Flik 37/P. The Austrians, after the heavy losses of their slow reconnaissance airplanes, were now using fighters equipped with cameras to check the movements of their enemies. Despite the not too remarkable speed of the Hanriot, Ticconi managed to catch it and shoot it down near Colbertaldo. The enemy pilot managed an emergency landing at S. Fior di Sopra airfield, destroying his airplane but surviving.

It was by this time more and more difficult to find enemy airplanes. Ticconi flew more missions, including many escort flights

Personnel of the 76ª Squadriglia in front of the Hanriot of Baracchini with the four aces. Ticconi is the first from the right, standing.

The Hanriot of *magg.* Lombard, commander of the VI Gruppo, showing his insignia, a shield in the national colors and the baton of command.

Ticconi in his Hanriot, showing his personal markings of black triangles.

for RE.8 reconnaissance airplanes of the RAF and launches of leaflets over enemy trenches, but he had only one combat, on 3 October, cut short by a gun jam.

His work in the final months of the war was not useless, however. Fighter squadrons were replacing old pilots, many of them tired and worn out, with newcomers coming from the flying schools. The experience of Ticconi was very useful in teaching them how to survive in the crucible of combat flying.

In October 1918 Ticconi was one of the millions who were hit by the epidemics of the "Spanish flu" that ravaged Europe. Despite his illness he had the will to return to his unit and fly a few more actions during the final offensive at the end of October 1918. One more decoration, a Bronze Medal, crowned his activity. In its citation two more victories are reported, the ones of 14 and 15 August, for a total of seven, although only six victories were credited to him in the final postwar list of aces.

After the armistice Ticconi was posted to the 81[a] Squadriglia da caccia, and he served there until the day of his death, 26 August 1919 when for unknown reasons as he was flying from Montecelio test airfield near Rome he crashed and was instantly killed, one of the many valiant Italian airmen who, having survived the war, lost their life in a long string of accidents in 1919.

The Ticconi family, years later, immigrated to the United States, thus no memory of this humble but valiant ace survives in Italy.

## Romolo Ticconi
## 6 confirmed victories

| Claim | Vict. # | Date | Sqn. | Aircraft | Hour | Place | Victim |
|---|---|---|---|---|---|---|---|
| 1 | 1 | 3.5.1918 | 76 | Hanriot | - | M. Cismon-Col Moschin | Two-seater |
| *[with ten. Nannini, Buzio, Censi, s.ten. Donati, 76ª Sq.] Korp. Franz Leo Sigl UNH/Oblt. i. d. R. Franz Weintritt WIA, Flik 16/D, Br.C.1 169.78* | | | | | | | |
| 2 | 2 | 25.6.1918 | 76 | Hanriot | - | Mareno di Piave | Fighter |
| *[with cap. Lega & ten. Scaroni, 76ª Sq.]* | | | | | | | |
| 3 | 3 | 25.6.1918 | 76 | Hanriot | - | Borgo Villa | Fighter |
| *[with cap. Lega, ten. Scaroni & serg. Mondini, 76ª Sq.] Maybe: Oblt. August SelingerUNH, Flik 42/J, Alb. DIII 153.202 FTL* | | | | | | | |
| 4 | 4 | 7.7.1918 | 76 | Hanriot | - | M. Valbella | Fighter |
| *[with ten. Scaroni, 76ª Sq.] Korp. titl. Zugsf. Oswald Bierlotter KIA, Flik 9/J, Alb. D.III 153.140* | | | | | | | |
| 5 | - | 7.7.1918 | 76 | Hanriot | - | Mount Spitz | Fighter |
| *Korp. Franz Pensl WIA, Flik 9/J, Alb. D.III 153.98* | | | | | | | |
| 6 | 5 | 14.8.1918 | 76 | Hanriot | - | Rasai | Fighter |
| *[with ten. Michetti, 76ª Sq.]* | | | | | | | |
| 7 | 6 | 15.8.1918 | 76 | Hanriot | - | Colbertaldo | Fighter |
| *Fw. Gustav Franz UNH, Flik 37/P, Phönix D.1 328.04* | | | | | | | |

# Profiles

One of the main reasons why the deeds of the Knights of the Air of World War I are still so fascinating is the range of colors of their aircraft and personal insignia. While the history of the main aces, particularly the German and British ones, are relatively well known, only recently the Austro-Hungarians have become the subject of research. For what concerns the Italians, excluding the universally known "prancing horse" of Baracca, which through the Ferrari racing cars is a symbol of national pride, very little is known about the airplanes and insignia of the other aces.

This chapter is trying to bring light to an unexplored subject, those personal and squadron insignia that reflect the character and feelings of a generation. They range from the four aces that Baracchini painted on his Hanriot, an obvious symbol of his aspiration to become ace of aces, to the funny ones, like a devil making gestures, or the simply practical ones, like a square or a triangle, with the only purpose of being recognizable, without any artistic pretension.

Some squadrons tolerated or approved personal insignia; others had unit emblems, like the red heart of the 77ª, or the griffin of the 91ª in 1918. Happy Hooligan, in Italian "Fortunello", Lucky, was very popular for obvious reasons, and in 1917 the pilots of the 80ª Squadriglia used it with many personal variations.

The color profiles are arranged according to the type of airplane, with four views of each fighter model, followed by side color profiles of the same type. Colors and insignia are reproduced on the basis of all published material and the archive research of the authors. It must be strongly stated that in this field there are many gaps of knowledge and lack of reliable documents, so these profiles include the personal interpretation of the author, and the reader is entitled to different opinions. In order not to spread confusion and to state a certainty that simply is not there, the sources are quoted each time.

Concerning colors, the author also supplies general references, because even when original relics have survived, one cannot be sure of how much the passing of years and the conditions of preservation have affected the quality of the colors.

## Insignia of Nationality

At the outbreak of hostilities with Austria-Hungary, the Italian air force dropped its pre-war insignia (the state shield, with an eagle and the shield of Savoy on the rudder, black circles and diagonal bands on the wings) and introduced colors similar to the ones of its new allies. Identification of aircraft was an immediate problem, but the main worry was not air-to-air recognition, as air combats were not even considered then. The only concern was to make aircraft nationality apparent to ground troops, and thus Italian nationality insignia were initially applied only to the bottom surface of wings. A circular dated 26 August 1915 specified that the lower surface of wings had to be painted with the national colors of Green, White and Red, to be painted also on rudders, like a flag. No mention was made of roundels, neither on fuselages nor on wing upper surfaces. Identification didn't really need colors, as the shape of Italian aircraft was markedly different from enemy ones. Most Italian aircraft were tailboom pushers (Farman, Caudron, Voisin, SP), while the Austrians used exclusively tractor biplanes. Aircraft identification however was yet to come, and ground observers directed the fire of artillery indiscriminately against friend and foe. The seaplanes of naval aviation, on the other hand, were liberally decorated with roundels in all possible positions for a specific reason, that Italian flying boats were very similar to the Austrian ones, and some of them, like the Macchi Lohners, were exact copies.

Roundels appeared only in late 1916 on the fuselages of the third batch of Nieuport 11s of Macchi production, but more months were to pass before roundels became commonplace. There was an exception for some new type, like the SAML and the Pomilio, tractor biplanes similar to enemy models that had roundels everywhere, including the nose.

French built aircraft were drawn from the production line for the French air force, so they arrived in Italy with roundels applied. The Spad 7 that first reached Italy in March 1917 usually had a roundel added on the fuselage also. The problem of the correct position of colors was not particularly appreciated: generally roundels had Green as the outside color, with the exception of the early

Spads and of other French-built aircraft. No other official instruction on insignia is known until late 1917, when British pilots deployed to Italy remarked that Italian reconnaissance airplanes carried no roundels on the top wings, or when they did, the order of colors could vary.

So the command issued an order to paint roundels on wings and fuselage, with Green outside and Red inside (the actual document doesn't seem to have survived) creating in later years the wrong opinion that until 1917 Italian aircraft had roundels with Red outside and that they inverted the colors in early 1918.

Besides the order and the requests of the allies, the fact remains that Italian individualism played its part also in the final year of the war, and many airplanes went on flying without wing roundels. It is doubtful to which amount the official order had effect, as it was by itself of doubtful interpretation, so much that a letter has survived in which the commander of a reconnaissance Squadriglia at the beginning of 1918 answered that he was ready to paint roundels, but he asked *"in which order"* the colors had to be applied.

## *Profile 1*

**Nieuport 11 serial no. Ni 1451, *cap.* Francesco Baracca, 2ª Squadriglia caccia, April 1916.**

This is a French built Nieuport 11 in pristine conditions. Its only addition is the Italian serial number on its fuselage, while the French number on the rudder was erased .It is not certain that the serial number was repeated on the port fuselage side too. Also typical for the Italians was the lack of roundels on wings upper surfaces. White color of the airplane is not confirmed; it could also be a very light tan, corresponding to natural fabric. On the top wing there is a Lewis with wooden stock and the typical arched gun mount used by French built aircraft. Baracca scored his first confirmed victory flying with this airplane on 7 April 1916. In early 1917 this airplane was still in service with the 70ª Squadriglia and then it was assigned to second-line units. It had a very long operational life considering the standards of the time.

## *Profile 2*

**Nieuport 11 serial no. Ni 2123, *serg.* Alvaro Leonardi, 80ª Squadriglia, May 1917.**

This is a Nieuport 11 built under French license by Nieuport Macchi at Varese. The Nieuport 11 was the first real fighter of the Italian air force. After the supply of some aircraft built in France, Macchi supplied a large quantity of them from mid 1916 to mid 1917. They differ from the French ones in many details, the most obvious being the different type of mount for the Lewis gun above the top wing. This aircraft belongs to the second Macchi production batch, with the color used also for the third lot. This color cannot be confirmed as absolutely exact: white & black photos only confirm that top and bottom surfaces had different colors. Some authors stated, without positive proof, that top surfaces were painted olive green: here the artist opted for a color corresponding to the one of the Nieuport 10 preserved in the Museo della Scienza e della Tecnica of Milan. We don't know how it was painted this way during restoration, but it is likely that at the time the managers of the Museum had the advice of veteran pilots, many of them being alive at the time. Also the Hanriot Hd.1 produced by Macchi in 1917/18 had this same color. This airplane shows standard nationality insignia under the bottom wing, on the engine cowling and on the rudder, and a fuselage roundel with Green outside. The smiling face of *"Fortunello"*, Happy Hooligan, a cartoon character then very popular in Italy, is painted over the roundel. Each pilot of this Squadriglia had a different variation of this subject. Leonardi scored the first confirmed victory with this airplane on 17 May 1917.

## *Profile 3*

**Nieuport 11 serial no. Ni 1611, *ten.* Ferruccio Ranza, 77ª Squadriglia, autumn 1916.**

A Nieuport 11 of Macchi production belonging to the first lot, with an overall light finish. Its only nationality markings are the colors below the bottom wings and on the rudder. This airplane is shown in a White finish, although other believe this color to be natural doped linen Cream. The use of White in Italy is confirmed by the fins of Caudron C 1077 preserved in the Museo della Scienza e della Tecnica of Milan, a contemporary airplane to the Nieuport and that in original photos shows a similar finish. Paintings of the famous contemporary artist Achille Beltrame also always show Nieuports in a white-light gray color instead of the so oft depicted light cream. Some recently discovered photo show that for a short period there was an attempt to apply mimetizations to the Nieuport 11 of 70ª and 77ª squadriglia. The desired mimetic scheme consisted of dark swirls made with a rather dry big brush. The color used is unknown but probably some sort of dark green. On the fuselage side there is the unit marking, a red heart, the personal badge of his first unit commander magg. Pier Ruggero Piccio.

## *Profile 4*

**Nieuport 11 serial no. Ni 1651, *ten.* Luigi Olivi, and *serg.* Mario Stoppani 76ª Squadriglia, autumn 1916.**

This Nieuport sports on its fuselage two diagonal Black stripes. Those were sometimes interpreted as in the national colors of Green and Red: Although there were Nieuports with such markings, in the case of the 76ª Squadriglia the bands indicated the *sezione* (flight) in the Squadriglia, one band for the first and two bands for the second *sezione*. The Black color of the bands was determined by research made over 40 years ago by the historian Rinaldo D'Ami.

## *Profile 5*

**Nieuport 11 serial no. Ni 1675, *serg.* Guglielmo Fornagiari (?), 78ª Squadriglia, 1916/1917.**

A Nieuport similar to the one of Profile 4 but showing a darker lower wing, most probably because it was re-covered with clear doped fabric. The fuselage insignia is not completely visible, what appears here is a reconstruction. The two dark circles may indicate the flight within the squadron.

### *Profile 6*

**Nieuport 11 serial no. Ni 1685, *ten.* Fulco Ruffo di Calabria, 70ª Squadriglia, winter 1916/1917.**

This airplane was pictured after a long period of operational life. Its frontal fuselage panels are natural metal, while all the back part of the fuselage is darker, maybe because its cloth covering was replaced. Visibility of the serial number is enhanced with a White shadowing. On the fuselage there is the first version of the black skull and crossbones, personal insignia of prince Ruffo. The machine gun above the top wing is a Colt and not the usual Lewis. Also Baracca equipped for some times its Ni 11 with a Colt gun.

### *Profile 7*

**Nieuport 11 serial no. Ni 1763, *ten.* Alessandro Buzio, 75ª Squadriglia, 1916/1917.**

Nieuport 11 of Macchi production with a large tricolor band on its fuselage. Probably it was meant to aid identification by ground observers that on the Asiago front at that time were not equipped with binoculars and had big difficult in the identifications of friends and foe. The original serial number, hidden under the band, was repainted further back, with non-standard characters. The top part of the front fuselage shows a different color, a feature appearing only on this airplane.

### *Profile 8*

**Nieuport 11 serial no. Ni 2125, *cpr.* Attilio Imolesi, 79ª Squadriglia, 1916/1917.**

The second production batch of Nieuports supplied by Macchi shows a different finish, as described in Profile 2. Some aircraft, like this one, have a smaller portion of the cowling painted with the national colors. The sloping white fuselage band indicates the first *sezione* of the 79ª Squadriglia.

### *Profile 9*

**Nieuport 11 serial no. Ni 2126, *cpr.* Marziale Cerutti, 79ª Squadriglia, 1916/1917.**

There are photographs of Cerutti in the cockpits of several Nieuports. This Ni. 11 of Macchi production carries a standard finish with the engine cowling completely painted in the national colors. The three sloping white fuselage bands indicate the third *sezione* of the 79ª Squadriglia.

### *Profile 10*

**Nieuport 11 serial no. Ni 2179, *s.ten.* Flavio Torello Baracchini, 81ª Squadriglia, spring 1917.**

Baracchini initially flew the Nieuport 11, a fighter already obsolete in the spring of 1917. This airplane appears in the standard Macchi finish, showing on the fuselage sides the "Black shield of D'Artagnan", personal emblem of the ace, and on the fuselage deck a child bothered by what seems a wasp. Baracchini flew the Nieuport 11 until 20 May 1917, and then, after his first confirmed victory, he flew the Type 17. He probably claimed the first confirmed victory flying this aircraft.

### *Profile 11*

**Nieuport 17 serial no. N 3139, *ten.* Fulco Ruffo di Calabria, 70ª/91ª Squadriglia, winter 1916/1917.**

By the end of 1916 Italian units received from France the first examples of the Nieuport 17, faster than the Ni. 11 and armed with a synchronized Vickers gun, which were assigned to all squadrons, and reserved for the best pilots. One of the many photo of this aircraft clearly show that the French roundels over the top wings were painted off. This aircraft shows the Black skull and crossbones emblem of Prince Fulco Ruffo di Calabria on both sides of the fuselage. It has a standards White-Silver finish, national colors below the wings and on the rudder. This aircraft arrived at the squadron on 2 December 1916 and Ruffo used it until April 1917. This emblem is even described in a combat report of an Austrian crew that had a combat with Ruffo on 28 February 1917, specifying that the attacking Nieuport showed *"a skull with crossed tibiae on its fuselage"*. As Vickers guns were scarce, Ruffo initially armed its airplane with a Lewis gun over the top wing.

### *Profile 12*

**Nieuport 17 serial no. Ni 3592, *s.ten.* Giannino Ancillotto, 80ª Squadriglia, October 1917.**

In the summer of 1917 fighter squadrons were finally largely re-equipped with this type, using Macchi-built aircraft. This one is an Italian built Ni. 17 in the typical Silver-White finish, as it appeared in October 1917. Italian Nieuport 17s carried standard color sections under the bottom wings and roundels on the fuselage and upper top wings. The 77ª and 80ª Squadriglia, belonging to the same Gruppo, applied the squadron emblem, a Red heart for 77ª and the "Star of Italy" for 80ª over the fuselage roundels. The star was White or Silver, while the roundel background seems to have been repainted solid Red. Not much paint was used for the star, which appears to be sloppily painted. Ancillotto scored his first victories on the Nieuport 17 in late October 1917. Presence of roundels over the top wings of this aircraft cannot be confirmed: some aircraft didn't carry them, although we don't know if they were not applied at the factory or if they were removed by the units.

### *Profile 13*

**Nieuport 17 serial no. N 2614, *cap.* Francesco Baracca, 70ª/91ª Squadriglia, winter 1916/ 1917.**

This is the first Nieuport 17 of French production supplied to the 70ª Squadriglia on 21 October 1916. Baracca used it until April 1917. It shows an overall Silver finish, or *"enduit metallisé"* as it was called in France, with the national colors below the wings and on the rudder. Most Nieuport 17s supplied from France didn't carry roundels above the top wings, while roundels appeared on all the fighters of French units such as the *"Escadrille de Venise"* based at Venice; so one may believe that the Italians had required not applying roundels or as is for this aircraft scraped them off. This is the first airplane known to sport the "prancing horse" emblem, in Black on both sides of the fuselage. So far no evidence was discovered of this emblem appearing on a Nieuport 11. Baracca introduced it as homage to his cavalry unit, the Piemonte Reale Regiment that had as its emblem a Silver prancing horse on a red field. On the rudder there is a patch covering a bullet hole. The original French serial number appears, together with the weight data, in French. Italian serials were not adopted for French built Nieuport 17s; only some French-built Hanriot Hd.1 and Nieuport 27 receive Italian military serial numbers.

## *Profile 14*

**Nieuport 17, serial no. N 2142, *cpr.* Attilio Imolesi, 79ª Squadriglia, spring 1917.**

Squadrons operating on the northern front of the Trentino were often called to send their pilots and aircraft as reinforcement to the front of the Isonzo when new offensives were in preparation. *Serg.* Imolesi was one of them. His Nieuport shows a personal insignia, a good luck charm horseshoe. This one is probably airplane serial N 2142, described as armed with both a Vickers and a Lewis gun over the top wing, with which Imolesi scored his first confirmed victory, together with *cap.* Baracca, on 24 April 1917.

## *Profile 15*

**Nieuport 17, serial no. unknown, *s.ten.* Flavio Torello Baracchini, 81ª Squadriglia, spring 1917.**

After his first successes, Baracchini also received one of the few available Nieuport 17s. According to his logbook, he first flew it on 21 May 1917. This is a French-built machine. The French serial number and technical data on the rudder are deleted. The personal emblem of Baracchini, a large Black shield, appears on the fuselage.

## *Profile 16*

**Nieuport 17, serial no. unknown, *serg.* Cosimo Rizzotto, 77ª Squadriglia, spring/summer 1917.**

The first commander of the 77ª Squadriglia was *maggiore* Piccio, who used a Red heart as his personal emblem. When he left that unit, the heart remained as a permanent unit insignia. On French-built Nieuport 17s it was applied directly over the light background of the fuselage, and when in the summer 1917 the Macchi-built machines arrived, sporting also a fuselage roundel, it was applied in a white circle obtained overpainting the fuselage roundel.

## *Profile 17*

**Nieuport 17, serial no. unknown, *serg.* Guido Nardini, 78ª Squadriglia, spring/summer 1917.**

The 78ª Squadriglia was one of the most colorful units of the Italian air force. In 1918 most of its fighters sported large pennants, of different color for each pilot. This tradition however began much earlier, in the summer of 1917, when a smaller pennant appeared on the fuselage of its Nieuports and Hanriots. Color is believed to be Red, but this cannot be confirmed as absolutely certain.

## *Profile 18*

**Nieuport 17, serial no. unknown, *maresc.* Ernesto Cabruna, 80ª Squadriglia, summer/autumn 1917.**

The caption of a photo describe this aircraft as that of *maresciallo* Cabruna, while visiting his old reconnaissance unit, the 29ª squadriglia, in Cavazzo Carnico. The Nieuport is a French built aircraft, without serial number on the tail. On the fuselage is visible the Red Star, the unit badge. Only the Vickers gun shortage justify the use of the armament of the Lewis gun mounted over the top plane. Cabruna was transferred to 80ª squadriglia in September or October 1917 and the photo was taken in this period, before the Italian retreat to the Piave.

## *Profile 19*

**Nieuport 17, serial no. unknown, *serg.* Antonio Reali, 79ª Squadriglia, summer/autumn 1917.**

*Serg.* Reali was a successful Nieuport pilot. This French-built aircraft is emblazoned with the *"fez"*, the soft cloth cap of the Bersaglieri troops. Probably it was not Reali's personal airplane, but the one of his commander, *ten.* Mazzini. Many pilots used emblems that recalled their original service or unit. Some squadron commanders didn't fly very much, and it often happened that other pilots of the squadron used their personal machine. This would explain why Reali was pictured in the cockpit of this Nieuport.

## *Profile 20*

**Nieuport 17, serial no. Ni 3603, *s.ten.* Carlo "Francis" Lombardi, 77ª Squadriglia, summer/autumn 1917**

This aircraft show how the unit's emblem of the red heart was applied on Macchi built Nieuport 17. Lombardi, was photographed in the cockpit of this aircraft in October 1917, before the disastrous Italian retreat . Several others pilots were photographed with this aircraft, in the same day and place, and it is doubtful that it was his personal mount.

## *Profile 21*

**Nieuport 27 bis, serial no. N 19750, *serg.* Marziale Cerutti, 79ª Squadriglia, autumn 1918.**

Together with *serg.* Reali, also of the 79ª Squadriglia, Marziale Cerutti was the most successful Nieuport 27 pilot, scoring most of his seventeen confirmed victories with this type. In the crisis after the retreat of autumn 1917, about one hundred Nieuport 27s arrived from France. These aircraft however were not new, and they had some defects, including some structural weakness. When a sufficient quantity of Hanriots was available, the Nieuport 27 was concentrated with the 79ª and 83ª Squadriglia that used them until the end of the war. This aircraft shows the French standard camouflage finish in five colors. It is not known if it carried roundels or color bands below the wings. Other photos of Italian Nieuport 27s show roundels under the bottom wings, apparently with Red outside. On the fuselage there is the "Ace of clubs" personal insignia of Marziale Cerutti, reproducing the Italian traditional playing cards. Similar insignia were used by Italian units in the Spanish Civil War and in World War II. The acrostic MIR was reported to stand for "Marziale Imperatore Romano" (Marziale Roman Emperor). Initially Italian Nieuport 27s kept their French serial numbers, but later, around mid 1918, they also received Italian serial numbers, usually applied on the fuselage and rudder. In this case the fuselage emblem hides the serial number.

## *Profile 22*

**Nieuport 17 bis, serial no. N 2628, *s.ten.* Flavio Torello Baracchini, 81ª Squadriglia, spring/summer 1917.**

Two recently discovered photograph proved the unknown presence of this variant of the Nieuport in the Italian air force. It is hard to discern the 17 bis variant from the 24 bis. The other known photograph, however, clearly show under the tailplane the script "TYPE 17" proving that this was one of the few Nieuport 17 bis build. The other aircraft received by the Italians had the serial number N 2629. The French almost always quickly provided Italy with a small number of all their new aircraft, as a suggestion for large orders. Baracchini had a further Lewis gun fitted on the top wing to increase firepower, probably a concession to his standing as an ace, because aircraft machine guns until mid 1918 were scarce in Italy. On the fuselage the Black shield appears again.

## *Profile 23*

**Nieuport 27, serial no. N 5907/11358, *serg.* Marziale Cerutti, 79ª Squadriglia, spring 1918.**

This aircraft show both the French and Italian serial number, respectively on the tail and on fuselage. Other that the squadriglia number below the cockpit, probably in black, the aircraft show a personal badge consisting in what seems the Italian flag with something like a shield and two bars in the central white field. The 79ª squadriglia was 78ª twin unit at this time, and this explain the badge in typical 78ª style. Of note is the unusual decoration on the aircraft wheel.

## *Profile 24*

**Nieuport 27, serial no. N 11339, *ten.* Flavio Torello Baracchini, 81ª Squadriglia, June 1918.**

This aircraft was not Baracchini's personal mount, but that of *serg.* Palpacelli. At the beginning of 1918 this unit was re-equipped with Nieuport 27 but in the following months started to receive the more effective Hanriot Hd.1. In June it was almost totally equipped with the Hd and photographic documentation show that even Palpacelli had a new Hanriot with his personal insignia painted on the fuselage sides. Another photograph taken some days later show a 81ª squadriglia line up showing Palpacelli's old mount equipped with Le Prieur rockets. In June 1918 the unit papers clearly show that Baracchini was the only pilot that attacked enemy drakens using rockets, or "turpilles", as the Italians called them. His log book clearly state that the aircraft he used (without mentioning the type) was equipped with only four of them, just like the Nieuport 27 of the photograph. It is evident that Baracchini flew his Hanriot Hd 1 (see profile number 46 ) for normal patrols, but used Palpacelli's old aircraft when he wanted attack drakens. The use of rockets is proved for at least four attacks performed on 19 and 21 June when he finally managed to shoot down in flames one of them over S. Pietro. The use of rockets however was not particularly successful, as they oft refused to ignite. He probably used this aircraft even on 22 June when while in search of kite balloons he attacked an enemy patrol *"with another Ni 27 of 79ª"* and claimed three of them shot down. Also on 25 June, when he was wounded by ground fire after claiming his last victory he was flying *"in search of drakens"*, and it could be possible that was also using one more time this aircraft.

## *Profile 25*

**Hanriot Hd.1, serial no. H.D 523 (?) N° 11411, *ten.* Flaminio Avet, 70ª Squadriglia, April-October 1918.**

The Hanriot Hd.1 was the main Italian fighter in 1918. The first machines arrived from France in July 1917, equipping first the 76ª and then the 78ª Squadriglia. Contrary to a widespread opinion, a large quantity of French-built Hanriots arrived to Italy, and deliveries went on even after Nieuport Macchi in Italy had begun mass production of this fighter. The exact number of Hanriots supplied by France is not known, but the authors estimate it at over 500. Hanriots of French production differ from the ones built by Macchi in many structural details and in the position of markings. The main differences are the rudder, which is bigger in French machines, and in different positions of the inspection panels. Photos show that French aircraft always have a lighter finish. An official report issued in July 1917 describes the new fighters as *"painted completely White"* and also a report of December 1917 describes the Hd of *cap.* Riva as *"White with a Red flag on its fuselage"*. The fuselage roundel of French aircraft is closer to the cockpit and larger than Macchi aircraft. The rudder appears to have an inverted disposition of colors, with Red leading, and there are color bands only below the bottom wings. This photo can be dated after 18 March 1918, when anti-aircraft fire damaged this airplane. Over the top wing little roundels patch up the holes in the fabric. The two bands over the top wing are described (without specifying which color) in a combat report of Mario Fucini of 17 May 1918. As this color is not known, the artist has chosen Red, considering that Avet was an officer in the Lancieri di Firenze Regiment, called *"the Red lancers"* from the color of their collar tabs. Vertical fuselage bands, together with an individual number, were a typical feature of the 70ª Squadriglia. The number of Squadriglia on the fuselage, close to the cockpit, and above the top wing, was the subject of a specific order presumably in November 1917, but some squadrons kept sporting a unit insignia instead of the number. This is one of the few French-built Hanriots to clearly show an Italian serial number, N° 11411 written in small digits behind the fuselage roundel, and it is supposed to match with French serial H.D 523. This particular aircraft had a long operational life, and it was still in charge of the 70° Squadriglia at the time of armistice.

## *Profile 26*

**Hanriot Hd.1, serial no. Hd 7549, *ten.* Mario Fucini, 78ª Squadriglia, spring/summer 1918.**

This airplane with its beautiful winged roundel decoration is a typical example of Hanriots of Macchi production. While the first

French-built aircraft arrived in Italian units in July 1917, the first Hd built in Italy arrived to the combat units only after the retreat across the Piave, in early November 1917. The Hanriots built by Macchi had a smaller rudder, a different location of inspection panels and different finish and position of roundels: all this details and the researches in Italian archives give us the possibility to discern easily between Italian and French built Hanriots. The overall yellow finish is based on a painting commissioned to an unknown artist by General Scaroni in the early 1970s. The painting, showing the famous combat between Scaroni himself and the Austro-Hungarian ace Kiss, show the ace's Hanriot in an overall deep bright yellow.

## *Profile 27*

**Hanriot Hd.1, serial no. Hd 18 – 6614, *serg.* Guido Nardini, 78ª Squadriglia, summer 1917.**

This is one of the earliest Hd.1 delivered to Italy in the summer 1917, which equipped first the 76ª and then the 78ª Squadriglia. These aircraft appear to carry roundels with Red outer ring, and from known photos sometimes they seem to carry roundels below the bottom wings instead of color bands. On the rudder there are both the French production number HD 18 and the Italian military serial number 6614. On its fuselage there are the Red (?) pennant of the 78ª Squadriglia and a little devil making gestures. The origin of this figure is unknown: it may come from a magazine illustration, or it may be from a kind of statue that was common in contemporary brothels. The Canadian ace William G. Barker, who flew later on the Italian front, had a very similar metal statue of a little devil attached to his Sopwith Camel. An upper view of this aircraft and marking details are depicted in the 78ª squadriglia's badge page.

## *Profile 28*

**Hanriot Hd.1, serial no. Hd 21 – 6647, *ten.* Mario Fucini, 76ª Squadriglia, January 1918.**

This is another Hd.1 of the first batch delivered from France in the summer 1917. It was pictured at Istrana airfield in early 1918 when its national markings were much modified. The fuselage roundel was repainted in a different size, maybe when it was modified with Green outer ring. On the fuselage and above the upper wing there is the squadron number in Black, as prescribed in autumn 1917. The Italian serial number 6647 was applied both on the fuselage and on the rudder. This airplane also has additional openings on its engine cowling ( a feature typical of Hanriots belonging to this unit) and a smaller rudder, cannibalized from a Macchi-built Hd.1. There are ribbons, possibly in the national colors of Red, White and Green, hanging from the rudder, to indicate Fucini's role of flight leader. The lightning motif appeared also on the fuselage of the Nieuport 11s of the Sezione Difesa of Grottaglie, of which Fucini was commander in the summer 1917. The color of the lightning is probable, but not sure.

## *Profile 29*

**Hanriot Hd.1, serial no. Hd 515, *ten.* Flavio Torello Baracchini, 76ª Squadriglia, March 1918.**

Deliveries of French-built Hanriots went on in 1918, but later machines differed in their national insignia, now having a Green outer ring. Possibly misinterpreting Italian instructions, the French inverted also the colors on the rudder that seem to have Red leading. Baracchini, as he returned to service after many months of hospital care for a bad wound in the face, using an insignia with four aces seems to show his wish to become the Italian "ace of aces". The number of Squadriglia appears on the fuselage and above the top wing, centrally. This machine had a short life, and was wrecked in landing a few days after its photograph was taken. The position of the four aces is inverted on the other side of the fuselage, while on the horizontal tailplane it follows the order of the left side of the fuselage, with two aces on the right and two on the left. Baracchini flew with at least three different machines carrying this emblem: when he was transferred to the 81ª Squadriglia in June 1918, the last one became the personal mount of *ten.* Censi. A upper view of this aircraft is depicted in the 76ª squadriglia's badge page.

## *Profile 30*

**Hanriot Hd.1, serial no. unknown, *cap.* Giulio Lega, 76ª Squadriglia, June 1918.**

Standard coloring with the pilot's emblem, formed by the collar tabs of his unit, the "Granatieri di Sardegna" and the grenade with a propeller superimposed. The number 76 on the fuselage is applied in negative over a dark band, supposed to be black, surrounding the cockpit. The emblem, as was customary in this Squadriglia, probably appears also on the tailplanes. See also the 76ª squadriglia's badge page for details.

## *Profile 31*

**Hanriot Hd.1, serial no. unknown, *serg.* Romolo Ticconi, 76ª Squadriglia, spring 1918.**

This airplane has the typical colors and insignia of the 76ª Squadriglia, with a Black triangle, personal emblem of the pilot, on the sides and top of the fuselage. Documents preserved at the USSMA confirm the badge color as being Black. Of note is the wheel taken from a Macchi build aircraft. See also the 76ª squadriglia's badge page.

## *Profile 32*

**Hanriot Hd.1, serial no. unknown, *cap.* Giulio Lega, 76ª Squadriglia, June/July 1918.**

A photograph taken in June 1918 show a simplified version of the insignia of the Black grenade on the fuselage sides. This aircraft was modified on the field to carry two machine guns. At a certain time it also had its engine cowling and wheels painted in a darker color, here interpreted as Green.

## *Profile 33*

**Hanriot Hd.1, serial no. HD 556 - 11451, *serg.* Antonio Chiri, 78ª Squadriglia, spring 1918.**

All through 1918 Chiri flew several Hd.1, all of them of French production and painted with a Black tail, described in many combat reports of other pilots. This airplane shows the French production number HD 556 on its rudder and the Italian military serial 11451 in small digits above the roundel.

## *Profile 34*

**Hanriot Hd.1, serial no. HD 606, *cap.* Antonio Riva, 78ª Squadriglia, spring 1918.**

A Red (?) pennant became the unit emblem of the 78ª Squadriglia in the summer of 1917. When Riva became commander of this squadron the pennant became bigger and each pilot had one of a different color. Red remained as the color for the personal airplane of the commander, Fucini had White, Brenta had Green, Mecozzi had Blue. Other pilots had pennants striped in more colors. A report dated 26 December 1917 described the airplane of Riva as *"white and carrying a Red flag on its fuselage"*.

## *Profile 35*

**Hanriot Hd.1, serial no. unknown, *s.ten.* Alvaro Leonardi, 80ª Squadriglia, autumn 1918.**

It is wrongly believed that Italian built Hanriots had the machine gun repositioned in the center of the fuselage. Actually all Macchi-built Hanriots had their Vickers gun located on the left side of the fuselage, and only a few French-built airplanes delivered in the second half of 1918 had the gun located centrally. One of them is this one, used by *s.ten.* Alvaro Leonardi, showing on its fuselage the Red star, emblem of the Squadriglia, painted over the roundel. According to a preserved document of the USSMA, the thin border of the circle around the star is Black. The individual number "13", Black bordered in Red, according to a statement of the ace's daughter, was his lucky number, also painted on top of the right upper wing, but in a different style. The port wheel is from a Macchi build Hanriot. See also the upper view of this aircraft on the badge section of profiles.

## *Profile 36*

**Hanriot Hd.1, serial no. 614, *serg.* Antonio Chiri, 78ª Squadriglia, autumn 1918.**

Although this may be confused with the similar airplane appearing in profile No. 33, this is a different, late production Hanriot armed with a Vickers gun located along the center line of the fuselage. The ace Cerutti in his combat report of 27 October 1918, quoted Chiri's airplane as an *"Hd marked with a Black tail"*.

## *Profile 37*

**Hanriot Hd.1, serial no. Hd. 6092, *serg.* Guglielmo Fornagiari, 78ª Squadriglia, autumn/winter 1917/1918**

One of the earliest Hd.1 of Italian production belonging to the 78ª Squadriglia, showing the position of insignia typical of Macchi. The only exception is the rudder, cannibalized from a French-built aircraft, showing number 60. The Squadriglia number appears on the fuselage and above the top wing. On the fuselage there is the pennant of the 78ª. The color of Fornagiari's pennant is unknown; here it is shown in Black, as the tone quality of the pennant appears darker than the Red and Green of the roundel and fin. It is not sure that this was the personal airplane of Fornagiari.

## *Profile 38*

**Hanriot Hd.1, serial no. Hd. 6112, *serg.* Alvaro Leonardi, 80ª Squadriglia, early 1918.**

In the dramatic situation after the breakthrough at Caporetto, when many aircraft were lost, there was no time for decoration of aircraft. A clear example is this Hanriot, again one of the earliest of Macchi production, used by Leonardi, without any personal insignia and curiously without roundels on the top wings.

## *Profile 39*

**Hanriot Hd.1, serial no. Hd. 6212, *ten.* Amedeo Mecozzi, 78ª Squadriglia, January 1918.**

Fucini recalls in his memories that Mecozzi used many insignia, and one of his first ones was a question mark. This Hanriot of early Macchi production shows the question mark and two bands around the fuselage. Blue was chosen as color of the circle and bands because a combat report of June 1918 describes Mecozzi's airplane as carrying "two blue bands on its fuselage". Also another Mecozzi's insignia was a Blue pennant, similar to the ones shown for Riva and Fucini.

## *Profile 40*

**Hanriot Hd.1, serial no. Hd. 6239, *s.ten.* Silvio Scaroni, 76ª Squadriglia, autumn/ winter 1917/ 1918.**

Scaroni claimed thirty victories, all of them flying a Hanriot. He was the pilot who best exploited the exceptional qualities of this small French fighter. This one is probably his first aircraft of this type. It shows, besides the usual insignia of nationality, number 76 in white on the fuselage and on the top wing. The personal insignia of the pilot, a white square bordered black, appears on the fuselage, fin and tailplanes. The engine cowling appears White-Silver, according to a painting of his aircraft commissioned by Scaroni himself, and of which the author of profiles have a color photograph, although there are no known photos to confirm this. Colored engine cowling seems to be used by flight and unit commanders of the 76ª and 81ª Squadriglia.

## *Profile 41*

**Hanriot Hd.1, serial no. Hd. 6252, *s.ten.* Alessandro Resch, 70ª Squadriglia, spring 1918.**

Resch scored his first victory flying a Voisin, and then he had to wait until 1918 to score again. His Hanriot shows the Squadriglia number on the sides of the fuselage and above the top wing, and an individual number 13. On the fin there is the collar tab of the Bersaglieri corps to which Resch belonged. The ailerons, headrest and the entire horizontal stabilizer are painted in a dark color, possibly the same of the collar tab, that is to say Crimson Red. The upper view of this aircraft is depicted in the badge section of the book.

### *Profile 42*

**Hanriot Hd.1, serial no. Hd. 6254, *s.ten.* Giorgio Michetti, 76ª Squadriglia, spring 1918.**

An aircraft of Macchi production with the number of the 76ª Squadriglia on the fuselage sides and above the top wing. The personal insignia is a White seahorse on a Black background. Documents preserved in the archives of USSMA confirm the colors of this insignia, which appears also on the tailplanes. A upper view of this aircraft is depicted in the 76° squadriglia's badge page.

### *Profile 43*

**Hanriot Hd.1, serial no. Hd. 6272, ten. Flavio Torello Baracchini, 76ª Squadriglia, spring 1918.**

Hanriot of Macchi production with the emblem of Baracchini positioned as on Hanriot HD 515. As markings of nationality on Macchi aircraft were different from French machines, the fuselage roundel is partly covered by the four aces, which appear also over the horizontal tailplanes.

### *Profile 44*

**Hanriot Hd.1, serial no. unknown, *s.ten.* Giorgio Michetti, 76ª Squadriglia, spring 1918.**

A line-up of Hanriots of the 76ª Squadriglia at Casoni in the spring of 1918 shows some aircraft with a very dark finish, irregularly applied over wings and fuselage. The airplanes of aces Scaroni and Michetti kept their personal insignia with reversed colors, to improve visibility. This profile is a presumed presentation of one of these aircraft. The seahorse is slightly different from the one appearing in profile 42.

### *Profile 45*

**Hanriot Hd.1, serial no. Hd. 7517, *ten.* Silvio Scaroni, 76ª Squadriglia, June/July 1918.**

Scaroni received this machine, probably the most famous Hanriot of them all, on 16 or 17 June 1918, having destroyed his former fighter in a forced landing due to engine failure on 15 June. Flying this Hd.1 Scaroni scored nine victories in three weeks, and then he was shot down and wounded on 12 July 1918. This fighter carries an armament of two machine guns, as conceived and tested by the pilot, and the Squadriglia number on the fuselage and above the top wing. The personal emblem, a White square with a Black border, appears also over the tailplanes.

### *Profile 46*

**Hanriot Hd.1, serial no. H.d. 11383(?), *ten.* Flavio Torello Baracchini, 81ª Squadriglia, June 1918.**

On 1 June 1918 Baracchini returned to the 81ª Squadriglia, where he had served originally in 1917. He left his Hanriot with the four aces to *ten.* Censi and took back as his personal insignia the Black shield. He obtained most of his final victories with this airplane until he was wounded on 26 June 1918. Initially it carried only one machine gun but later, following the example of Scaroni, a second gun was fitted on the right side of the fuselage. This airplane at some point appears as having a darker engine cowling and wheels, possibly Green. One could guess that flight leaders within the VI Gruppo sported engine cowlings of different colors (Green, Red and White). The serial number, probably Hd 11383, disappears under the shield and it was not repainted elsewhere.

### *Profile 47*

**Hanriot Hd.1, serial no. Hd. 19209, *serg.* Aldo Bocchese, 70ª Squadriglia, autumn 1918.**

A Macchi production aircraft with the individual number 17 in Black. The headrest fairing is colored, possibly Red. The pilots of the 70ª Squadriglia didn't fear to use the numbers 13 and 17, considered bad luck omens in Italy. This airplane was delivered to the Squadriglia on 8 October 1918.

### *Profile 48*

**Hanriot Hd.1, serial no. Hd. 19254, *ten.* Leopoldo Eleuteri, 70ª Squadriglia, summer 1918.**

A report from an observation post describes a combat in which a Sopwith shot down an enemy aircraft. The commander of the 70ª Squadriglia explained that it was not a British Sopwith Camel but an Hd.1 *"by Macchi, spotted like the British ones and carrying white bands and personal number on its fuselage"* of pilot Eleuteri. Obviously this Hanriot was painted Dark Olive Green like the Camels. If the observer made a mistake, this kind of finish must have been rather unusual. What is certain is that French and Italian Hanriots could be distinguished even from a distance. For instance a pilot who witnessed from afar the shooting down of *serg.* Nava, over the Montello on 19 June 1918, described it in his report as an *"Hd of French production"*. This airplane sports the colored bands typical of the 70ª Squadriglia, the unit number 70 White and the individual number 4, also White, behind the roundel, and a fin that appears in varnished linen, maybe the two "White" bands reported by Avet. Serial number was erased, but it is believed to be 19254.

### *Profile 49*

**Hanriot Hd.1, serial no. unknown, *ten.* Mario Fucini, 78ª Squadriglia, autumn 1918.**

This Hanriot of Macchi production features one of the few examples of camouflage applied in the field. It has a finish of irregular bands, maybe of different colors, thicker on the fuselage and upper wings, softer on the lower wings. The stencil ALZARE (lift here) on the fuselage is partly overpainted. This airplane has an armament of two guns. The number 78 in black appears on the fuselage; later on it received a thin White border to improve its visibility. The White flag (Fucini called it with the heraldry term of "Silver pennant") shows 12 little black skulls, so the original photo can be dated after 28 August 1918, when Fucini claimed his 12th victory.

### *Profile 50*

**Hanriot Hd.1, serial no. Hd. 20960, *serg.* Cosimo Rennella, 78ª Squadriglia, autumn/winter 1918.**

Hanriots produced in Italy received a camouflage finish only close to the end of the war. This finish appears to be formed by patches of different colors painted over the yellowish basic finish, and then spotted with a lighter paint. The finish appearing here is a general reconstruction, because the new camouflage varied considerably among the different aircraft. Hanriot serial 20960 was probably delivered and photographed after the end of the war, as it doesn't show any armament nor any personal or Squadriglia emblem.

## *Profile 51*

**Spad 7, serial no. S 1420, *ten.* Antonio Cabruna, 77ª Squadriglia, summer/autumn 1918.**

The Spad 7 was a most important fighter of the Italian air force in the Great War. It was delivered to combat units since April 1917 but it was not built under license in Italy, possibly because it had a very complex structure that did not fit with the construction capacities of Italian aviation industries. Its main user was the 91ª Squadriglia, but other units used it too, particularly the 77ª Squadriglia, that in the second half of 1917 had a mixed complement of Nieuport 11, 17 and Spad. The aircraft shown here shows the typical factory finish of Spads, with natural doped linen, while metal parts are painted in a similar darker color, which could vary from light Tan to a rather darker Brown. This profile is based on the colors of the original Spad of Cabruna as photographed before its restoration, but one should consider that the passing of years may have markedly altered the tonality of pigments. This airplane here shows the insignia it carried during wartime. It still shows on its top wings roundels with a Red outer ring but strangely the fuselage roundel, hidden by the emblem, had a Green outer ring that is visible though the fading White paint. There are the three colored sections below the bottom wings, with natural doped fabric instead of the central White section. The fuselage roundel, overpainted in White, forms the background for the unit emblem, the Red heart. Cabruna added his personal insignia on both sides of the fuselage, the shield of his hometown Tortona. Over the top wing there is his individual number, 13 in Roman numerals. His good friend Leonardi of the 80ª Squadriglia, the twin unit in the XIII Gruppo, also had 13 as personal number. After the end of the war nine Black crosses were painted on the spine of the fuselage to indicate the victory claims of ace Cabruna. As they don't appear in wartime photographs, they are not reproduced here. The lateral panels of the engine cowling were often removed during hot periods for better engine cooling.

## *Profile 52*

**Spad 7, serial no. S 4689, *s.ten.* Luigi Olivari, 91ª Squadriglia, May 1917.**

The 70ª was the first Squadriglia to receive the Spad, aircraft serials 4688 and 4689 were ferried by Ruffo and Olivari. When the 91ª Squadriglia was created, on 1 May 1917, the best pilots and all the Spads were assigned to his new unit. The one shown here is a Spad as it was delivered to the front, with French-style roundels over the top wing, with Red outer ring. Italians added fuselage roundels and colored sections below the bottom wings. These aircraft also retained their French serial number. The fuselage roundel was probably added as a further identification for a new and unusual airplane appearing at the front.

## *Profile 53*

**Spad 7, serial no. S 4702, *cap.* Francesco Baracca, 91ª Squadriglia, May 1917.**

This airplane shows the standard finish of Italian Spads, with the Black prancing horse of Baracca on the fuselage. The design of the horse's mane is more detailed than the one appearing on the Nieuport 17 of Profile 13. Baracca used several Spads: in the photograph from which this profile is derived the serial number is not visible. The chosen one, S 4702, is the serial of the first Spad he used, that arrived at the 91ª Squadriglia on 7 May 1917 and was destroyed in an accidental fire on the ground on 7 July, after Baracca had scored his 13th victory.

## *Profile 54*

**Spad 7, serial no. S 4703, *ten.* Fulco Ruffo di Calabria, 91ª Squadriglia, May 1917.**

When the Spad arrived, the victories of Ruffo increased considerably. In May and June 1917 he was always in action, at an outstanding operative rate, and he often returned to base with a damaged aircraft. His fighter at that time sported the standard finish of Italian Spads, with roundels over the top wings and fuselage, and colored sections below the bottom wings. On this airplane Ruffo painted his Black skull over the fuselage roundel. It differs only in small details from the skull that previously appeared on his Nieuport 17. It was reported that Ruffo, an easy-going and kind prince, made an exception to his democratic attitude in the fact that he used his authority to forbid any other Italian pilot to use the skull as his personal insignia. The serial number shows that this is the Spad that was written off due to the battle damage sustained on 20 May 1917, after having brought its pilot safely home. Other aircraft used by Ruffo were S 4702 of Baracca, once, and mainly S 4705 in May and S 5408 in July 1917.

## *Profile 55*

**Spad 7, serial no. unknown, *s.ten.* Luigi Olivari, 91ª Squadriglia, spring/summer 1917.**

It was reported that Baracca, Tacchini and Olivari were called "the tricolor patrol", maybe because for a short period they had the noses of their aircraft painted in the national colors: White Baracca, Red Olivari and Green Tacchini. There is no proof to confirm this, but Olivari did use a Spad 7 with its engine cowling painted dark, and it is believed that the color was Red. An enemy pilot captured in that period said that one of the most feared enemies was the one with the "Red nose" which would confirm this option. The only known serial number of a Spad flown by Olivari is S 4700.

### *Profile 56*

**Spad 7, serial no. S 4691, *s.ten.* Ferruccio Ranza, 91ª Squadriglia, spring/summer 1917.**

In his first period with the 91ª Squadriglia Ranza introduced his personal insignia, this beautiful owlet. This airplane arrived at the unit on 1 May 1917 and remained operational until November. Unfortunately there is a single photo of this emblem, and as the lower part of it is not visible, one cannot be sure if the bird also had a tail or if it rested on a branch. This artist's choice was not to paint what has not been seen. Other serial numbers of Spads flown by Ranza in the summer of 1917 were S 1307, S 4704, S 4789, S 4883, S 5491 and in November 1917 S 6367.

### *Profile 57*

**Spad 7, serial no. S 4695, *ten.* Bartolomeo Costantini, 91ª Squadriglia, July 1917.**

Costantini adopted a very plain emblem, while other pilots of the squadron had more elaborate designs. Probably he used the Black pennant a long time, up to the end of the war, despite the fact that in 1918 many aircraft of the 91ª Squadriglia sported the Black griffin on the fuselage. *Ten.* De Bernardi, who was assigned to the 91ª Squadriglia in the summer 1918, is reported as having flown a Spad decorated with a Black flag, so he probably inherited one of the aircraft of Costantini. In November 1917 Costantini flew frequently with Spad S 5398.

### *Profile 58*

**Spad 7, serial no. S 4699, *ten.* Luigi Sabelli, 91ª Squadriglia, summer/autumn 1917.**

The fighter of Sabelli was specifically described as *"the Spad with the ladder"* in a combat report. We don't know why the pilot adopted this particular emblem. Sabelli was shot down in combat by German fighters in October 1917 as he was flying Spad serial S 4699. It seems that from June 1917 to the day of his death he flew exclusively this aircraft. The photo from which this profile is based have a caption that described it as being taken in Casoni airfield, but is probably a mistake, confusing Sabelli's badge with the same one used by Ranza later on. In fact Ranza on 21 November 1917 shot down a German two-seater over Casoni airfield, but the aircraft serial number is known to be S 6367 and not this one.

### *Profile 59*

**Spad 7, serial no. S 1544 - 4707, *ten.* Giuliano Parvis, 91ª Squadriglia, summer/autumn 1917.**

Also the fighter of Parvis was specifically described in a combat report as *"the Spad with the Black crescent"*. The color of this emblem is thus confirmed, and Black paint was indeed available in all units, and particularly the 91ª Squadriglia. On the rudder there are two serial numbers: possibly the rudder was cannibalized from another Spad and instead of removing the old serial 1544 the correct one for this Spad, 4707, was painted below the older one. This airplane was delivered to the squadron on 28 September 1917 and Parvis used it extensively in November 1917. Other airplanes that he used in the summer of 1917 were S 4690 and S 4697.

### *Profile 60*

**Spad 7, serial no. unknown, *s.ten.* Flavio Torello Baracchini, 76ª Squadriglia, July-August 1917.**

While the 91ª Squadriglia was the only unit completely equipped with the French fighter, also other fighter units received a few Spad aircraft. One of them was the 76ª Squadriglia, which had at least three of them in summer 1917, one being used by the ace Baracchini, who at that time was the only pilot who could compare with the aces of the 91ª Squadriglia and compete with them for leadership. When on 3 August 1917 *cap.* Baracca shot down an aircraft together with Baracchini he described the other fighter as *"Spad showing the number 76 on its fuselage"*. The use of the number of Squadriglia on the fuselages of airplanes began around July 1917 at the 76ª and has remained as a constant of Italian Air Force aircraft to this day. Baracchini converted on the Spad at the 91ª and he started using it in operations since 16 July, when he was shifted from the 81ª to the 76ª Squadriglia. He flew 43 combat sorties with this type, claiming 4 or 5 victories. The proportion of the rings of the roundel is different from common practice.

### *Profile 61*

**Spad 7, serial no. unknown, *s.ten.* Carlo "Francis" Lombardi, 77ª Squadriglia, autumn 1917?**

This Spad shows the emblem of the 77ª Squadriglia painted in an unusual way. Photographic evidence shows that the first Nieuports of the 77ª had a Red heart painted directly on the fuselage, with no background. Later on, when Nieuports of Italian production arrived, in the summer of 1917, it was painted over a White circle painted over the fuselage roundel. This practice was extended to the Spad 7s of the unit. This is one of the few cases known so far of the heart emblem painted in an unusual way. In this case also the proportion of the rings of the roundel are different than the ones commonly used.

### *Profile 62*

**Spad 7, serial no. unknown, *ten.* Luigi Sabelli, or *s.ten.* Ferruccio Ranza 91ª Squadriglia, summer/autumn 1917.**

This aircraft shows another version of Sabelli's badge, with the latter painted larger than that of profile 58. It is not known when this aircraft was photographed: it could be either another Sabelli's aircraft or a Spad flown in autumn 1917 by Ranza, who adopted this badge after his close friend's death in combat.

### *Profile 63*

**Spad 7, serial no. S 6334, *ten.* Guido Novelli, 91ª Squadriglia, November 1917/Spring 1918.**

There have been doubts whether the airplane with the "ace of Clubs" was the one of Novelli, particularly considering that the clover was indicated in a document as the personal insignia of Adriano Bacula. Data in the archives however confirm that Novelli constantly used this airplane, serial S 6344, in November 1917. In 1918 Novelli adopted a white cowl as personal insigna and it is possible that his old aircraft and badge, was then used by Bacula.

## *Profile 64*

**Spad 7, serial no. unknown, *ten. col.* Pier Ruggero Piccio, 91ª Squadriglia, 1918.**

Despite his role as commander of Italian fighter forces, *ten. col.* Piccio is not shown in many photographs. This profile is derived from the only known image of the ace in the cockpit of a Spad. What is know about his insignias is that when at the 77ª Squadriglia he used a Red heart as his personal emblem, which later became the unit insignia. Ranza in the early Sixties recalled the insignia of Piccio as "a Red heart and a diagonal red band on the fuselage" but so far no photographic evidence of these has been found. Scaroni wrote that he met Piccio during a sortie in the spring - summer 1918 and to have identified him thanks to his emblem, without describing it. This Spad shows the griffin, which was introduced in 1918 as unit insignia of the 91ª Squadriglia. We don't know if there were other emblems or Roman numerals on the rest of the fuselage. The lateral panel of the engine cowling is removed to improve cooling.

## *Profile 65*

**Spad 7, serial no. unknown, *ten.* Ferruccio Ranza, 91ª Squadriglia, May 1918.**

After the death of Sabelli, killed in air combat, Ranza took over his insignia, the ladder, as homage to his friend. The profile is drawn from a photo in a hot day of 1918, with the lateral panes removed. The roundel appears with a Green outer ring. An oil painting belonging to the Ranza family shows the Spad of the ace with a Black ladder and roundels with Green outer ring.

## *Profile 66*

**Spad 7, serial no. unknown, *cap.* Fulco Ruffo di Calabria, 91ª Squadriglia, March 1918.**

Even after the introduction of the Spad 13, the earlier Spad 7 was extensively used at the 91ª Squadriglia until the end of the war. This airplane, pictured in March 1918, carries no fuselage roundel, and it is adorned with a Black skull, personal insignia of Ruffo di Calabria, in a more elaborate version. It is not known if on the right side of the fuselage there was the skull emblem, or rather the Black Griffin of 91ª.

## *Profile 67*

**Spad 7, serial no. unknown, *s.ten.* Carlo "Francis" Lombardi, 77ª Squadriglia, 1918**

This Spad flown by Lombardi shows the emblem of the 77ª Squadriglia in its most common version, adopted in summer 1917 and used until the end of the war. The pilots of 77ª generally had no personal insigna on the fuselage sides of their aircraft.

## *Profile 68*

**Spad 7, serial no. unknown, *ten.* Michele Allasia, 77ª Squadriglia, spring 1918.**

A variation of the Red Heart insignia of the 77ª Squadriglia decorates the fuselage of the Spad 7 of Allasia. On the clear ribbon around the heart something is written, unfortunately not clear enough for interpretation.

## *Profile 69*

**Spad 7, serial no. S 2489, *cap.* Francesco Baracca, 91ª Squadriglia, November 1917.**

This is a Spad known to be used by Baracca on 7 December 1917 for scoring his 30th victory over Monte Kaberlaba. There is no further mention of its use in the squadron documents until 16 October 1918 when it was flown by Olivero. It remained in squadron service in 1919 too and then on 19 June 1921 it was donated to the Museo Baracca in Lugo di Romagna, decorated with Baracca's markings, the Black prancing horse on the left side and the Black griffin on the right. Further back on the fuselage there is, again in black, the individual number ten in roman characters. This Spad restored by G.A.V.S. association, is today preserved in the newly reopened Baracca Museum.

## *Profile 70*

**Spad 7, serial no. 139?, *ten.* Flaminio Avet, 70ª Squadriglia, October 1918.**

In late summer 1918 also the 70ª Squadriglia received some Spad 7. Clashes with enemy aircraft were rare at that time, and the Hanriot lacked the speed to intercept the new fast enemy two-seaters. This is one of the very few Italian Spad 7 with a camouflage finish. On its fuselage there is the roundel, the Squadriglia number in White and the vertical stripes as personal marking of the pilot that continue on the tailplanes. Roundels over the top wing are smaller than usual, and don't overlap the ailerons. The French construction number on the rudder was erased above the Red stripe: according to squadron documents that number could be either S 1398 or S1390.

## *Profile 71*

**Spad 13, serial no. S 2445, *magg.* Francesco Baracca, 91ª Squadriglia, April 1918.**

The Spad 13 of Baracca was usually presented wrongly with an unlikely solid Green finish. There never were Spads with that finish, as all Italian Spad 13 had the typical French camouflage scheme with five colors. Many French industries took part in the mass production of this fighter, and each one had its different way of applying camouflage colors and roundels: in this case the disposition of the color patches shows an aircraft produced by Bleriot. The airplane of Baracca additionally carries a fuselage roundel with Red outer ring, to improve visibility, while there are no roundels on the top wings. The lower surfaces of the bottom wings are here shown with the required color sections, but it cannot be determined whether they carrier roundels instead. On the inner strut there is a small pennant, maybe in metal, of unknown color. On the left side of the fuselage there is the *"Cavallino Rampante"* (prancing horse) insignia of the ace, inside a White cloud. According to the recollections of General Ranza and to known photographs, it is believed that on the right side there was the insignia of the griffin, squadron emblem of the 91ª since April 1918. This fact however cannot be positively confirmed. Lately some historians have guessed that the prancing horse was Red. Debate on this question has raged in Italy, but these authors are convinced that Black was the correct color. Serial number S 2445 is not certain. If it were so, this would be an aircraft delivered to the squadron on 14 February 1918, together with four other ones. This aircraft was turned over to an aviation park on 21 May 1918. Baracca was killed on 19 June 1918 and he probably was flying S 5382, a Spad 7 with 180 hp engine.

## *Profile 72*

**Spad 13, serial no. S 7206, *magg.* Francesco Baracca, 91ª Squadriglia, October 1917 - February 1918.**

This aircraft is the first Spad 13 delivered to Italy. The 91ª Squadriglia took it on charge on 4 September 1917, calling it "Spad 200 hp" in its documents. It became the personal aircraft of the commander, who used it, together with a Spad 7, during the hard combats after Caporetto, scoring many victories. On 25 October the gunner of a German two-seater hit it and brought it to a forced landing. This Spad was repaired and it remained in service until 22 February 1918, when it was severely damaged inside a hangar during a night bombing raid. Up to that time it was the only Spad 13 in service in the Italian air force. It is an early production aircraft, supposedly with rounded top wing tips. It is in natural linen finish, with the front section painted in a similar color that could vary in its tonality: this presentation is based on the colors of the Spad 7 of French ace Guynemère, before its restoration. When it was photographed this airplane didn't show on its fuselage the prancing horse, but only a roundel. Baracca scored at least four victories flying this Spad, always against German aircraft: three DFW C.5 reconnaissance airplanes, two on 22 October and one on 15 November 1917, and an Albatros fighter on 23 November.

## *Profile 73*

**Spad 13, serial no. unknown, *serg.* Guido Nardini, 91ª Squadriglia, July 1918.**

Since April 1918 the 91ª Squadriglia introduced the griffin as unit emblem. Each airplane of the squadron also carried an individual number in Roman numerals on the sides and deck of the fuselage. This Spad shows the standard finish, but in order to increase visibility of the griffin the lighter part of the camouflage on the rear fuselage was increased. Nardini destroyed this airplane in a forced landing due to engine failure (a common occurrence) on 14 July 1918. The Spad was reputed for its structural strength, and the pilot got out unharmed from the wreck. The serial number of this airplane is not known.

## *Profile 74*

**Spad 13, serial no. unknown, *cap.* Fulco Ruffo di Calabria, 91ª Squadriglia, September 1918.**

After the death of Baracca Ruffo became commander of the 91ª Squadriglia until September 1918, and then he became commander of the XVII Gruppo. A photograph dated 19 September 1918 shows him with other pilots next to this Spad 13 decorated with a Black skull. To improve its visibility the side of the fuselage received a light coating of White. The side panels of the cowling are removed. On the tricolor stripes on the rudder no serial number is visible. Like the Spad 13 of Baracca, this aircraft also doesn't carry roundels over the top wings. Again, it is not known if on the right side there was the skull emblem or rather the griffin. The fact that all known photograph are taken from the left side suggest that the personal emblem of the skull was only painted on the left side.

## *Profile 75*

**Spad 13, serial no. unknown, *cap.* Bartolomeo Costantini, 91ª Squadriglia, autumn 1918.**

This airplane carries a standard finish and its griffin, the squadron emblem, appears inside a small White cloud. The number 3 in Roman numerals on the fuselage sides and over the spine has a thin White shading. Costantini scored his last two confirmed victories flying a Spad 13. His victims were the commander of Flik 40/P, *Oblt.* Stillmungus, KIA, and the ace Friedrich Hefty, of Flik 42/J, who saved his life jumping with a parachute.

## *Profile 76*

**Macchi M.5, serial no. M 7256, *Ten. di Vascello* Orazio Pierozzi, 261ª Squadriglia, spring/summer 1918.**

The Macchi M.5 was probably the best seaplane fighter of the war. The most successful Italian naval ace was Orazio Pierozzi and this was his personal aircraft. The Macchis came out of the factory at Varese with the fuselage in natural varnished wood above, and White waterproof paint below. Wings initially were natural doped fabric, but later on, according to documents, they were painted Olive Green above. As Austrian seaplanes had similar shapes, all though the war both sides abounded in roundels and Black crosses for identification. This seaplane shows roundels with Green outer ring on the top wings and fuselage, and the national colors also appear in bands on the rudder and below the wings. The 260ª Squadriglia even added further roundels below the bottom wings. The Macchi of Pierozzi, serial number M 7256, features the distinguishing pattern of the 261ª Squadriglia, a line of White triangles along the fuselage, his individual number 18 and an insignia of a dog's head biting an Austrian seaplane. The engine cowling was often removed to improve cooling. Pierozzi scored most of his victories with this fighter.

## *Profile 77*

**Macchi M.5, serial no. M 7087, *Ten. di Vascello* Orazio Pierozzi, 255ª Squadriglia, autumn 1917.**

This is a Macchi M.5 of the initial production batch, with early faired lateral floats. The engine support and its frontal strut both have their fairings. On the fuselage there is the individual number "4".

## *Profile 78*

**Macchi M.5, serial no. M 7242, *Ten. di Vascello* Federico Martinengo, 260ª Squadriglia, spring/summer 1918.**

This Macchi shows the red and green fuselage bands indicating

260ª Squadriglia. The engine cowling is removed. Individual markings are the number "1" and a winged rat. Many aircraft of the 260ª carried roundels under the bottom wings, but we cannot be sure if this particular aircraft actually carried them.

## *Profile 79*

**Macchi M.5, serial no. M 7289 (?), *S.ten. di Vascello* Umberto Calvello, 260ª Squadriglia, spring/summer 1918.**

Calvello, also a pilot with the 260ª Squadriglia, had a Macchi M.5 similar to the one of Martinengo, marked with a "2" and with one more "Fortunello" walking with a suitcase with a patriotic motto: "marciare non marcire" (March, don't rot!) and "Ocio! Ocio fiol d'un can", (Watch out, son of a dog) in the dialect of Venice. Also in this case the engine cowling is removed, but not the fairing of the front engine strut. Only the first three digits of the serial number are visible, from known data full serial can be only either M 7283 or M 7289.

## Squadron and personal markings

The beauty of some personal emblem cannot be fully appreciated on the profiles depicted: we have chosen to show some of them in a larger size.

***Page A shows the markings of 78ª Squadriglia aces:***

A1 – A2 )
Personal emblem of *serg*. Guido Nardini, Hanriot Hd.1 (see profile number 27). The upper view shows the early production Hanriots with roundels having the red at the outside of the cockade.

A3 )
Squadron emblem of a pennant carried by 78ª Squadriglia Nieuports and Hanriots Hd.1 on summer 1917, common to all aircraft. (See profile number 17 and 27 )

A4 )
Personal marking of *serg*. Fornagiari, Hanriot Hd.1, painted over the fuselage roundel. (See profile number 37)

A5 )
Personal marking of *capitano* Antonio Riva, Hd.1. (See profile number 34)

A6 )
Personal marking of *ten*. Amedeo Mecozzi, Hd.1. (See profile number 39)

A7 - A8 - A9 )
Upper view and personal marking of *ten*. Mario Fucini, Hd.1, armed with twin machine guns. (See profile number 49)

***Page B shows the markings of 76ª Squadriglia aces:***

B1 - B2 )
Personal markings of *ten*. Flavio Torello Baracchini, Hanriot Hd.1. Of note are the squadron number on the center upper wing and the four aces on the tailplanes. (See profile number 29 )

B3 - B4 )
Personal markings of *capitano* Giulio Lega, Hanriot Hd.1, applied even on the upper tailplanes (See profile number 30).

B5 )
Personal markings of *serg*. Romolo Ticconi, applied over the fuselage, instead of over the tailplanes. (See profile number 31)

B6 - B7 - B8 )
Personal markings of *s.ten*. Giorgio Michetti, Hd.1. The seahorse is reversed on the tailplanes and on the left side of the fuselage. (See profile number 42 and 44 )

B9 - B10 )
Personal markings of *s. ten*. Silvio Scaroni, Hd.1, applied even over the tailplanes, in typical squadrons style. (See profile number 40)

***Page C shows the markings of several other Squadriglia aces:***

C1 )
Personal marking of *s. ten*. Flavio Torello, Baracchini Nieuport 11, painted over the fuselage deck (See profile number 10)

C2 )
Personal marking of *serg*. Attilio Imolesi, Nieuport 17 (See profile number 14)

C3 – C4 )
Personal marking of *ten*. Alvaro Leonardi, Hd.1.Of note is the different style of number 13 on the upper wing when compared with the more elaborated version applied on the fuselage side. (See profile number 35)

C5)
Personal marking of *serg*. Michele Allasia, Spad 7 . The writing on the band is not known (See profile number 68)

C6 –C7 )
Personal marking of *ten*. Antonio Resch, Hd.1. As common for squadron practice the squadron number 70 and personal number 13 are repeated on the upper wing too. The colored ailerons and tailplanes are probably flight leader markings.(See profile number 41)

C8 )
Personal marking of *serg*. Antonio Reali, Nieuport 17. The oil from the engine have partially scraped off the red paint of the marking. (See profile number 19)

***Page D show the markings of 91ª Squadriglia aces:***

D1 – D2 – D3 )
The evolution in detail of the personal markings of *maggiore* Francesco Baracca is easily visible here, where the black prancing horse is shown from his first version applied on the Nieuport 17, and then on Spad 7 and Spad 13. (See profiles number 13, 53 and 71)

D4 – D5 – D6 – D7 –D8 )
The evolution in detail of the personal markings of *capitano* Fulco Ruffo di Calabria is easily visible here, where the black skull and crossbones insignia is shown from his first version applied on the Nieuport 11 and then on Nieuport 17, Spad 7 and Spad 13. (See profiles number 6,11,54,66 and 74 )

D9)
Personal marking of *ten*. Guido Novelli, Spad 7 (See profile number 63)

D10)
Personal marking of *s. ten*. Ferruccio Ranza, Spad 7 (See profile number 56 )

D11)
The Black griffin, emblem of the 91ª Squadriglia in 1918 (See profiles number 64, 73 and 75 )

1

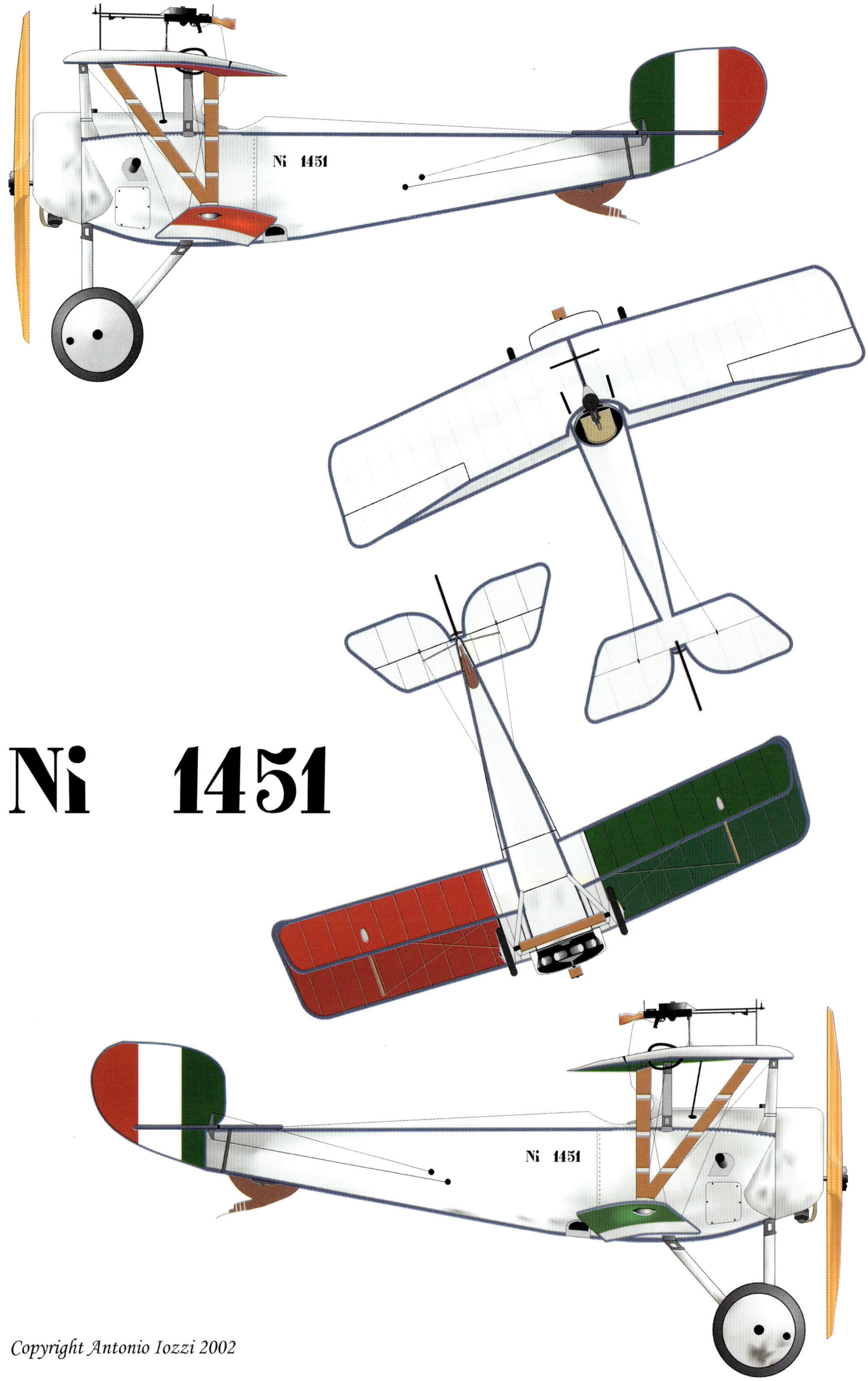

2

3

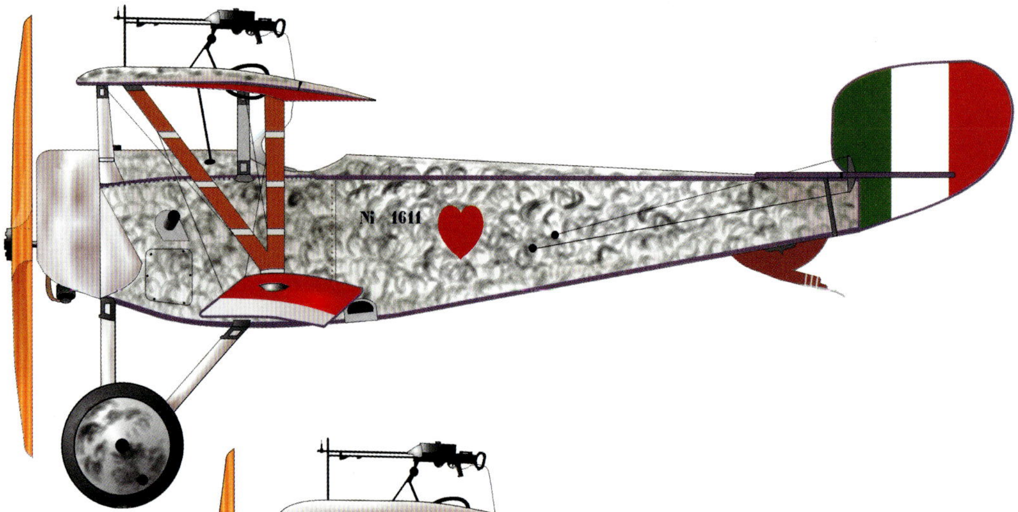


4

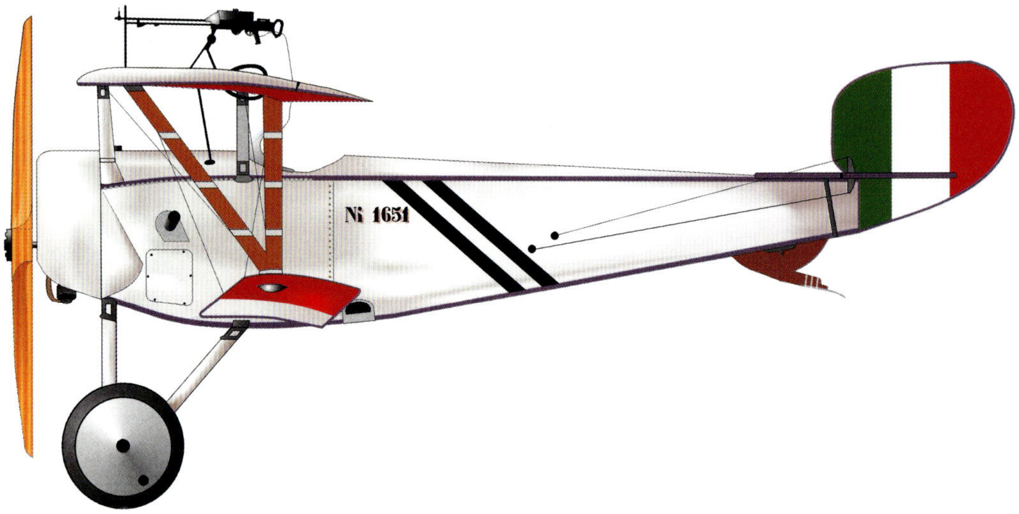


5

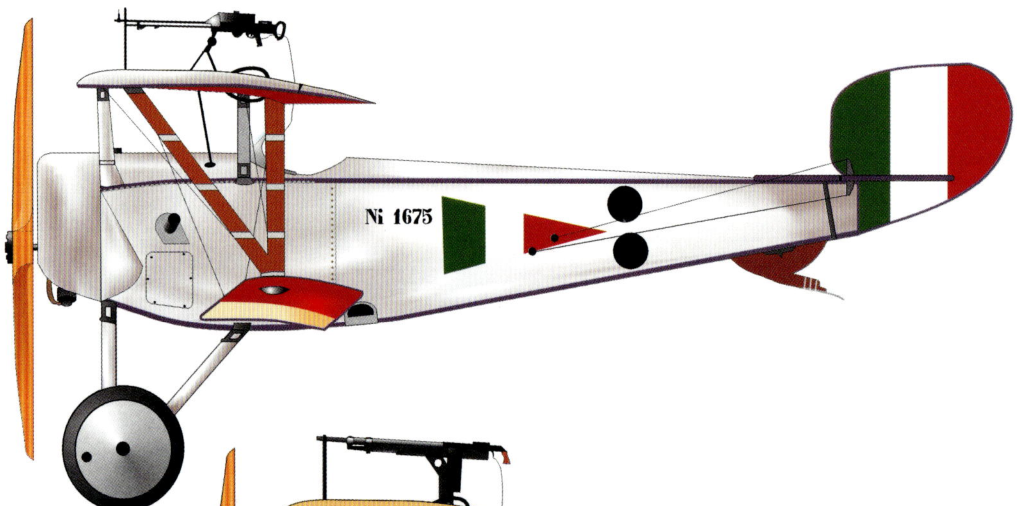


6

7

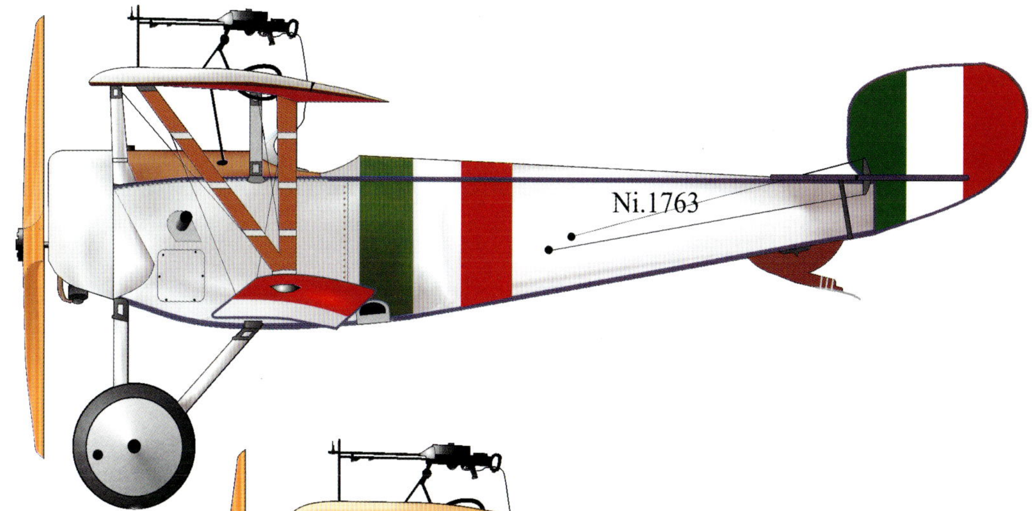


8

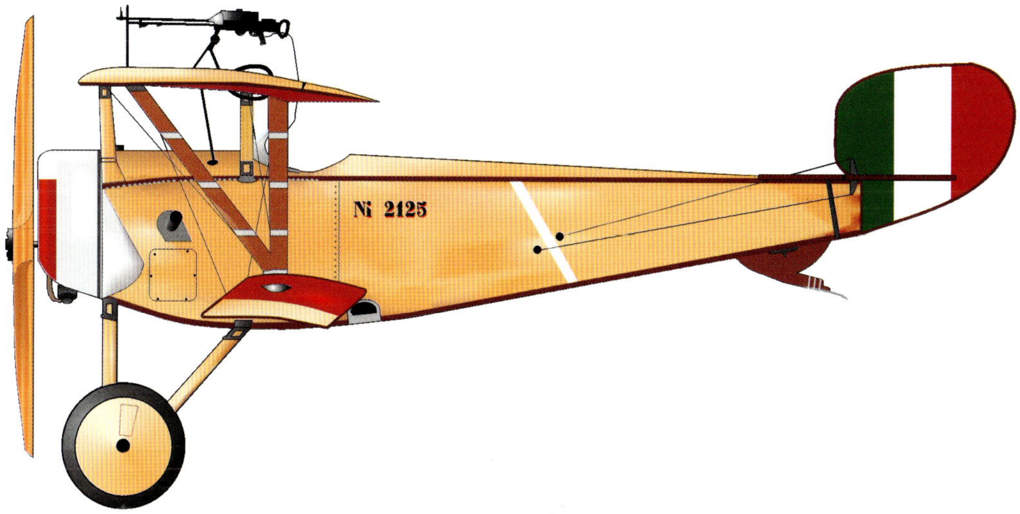


9

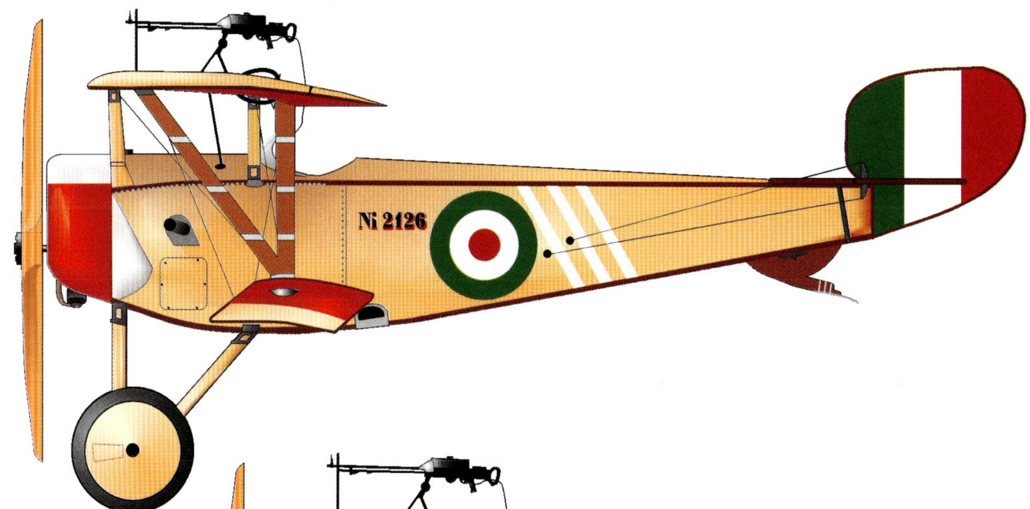


10

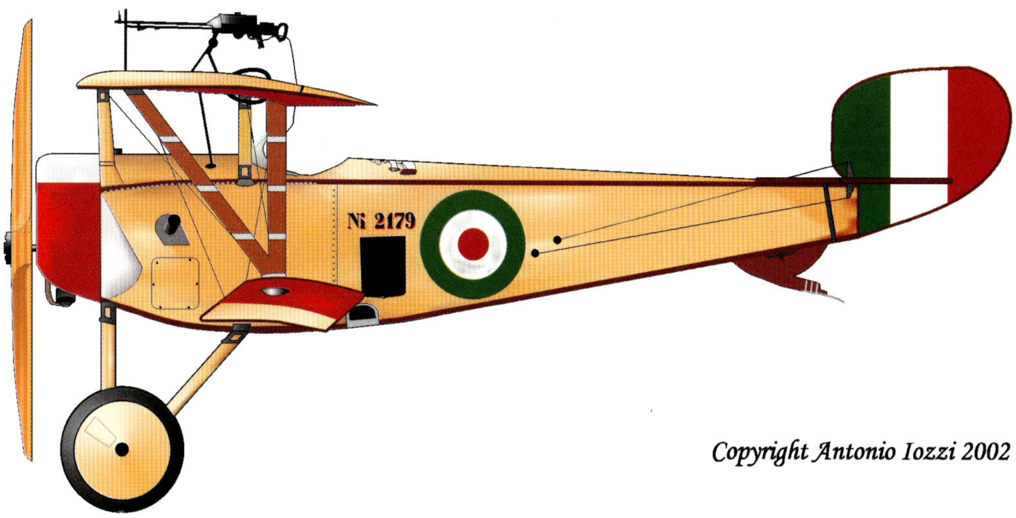

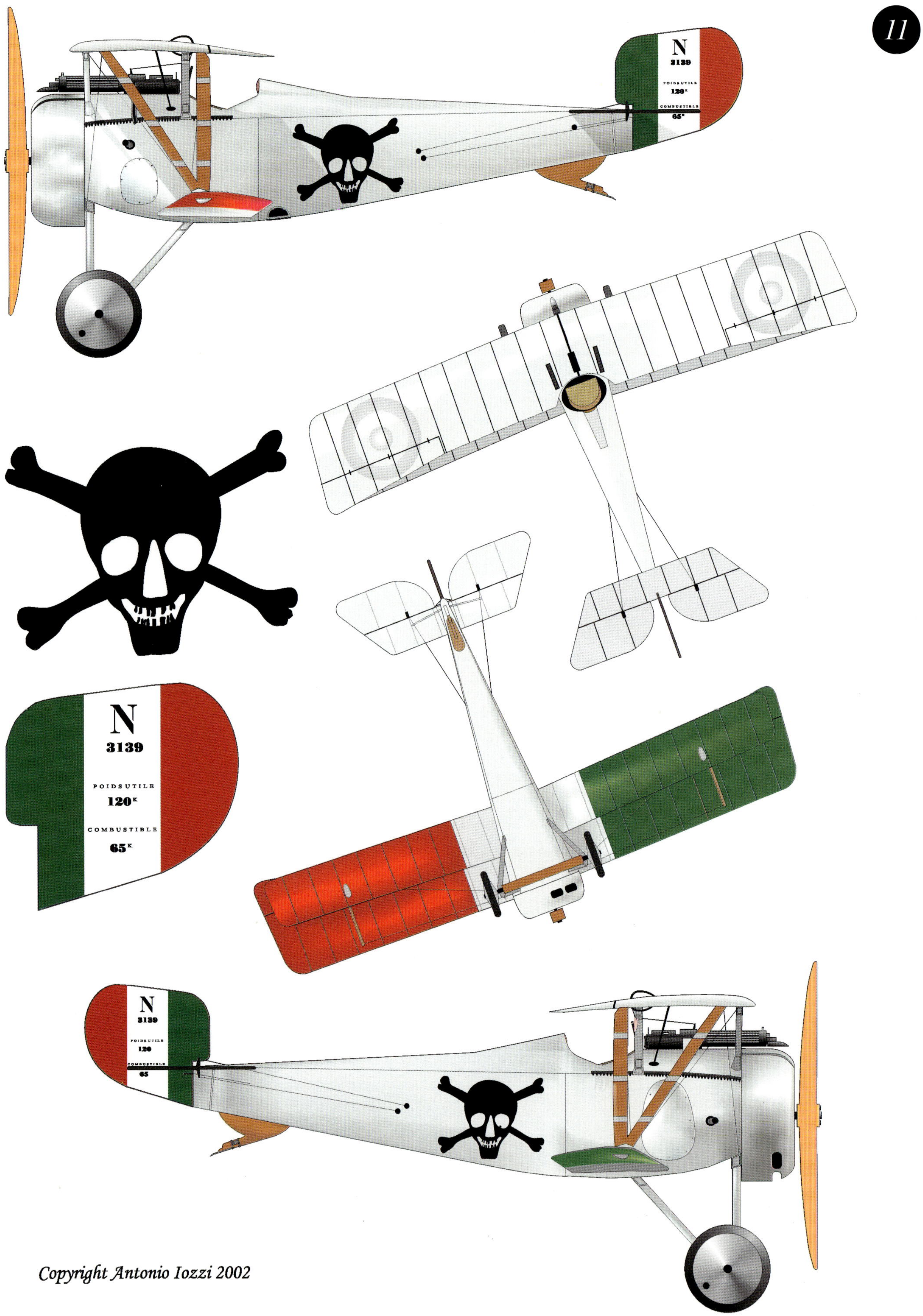
11
N
3139
POIDSUTILE
120K
COMBUSTIBLE
65K

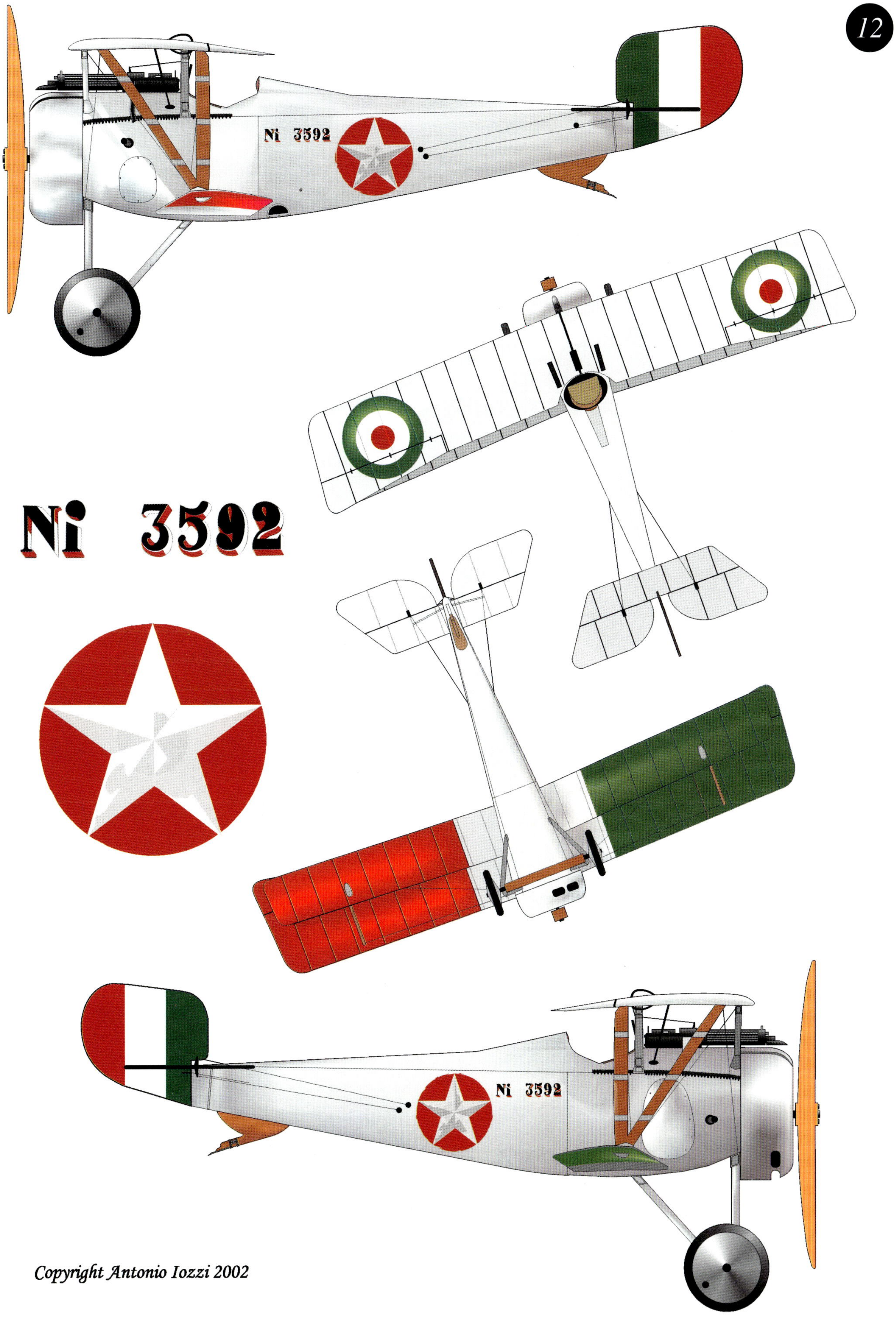
12
Ni 3592
Ni 3592
Ni 3592

13

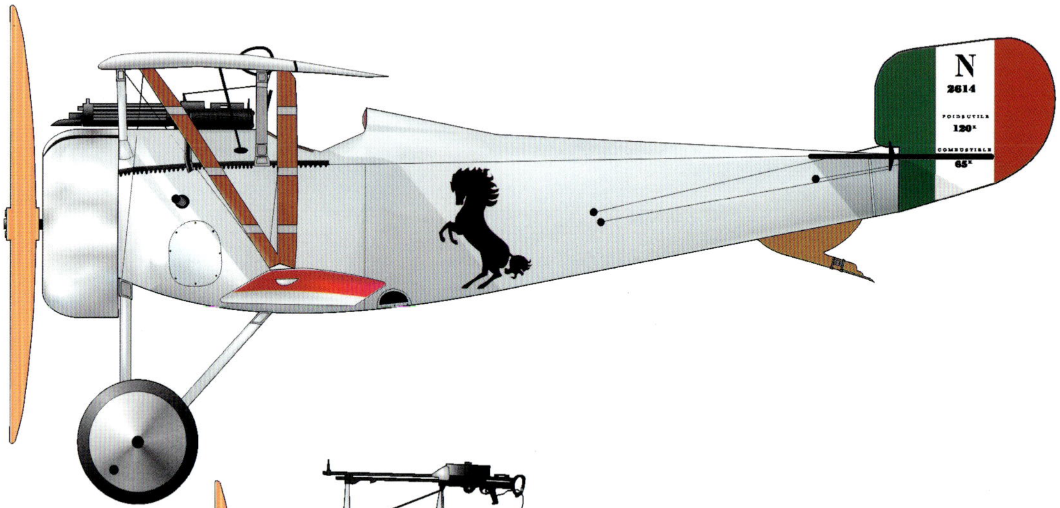


14

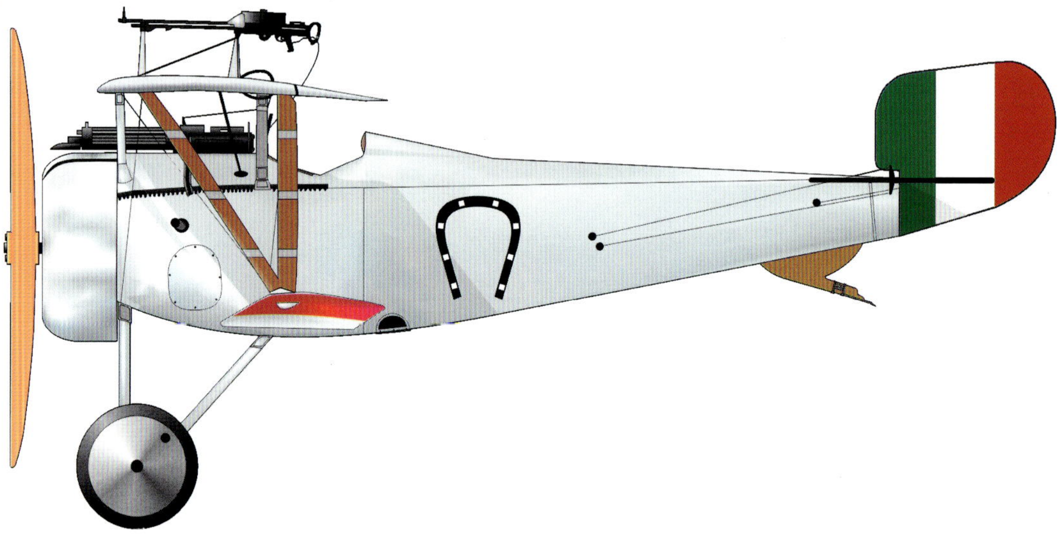

15

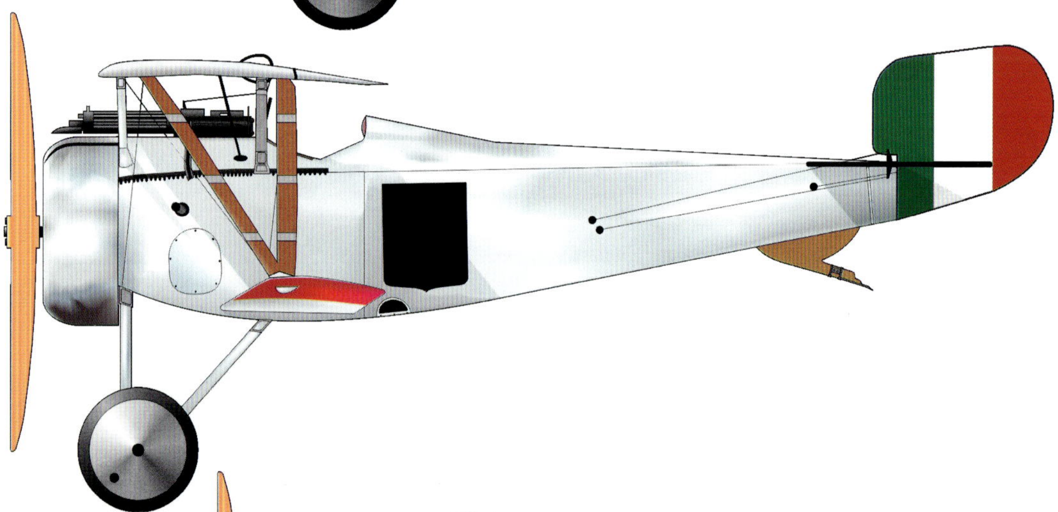

16

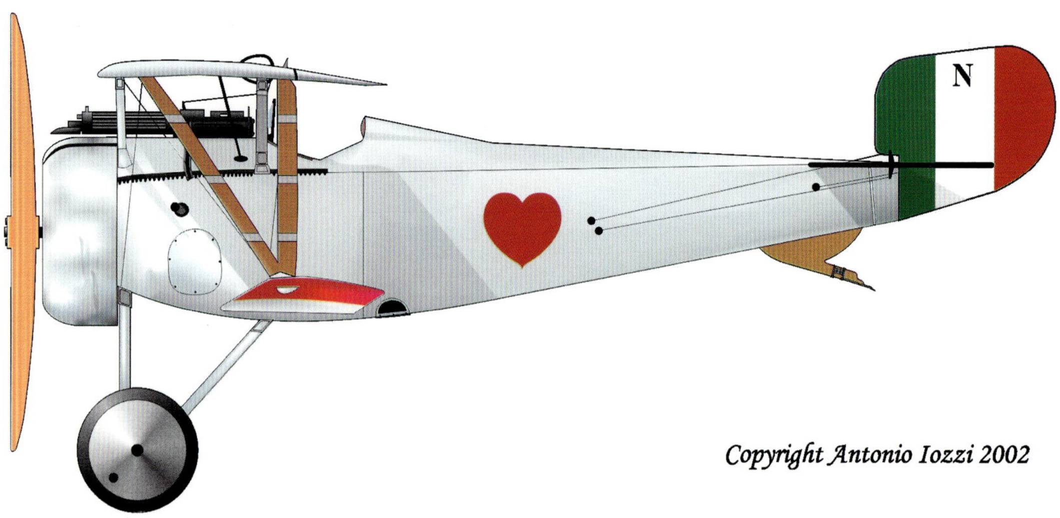

17

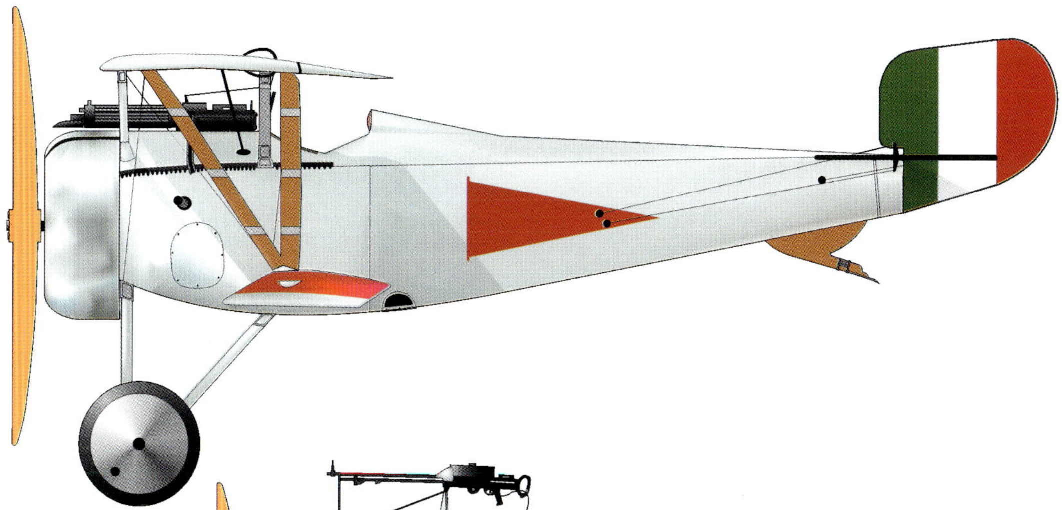

18

19

20

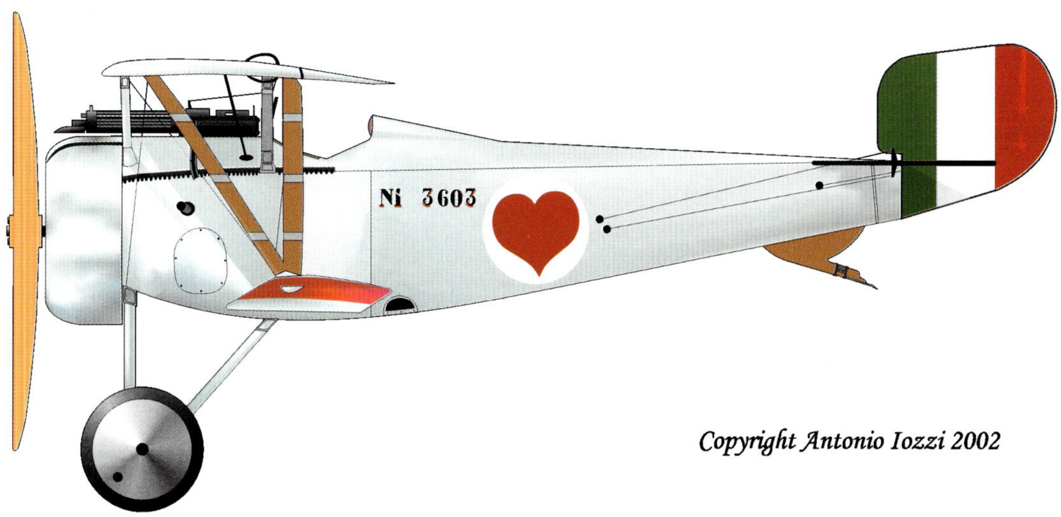

21

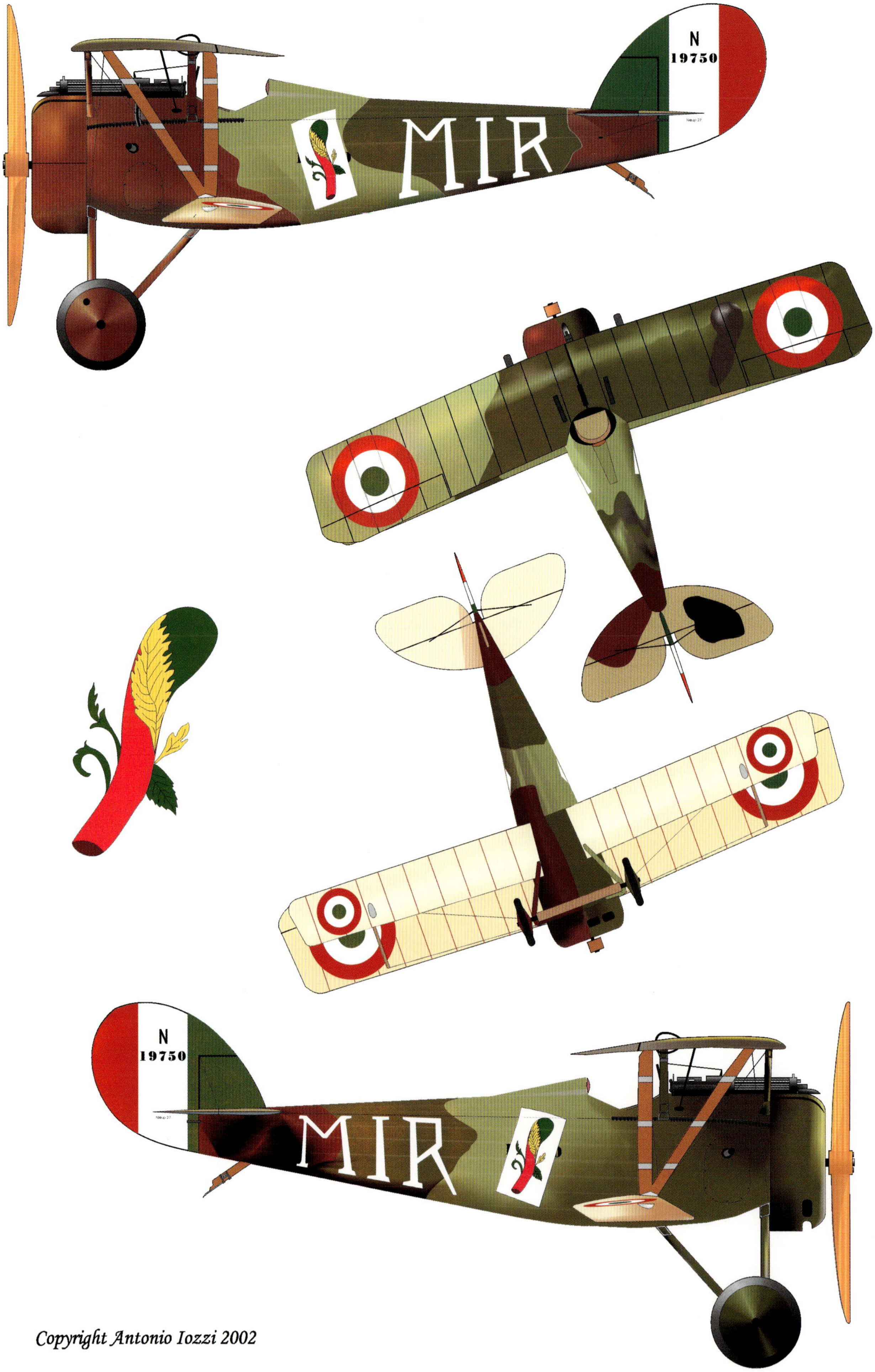

22

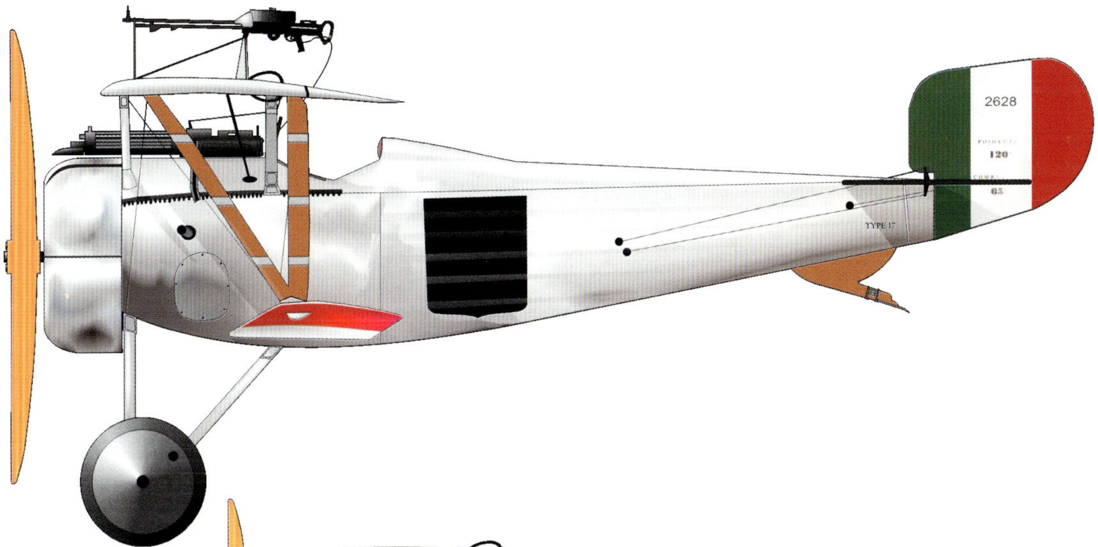


23

24

25

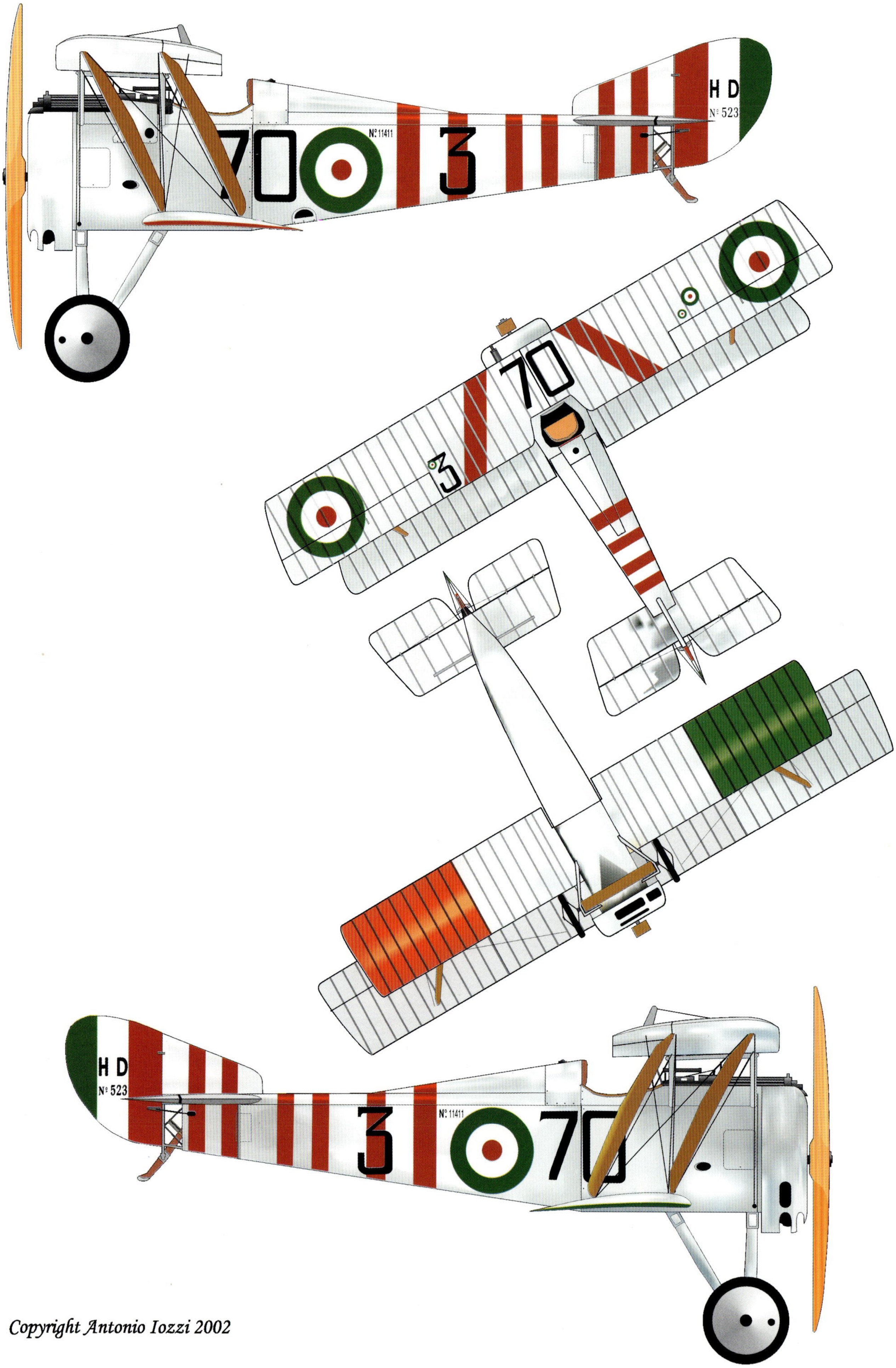

26

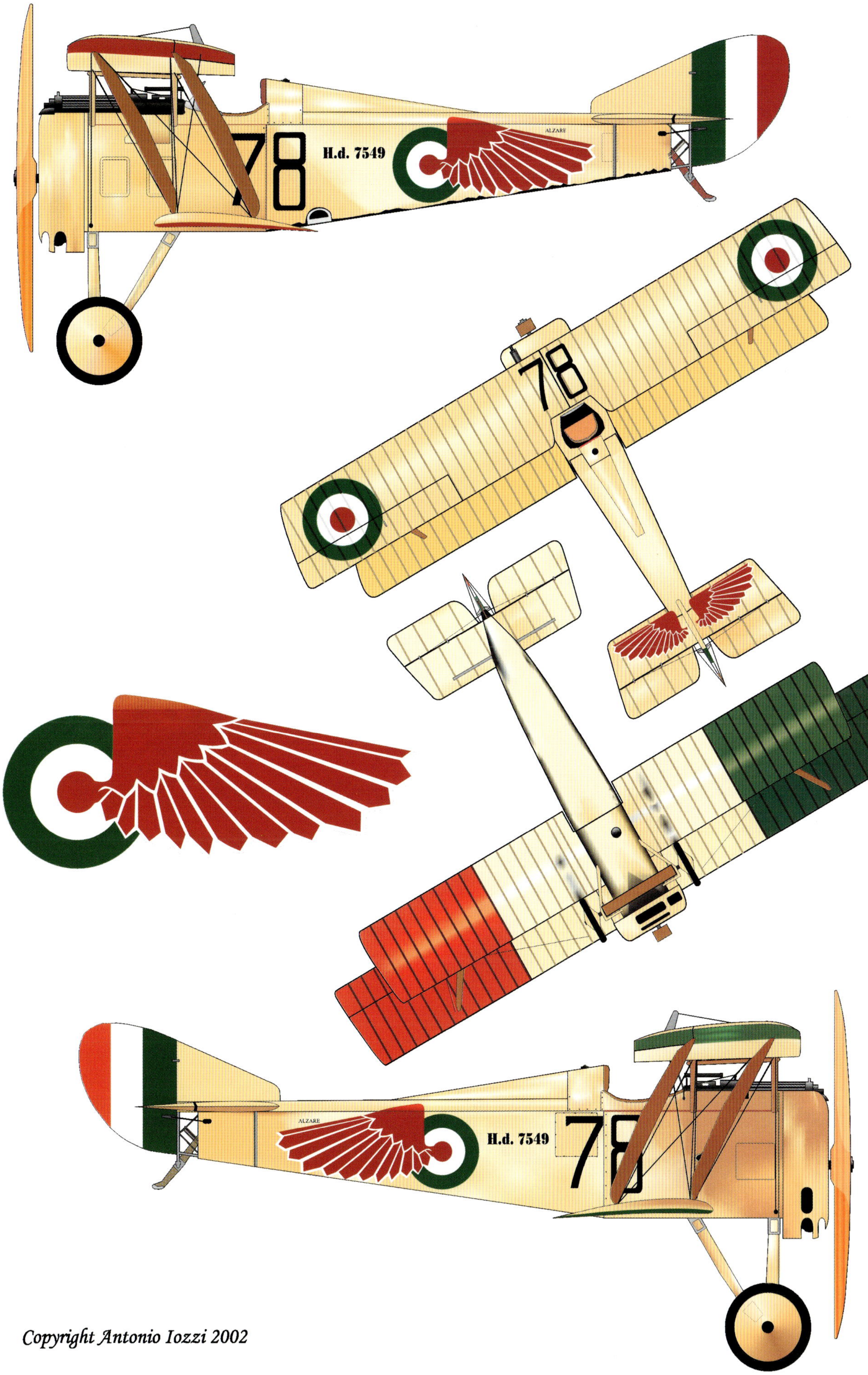

27

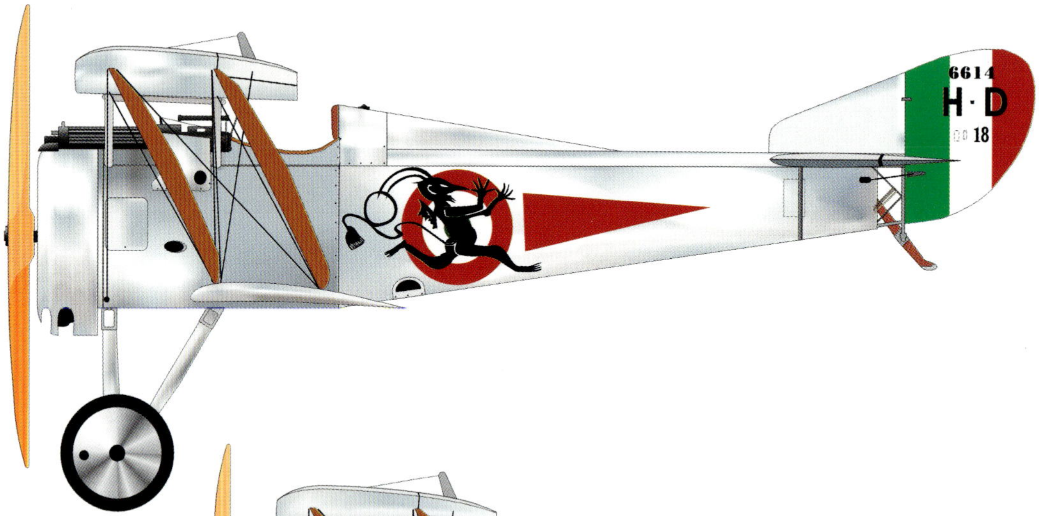


28

29

30

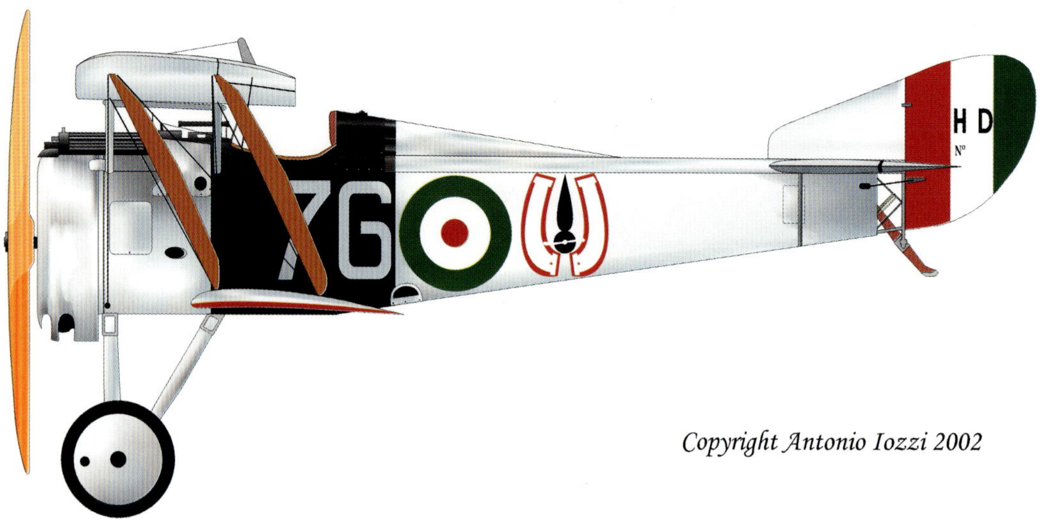

31

32

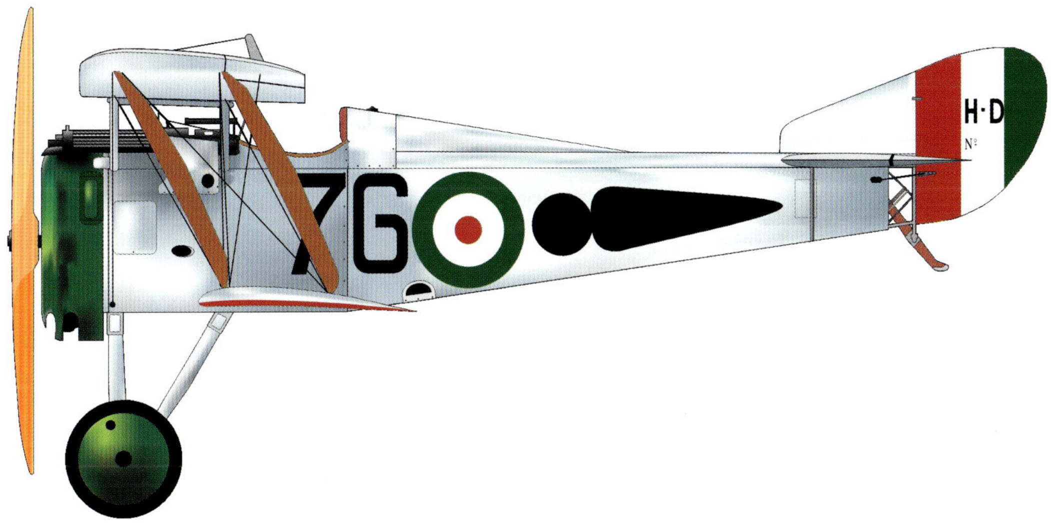


33

34

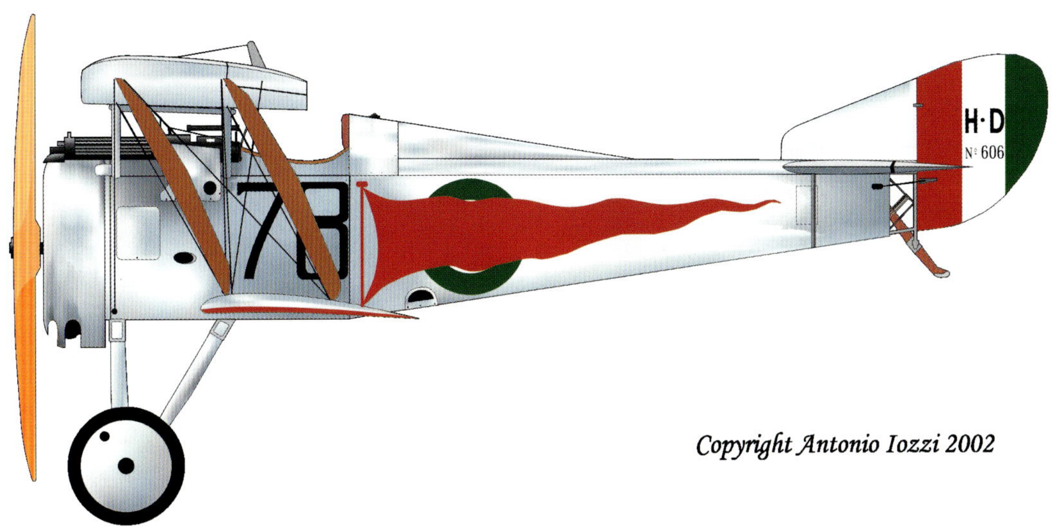

35

36

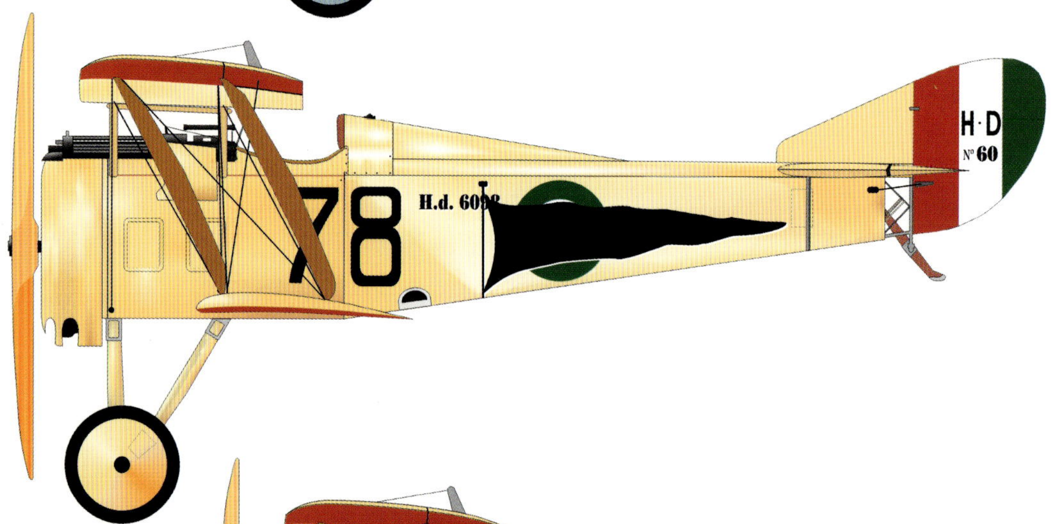


37

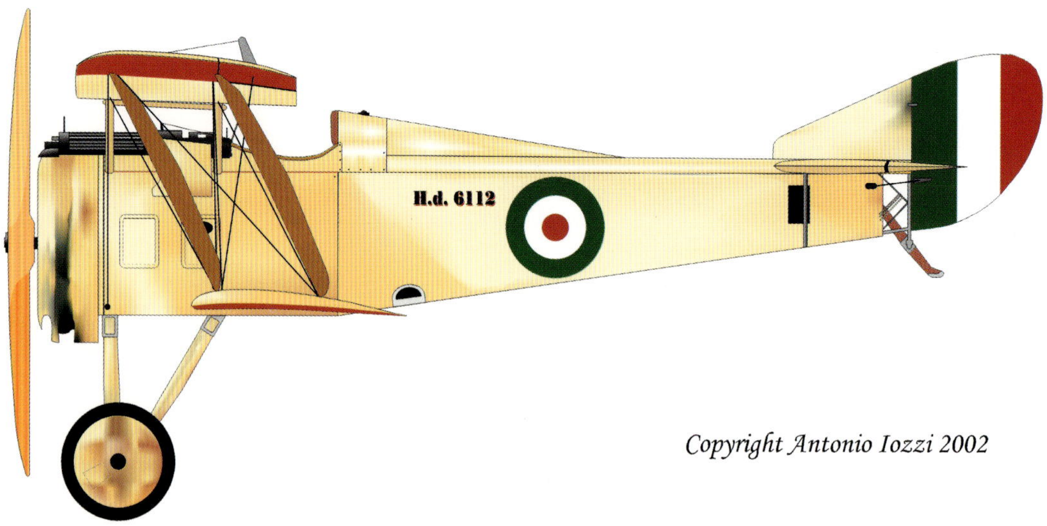


38

39

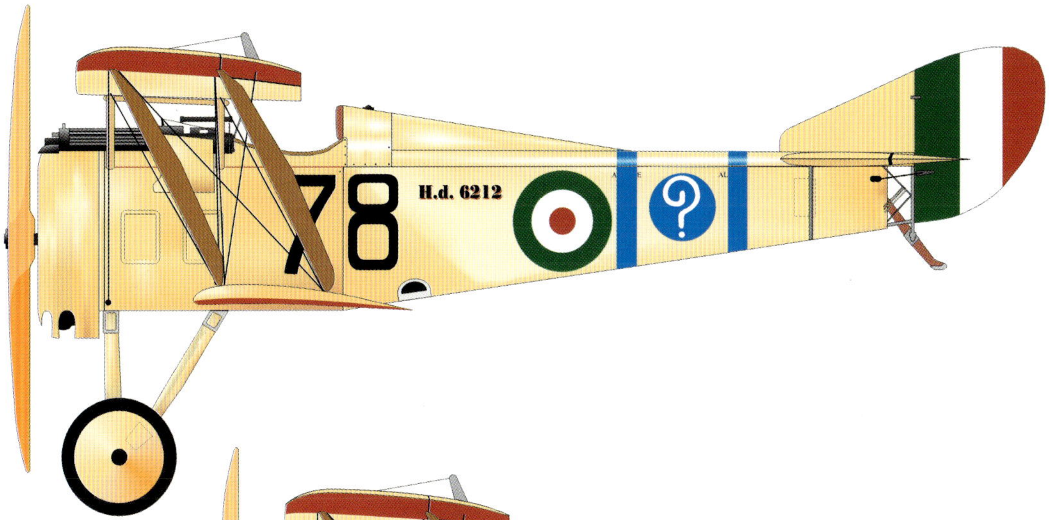
78
H.d. 6212

40

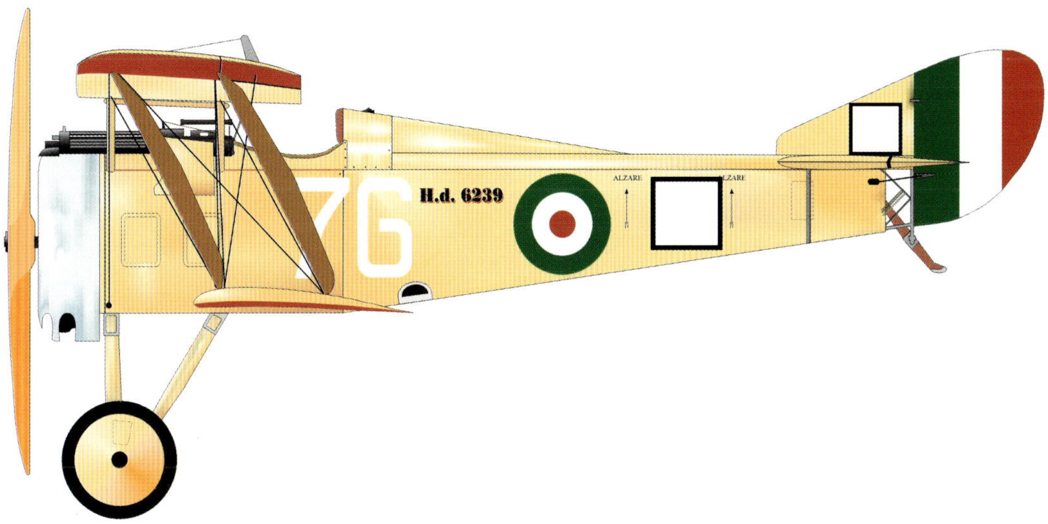
76
H.d. 6239
ALZARE

41

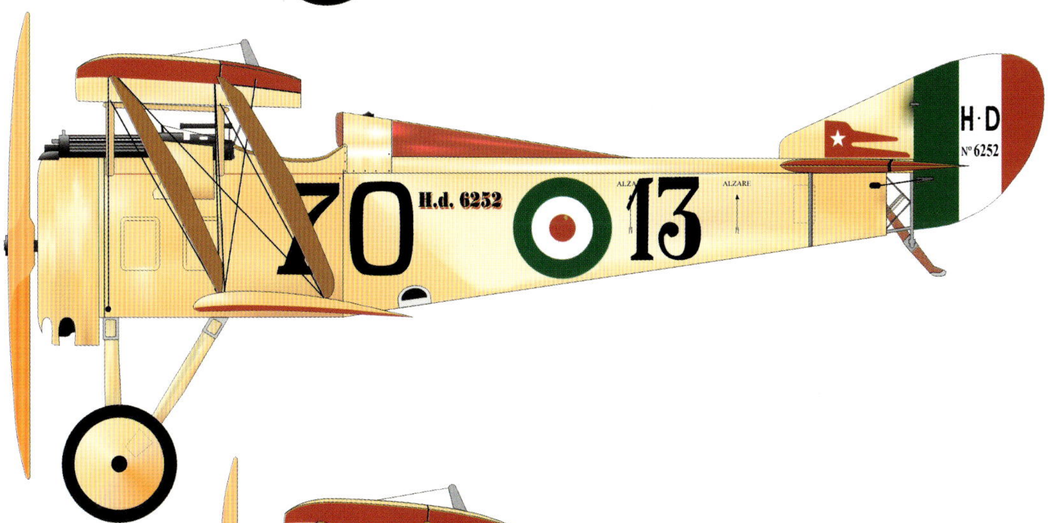
70
H.d. 6252
13
ALZARE
H·D
N° 6252

42

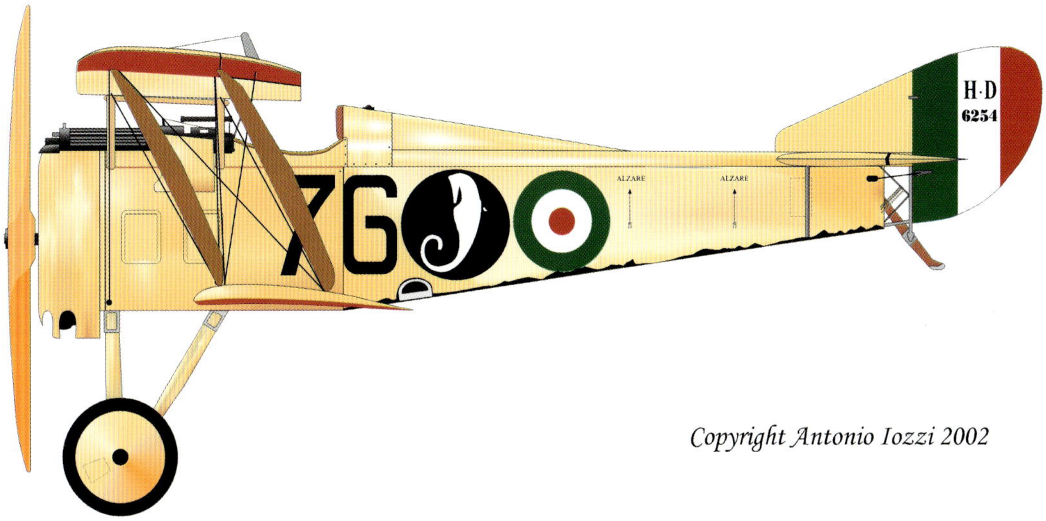
76
ALZARE
H·D
6254

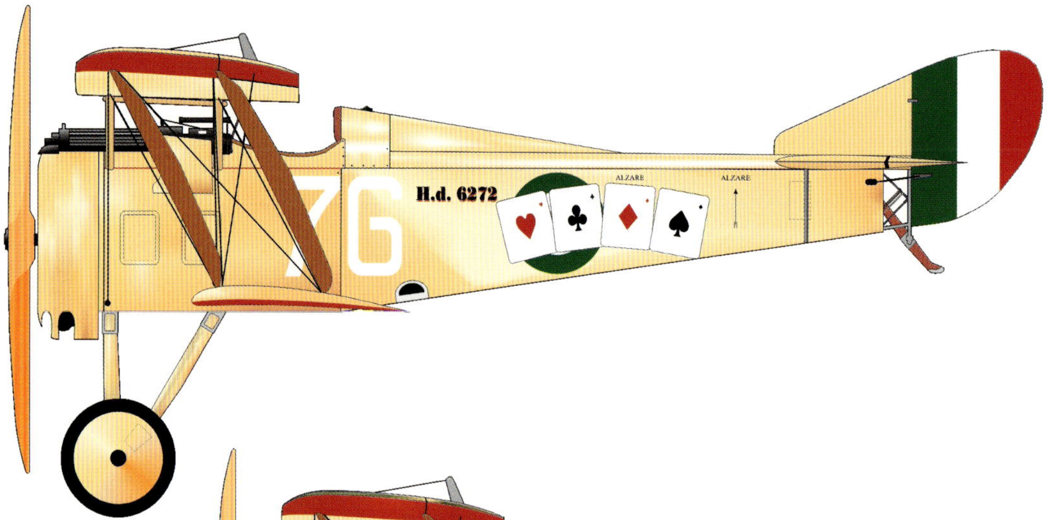


43

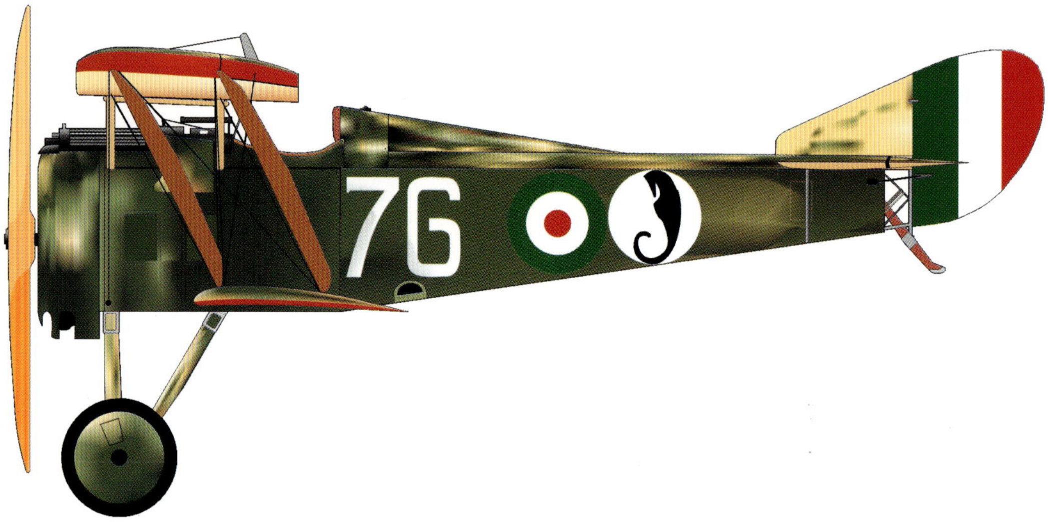


44

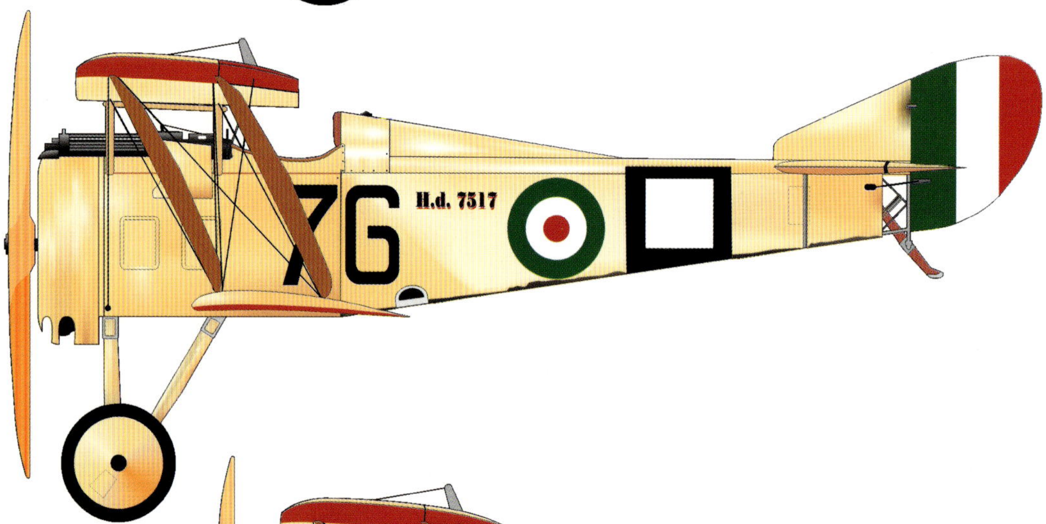


45

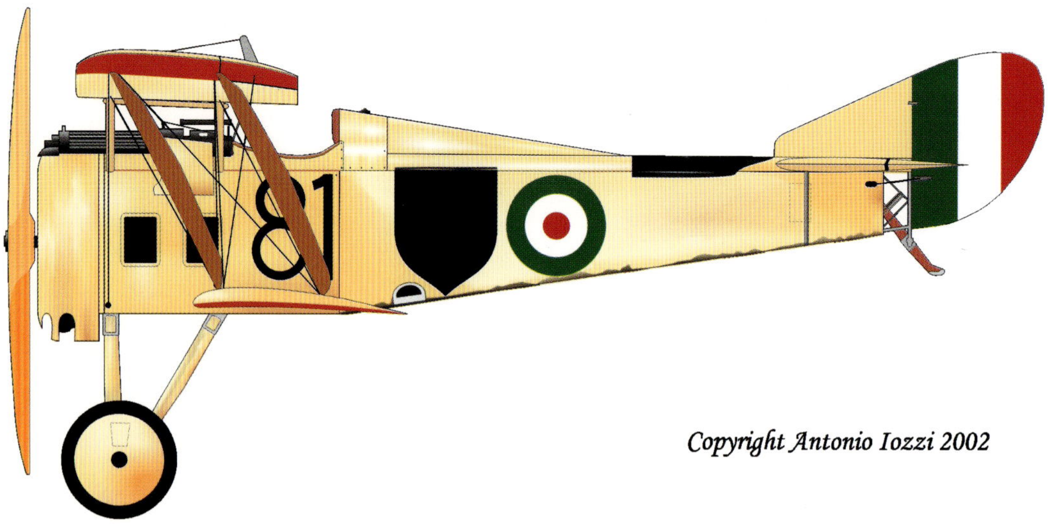


46

47

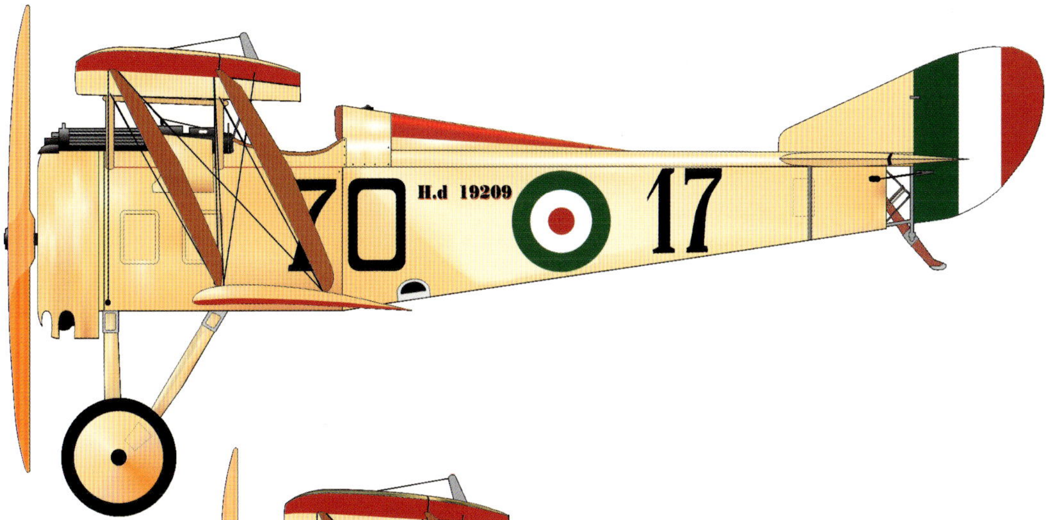


48

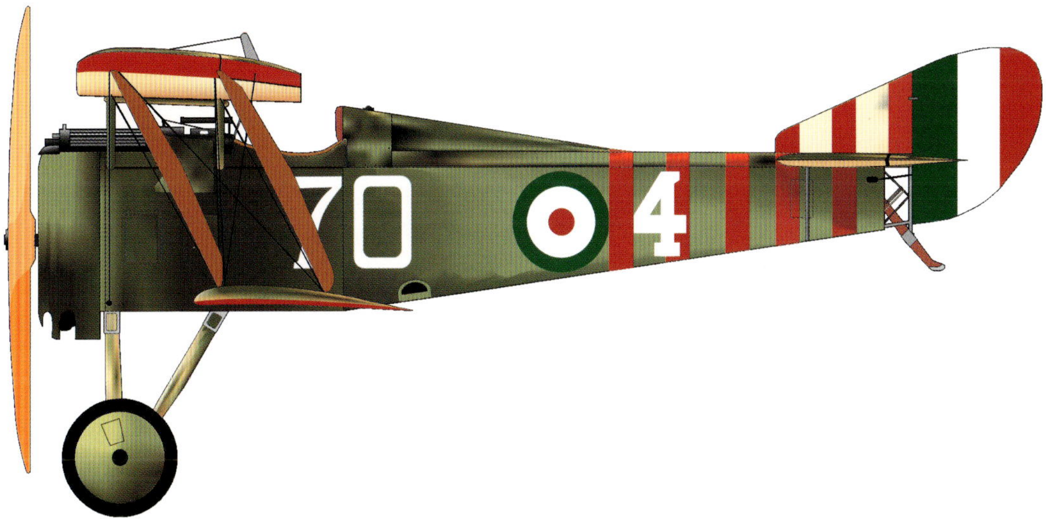


49

50

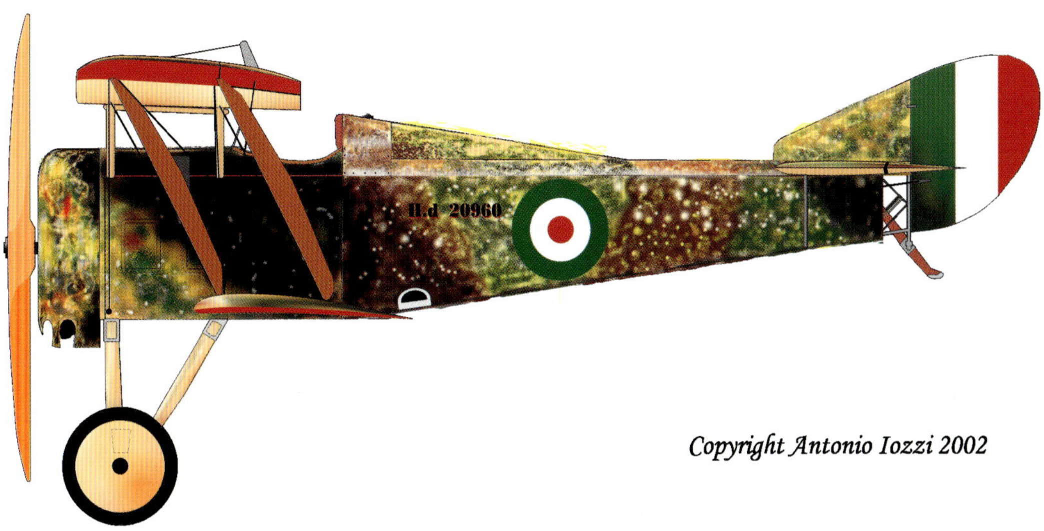

51

52

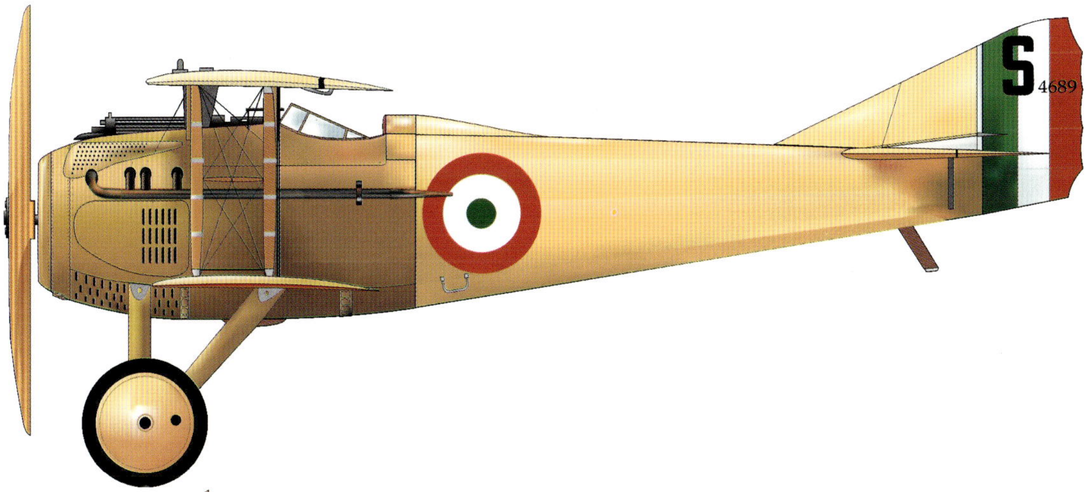


53

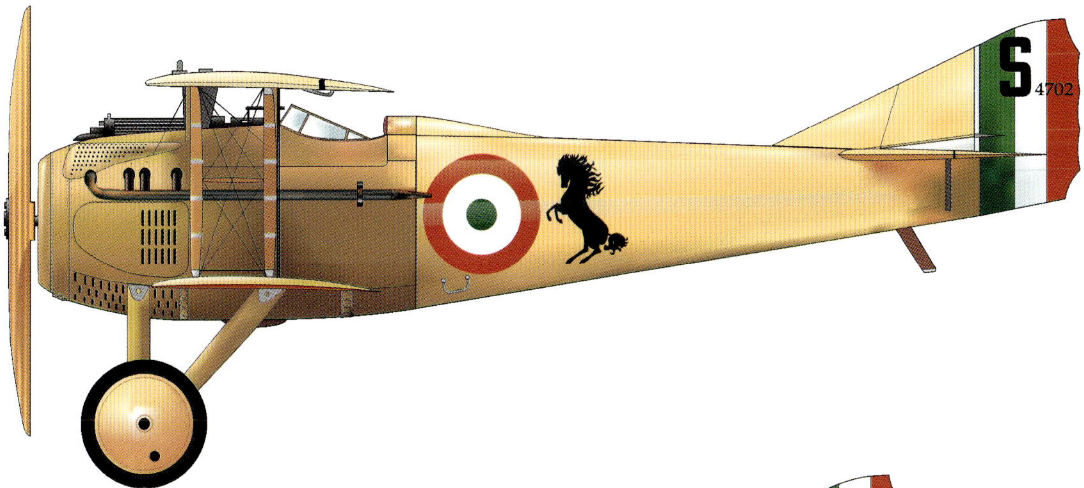

54

55

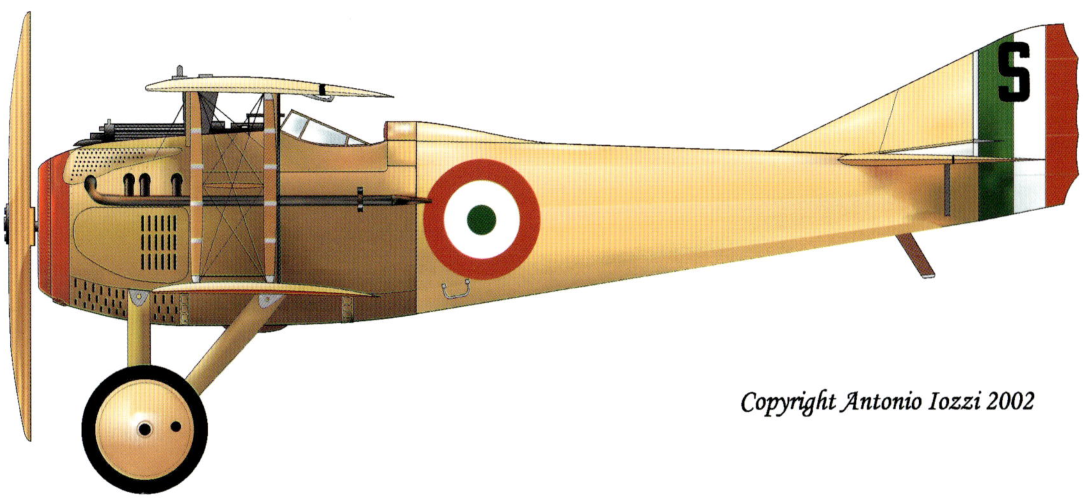

56

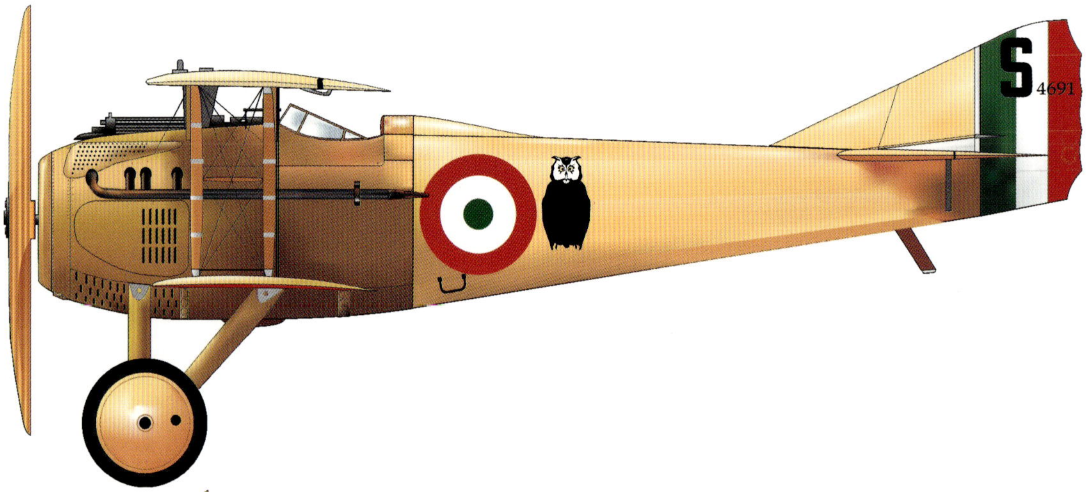


57

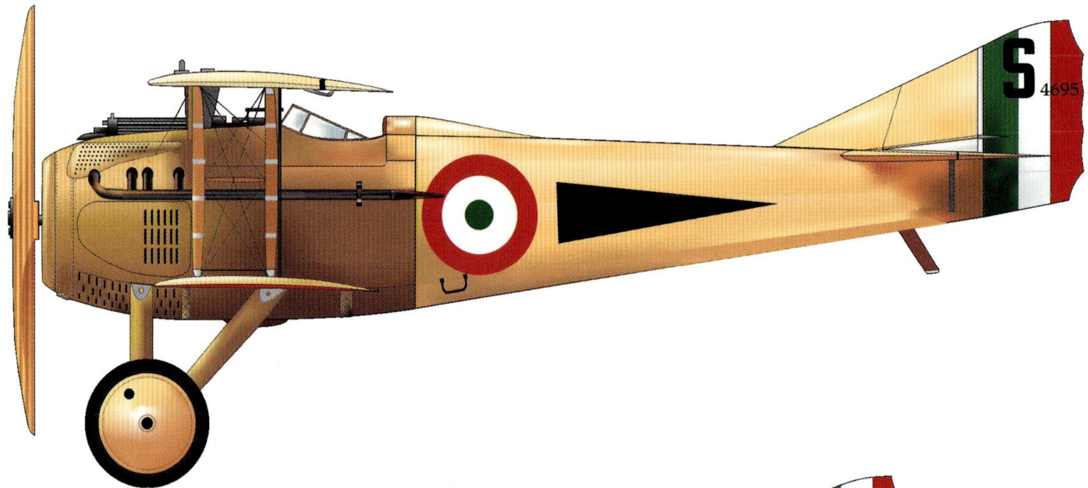


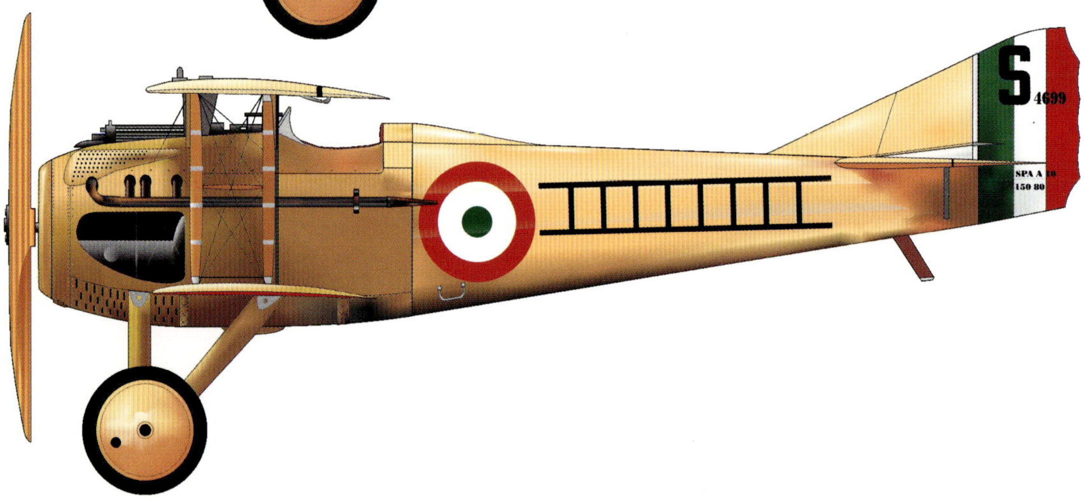


58

59

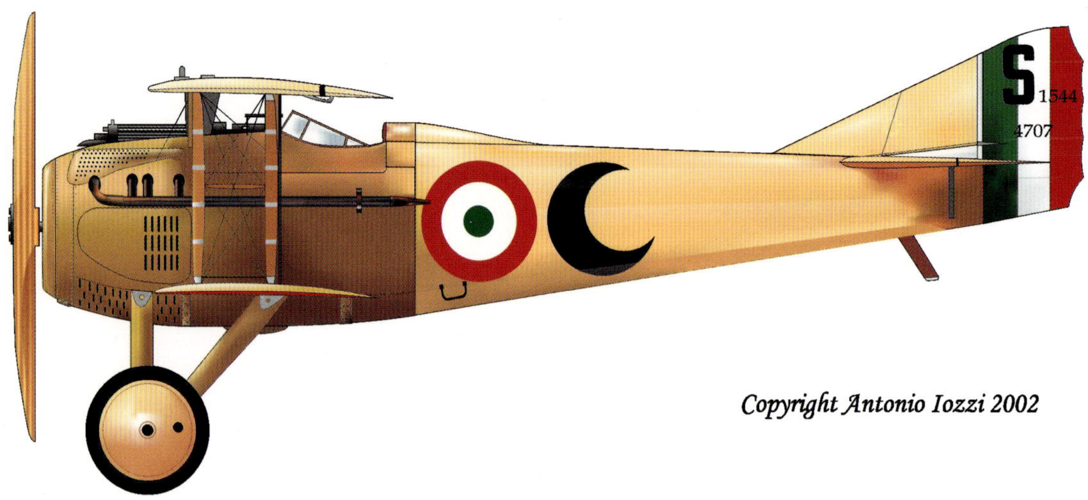

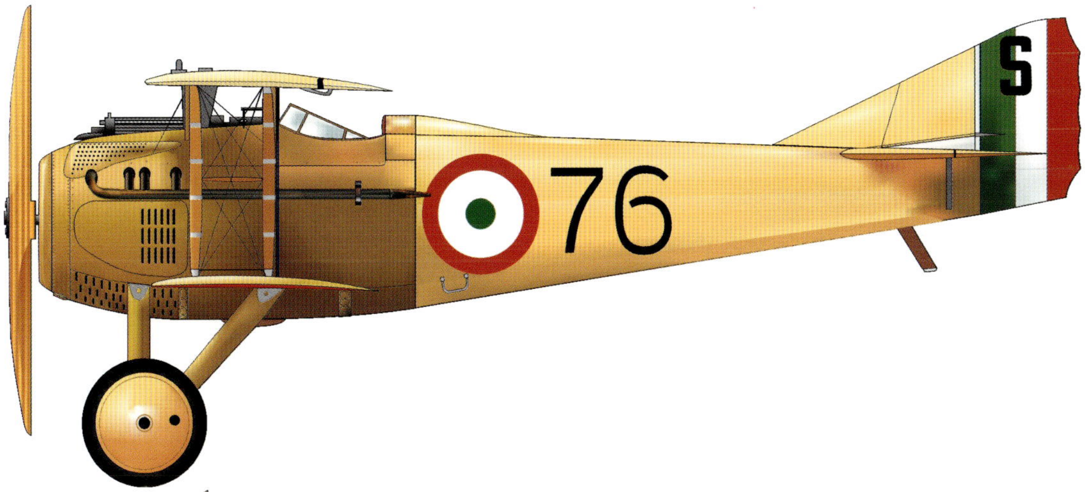


60

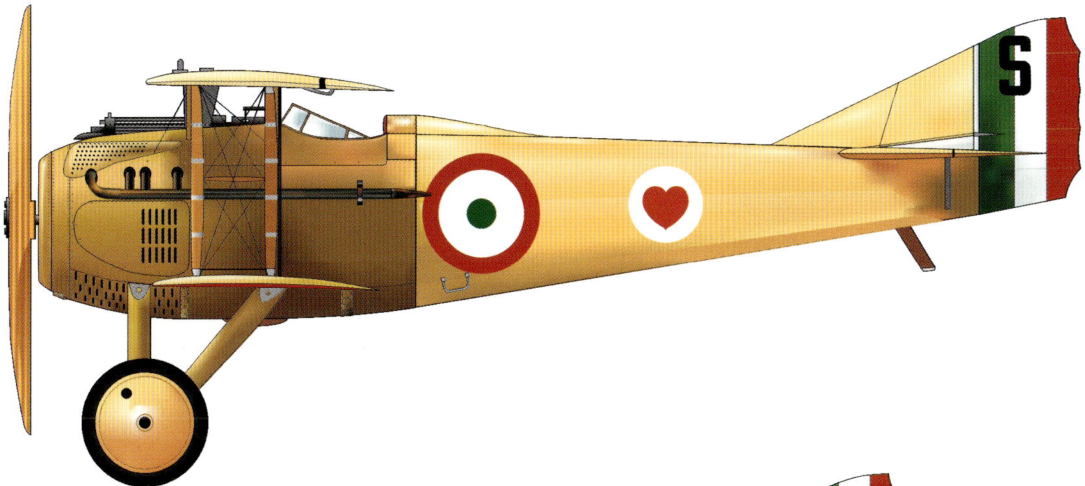


61

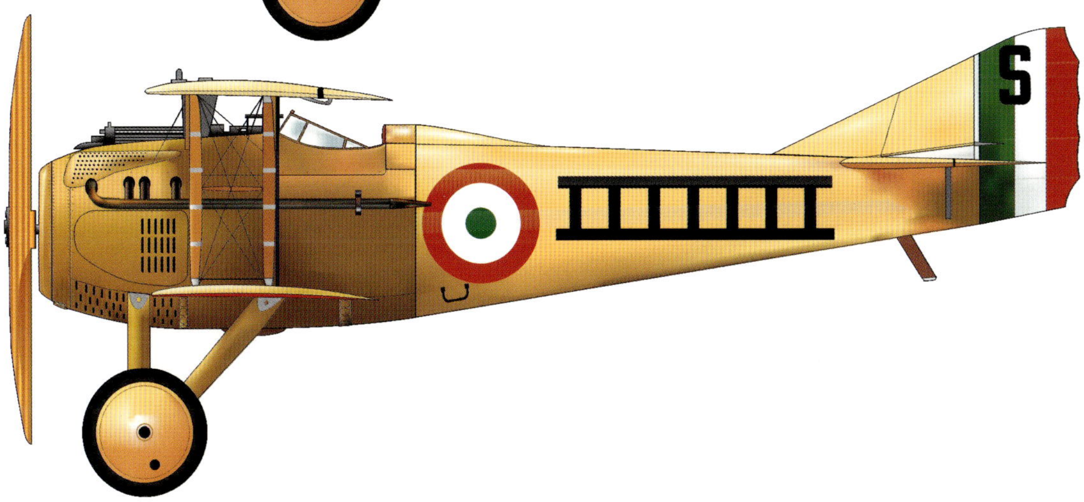


62

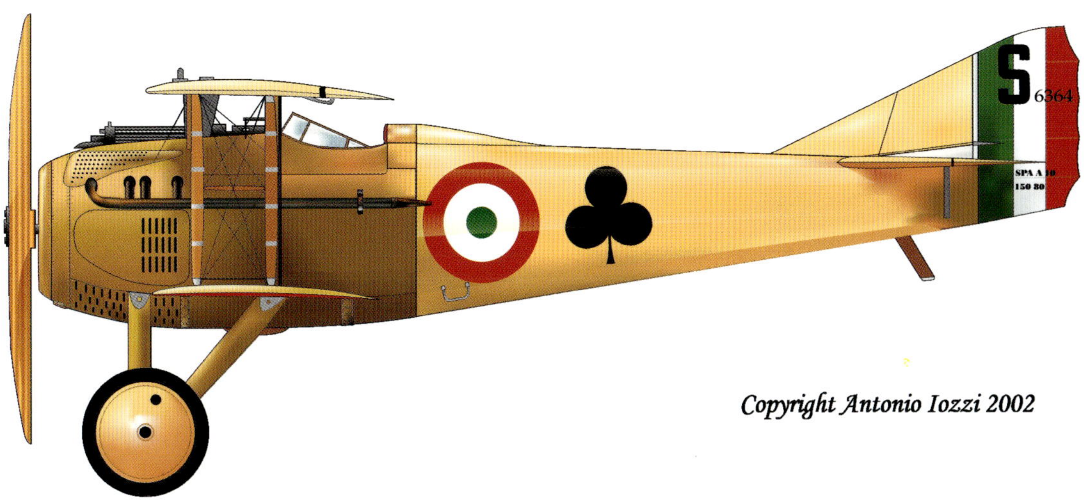


63

64

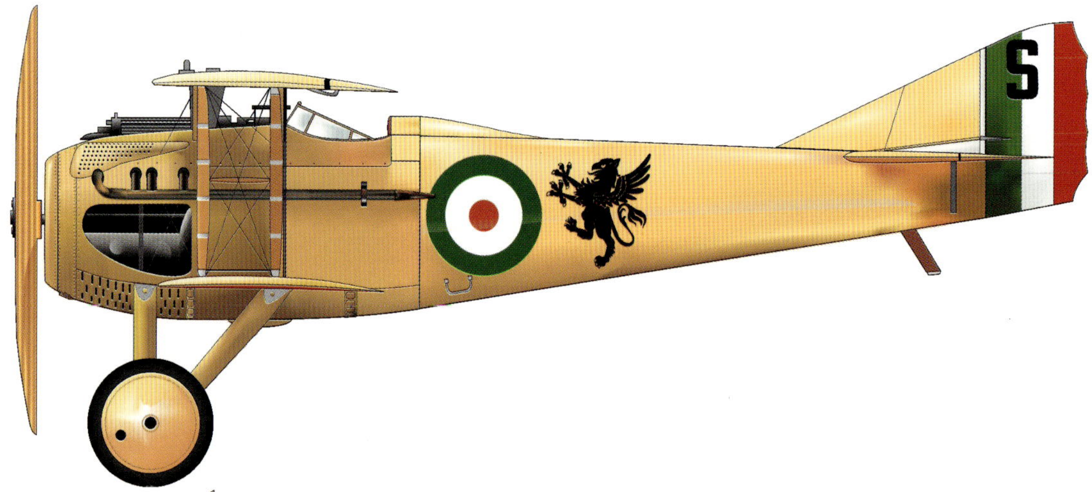

65

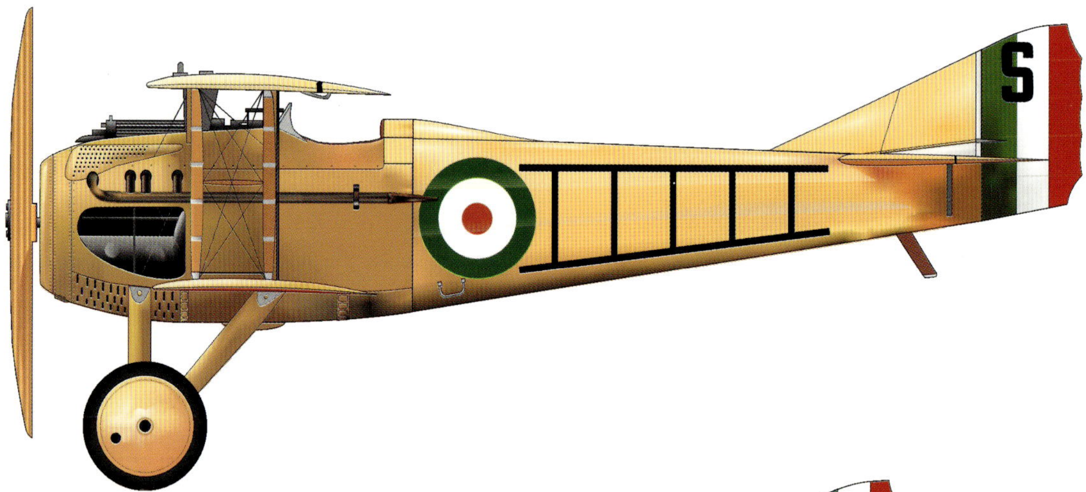

66

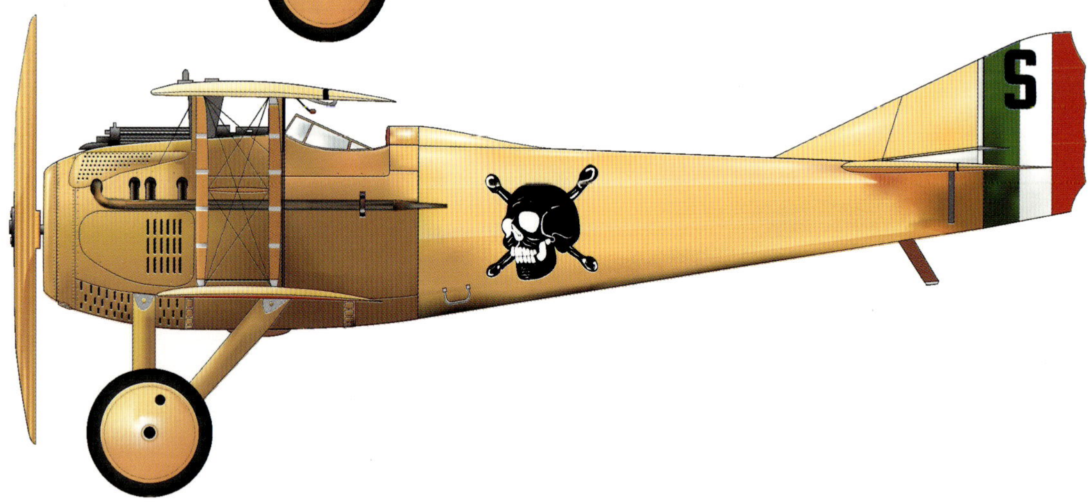

67

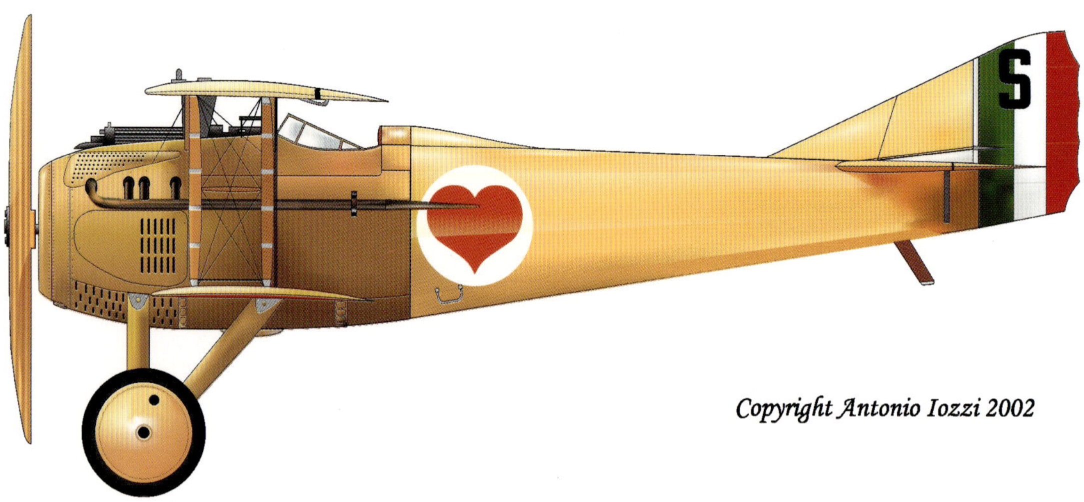

68

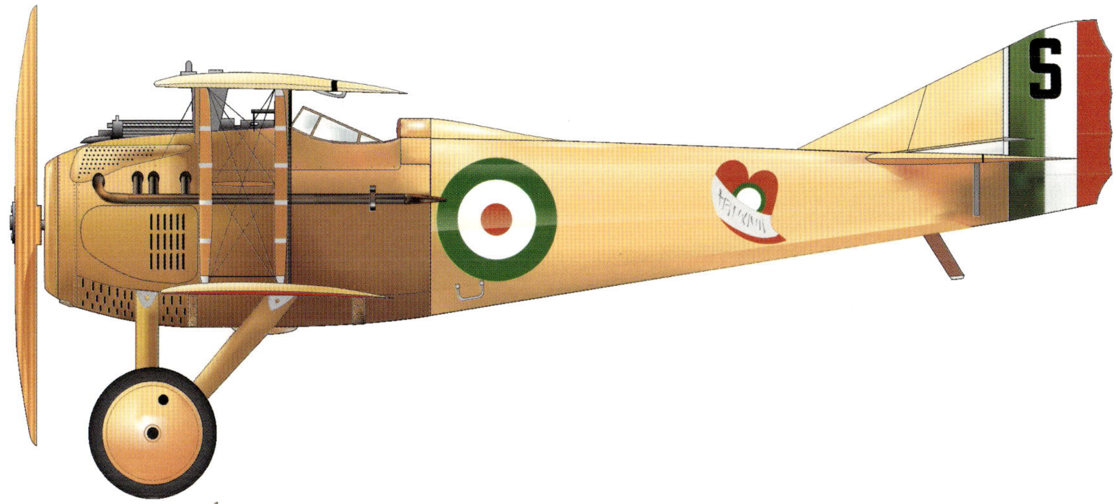


69

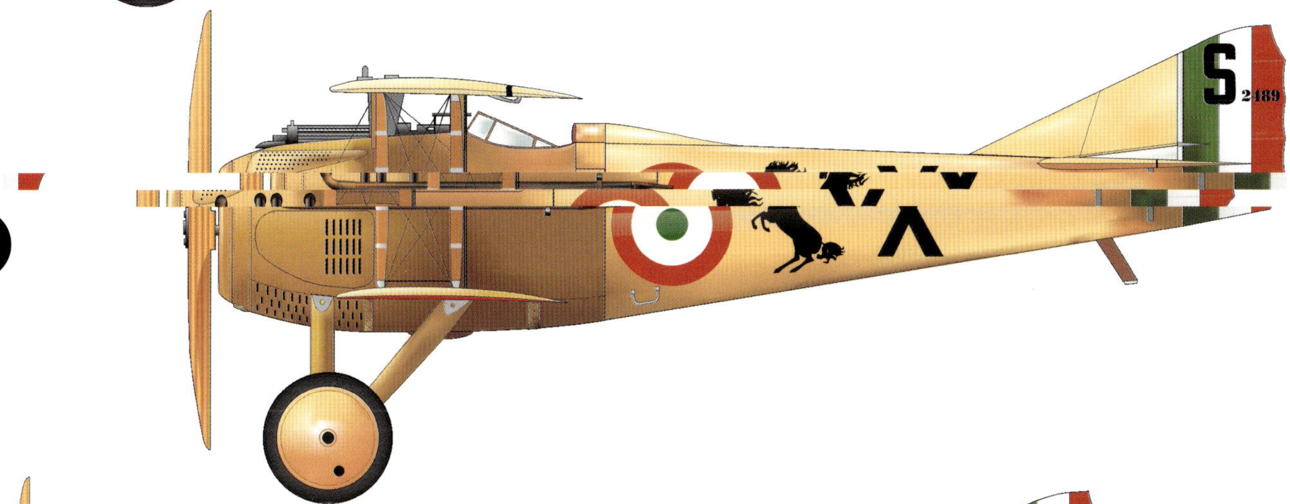


70

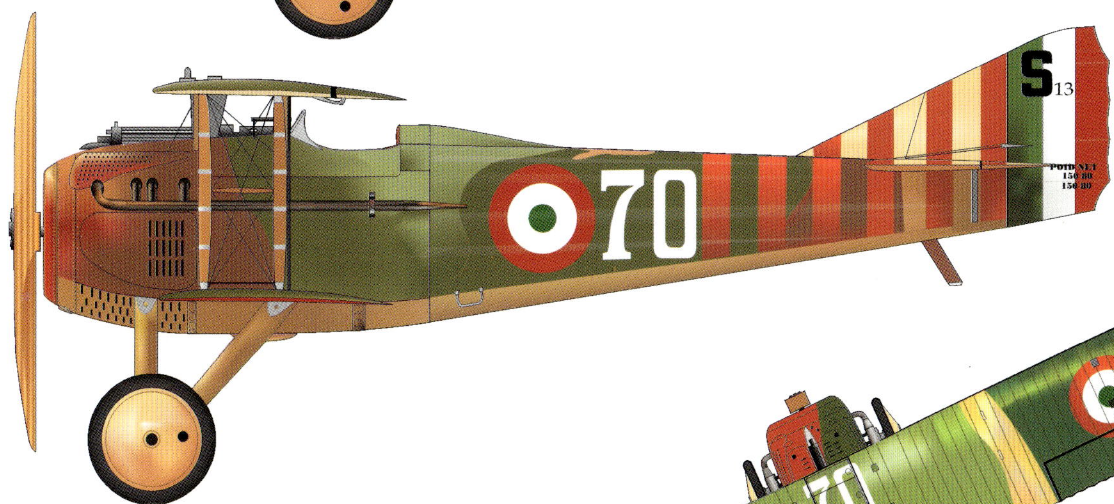

71

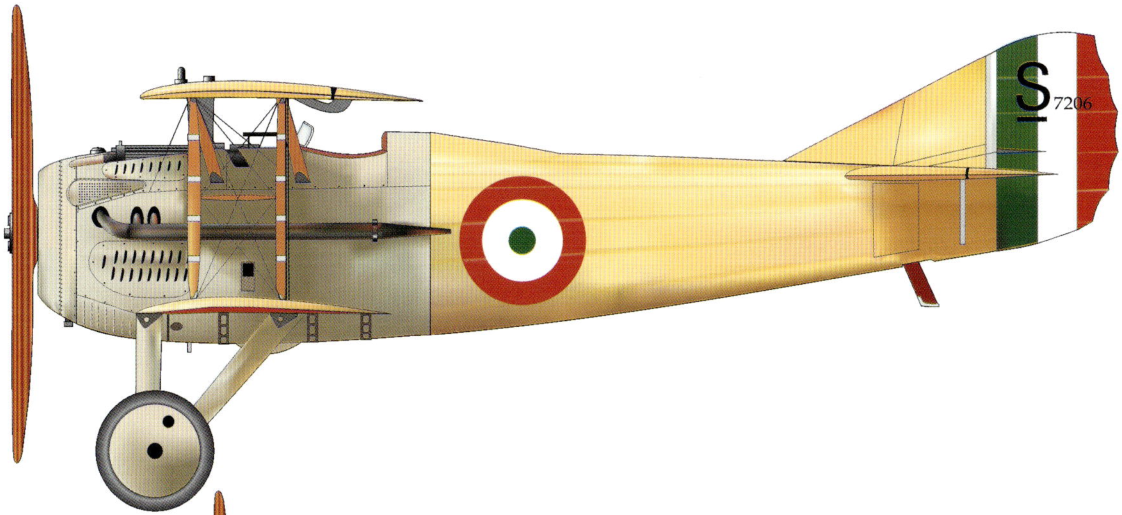


72

73

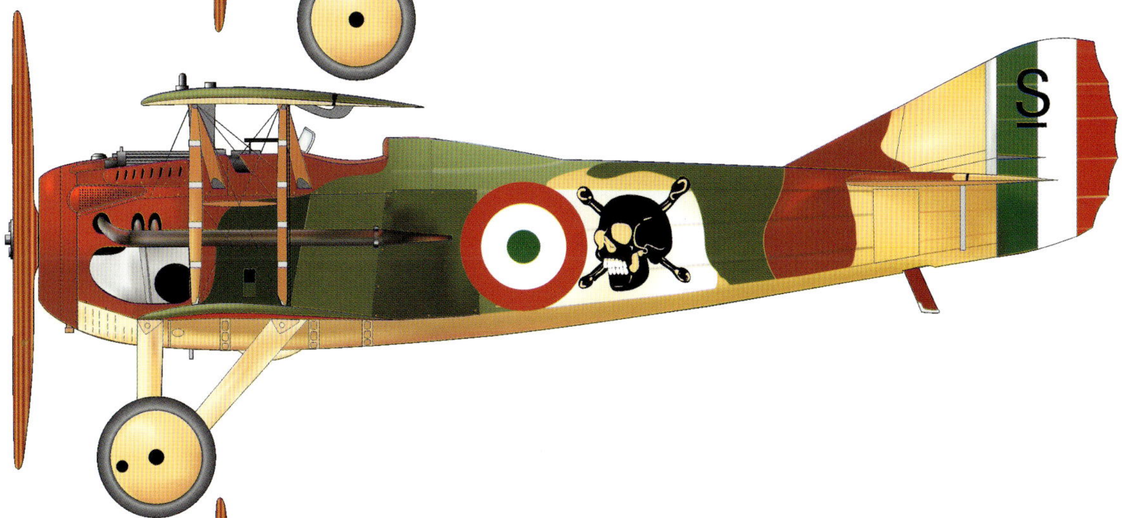


74

75

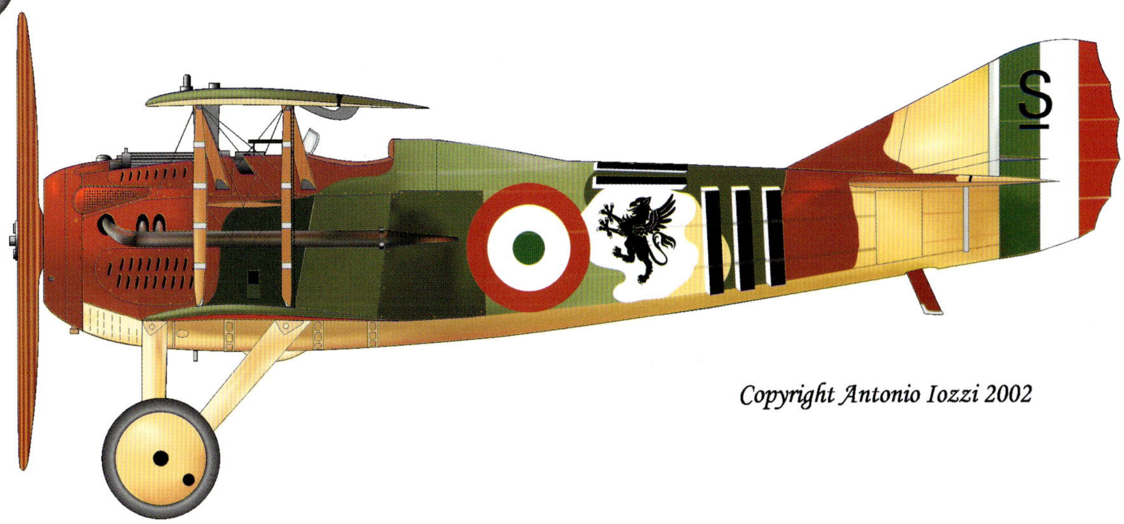

76

77

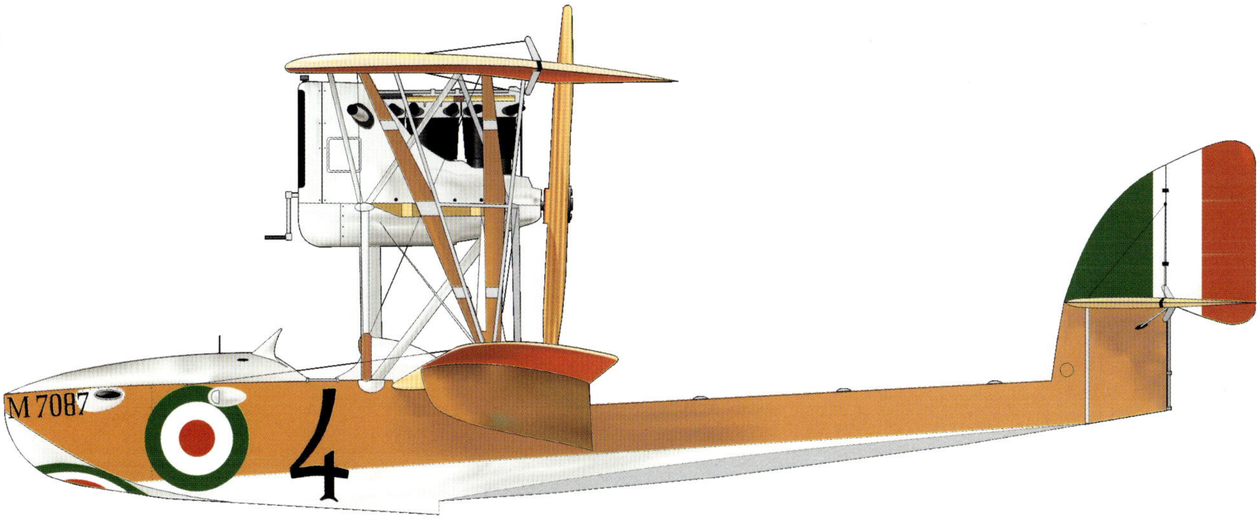
M 7087
4

78

M 7242
1

79

M 7289
2

OCIO! OCIO
FIOL D'UN CAN
MARCIARE
NON
MARCIRE
2

A1

A2

A3

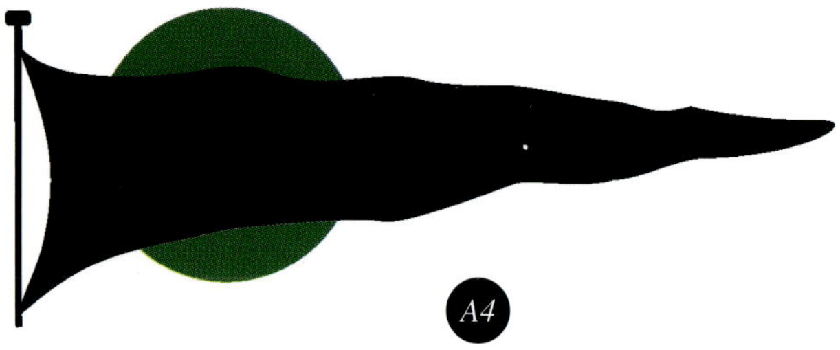

A4

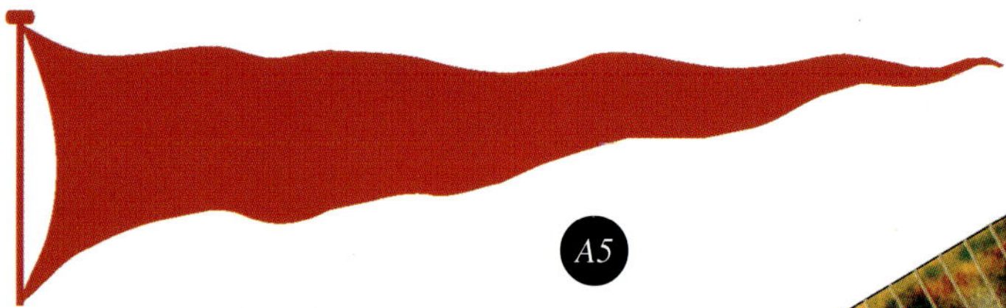

A5

A6

A9

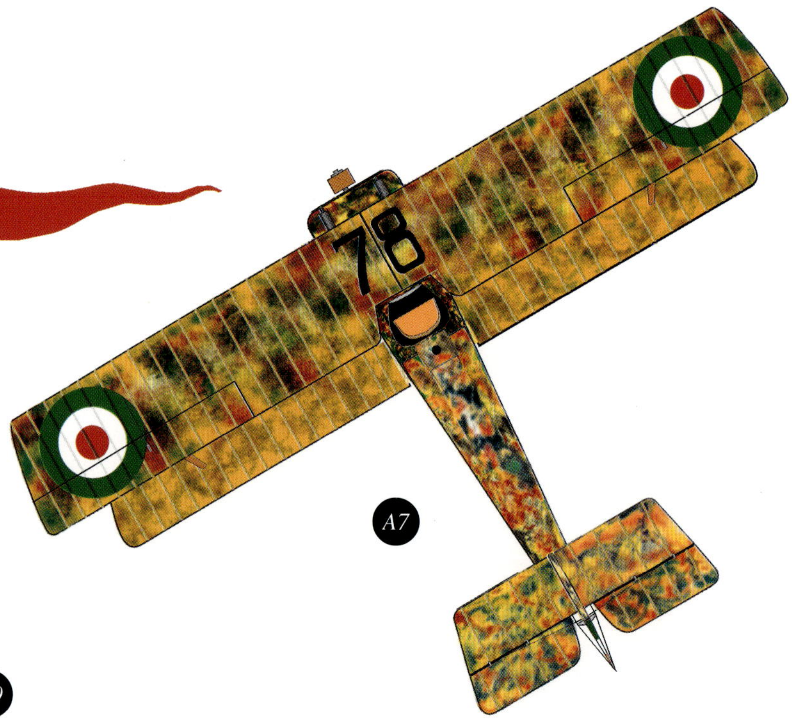


A7

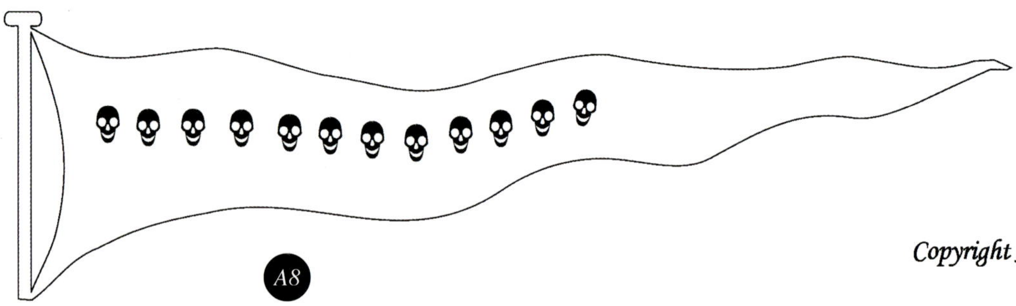

A8

B1

B2

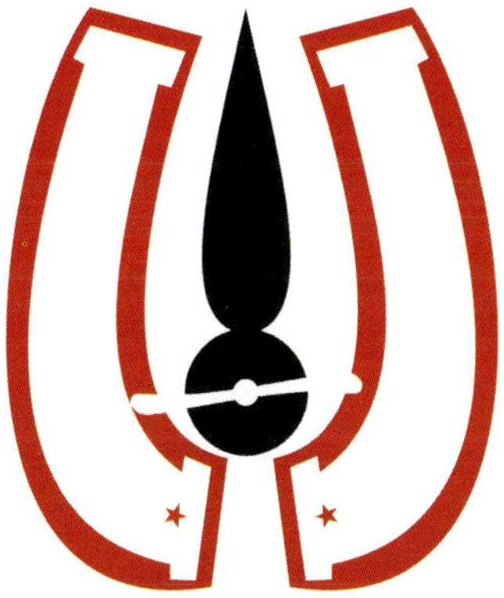

B3

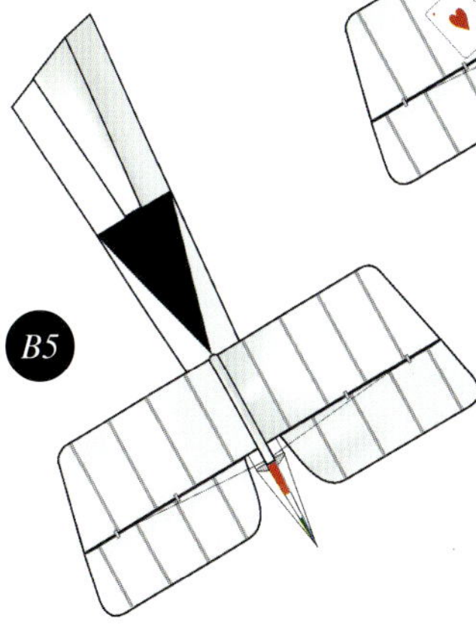

B5

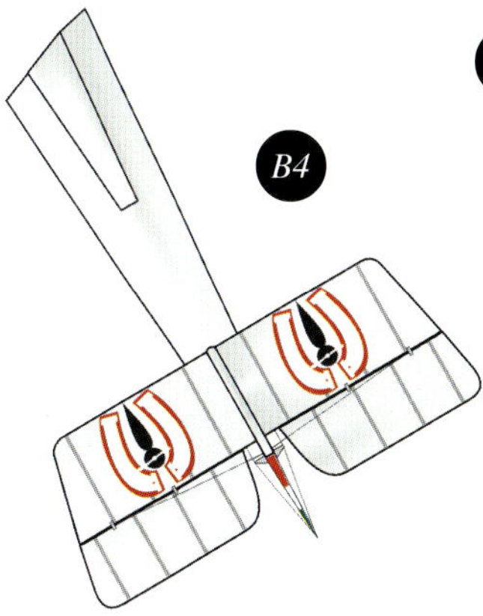

B4

B6

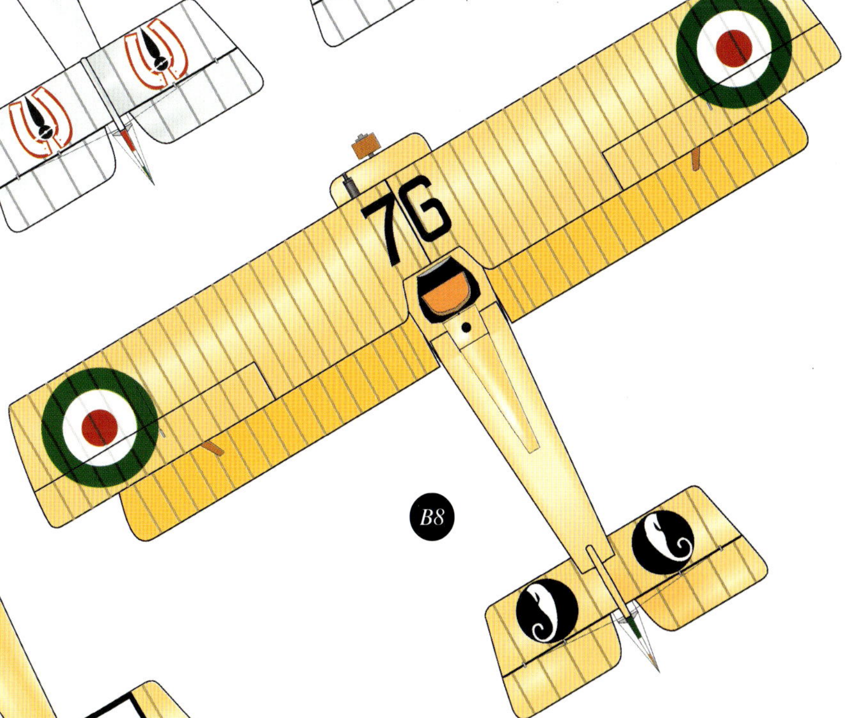

B8

B7

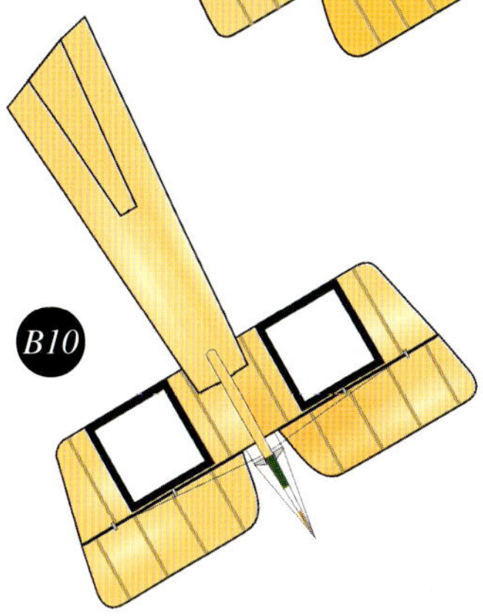

B10

B9

C1

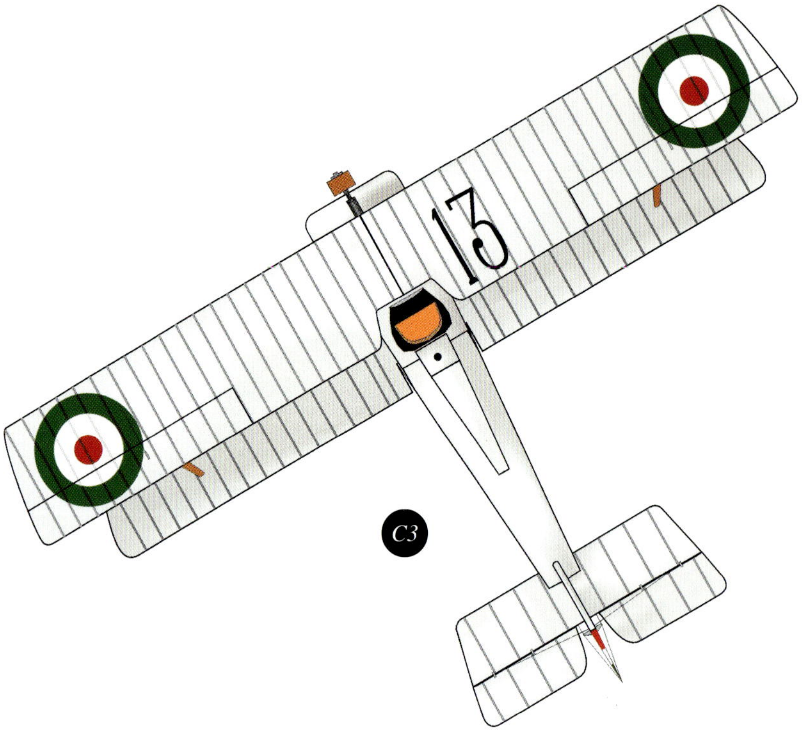


C3

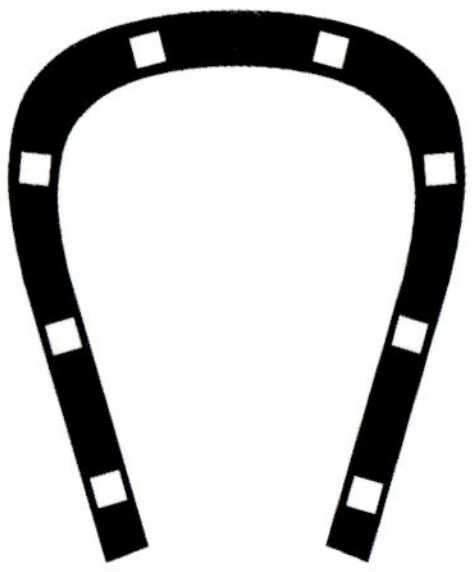

C2

C4

C5

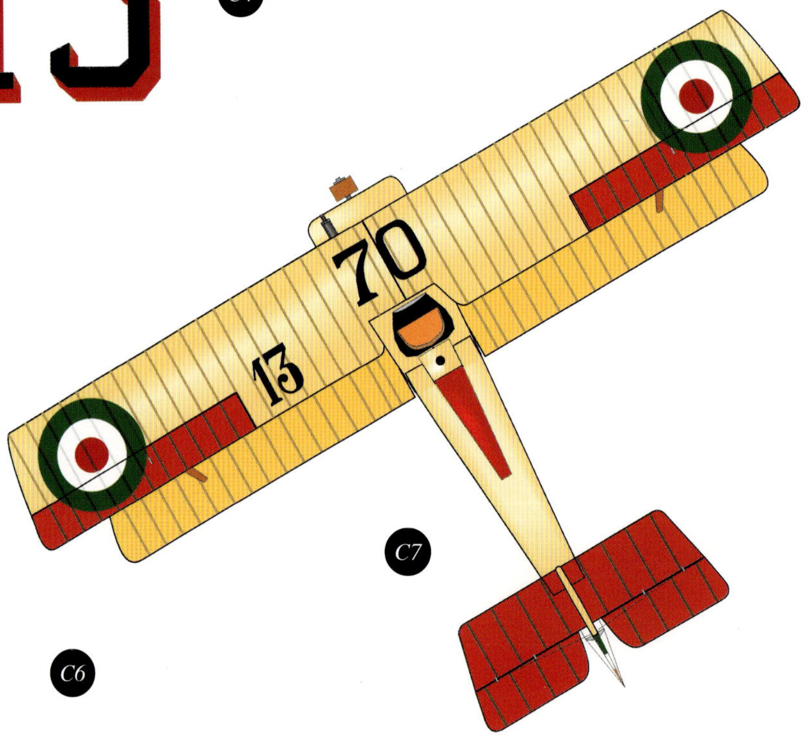


C7

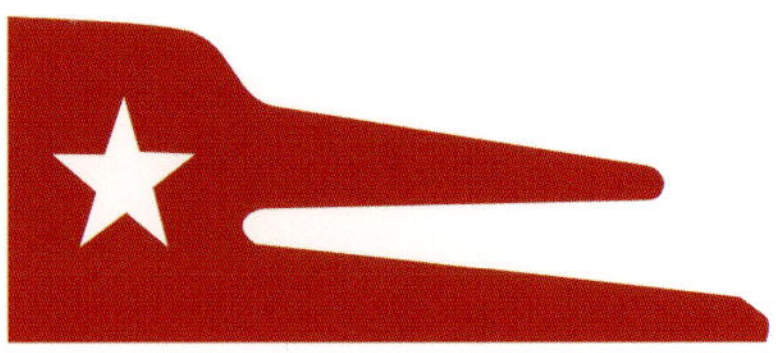

C6

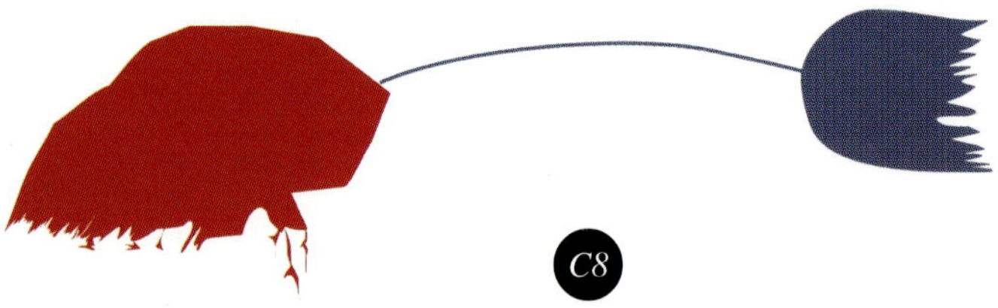

C8

D1

D2

D3

D4

D5

D6

D7

D8

D9

D10

D11

# Abbreviations

Alb. Albatros
Asp. Aspirante
B.G. Bomber Geschwader
Br. Brandenburg
Brig. Brigadiere
Cap. Capitano
Capt. Captain
Col. Colonnello
Cpr. Caporale
DOW Dead of wounds
EA Enemy aircraft
FA Flieger Abteilung
FA (A) Flieger Abteilung (Armée)
Fähnr. Fähnrich
FlG Flieg Geschwader
Flgst. Fliegergast
Flik Flieger Kompanie
FP Feld Pilot
Frglt. Fregattenleutenant
FTL Forced to Land
Fw. Feldwebel
Gefr. Gefreiter
GM Guardiamarina
Hptm. Hauptmann
HMS His Majesty's Ship
i.d.R. in den Reserve
Ing. Ingegnere
Jasta Jagd Staffel
JCS Joint Chiefs of Staff
KIA Killed in action
KIC Killed in crash
Korp. Korporal
Lieut. Lieutenant
Lschlt. Linienschiffleutnant
Lt. Leutnant
Magg. Maggiore
Maj. Major
Marò Marinaio
MIA Missing in Action
M.llo Maresciallo
Oblt. Oberleutnant
Offstv. Offizierstellvertreter
OOC Out Of Control
RAF Royal Air Force
Res. Reserve
RFC Royal Flying Corps
RHBZ Reihenbild Zug
Rittm. Rittmeister
R.N. Regia Nave
Serg. Sergente
SFS Seeflugstation
Sold. Soldato
Sq. Squadriglia
Stfw. Stabsfeldwebel
St.b. Staabsbootmann
S.ten. Sottotenente
STV Sottotenente di Vascello
T.col. Tenente colonnello
Ten. Tenente
TV Tenente di Vascello
Uffz. Unteroffizier
UNH Unhurt
V.M. Valor Militare
WIA Wounded in action
WIC Wounded in crash
Zugsf. Zugsführer

# Bibliography


Anon., *Albo d'Oro dell'Aeronautica*, Provveditorato Generale dello Stato, Roma 1928.

Anon., *Air Aces of the 1914-1918 War*, Harleyford, Letchworth 1959.

Alegi Gregory et al., *In volo per Vienna*, Museo Caproni Trento, Trento 1993.

Alegi Gregory, *Ansaldo SVA 5*, Albatros Production Ltd, Berkhamsted (U.K.), 1993

Alegi Gregory, Falessi Cesare, *L'asso degli Assi*, Bariletti, Roma 1992.

Aliperta Gaetano, *Memorie di volo e di guerra*, Arti Grafiche Favia, Bari s.i.d.

Banfield Goffredo de, *L'aquila di Trieste*, Lint, Trieste 1984.

Bignozzi Giorgio, Catalanotto Baldassarre, *Storia degli aerei italiani*, Edizioni Cielo, Roma 1962.

Bompiani G., Prepositi C., *Le ali della guerra*, Mondadori, Milano 1931.

Briganti Alberto, *Oltre le nubi il sereno*, Nuovo Studio Tecna, Roma 1994 (2° ed.).

Bruce J.M., *War Planes of the First World War - Fighters Volume 5*, Mac Donald, London 1972.

Bruce J.M., *Nieuport Aircraft of World War One*, Arms and Armour Press, London 1988.

Bruce J.M., *Nieuport Fighters*, Albatros Productiona Ltd., Berkhamsted (U.K.), 1993.

Bruce J.M., *Hanriot HD 1*, Albatros Production Ltd., Berkhamsted 1988.

Camurati Gastone, *Aerei Italiani 1914-1918*, Rivista Aeronautica, Roma 1973.

Cartosio Tomaso, *Vita eroica di Ernesto Cabruna*, Stato Maggiore Aeronautica - Ufficio Storico, Roma 1972.

Cavara Otello, *Voli di guerra*, F.lli Treves, Milano 1918.

Cavigioli Riccardo, *L'aviazione Austro Ungarica sulla fronte italiana 1915-1918*, Milano 1934.

Cavina Giovanni, *Gorizia – Caporetto – Montello – Vittorio Veneto*, Edizioni Lega, Faenza 1967.

Connors John F., *SPAD fighters in action*, Squadron/Signal Publications, Carrollton Texas 1989.

Contini Luigi, *L'aviazione italiana in guerra*, Marangoni, Milano 1932.

Contini Luigi, *Francesco Baracca, l'asso Italiano*, Marangoni, Milano 1933.

Davilla James & Soltan Arthur, *French Aircraft of the First World War*, Flying Machine Press, Stratford 1997.

De Bernardis E., *La guerra nell'aria*, Edizioni Mediolanum, Milano 1935.

De Martino Basilio, *Ali sulle trincee*, Stato Maggiore dell'Aeronautica – Ufficio Storico, Roma 1999.

De Martino Basilio, *Guerra Aerea*, Rossato, Valdagno 1999.

Evangelisti Giorgio, *Una vita in gioco*, Editoriale Olimpia, Firenze 1993

Fabi Lucio, *Gente di Trincea*, Mursia, Milano 1994.

Fabi Lucio, *La prima guerra mondiale 1915-1918*, Editori Riuniti, Roma 1998.

Fabris Eugenio et al., *Ali in Pace e in Guerra 1903-1930*, Circolo Culturale Filatelico Numismatico, Fiorenzuola d'Arda, 1996.

Falessi Cesare, Pagliano Maurizio, *Storia fotografica dell'aviazione italiana*, Newton Compton, Roma 1984.

Ferrari Atlantico, *L'asso di cuori Guido Keller*, Cremonese Editore, Roma 1933.

Ferrari Paolo et al., *La Grande Guerra Aerea*, Rossato, Valdagno 1994.

Franks N., Guest R. & Alegi G., *Above the Warfronts*, Grub Street, Londron 1997

Franks N., Bailey F. & Duiven R., *The Jasta War Chronology*, Grub Street, London 1998

Franks N., Bailey F. & Duiven R., *Casualties of the German Air Service*, Grub Street, London 1999.

Fucini Mario, *Voli sul nemico*, Bemporad, Firenze 1933 (3° ed.).

Gentilli R., Varriale P., *I reparti dell'aviazione italiana nella Grande Guerra*, Stato Maggiore dell'Aeronautica – Ufficio Storico, Roma 1999.

Giuliotti Adolfo, *Baracchini*, Edizioni Ardita, Roma 1934.

Grassi Pietro, *La Medaglia d'Oro al Valor Militare Fulco Ruffo di Calabria,* Padova 1990.

Grosz P.M., *Albatros D.III (OEF)*, Albatros Production Ltd., Berkhamssted (U.K.)

Grosz P.M., *Aviatik D.I*, Albatros Production Ltd., Berkhamsted (U.K.)

Grosz P., Haddow G., Schiemer P., *Austro-Hungarian Army Aircraft of World War One*, Flying Machine Press, Mountain View, Ca. 1993.

Hauke E., Schroeder W., Tötschinger B., *Die Flugzeuge der K.u.K. Luftfhartruppe und Seeflieger 1914-1918*, H. Weishaupt Verlag, Graz 1988.

Höfling Rudolf, *Albatros D-II*, Flugzeug Publikations GmbH, Illertissen 2001.

Lamberton & Cheesman, *Fighter Aircraft of the 1914-1918 war*, Harleyford, Letchworth 1960.

Laredo di Mendoza S., Russo A., *Ali e squadriglie*, Impresa Editoriale Italiana, Milano 1933.

Lenotti Wolfram, *Rot-Weiss-Rot zur Luft*, Osterreichische Luftfahrverlag, Wien 1957.

Lombardi Francis, *Gli amici di Marcon*, Associazione Arma Aeronautica, Vercelli 1979.

Lombardi Francis, *Tre raids del 1930*, Gallo, Vercelli 1980.

Longhi Fiorenzo, *Piloti ed aerei italiani dal Pacifico alle Ande*, Longhi, Belgirate 1996.

Longoni Maurizio, *Nieuport Macchi 11 & 17*, Intergest Milano, 1976.

Ludovico Domenico, *Aviatori Italiani da Roma a Tokyo nel 1920*, Etas Kompass, Milano 1970.

Maddalena Umberto, *Lotte e vittorie sul mare e nel cielo*, Mondadori, Milano 1930.

Mancini Luigi, *Grande enciclopedia aeronautica*, Edizioni "Aeronautica", Milano 1936.

Maravigna Pietro, *Guerra e Vittoria*, UTET, Torino 1927.

Meindl K., Schroeder W., Tötschinger B., *Hansa Brandenburg D1 1916-18*, OFH Nachrichten, Wien 1995.

Meindl K., *Luftfahrtruppen Verlustliste*, Unpublished manuscript.

Milanesi Guido, *Albatros*, Alfieri e Lacroix, Milano 1920

Milanesi Guido, *Le aquile*, Ceschina, Milano 1929.

Molfese Manlio, *L'aviazione da ricognizione italiana durante la guerra europea*, Provveditorato Generale dello Stato, Roma 1925.

Monelli Paolo, *La Nostra Guerra 1915-'18*, Touring Club Italiano, 1968.

Montù Carlo, *La nostra aviazione*, Lattes, Torino 1916.

O'Connor, Martin, *Air Aces of the Austro-Hungarian Empire*, Champlin Fighter Museum Press, Falcon Field (Arizona) 1986.

Perissinotto Marino, *Giannino Ancillotto un eroe sandonatese*, Città di San Donà di Piave, 1995

Peter Ernest, *Die K.u.K. Luftschiffer und Fliegertruppe Osterreichs-Ungarns 1794-1919*, Motorbuch Verlag, Stuttgart 1981.

Pieri Piero, *L'Italia nella Prima Guerra Mondiale*, Einaudi, Terza edizione aumentata, Torino 1968

Pieropan Gianni, *1914-1918 Storia della Grande Guerra sul fronte italiano*, Mursia, Milano 1988

Porret Daniel, *Les As français de la Grande Guerre*, Service Historique de l'Armée de l'Air, Paris 1983

Porro Felice, *La guerra nell'aria 1915-1918*, Corbaccio, Milano 193X (4° ed., Edizioni Mate, Milano 1965).

Pozzi Sandro, *Guido Keller l'ardito del cielo*, Edizioni Mediolanum, Milano 1933.

Ralph Wayne, *Barker VC*, Doubleday, Toronto 1997

Rebora Enrico, *Appunti sul contributo lombardo alla storia dell'aviazione italiana*, Rivista Aeronautica, Roma 1968

Rebora Enrico, *I precedenti del volo su Vienna*, Uffico Storico SMA, Roma 1974.

Ufficio Storico dell'Aeronautica, *Cronistoria dell'Aeronautica Militare Italiana, 8 Voll.*, Stato Maggiore Aeronautica - Ufficio Storico, Roma 1968.

Ufficio Storico dell'Aeronautica, *Istruzione di allievi piloti americani in Italia nella Prima guerra Mondiale*, Stato Maggiore Aeronautica - Ufficio Storico, Roma 1968

Ufficio Storico della Regia Marina, a cura dell'ammiraglio Silvio Salza, *La Marina italiana nella Grande Guerra*, 8 Voll., Vallecchi, Firenze 1942.

Scaroni Silvio, *Impressioni e ricordi di guerra aerea*, Società Tipografico Editrice Nazionale, Torino 1922,

Scaroni Silvio, *Battaglie nel cielo*, Longanesi, Milano 1971.

Schiemer Peter *Die Albatros (Oeffag)-Jagdflugzeuge der K.u.K Luftfhartruppen*, H. Weishaupt Verlag, Graz.

Schroeder Walter, *Die Oesterreichischen Hansa-Brandenburg C1*, OFH Nachrichten, Vienna 1980

Schroeder Walter, WKF, OFH Nachrichten, Wien 1980

Schupita Peter, *Die K.u.K Seeflieger*, Bernard & Gaefe Verlag, Koblenz 1983.

Sclerandi Bernardo, *Aviazione 1900-1918*, Curcio Periodici, Roma.

Straulino Luigi, *Messaggi tra aviatori nemici*, Stato Maggiore Aeronautica - Ufficio Storico, Roma 1987.

Tamozzo Dina, *La prima guerra mondiale a Castelgomberto e Sovizzo*, 1998.

Tesar Petr Aharon, *Albatros D.II & D.III Oeffag*, JaPo, Hradec Kràlové 1998

Tonicchi G., *La scuola idrovolanti della Regia Marina di Bolsena 1917-1921*, Tarquinia s.i.d.

Tötschinger B. & Schroeder W., *Handbuch der K.u.K. Luftfahrtruppe 1914-1918*, OFH Nachrichten, Wien 1982.

Tötschinger B. & Schroeder W., *Phoenix Marine Jagdeinsitzer*, OFH Nachrichten, Wien 1982

Tötschinger B. & Schroeder W., *Hansa Brandenburg C1 1916-18*, OFH Nachrichten, Wien 1992

Tötschinger B. & Schroeder W., *Luftsiege der Flik 1*, OFH Nachrichten, Wien 1989

Van Wyen Adrian, *Naval Aviation in World War I*, Chief of Naval Operation, Washington 1969

Various authors: *Air Aces of the 1914-1918 War*, Harleyford, Letchworth 1959

Various authors, with Gregory Alegi, *In volo per Wien*, Museo Caproni Trento, Trento 1993

Various authors, with Paolo Ferrari, *La Grande Guerra Aerea*, Rossato, Valdagno 1994

Vergnano Piero, *Origini dell'aviazione in Italia 1783-1916*, Intyprint, Genova 1964

Weber Fritz, *Tappe della disfatta*, Corticelli, Milano 1934.

**Magazines**

*Aerofan, Air Enthusiast, Air History, Ali Antiche, Ali Nuove, Avions, Cross & Cockade International, Icare, Il Cielo, JP 4 Mensile d'Aeronautica, Domenica del Corriere, O.F.H. Nachrichten, Over the Front, Rivista Aeronautica, Windsock, World War One Aero.*

**Official sources of images**

Fototeca Stato Maggiore Aeronautica, Fototeca Ufficio Storico Stato Maggiore Esercito, Fototeca Ufficio Storico Marina Militare, Istituto di Storia e Cultura dell'Arma del Genio, Museo Storico dell'Aeronautica Militare, Associazione Nazionale delle Medaglie d'Oro, Collections Service Historique de l'Armée de l'Air.

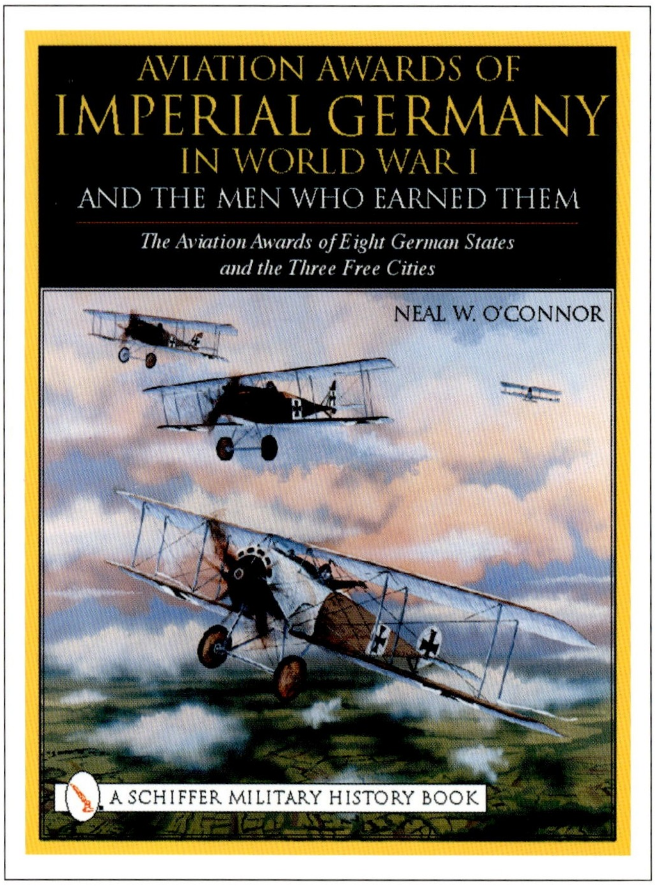


# AVIATION AWARDS OF IMPERIAL GERMANY IN WORLD WAR I
## AND THE MEN WHO EARNED THEM
### *Vol. VII: The Aviation Awards of Eight German States and the Three Free Cities*

Neal W. O'Connor

This volume is the last of Neal O'Connor's series covering the air awards and pilots of Imperial Germany – the focus of this book being the remaining German states not covered in previous volumes. The eight German states appear first in alphabetical order: thus we find the Duchy of Brunswick leading off the discussion; the Grand Duchy of Hesse; the Principality of Hohenzollern; the Principalities of Lippe and Schaumburg-Lippe; the Grand Duchies of Mecklenburg-Schwerin and Mecklenburg-Strelitz; and the Principality of Waldeck; and finally, the three Free Hanse Cities of Bremen, Hamburg and Lübeck. Among the forty-three biographies included in this volume are: *Oblt.* Ernst *Freiherr* von Althaus, *Lt.* Erwin Böhme, *Oblt.* Maximilian von Cossel, *Lt.* Ludwig Fritz Dornheim, *Lt.* Emil Flörke, *Offz.*-Stv. Ludwig Friedewald, *Rittm.* Hans-Achim von Grone, *Major* Wilhelm Haehnelt, *Lt.* Franz von Kerssenbrock, *Lt.* Günther Ludeke, *Lt.* Friedrich Niederhöfer, *Lt.* Otto Parschau, *Rittm.* Manfred *Freiherr* von Richthofen, *Lt.* Karl-Emil Schäfer, *Lt.* Carl-August von Schoenebeck, *Lt.* Günther Schroth, *Lt.* Otto-Martin *Graf* von Schwerin, *Offz.-Stv.* Gottlieb Vothknecht, *Lt.* Friedrich-Wilhelm Wichard, and *Oblt.* Heinz-Hellmuth von Wühlisch.
Size: 8 1/2" x 11" • over 700 black and white photographs • 528 pages
ISBN: 0-7643-1626-5 • hard cover • $69.95

# The Italian front : November 1917 October 1918

- Italian airfields
- Austro-Hungarian airfields
- City
- Front lines

Map 1 : 1.000.000

0 km 50 km

0 km 10 km 20 km 30 km 40 km 50 km